Handbook of Experimental Pharmacology

Volume 149

Springer

Berlin
Heidelberg
New York
Barcelona
Hong Kong
London
Milan
Paris
Singapore
Tokyo

Obesity: Pathology and Therapy

Contributors

P. Björntorp, S.R. Bloom, J.E. Blundell, C. Bouchard, C.F. Burant, L.A. Campfield, J.F. Caro, Y.C. Chagnon, M. DiGirolamo, V. Drapeau, J. Dwyer, B.J. Goldstein, J.G. Granneman, J.C.G. Halford, B.C. Hansen, J.B. Harp, M.D. Hirvonen, R.E. Keesey, J.W. Kolaczynski, R. Konikoff, Y. Ma, K.G. MacDonald, D.E. Moller, L. Pérusse, F.X. Pi-Sunyer, W.J. Pories, M. Rossi, D.M. Smith, F.J. Smith, J. Stevens, M.E. Trautmann, A. Tremblay, A.A. Truett, L.H.T. Van der Ploeg, D.B. West, B. York

Editors:

D.H. Lockwood and T.G. Heffner

Springer

Dean H. Lockwood, M.D.
Vice President
Department of Clinical Research
Parke-Davis Pharmaceutical Research
Division of Warner Lambert Company
2800 Plymouth Road
Ann Arbor, MI 48105
USA

Thomas C. Heffner, Ph.D.
Sr. Director, Neuroscience Therapeutics
Parke-Davis Pharmaceutical Research
Division of Warner Lambert Company
2800 Plymouth Road
Ann Arbor, MI 48105
USA
e-mail: Thomas.Heffner@wl.com

With 60 Figures and 42 Tables

ISBN-13: 978-3-642-64070-4 e-ISBN-13: 978-3-642-59651-3
DOI: 10.1007/978-3-642-59651-3
Library of Congress Cataloging-in-Publication Data
Obesity: pathology and therapy/contributors, P. Björntorp ... [et al.]; editors D.H. Lockwood and T.G. Heffner.
p. cm. – (Handbook of experimental pharmacology; v. 149)
Includes bibliographical references and index.
ISBN-13: 978-3-642-64070-4
1. Obesity. I. Björntorp, Per. II. Lockwood, Dean H. III. Heffner, T.G. (Thomas G.), 1949– IV. Series.
QP905.H3 vol. 149
[RC628]
615'.1 s – dc21
[616.3'98] 00-044025

Springer-Verlag Berlin Heidelberg New York
a member of BertelsmannSpringer Science+Business Media GmbH

Softcover reprint of the hardcover 1st edition 2000

Cover design: design & production GmbH, Heidelberg

Typesetting: Best-set Typesetter Ltd., Hong Kong

SPIN: 10666361 27/3020-5 4 3 2 1 0

Preface

Obesity is a serious medical problem that affects millions of people, especially in Western societies. Although long considered a complicating factor in a variety of diseases, there is now widespread agreement that obesity itself should be classified and treated as a disease and that it has important consequences for personal health, quality of life and cost to society. Understanding obesity and the means of treating it have been hampered in the past. There have been misperceptions that obesity is a behavioral disorder and that its treatments provides only cosmetic benefits. Pharmacologic approaches to treatment have suffered from problems of limited efficacy, reduced activity upon chronic use, and serious side effects, including abuse liability, cardiac disease, hypertension, and respiratory complications. Finally, there has been a proliferation of consumer and natural products with unproven benefits. This book attempts to address both the problems associated with obesity and the approaches to treating it.

In the first section devoted to pathology, Drs. DIGIROLAMO, HARP, and STEVENS elaborate in Chap. 1 on how obesity and its medical complications develop. As described by Dr. PI-SUNYER in Chap. 2, obesity is a disease seen most often in affluent Western societies and is associated with the aforementioned medical problems, as well as Type II diabetes mellitus and gallbladder disease. Drs. CHAGNON, PÉRUSSE, and BOUCHARD review the human genetics of obesity in Chap. 3, and Drs. GOLDSTEIN and KOLACZYNSKI review the compelling evidence of the important role that obesity plays in the development of Type II diabetes in Chap. 4

Greater awareness of the consequences of obesity, as well as its recognition as a disease, has stimulated research aimed at understanding its etiology and developing new treatments. The second section of this book is devoted to what is known about the pharmacology and treatment of obesity. The role of fat metabolism is described by Dr. BJÖRNTORP in Chap. 5, while in Chap. 6, Drs. HIRVONEN and KEESEY elaborate on the concept of body weight set point and its use in the understanding of (as well as its implications for) body weight regulation. Pharmacologic approaches to treating obesity have included serotonergic drugs and other agents, as described by Drs. HALFORD and BLUNDELL in Chap. 7 and by Drs. CAMPFIELD and SMITH in Chap. 8. Dietary approaches to the treatment of obesity are reviewed by Drs. DWYER and KONIKOFF in Chap. 9 and by Drs. Driapeau and TREMBLAY in Chap. 10. Use of surgical treatments are discussed by Drs. SMITH, PORIES, and MACDONALD in Chap. 11.

The third section of this book is devoted to current pharmacologic targets for obesity. Leptin and its critical role in adipocyte-to-brain signaling is described by Drs. CARO and TRAUTMANN in Chap. 12. In Chap. 13, Drs. ROSSI and BLOOM review the important role of central nervous system neuropeptides in feeding, while Dr. GRANNEMAN describes the concept of β3 adrenergic receptors as a target for obesity treatment in Chap. 14. Finally, Dr. BURANT describes the insulin sensitizers in Chap. 15.

Strategies for developing future targets for antiobesity agents are described in the fourth and final section of the book. In Chap. 16, MOLLER and VAN DER PLOEG suggest important future directions for identifying new targets for obesity treatment based on genetic and transgenic approaches. Drs. WEST, MA, TRUETT, and YORK describe approaches for identifying new genes involved in obesity in Chap. 17. Dr. HANSEN reviews the importance of nonhuman primates as experimental models of obesity in Chap. 18.

It is our hope that this volume will provide a useful review of our present understanding of obesity, its etiology, and the future directions in the development of effective treatments of it.

D. LOCKWOOD and T. HEFFNER, EDITORS

List of Contributors

BJÖRNTORP, P., Department of Heart and Lung Disease, Sahlgren's University Hospital, University of Göteborg, S-413 45 Göteborg, Sweden
e-mail: Per.Björntorp@hjl.gu.su

BLOOM, S.R., Endocrine Unit, Imperial Collge of Science, Technology and Medicine, Hammersmith Hospital, DuCane Road, London W12 0NN, United Kingdom
FAX: (208) 383-3142

BLUNDELL, J.E., Biopsychology Group, School of Psychology, University of Leeds, Leeds LS2 9JT, United Kingdom
e-mail: johneb@psychology.leeds.ac.uk

BOUCHARD, C., Executive Director, Pennington Biomedical Research Center, 6400 Perkins Road, Baton Rouge, Louisiana 70808-4124, USA
e-mail: bouchac@mhs.pbrc.edu

BURANT, C.F., Cell Biology, Parke-Davis Pharmaceutical Research, 2800 Plymouth Road, Ann Arbor, MI 48105, USA
e-mail: Charles.Burant@wl.com

CAMPFIELD, L.A., Department of Metabolic Diseases, Hoffman-La Roche Inc., Nutley, NJ 07110, USA
FAX: (973) 235-8128

CARO, J.F., Lilly Research Laboratories, Eli Lilly and Company, Lilly Corporate Center, Indianapolis, IN 46285, USA

CHAGNON, Y.C., Physical Activity Sciences Laboratory, Kinèsiologie, PEPS, Laval University, Ste-Foy, Quèbec G1K 7P4, Canada
e-mail: Yvon.Chagnon@kin.msp.ulaval.ca

DiGirolamo, M., Department of Medicine, Division of Gerontology and Geriatric Medicine, Emory University School of Medicine, Woodruff Memorial Research Building, 1639 Pierce Drive, Atlanta, GA 30322, USA
e-mail: mdigi01@emory.edu

Drapeau, V., Physical Activity Sciences Laboratory, Division of Kinesiology, PEPS, Laval University, Ste-Foy, Québec G1K 7P4, Canada
e-mail: vicky.drapeau@kin.msp.ulaval.ca

Dwyer, J., Tufts Medical School and Tufts School of Nutrition, Science and Policy and Frances Stern Nutrition Center, New England Medical Center, Hospital Boston, MA 02111, USA
e-mail: jdwyer1@lifespan.org

Goldstein, B.J., Division of Endocrinology, Diabetes and Metabolic Diseases, Jefferson Medical College of Thomas Jefferson University, 1020 Locust St, Room 349, Philadelphia, PA 19107, USA

Granneman, J.G., Department of Psychiatry and Behavioral Neuroscience, Wayne State University School of Medicine, 540 East Canfield, Detroit, MI 48201, USA
e-mail: James.Granneman@wl.com

Halford, J.C.G., Department of Phsychology, University of Liverpool, Eleanor Rathbone Building, Liverpool L69 7EA, United Kingdom
e-mail: j.c.g.halford@liverpool.ac.uk

Hansen, B.C., Obesity and Diabetes Research Center, University of Maryland School of Medicine, 10 South Pine Street, MSTF # 6-00, Baltimore, MD 21201, USA
e-mail: bchansen@aol.com

Harp, J.B., Department of Nutrition and Medicine, University of North Carolina at Chapel Hill, CB#7400, McGavran-Greenberg Hall, Chapel Hill, NC 27599, USA
e-mail: jharp@sph.unc.edu

Hirvonen, M.D., University of Wisconsin, Brogden Psychology Building, 1202 W. Johnson Street, Madison, WI 53706, USA
e-mail: mdhirvonen@facstaff.wisc.edu

Keesey, R.E., University of Wisconsin, Brogden Psychology Building, 1202 W. Johnson Street, Madison, WI 53706, USA
e-mail: rekeesey@facstaff.wisc.edu

Kolaczynski, J.W., Novo Nordisk Pharmaceuticals, Inc., 100 Overlook Center, Suite 200, Princeton, NJ 08540-7810, USA
e-mail: jkol@nnpi.com

Konikoff, R., Tufts University, School of Nutrition, Science and Policy and Frances Stern Nutrition Center, New England Medical Center, Hospital Boston, MA 02111, USA

Ma, Y., Parke-Davis Laboratory for Molecular Genetics, 1501 Harbor Bay Parkway, Alameda, CA 94502, USA

MacDonald, K.G., Department of Surgery, Rm. #204, Brody School of Medicine at East Carolina University, Greenville, NC 27858, USA
e-mail: macdonaldk@mail.ecu.edu

Moller, D.E., Departments of Molecular Endocrinology and Metabolic Disorders, Merck Research Laboratories, Rahway, NJ 07065, USA
e-mail: david_moller@merck.com

Pérusse, L., Physical Activity Sciences Laboratory, Kinèsiologie, PEPS, Laval University, Ste-Foy, Québec G1K 7P4, Canada

Pi-Sunyer, F. X., Obesity Research Center, St. Luke's-Roosevelt Hospital Center, 1111 Amsterdam Avenue, New York, NY 10025, USA
e-mail: Fxp1@columbia.edu

Pories, W.J., Department of Surgery, Rm. #204, Brody School of Medicine at East Carolina University, Greenville, NC 27858, USA
e-mail: poriesw@mail.ecu.edu

Rossi, M., Endocrine Unit, Imperial Collge of Science, Technology and Medicine, Hammersmith Hospital, DuCane Road, London W12 0NN, United Kingdom

Smith, D.M., Department of Surgery, Rm. #204, Brody School of Medicine at East Carolina University, Greenville, NC 27858, USA

Smith, F.J., Department of Metabolic Diseases, Hoffman-La Roche Inc., Nutley, NJ 07110, USA

Stevens, J., Departments of Nutrition and Epidemiology, University of North Carolina at Chapel Hill, Chapel Hill, NC 27599, USA

TRAUTMANN, M.E., Lilly Research Laboratories, Eli Lilly and Company, Lilly Corporate Center, Indianapolis, IN 46285, USA
e-mail: trautman_Michael@lilly.com

TREMBLAY, A., Physical Activity Sciences Laboratory, Division of Kinesiology, PEPS, Laval University, Ste-Foy, Québec G1K 7P4, Canada
e-mail: angelo.tremblay@kin.msp.ulaval.ca

TRUETT, A.A., Pennington Biomedical Research Center, 6400 Perkins Road, Baton Rouge, LA 70808, USA

VAN DER PLOEG, L.H.T., Departments of Obesity Research and Metabolic Disorders, Merck Research Laboratories, Rahway, NJ 07065, USA

WEST, D.B., Parke-Davis Laboratory for Molecular Genetics, 1501 Harbor Bay Parkway, Alameda, CA 94502, USA
e-mail: david.west@wl.com

YORK, B., Pennington Biomedical Research Center, 6400 Perkins Road, Baton Rouge, LA 70808, USA

Contents

Section I: Pathology

CHAPTER 1

Obesity: Definition and Epidemiology

Section I
Pathology

CHAPTER 1

Obesity: Definition and Epidemiology

M. DiGirolamo, J. Harp, and J. Stevens

A. Introduction

Obesity, defined by an excess of body fat, is a prevalent disorder in the Western world. In the United States, it has been estimated that one out of three adults is obese, and, in some strata of the population, one out of two (Wickelgren 1988). Obesity is a multifactorial condition, where genetic, environmental, behavioral, and social-economic conditions influence and determine body weight and adipose tissue mass and distribution (Bray 1992).

The relevance of obesity to health derives from the observation, over time and in many countries, that obesity predisposes to, or aggravates, many clinical conditions such as high blood pressure, hyperlipidemia, diabetes, gout, atherosclerotic heart disease, restrictive lung disease, gallbladder disease, degenerative arthritis, and infertility (Anderson and Kannel 1992; Matsuzawa et al. 1992; Seidell 1998). Furthermore, the obese state predisposes to, or aggravates, psychological maladaptation, poverty, negative employment practices, and adverse conditions and rates of health insurance. This chapter will deal initially with the problem of diagnosing obesity, defining limits of normal body fat and body weight, adipose tissue physiology, and with characterization of the various etiologies of obesity. The second part of this chapter will show the prevalence of obesity in the United States and around the world. Demographic and lifestyle factors associated with obesity are reviewed.

B. Historical Aspects

Obesity has been known to exist for many thousands of years. Occasional artifacts show obese human forms, mostly women. Some prehistoric stone statues (some over 3000 years old) have been found with typical characteristics of gynoid obesity (Brown 1992). In Malta, a headless statue with significant obesity has been unearthed. The Venus of Willendorf has received wide attention in obesity circles. Figure 1 shows the Venus of Dolni Vestonice which originated in Moravia. This small statue was modeled in clay and then fired. Both statues of Venus show features of moderate to severe obesity. Perusal of art (sculpture, drawings, and paintings) over the centuries reveals that, in certain

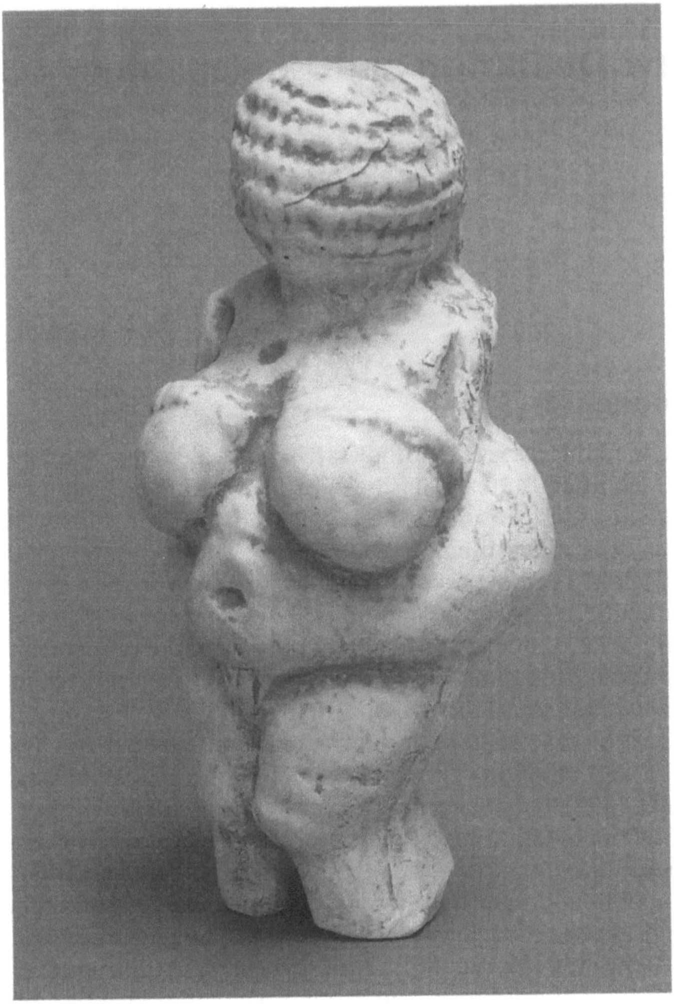

Fig. 1. The Venus of Dolni Vestonice

periods of history, and in different societies, obesity was not only accepted, but desirable. The Renaissance paintings of Titian and Rubens with well-endowed, voluptuous features, come to mind. More recently, the sculptures and distinctive paintings of Columbia-born artist Fernando Botero show exaggerated human and animal forms of obesity. Thus, over time, style and fashion have accompanied, and possibly influenced, nutrient intake to mold the prevailing image.

There is no question that obesity prevalence is increasing in the present century. The Pima Indians, who were mostly lean at the turn of the century, now present with a 70%–80% rate of obesity (Price et al. 1993). In many

Table 1. Classification of overweight

Classification	% of Ideal Body Weight	BMI (kg/m^2)
Underweight	<90	<18.5
Normal Range	90–109	18.5–24.9
Overweight	≥110	25.0 or higher
Pre-obese	110–119	25.0–29.9
Obese class I	120–149	30.0–34.9
Obese class II	150–199	35.0–39.9
Obese class III	≥200	40 or higher

The classification according to BMI is based on data in INTERNATIONAL OBESITY TASK FORCE (1998); FLEGAL et al. (1998). The classification based on percent of ideal body weight is based on data in METROPOLITAN LIFE INSURANCE COMPANY (1983). The two classifications may be close but they have not been matched as yet.

European countries of the western world, obesity is now found in ~15% of men and 22% of women, and 1%–2% of the population suffers from massive obesity (SEIDELL and FLEGAL 1997).

C. Definition of Obesity

Obesity is defined as a clinical condition characterized by an excessive accumulation of body fat. Since the direct or indirect measurement of body fat requires specialized techniques and is beyond the reach of most clinical settings, definition of normal (desirable) body weight and excess of body weight and body fat relies on surrogate measures such as the ones shown in Table 1. This table shows normative values related to desirable or excessive body weight; it also includes the index of body mass (BMI, body mass index = kg/m^2) and related variations in different classes of obesity, according to a classification by the International Task Force on Obesity (INTERNATIONAL OBESITY TASK FORCE 1998). Once normative values of body fat content in relation to BMI become established, it will be possible to define body fat in excess of normal values that are approximately 12%–22% in adult men and 21%–32% in adult women (KEYS et al. 1972). Obese people are often overweight, but caution is needed as some overweight people (i.e., weight lifters, football players) are not obese. Conversely, some individuals who are not overweight, but are sedentary, may have excessive body fat.

Conceptually, most researchers would agree that percent body fat is the variable of interest. However, percent body fat is not easily measured in large populations. Body weight, on the other hand, can be very easily and inexpensively measured in a variety of settings and is easily understood by the public.

The difficulty with weight is that it must be adjusted for height. In early work, researchers examined the effect of weight-for-height using several categories of height in increments of one inch, while more recently, ratios of weight and height have been used to express weight relative to height. Ratios of weight and height have been calculated using height raised to some power. Currently, the most popular form for this calculation is weight in kilograms divided by the square of height in meters (Keys et al. 1972; Quetelet 1981). This form has become so popular that, in the absence of definition, it is assumed to be the form used to calculate body mass index (BMI). BMI, calculated in this manner, has been shown to be related to percent body fat. In one study, a correlation coefficient of approximately 0.7 was found in adults (Wellens et al. 1996), whereas in another study a correlation coefficient of 0.96 was found (Zumoff et al. 1990). Some studies have shown that BMI is slightly negatively correlated with height, but since percent body fat is also slightly negatively correlated with height in several populations, this is not necessarily a weakness.

D. Criteria for Defining Overweight and Obesity

Obesity and overweight are two frequently used terms; however, definitions and the measures and cutpoints used to define these terms have varied. Obesity has been defined as excess adipose tissue whereas overweight has been defined as a body weight for height in excess of given reference values (Kuczmarski et al. 1997).

In the United States, the 1959 or the 1983 Metropolitan Life Insurance (MLI) Height and Weight tables have been used to provide cutpoints for overweight by many investigators (Metropolitan Life Insurance Company 1983). These widely quoted tables show a range of desirable weight, defined as the weight-for-height associated with the lowest mortality. However, in the 1985 NIH Consensus Conference on Obesity, the 85th percentile values of BMI for men and women aged 20 through 29 from NHANES II gained popularity as cutpoints for overweight. These cutpoints, based solely on the distribution of BMI within the described population were 27.8 for men and 27.3 for women. They are approximately equivalent to 124% of desirable weight for men and 120% of desirable weight for women defined as the mid-point range of weights for a medium frame from the 1983 Metropolitan Life Insurance (MLI) Height and Weight tables, after appropriate adjustments for clothing and shoes.

In 1995, the World Health Organization (WHO) recommended using BMI cutpoints of 25, 30, and 40 to define grade 1, grade 2, and grade 3 overweight (International Obesity Task Force 1998). Soon after the WHO report, the International Obesity Task Force (IOTF) suggested a somewhat similar classification; however, an addition cutpoint of a BMI at 35 was suggested (see

Table 1). Using these cutpoints, overweight is defined by a BMI of 25 or greater, while obesity is defined by a BMI of 30 or greater. In 1988, a U.S. government consensus conference adopted the WHO recommendation of 25 and 30. According to these criteria, 54% of the U.S. adult population is overweight (BMI > 25) and 22% is obese (BMI > 30) (FLEGAL et al. 1998).

E. Assessment of Body Fat and Degree of Obesity

Having tools to assess body fat accurately has taken on greater importance with the development of guidelines that recommend increasingly more aggressive therapies for obesity of increasing severity. An ideal diagnostic test of obesity would precisely and accurately quantify and localize adipose tissue stores. The test would be easy to perform, have low risk, low cost, and high predictive value to distinguish individuals with medically significant obesity from those without. The measure would be sensitive enough to detect clinically important changes in body fat that occur with weight loss and weight gain. Table 2 shows how body fat accumulation, in adult men and women, influences body composition and caloric storage.

The following sections will review the traditional and newer methods of measuring body fat and body composition in the context of their usefulness as tools for the assessment of obesity.

Table 2. Body fat accumulation influences body composition and caloric storage in adult men and women

	Lean Active	Moderately Obese Inactive	Severely Obese Sedentary
Female			
Height	5′5″	5′5″	5′5″
BWT	125 lb (100% IBWT)	180 lb (144% IBWT)	290 lb (232% IBWT)
BMI	20.8	30	48.3
% Fat	22	39.7	55
Kg Fat	12.5	32.5	72.5
Kcal in fat storage	101 000	263 200	584 500
Male			
Height	5′11″	5′11″	5′11″
BWT	172 lb (100% IBWT)	227 lbs (132% IBWT)	337 lbs (196% IBWT)
BMI	24	31.7	47
% Fat	12.9	29.2	45.8
Kg Fat	10.1	30.1	70.1
Kcal in fat storage	81 800	243 800	567 800

BMI, body mass index; BWT, body weight; IBWT, ideal body weight.

I. Body Composition

The traditional measures of body fat are based on the premise that total body weight is the sum of two compartments; fat mass and fat-free mass (FFM). Direct measures of either FFM, total body water, or body density are used along with total body weight to estimate the absolute and relative amount of body fat. In the last decade, more sophisticated methods have been developed to separate total body weight into four compartments: fat mass, body cell mass, extracellular water, and skeletal mass (Pierson et al. 1997).

II. Underwater Weighing – Hydrostatic Densitometry (UWW)

For many years UWW was considered the "gold standard" for the determination of body fat. UWW is based on the fact that, at normal body temperature, body fat has a density of $0.901\,kg/dm^3$ and FFM of $1.097\,kg/dm^3$ (Siri 1969). The recordings are performed while the subject is submerged underwater in a large tank. Air in the lungs and gut alters the buoyancy of the subject and introduces error in the measurements of body density. Since the lung contains the largest source of air, residual lung volume is measured and subtracted using the nitrogen dilution technique, for example. The measured body density is used to calculate body fat from the formula: body density = mass/volume, or = (weight in air)/[(weight in air minus weight in water)/(density of water minus residual lung volume)], then percentage body fat = (4.95/body density) –4.501 × 100. The margins of error of the measurements of body density are ~0.003 g/cc for children and 0.008 g/cc for adults, or about 2–3 kg of body fat for the average lean adult (Brozeck et al. 1963). Although several formulae have been suggested, there is uncertainty about the exact formula that should be used to calculated percent fat from the measured variables. An additional limitation of UWW is that obese subjects have difficulty staying submerged because they are too buoyant or are resistant to the thought of having to go underwater. The technical and space requirements of UWW restrict its use primarily to research centers and sports medicine facilities.

III. Total Body Water (TBW)

Estimations of body fat using TBW are based on the assumption that adipose tissue is anhydrous. Water content of adipose tissue decreases with increasing obesity and varies from 25% of the wet weight of adipose tissue in lean rats to 9% in obese rats (DiGirolamo and Owens 1976). Fat-free tissue contains approximately 73% water. Total body water is measured using deuterium, tritium, or ^{18}O isotopic dilution. The subject ingests or is injected with the isotope and the levels measured in the blood. Fat mass is calculated by the following formula: FFM (kg) = TBW (L)/0.73, body fat (kg) = body weight (kg) minus FFM (kg). TBW has a small margin of error (0.2%), but variations in hydration status may alter results (Pierson et al. 1997). Since obesity is a state of fluid retention, this technique is not optimal for measurement of fat mass in the obese.

IV. Total Body Potassium (TBK)

Based on the fact that approximately 98% of total body potassium stores are localized in fat-free tissue, total body potassium measures FFM (PIERSON et al. 1997). Using the two compartment model, fat mass is calculated as the difference between total body weight and fat-free weight. TBK can be measured by whole-body counting of ^{40}K, a naturally occurring isotope, or by isotopic dilution measurements of administered ^{42}K. Whole body counting is a highly technical measure that requires specialized equipment that is not amenable to the clinical setting. Increased tissue mass attenuates the detection of ^{40}K, thus obesity decreases the counting efficiency (COLT et al. 1981). Error in measurements may also be introduced by alteration in potassium that occur with ill health, age, and altered nutritional states (PIERSON et al. 1974).

V. Bioelectrical Impedance Analysis (BIA)

BIA is based on the fact that total body water and FFM have rich electrolyte concentrations and electrical conductance through these tissue is high relative to fat. Estimates of body fat are derived from the resistance and reactance values using electrodes placed on the wrists and ankles to generate a small unnoticeable current through the body. BIA is gaining greater acceptance because it can be used to measure body fat in a large number of subjects in many settings with minimal risk, time, and cost. Measurements of resistance and reactance are reproducible if done correctly, but many variables can alter the accuracy and precision of the measurements: body position, hydration status, food intake, ambient air, skin temperature, recent physical activity, and conductance of the examination table (DEURENBERG 1996). BIA has limitations in the measurements of body fat in the severely obese. An NIH consensus conference concluded that BIA is useful in mild-moderate obesity, but that specific well-defined procedures for performing BIA are not practiced (SCHOELLER 1996).

VI. Ultrasound

Measurement of adipose tissue with ultrasound has many attractive features. It is direct, uses no radiation, and is clinically available in many inpatient and outpatient settings. Although acceptable results have been obtained for measurements of subcutaneous fat, data are conflicting for the assessment of intraabdominal depots. In one study, it was determined that ultrasound was a good measure of subcutaneous adipose tissue, but not intraabdominal fat when compared to computerized tomography (CT) and magnetic resonance imaging (MRI) (BELLISARI et al. 1993). In contrast, two other studies found a good correlation between visceral fat measured by ultrasound and CT (Armellini et al. 1994; Tornaghi et al. 1994). One advantage of ultrasound over other more conventional methods, such as anthropometry, is in the clinical evaluation of subcutaneous fat in the severely obese where skinfold determinations are problematic.

VII. Computerized Tomography (CT)

Computerized tomography is the most accepted method to measure total and regional body fat. This technology is based on the fact that transmission of an X-ray beam has different attenuation coefficients among fat-free mass, bone, and adipose tissue that can be quantified using computer analysis (Plourde 1997). Abdominal visceral fat can be assessed separately from subcutaneous fat, which is highly desirable for assessing the metabolic risk of obesity. This method is considered more accurate than underwater weighing, anthropometry, and TBK since it also takes into account the body cell mass and extracellular water compartments, unlike the other methods that use the two compartment model (Pierson et al. 1997). The reproducibility of CT is good, 0.6%–1.4%, and its validity has been documented with cadaveric studies (Snyder et al. 1984; Van der Kooy and Seidell 1993). Single slice measurements at L4–L5 (designed to cut radiation exposure, cost, and time) have been found to be adequate for the assessment of total abdominal and visceral adipose tissue mass. CT has also been used successfully to detect regional changes in fat mass with weight loss in women and men (Chowdhury et al. 1993; Stallone et al. 1991). Practical limitations of CT include the high cost and radiation exposure. The procedure can not be performed in the severely obese because many of the CT tables have a 350lb weight limit.

VIII. Magnetic Resonance Imaging (MRI)

MRI has some advantages over CT. It does not expose the person to radiation and it can assess tissues in many planes. Adipose tissue appears as a distinct bright area contrasting with surrounding areas on T-1 weighted images (Plourde 1997). The difference between body fat assessment with MRI and cadaveric dissection was 3% in one study (Abate and Garg 1995). Drawbacks to MRI are the high cost and technical difficulty in the interpretation of results.

IX. Dual Energy X-ray Absorptiometry (DEXA)

DEXA is a safe and quick, widely available, method used to assess a patient's risk for osteoporosis that has also been used to assess body composition (Laskey 1996). DEXA measurement of fat mass is based on the fact that the instruments generate X-ray beams and measure the attenuation of the beams at two different energies. Bone mineral content and soft tissue composition are calculated in the area tested and fat mass estimated. The assessment of body composition is most precise in young healthy subjects and is less reliable in the obese and osteoporotic subject. DEXA can overestimate body fat in individuals with low bone density such as in postmenopausal women. Some consider this method to be the best technique to measure body composition in children (LAPILLONNE et al. 1997). Another limitation is that DEXA is not a good measure of intraabdominal fat either alone or when combined with anthropometry (VAN DER KOOY and SEIDELL 1993).

X. Neutron Inelastic Scattering

This newer method is different in that it measures fat mass more directly than some of the previous methods. The instrument measures total body carbon and oxygen. Fat contains most of the total body carbon (ARMELLINI et al. 1994). This new technique suffers many of the limitations of other highly technical expensive measures of body composition.

XI. Anthropometry

1. Body Mass Index (BMI)

The BMI is based on the one compartment model representing both fat mass and lean body mass; it is designed to assess weight conveniently, independent of stature. Overweight implies an increase in BMI without discriminating whether this increase is due to increased body fatness or lean tissue mass. BMI is highly correlated ($r = 0.7$) with percentage body fat (WELLENS et al. 1996). A study of 300 men and 77 women determined the sensitivity, specificity, and predictive value of BMI as an index of obesity compared to measurements of body fatness obtained with hydrodensity. To achieve a sensitivity of 91% and specificity of 47%, the BMI cut off value corresponding to 25% body fat in men and 30% in women was 24.5 kg/m^2 in men and 22 kg/m^2 in women (ZUMOFF et al. 1990). Another group found that 25% body fat in men and 33% in women correlated with a BMI of 25 kg/m^2 in men and 23 kg/m^2 in women (HORTOBAGYI et al. 1994). These BMI levels are similar to the IOTF for overweight (FORBES et al. 1988) and lower than the cutoff for obesity (HOOGWERF and NUTTALL 1984).

2. Weight and Height Tables

Metropolitan Life actuarial tables have essentially been replaced by BMI in clinical and epidemiological settings. These tables give gender specific weight, height, and frame size based on mortality experience (METROPOLITAN LIFE INSURANCE CO 1988).

3. Circumferences

Waist and hip circumferences have gained widespread popularity in the research setting as surrogates of intraabdominal and gluteal fat, respectively.

Unfortunately the technique for obtaining waist and hip circumferences is not standardized. Waist circumference can be measured at the umbilicus, at the minimal abdominal circumference, and midway between the lower rib margin and the iliac crest. A waist circumference of 95 cm in both men and women, and a waist to hip ratio (WHR) of 0.94 in men and 0.88 in women corresponds to a visceral adipose tissue area of 130 cm^2 (Goodman-Gruen and Barrett-Connor 1996). After age 80, WHR is a poor measure of abdominal obesity, and waist is better. There has been much debate over the use of WHR or waist alone (Lemieux et al. 1996) in the assessment of regional fat distribution and metabolic risk, but no consensus has been reached on the best measure to use, if any, in the clinical setting.

4. Skinfold Thickness

Skinfold determinations estimate total body fat. Special calipers are used (Harpenden, Morse, Holtain-Harpenden, and Lange) to obtain measurements at multiple sites: midtriceps, inferior tip of scapula, iliac crest, and abdomen. This measurement is based on the fact that skin is 0.5–2 mm thick, and subcutaneous fat is the rest (Forbes et al. 1988). This method has limitations because the most accepted reference tables for calculating body fat from skinfold determinations are derived from a Scottish population (Durnin and Womersley 1974). The application of skinfolds obtained in this group to others has not been established. Obtaining reliable measurements of skinfolds requires an individual trained, and preferably certified, in the technique. Comparisons of results from skinfold thickness with results from CT show that as subcutaneous thickness increases, the calipers underestimate the true value of subcutaneous fat (Bray 1989). This limitation along with the added technical difficulty of obtaining skinfold measurements in individuals with large amounts of subcutaneous fat diminish the usefulness of this method in assessing body fat in the severely obese.

F. Practical Considerations in the Assessment of Body Fat

There are many methods, both new and old, to assess body fat. None has proven to be an ideal diagnostic test of obesity. Some of these methods are either too costly, too invasive, or not sufficiently validated through the extreme of body weight to be used widely in the assessment of obesity.

Relative (percent of ideal) body weight and BMI are generally used to estimate degree of overweight and obesity in the clinical setting. Underwater weighing, bioelectrical impedance, or dual energy X-ray absorptiometry best estimate absolute or relative body fat mass. Circumferences and skinfold thickness can give an approximation of regional fat distribution with little cost and risk. If more precise estimates are needed, to separate abdominal fat into visceral and subcutaneous regions, then CT scan at L4–L5 is best.

G. Anatomy and Physiology of Adipose Tissue

The first review on the physiology of adipose tissue appeared in 1948 (SHAPIRO and WERTHEIMER 1948). Prior to that, adipose tissue was considered to be a specialized connective tissue filled with droplets of fat. The tissue was known to have a small number of capillaries and no nerves. The fat stores were believed to be passive in nature and devoid of an important metabolic role. The only recognized function was insulation to trauma and heat for the body.

The last 50 years have revolutionized our understanding of adipose tissue, its anatomy, regional differences, blood flow, innervation, and the critical processes of nutrient arrival, processing, and storage, coupled with lipid breakdown and mobilization, and the exquisite endocrine regulation of these metabolic processes (DIGIROLAMO et al. 1998; DIGIROLAMO et al. 1992).

I. Anatomy of Adipose Tissue

White adipose tissue is present in several regions of the body. About 50% of it is present in the subcutaneous region, and the rest is present around endocrine organs (gonads, adrenal, kidney), inside the abdominal cavity, and along muscle fascias. The specialized cells, called adipocytes, are supported by connective stromal-vascular tissue. Vascular blood flow feeds the adipocytes through disseminated capillaries (some investigators believe there is at least one capillary per cell; CRANDALL et al. 1997), and a vast network of nerve endings has been recently described (REBUFFÉ-SCRIVE 1991).

The adipose tissue composition varies with stage of development and with adipose region. In young animals, about 70% of wet weight is fat, 25% is water, and 5% is protein material. When animals, including human, reach sexual maturity, the fat content is about 85%–92% of wet weight, water is 8%–10%, and defatted protein residue is about 1%–3% (DIGIROLAMO and OWENS 1976).

One of the major developments in the understanding of adipose tissue anatomy and physiology has been the clarification of major differences in the anatomy, blood flow, composition, and metabolic role of adipose tissue in different regions of the body. In rodents, retroperitoneal and gonadal fat has the highest lipid content (90%–92%) and the least defatted protein content (1%–2%). Intra-abdominal mesenteric tissue contains less fat (80%–82%) and more protein (5%–6%), whereas the subcutaneous fat contains the least fat (50%–60%) and the highest content of protein (10%–12%) and water (25%–35%) (DIGIROLAMO et al. 1998). In the human, significant regional differences in body fat deposition have been noted. The upper body segment obesity, seen frequently in the male (android), has increased body fat accumulation on the trunk, around the waist and intra-abdominally. This type of body fat distribution has been called "apple-like", central, or abdominal obesity. The lower body segment obesity, seen frequently in the female (gynoid), has increased body fat accumulation in the thighs and buttocks. This type of body fat distribution has been called "pear-like" or peripheral obesity (Vague et al. 1969).

Although gynoid obesity is seen more frequently in women, and the android form in men, there are occasional women with the "apple-like" type of obesity, and men with the "pear-like" appearance. A large body of evidence has indicated that the upper body segment, android, or abdominal, obesity is associated with major health risk factors, such as hyperlipidemia, hypertension, and diabetes (Anderson and Kannel 1992). Furthermore, these health risks correlated better with abdominal "visceral" fat deposition than with abdominal "subcutaneous" deposition (Kissebah 1996).

The blood flow regulation also varies with adipose region, being highest in the mesenteric tissue, intermediate for the gonadal and retroperitoneal fat, and lowest for the subcutaneous tissue. Food deprivation increases the blood flow to adipose tissue, partly due to the necessity of removing the products of fat breakdown and mobilization. Increment in blood flow with fasting are highest for mesenteric adipose tissue and lowest for subcutaneous tissue (Kowalski et al. 1997).

The main cellular components of adipose tissue are the adipocytes, specialized cells which house the accumulated triglyceride in near-spherical cells containing a selective membrane and a small rim of cytoplasm rich in mitochondria.

II. Physiology of Adipose Tissue

The major known adipose tissue functions are listed in Table 3. White adipose tissue is the major storage compartment of nutrients in the body. In a lean

Table 3. The major known adipose tissue functions

Non-metabolic	Protection of skin, skeleton, muscles, nerves, and blood vessels from trauma
	Insulation from heat loss
Metabolic	Uptake of circulating nutrients, synthesis and storage of triglycerides
	Glucose conversion to lactate and lactate release
	Lipolysis of stored triglycerides and mobilization of glycerol and free fatty acids into the circulation
	Removal of circulating lipoprotein-triglycerides by the action of lipoprotein lipase
	Binding and response of endocrine agents capable of affecting the above metabolic activities
Others	Storage and metabolism of lipid-soluble substances
	Aromatization of androgens to estrogens
	Synthesis and release of leptin, a new hormone with multiple effects, which include informing the hypothalamus of peripheral energy storage

individual, 80,000–120,000 Kcal are stored in adipose tissue (see Table 2). This value can be increased by a factor of 5–8 in obese people. The storage capacity of the adipose organ is due to the limited water content (usually <10% of wet tissue) and to the enhanced caloric density of lipid storage (1 g = 9 Kcal).

Adipose tissue participates in the daily metabolic exchanges by actively removing circulating nutrients (glucose, fatty acids, lactate) after a meal and also releasing nutrients (glycerol, fatty acids, lactate, etc.) into the circulation in the fasted state (see Fig. 2).

Glucose uptake by fat cells is facilitated by glucose transporters that exist on the cell surface and also by additional glucose transporters moved to the cell surface by the action of insulin (KONO 1985). The glucose is rapidly phosphorylated and utilized as an energy source. The main metabolic products of glucose in a fat cell are: carbon dioxide, α-glycerophosphate, fatty acids (the latter two are esterified as triglycerides), and lactate and pyruvate. Conversion of glucose to lactate is enhanced in large fat cells from obese subjects and after food deprivation (NEWBY et al. 1990). Only a small amount of glucose (<3%) is converted to glycogen and stored as such. Lipolysis, i.e., breakdown of stored triglycerides to glycerol and fatty acid is promoted by triglyceride lipase, activated by lipolytic hormones (CUSHMAN and SALANS 1973).

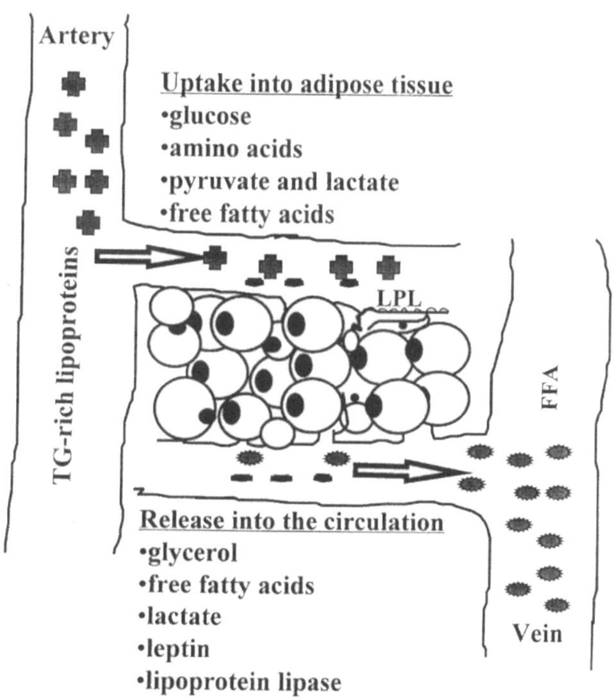

Fig. 2. The participation of adipose tissue in daily metabolic exchanges

Insulin has a dual role, anabolic and anticatabolic, in fat cell metabolism. By stimulating glucose transport and utilization, insulin favors synthesis and storage of triglycerides. By blocking the lipase stimulated by lipolytic hormones, insulin prevents or reduces lipolysis. In a broader sense, adipose tissue participates in important metabolic exchanges with other organs such as muscle, liver, and pancreas. During fasting, the liver receives from adipose tissue products such as glycerol and lactate for gluconeogenesis and free fatty acids for synthesis of triglyceride-rich lipoproteins. In view of greater capacity for lactate production in mesenteric adipose tissue and direct flow of mesenteric tissue to the portal vein and the liver, the contribution by visceral and mesenteric adipose tissue to hepatic processes of gluconeogenesis and glycogenesis may be greater than that from other adipose tissue regions (McGarry and Foster 1979). During fasting, muscle receives fatty acids and lactate generated in the fat cells.

In the fed state, chylomicrons produced in the intestinal tract are cleared from the circulation by the action of an enzyme, lipoprotein lipase, which is produced in the fat cells and is released to the endothelium where it hydrolyzes the triglyceride-rich lipoproteins. The resulting free fatty acids are taken up by the fat cells. In obesity, adipose tissue mass increases by a factor of 3–10. Excessive fat deposition is associated with increased perfusion requirements of the tissue. Consequently, cardiac output and plasma volume increase. Hypertension is a frequent concomitant of obesity (Alexander 1970).

Excessive accumulation of fat in the body, particularly in the visceral area, is associated with hyperlipidemia, carbohydrate intolerance, and diabetes. The precise mechanisms by which obesity leads to diabetes, the type II, non-insulin dependent, are not clear but appear related to a combination of peripheral insulin resistance, coupled with excessive demands of pancreatic secretion of insulin. Additionally, excessive production of free fatty acids and of lactate by enlarged fat cells of obese subjects may contribute to the carbohydrate intolerance of obesity by providing the muscle with alternate sources of fuel that can be utilized by the muscle preferentially to glucose (DiGirolamo et al. 1989; DiGirolamo et al. 1992).

H. Etiologic Classification of Obesity

Although many strides have been made in understanding the regulation of food intake and utilization, and in the treatment of obesity, considerably less is known about the etiology and pathogenesis of most cases of obesity. Table 4 is an etiologic classification scheme for rare and more common forms of obesity.

The genetics of obesity will be covered in chap. 10. The endocrine causes of obesity are exceedingly rare, but must be entertained during the clinical evaluation of obesity. One study found in patients with primary hypothy-

Table 4. Etiologic Classification of obesity

Genetic	Nutritional
Rare Syndromes	High fat
Leptin deficiency	Low fat/high calories
Convertase deficiency	Nutrient Storage
Prader-Willi	Triglyceride storage disease
Alström	Glycogen storage disease
Biemond	Drug-induced weight gain or obesity
Laurence-Moon-Bardet-Biedl	Insulin
Downs	Sulfonylureas
More common	Glucocorticoids
Familial obesity with recognizable patterns	Antipsychotics
Hypothalamic lesions	Depoprovera
Craniopharyngioma	Inactivity
Post-surgical	Post-operative
Trauma	Post-trauma
Tumor invasion	Aging
Empty sella syndrome	Industrialization
Inflammation	
Endocrine	
Insulinoma	
Cushing's syndrome	
Polycystic ovaries	
Growth hormone deficiency	
Pseudohypoparathyroidism	
Hypothyroidism	
Hypogonadism	

roidism that thyroxine replacement had no significant effect on body weight 1 year after starting therapy (PEARS et al. 1990). In a second study of hypothyroid subjects, thyroxine treatment led to a small decline in mean body weight over the first 6 months, but body weight returned to pretreatment levels by 24 months (HOOGWERF and NUTTALL 1984). These studies demonstrate the limited long-term effect of thyroid hormone replacement in the treatment of obesity.

The hypothalamus is the site for regulation of food intake. In animal models, lesions of the hypothalamus can cause obesity. In humans, lesions of the hypothalamus have been associated with obesity. Several common drugs have been associated with weight gain, including antidiabetic agents such as insulin and sulfonylureas, glucocorticoids, and centrally acting agents such as phenothiazines. More recently, Depo-Provera, a drug used widely as a female contraceptive, has been associated with significant amounts of weight gain. One study of adolescent females using depot medroxyprogesterone acetate for contraception found weight gain in 27% of patients (MATSON et al. 1997). The average weight gain was 6.0 kg at 11 months and 9.0 kg at 17 months.

The underlying cause of obesity in the majority of patients may be one of excess energy intake, inadequate energy expenditure, or a combination of both. In practice, limiting food intake and encouraging increased physical activity has been largely ineffective in long term weight loss and maintenance. The Vermont overfeeding study showed that normal weight volunteers who were overfed required 2700 kcal/m^2 to maintain their increased weight while spontaneously obese individuals required 1100 to 1400 kcal/m^2 to maintain their weight (Sims et al. 1973). More recently, it was reported that metabolic rate decreases with weight loss and increases with weight gain, but in obese individuals these compensatory mechanisms are skewed and favor maintenance of the obese state (Lewis et al. 1994). These studies show that endogenous defects in the handling of metabolic fuels may predispose the obese individual to gain weight and to resist weight loss.

Because of our limited understanding of the etiology and pathogenesis of obesity, many classification criteria have been proposed, but there continues to be no consensus reached regarding the best etiologic classification to aid in the clinical management of obesity. Given the rapid pace of new advances in the field of obesity and regulation of food intake, it is only a matter of time before there are consensus diagnostic criteria and etiologic classifications for obesity. Ideally, these criteria will match our improved understanding of the pathogenesis of obesity and its relationship to risk factors and obesity-related diseases.

I. Epidemiology of Obesity

The description of population trends in obesity or overweight depend upon the measures and cutpoints used. Generally, trends in populations have been described using BMI, but the cutpoints used to define obesity and/or overweight have varied across studies. In the discussion below, we rely most heavily on the cutpoints recommended by the International Task Force on Obesity (Table 1) (International Obesity Task Force 1998).

I. Observations in the United States

National surveys have provided cross sectional descriptions of BMI across time, gender, age and ethnic groups. In addition to these demographic factors, other factors such as educational status, marital status, parity, smoking, and physical activity have been shown to be associated with the prevalence of obesity. Prevalence estimates from both phases of the NHANES III (1988–1994) showed that over both genders and all race-ethnic groups 54.4% of Americans were overweight (BMI $\geq$ 25.0) and 22.5% were obese (BMI $\geq$ 30.0) (Flegal et al. 1998). The total costs attributable to obesity in the United States in 1995 has been estimated at $99.2 billion dollars (Wolf and Colditz 1998). Approximately $51.64 billion of those dollars were direct medical costs. This is approximately 5.7% of the U.S. health expenditure nationally.

1. Secular Trends

Using the 85th percentile cutpoints from the NHANES II study described above, KUCZMARSKI et al. 1997 showed a startling increase in the prevalence of overweight of 8% between NHANES II (1976–1980) and the first phase of NHANES III (1988–1991) to 33% of US adults 20 years of age or older. FLEGAL et al. 1998 have more recently summarized the trends in obesity over two decades using the International Obesity Task Force cutpoints and both phases of NHANES III, as well as data from previous national health surveys (Table 5). There was little increase over the survey periods in pre-

Table 5. Secular trends in the prevalence of obesity and overweight in the United States

	NHES I (1960–62)	NHANES I (1971–74)	NHANES II (1976–80)	NHANES III (1988–94)
Women				
Pre-obesity (25.0 ≤ BMI ≤ 29.9)				
Crude[a] (≥20 years)	25.3	24.3	24.8	25.7
Age adjusted (20–74 years)	23.6	23.6	24.3	24.7
Class I obesity (30.0 ≤ BMI ≤ 34.9)				
Crude (≥20 years)	11.2	10.9	10.4	14.5
Age adjusted (20–74 years)	10.4	10.5	10.2	14.2
Class II obesity (35.0 ≤ BMI ≤ 39.9)				
Crude (≥20 years)	3.6	3.9	4.3	6.6
Age adjusted (20–74 years)	3.3	3.8	4.2	6.8
Class III obesity (BMI ≥ 40.0)				
Crude (≥20 years)	1.4	1.9	2.1	3.8
Age adjusted (20–74 years)	1.3	1.9	2.1	3.9
Total overweight (BMI ≥ 25)				
Crude (≥20 years)	41.5	41.0	41.6	50.6
Age adjusted (20–74 years)	38.6	39.8	40.8	49.6
Total Obese (BMI ≥ 30)				
Crude (≥20 years)	16.2	16.7	16.8	24.9
Age adjusted (20–74 years)	15.0	16.2	16.5	24.9
Men				
Pre-obesity (25.0 ≤ BMI ≤ 29.9)				
Crude (≥20 years)	38.3	41.5	39.2	39.9
Age adjusted (20–74 years)	37.8	41.1	39.1	39.4
Class I obesity (30.0 ≤ BMI ≤ 34.9)				
Crude (≥20 years)	9.1	9.7	10.0	14.3
Age adjusted (20–74 years)	8.8	9.5	10.0	14.6
Class II obesity (35.0 ≤ BMI ≤ 39.9)				
Crude (≥20 years)	1.2	1.7	1.9	3.5
Age adjusted (20–74 years)	1.3	1.7	1.9	3.6
Class III obesity (BMI ≥ 40.0)				
Crude (≥20 years)	0.3	0.6	0.4	1.8
Age adjusted (20–74 years)	0.3	0.6	0.4	1.8
Total overweight (BMI ≥ 25)				
Crude (≥20 years)	48.9	53.5	51.5	59.5
Age adjusted (20–74 years)	48.2	52.9	51.4	59.4
Total Obese (BMI ≥ 30)				
Crude (≥20 years)	16.6	12.0	12.3	19.6
Age adjusted (20–74 years)	10.4	11.8	12.3	20.0

Adapted from FLEGAL et al. 1998.

[a] Crude (non-age-adjusted) total includes ages 20–79 years for NHES I, ages 20–74 years for NHANES I and NHANES II, and ages ≥20 years (no upper age limit) for NHANES III.

Table 6. Prevalence of pre-obesity and obesity by gender and ethnic groups in NHANES III (1988–94) and HHANES (1982–84)

	Women (%)	Men (%)
Non-Hispanic White		
Pre-obese (BMI 25.0–29.9)	23.0	39.6
Obese (BMI 30.0)	22.4	20.0
African-American		
Pre-obese (BMI 25.0–29.9)	29.2	36.2
Obese (BMI 30.0)	37.4	21.3
Mexican-American		
Pre-obese (BMI 25.0–29.9)	33.4	44.0
Obese (BMI 30.0)	34.2	23.1

Age-adjusted prevalences for men and women aged 20–74 years.

obesity (BMI 25.0–29.9) in men and women ages 20–74 years. However, the prevalence of obesity (BMI $\geq$ 30) increased between the last two surveys from 16.5% to almost 25% in women and from 12% to 20% in men.

2. Gender

Weight-height indices show that American women have a higher prevalence of obesity than men. As shown in Table 6, the gender discrepancy tends to be larger in minority populations than in Caucasians. Also, it should be noted that pre-obesity is more prevalent in men than in women.

3. Ethnicity

African-American and Mexican-American women have higher levels of both pre-obesity and obesity than White women. African-American and Mexican-American men have a higher prevalence of obesity than white men; however, white men are somewhat more likely to be pre-obese than African-American men (39.6% vs 36.2% in the NHANES III survey). Obesity is more prevalent among American Indians compared to Caucasians, however, prevalence varies among tribes (Broussard et al. 1991; Price et al. 1993; Welty et al. 1995). Asian Americans tend to weigh less than Caucasian Americans (Klatsky and Armstrong 1991; Kumanyika 1994; Wang et al. 1996) although there is evidence that third generation Asian Americans tend to have weights similar to those of Caucasian Americans (Klatsky and Armstrong 1991).

4. Age

Both cross-sectional and longitudinal data show that weight tends to increase between early and middle adulthood. Figure 3 shows the prevalences of pre-

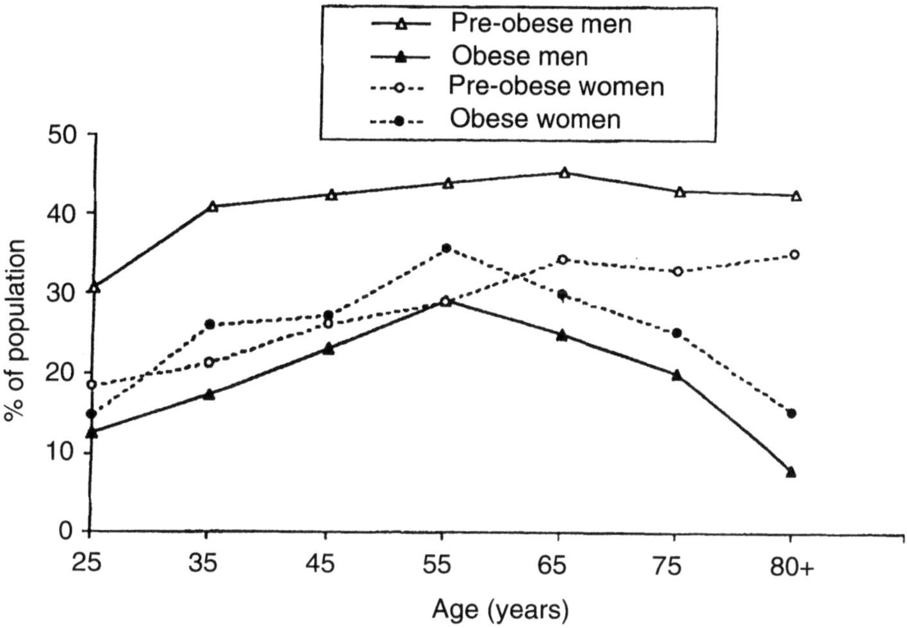

Fig. 3. The prevalence of preobesity and obesity in men and women in NHANES III by age categories

obesity and obesity in men and women in NHANES III by age categories. The prevalence of preobesity tended to increase until about 65 years of age and obesity increased until approximately age 55. Longitudinal data from a population-based cohort recruited from four U.S. communities as part of the ARIC (Atherosclerosis Risk in Communities) Study showed that between early adulthood (age 25) and middle age (45–64 years) most participants gained substantial amounts of weight (Stevens et al. 1998). Approximately 53% of white women, 79% of African-American women, 46% of white men, and 49% of African-American men gained more than 10 kg over that time period. Less than one quarter of all participants maintained their weight within 4 kg.

5. Diet

On average, obese individuals consume more calories than do thin individuals, although correlations between weight and reported caloric intake are often weak. Research of this topic is plagued by poor methodology to measure food intake in epidemiologic studies of free-living individuals. Also, the tendency of obese individuals to under-report their food intake (Schoeller 1995) biases associations between body weight and energy intake. Another reason for the difficulty in establishing associations between energy intake and weight is that only small deviations in energy balance sustained over time are necessary to produce large difference in body weight in the long term (Seidell and Flegal 1997). The methodological errors inherent in the determination of energy intake may be too large to allow examinations of the associations between energy intake and body weight in epidemiologic studies.

Recently, there has been controversy in regard to the effect of dietary fat on caloric intake and body weight (Popkin 1998; Willett 1998). There is evidence to indicate that subjects tend to consume more calories when the diet is higher in fat, and this can result in heavier body weights. Nevertheless, many of the long-term studies that show associations between fat intake and body weight are confounded by differences in economic development and physical activity level.

6. Physical Activity

The literature on associations between body weight and physical activity is also limited by methodology. It is difficult to measure accurately physical activity using methods appropriate for epidemiologic work. Nevertheless, several studies have shown that sedentary individuals are heavier than those who are physically active (Bouchard et al. 1993; Gortmaker and Dietz 1990). Very physically active individuals are rarely obese (Williams 1997).

7. Associations with Other Factors

Smokers have lower body weights than former smokers and never smokers (Flegal et al. 1995). Body weight usually declines with the initiation of habitual smoking and increases with cessation of smoking. The association between alcohol intake and body weight has been controversial; however, recent studies indicate a positive association between alcohol intake and body weight after controlling for confounding by cigarette smoking (Mannisto et al. 1996; Williamson et al. 1987). The prevalence of obesity is higher in Americans with lower levels of education (Sobal and Stunkard 1989). The number of children to whom a woman gives birth is weakly correlated with body weight, with increases likely to be less than 1 kg per pregnancy (Cederlof and Kaij 1970; Forster et al. 1986; Lewis et al. 1994).

II. International Trends in the Prevalence of Obesity

1. European Countries

Seidell (1997) has reviewed the prevalence of obesity (BMI $\geq$ 30) among European centers participating in the WHO-MONICA study between 1983 and 1986. Table 7 shows some of the results from that study. The average prevalence in the European countries studied was 15% in men and 22% in women. The populations measured were not necessarily representative of the countries in which they lived. Nevertheless, these data illustrate great variability

Table 7. Prevalence of obesity (BMI > 30 kg/m^2) in European WHO-MONICA populations (age-standardized prevalence 35–64 years)[a]

Part of Europe	Country	Center	Prevalence	
			Men	Women
Northern	Denmark	Glostrup,	11	10
	Finland	Kuopio Province	18	19
	Finland	North Karelia	17	24
	Finland	Turku-Loima	19	17
	Iceland	Iceland	11	11
	Sweden	Gothenburg	7	9
	Sweden,	Northern Sweden	11	14
Western	UK	Glasgow	11	16
	UK	Belfast	11	14
	Germany	Bremen	14	18
	Germany	Rhein-Neckar	13	12
	Germany	Augsburg	18	15
	Germany	Augsburg (rural)	20	22
	Belgium	Luxembourg Province	13	18
	Belgium	Ghent	11	15
	France	Lille	14	19
	France	Toulouse	9	11
	France	Strasbourg	22	23
	Switzerland	Ticinio	20	15
	Switzerland	Vaud-Fribourg	13	13
Eastern	Russia	Novosibirsk	14	44
	Russia	Moscow	13	33
	Lithuania	Kaunas	22	45
	East Germany	Halle County	18	27
	East Germany	Karl-Marx-Stadt	14	19
	East Germany	Cottbus County	17	23
	East Germany	'rest of DDR'	17	21
	Poland	Warsaw	18	26
	Poland	Tarnobrzeg	13	32
	Czech Rep.	Czech Republic	21	32
	Romania	Bucharest	20	31
	Serbia	Novi Sad	17	29
Southern	Spain	Catalonia	9	24
	Italy	Area Brianza	11	15
	Italy	Friuli	16	19

[a] Seidell 1997.

across countries. The lowest prevalence of obesity was found in Gothenburg, Sweden (7% in men, 9% in women) and the highest prevalence in Kaunas, Lithuania (22% in men, 45% in women).

2. Lower- and Middle-Income Countries

The prevalence of obesity tends to be lower in less well-developed countries than in high-income countries. Nevertheless, obesity rates are rising in lower-

Table 8. Estimated world prevalence of obesity

	Population aged 15+ in millions	Prevalence of obesity (%)	Approximate estimate of number of obese subjects in millions (midpoint)
Established market economies	640	15–20	96–128 (112)
Former socialist economies	330	20–25	66–83 (75)
India	535	0.5–1.0	3–7 (5)
China	825	0.5–1.0	4–8 (6)
Other Asia and Islands	430	1–3	4–12 (8)
Sub-Saharan Africa	276	0.5–1.0	1–3 (2)
Latin American and Caribbean	280	5–10	14–28 (21)
Middle Eastern crescent	300	5–10	15–30 (22)
World	3616		(251)

Population size and regions taken from Murray and Lopez 1996.

and middle-income countries (Popkin 1998). During the last several decades, a large number of countries that had formerly faced significant problems of malnutrition and famine have attained adequate food supplies. This has been accompanied by changes in diet composition. At the same time, there have been notable reductions in both occupational and non-occupational physical activity. Presumably, as a result of these two changes, the prevalence of obesity has increased over the last decade in virtually all countries for which measurements are available.

3. Estimated World Prevalence of Obesity

Murray and Lopez (1996) have attempted to quantify the number of obese adults in the world (Table 8). There is considerable uncertainty in their estimates due to the lack of studies of representative samples of many of the world's populations. The midpoints of the estimates add to slightly more than 250 million people, which is approximately 7% of the total adult world population. In many countries the prevalence of pre-obesity (BMI 25–29.9) is two to three times as large as the prevalence of obesity. Therefore, it is plausible that as many as one billion people are overweight (BMI ≥ 25) (Seidell 1997).

J. Conclusions

Obesity is and continues to be a major health problem in the Western World and a growing problem in Third World countries in which improvements in food availability have been achieved. Awareness of obesity and associated health risks will hopefully lead to measures by individuals, governments, and industry to influence the development of obesity, to limit its negative effects, and possibly to prevent its onset by a combination of judicious eating and steady physical activity (Hill and Peters 1998).

Continuing research is needed both in the definition of obesity (particularly in relation to degree and location of excess adiposity) and in the epidemiology of obesity (particularly in the link of obesity to major health hazards and world prevalence). It seems almost paradoxical that, at the time in history when efforts are made to reduce famine and its consequences worldwide, major health risks are uncovered, in the western world and in some third world countries, that are linked to unlimited food availability, excessive food ingestion, and enhanced fat accumulation in the body.

References

Abate N, Garg A (1995) Heterogeneity in adipose tissue metabolism: causes, implications, and management of regional adiposity. Prog Lipid Res 34:53–70
Alexander JK (1970) Cardiovascular effects of obesity. Med Counterpoint 15–29
Anderson KM, Kannel WB (1992) Obesity and disease. In: Björntorp P, Brodoff BN (eds) Obesity. Lippincott, Philadelphia, pp 465–473
Armellini, F, Zamboni M, Castelli S, Micciolo R, Mino A, Turcato E, Rigo L, Bergamo-Andreis IA, Bosello O (1994) Measured and predicted total and visceral adipose tissue in women. Int J Obes Relat Metab Disord 18:641–647
Bellisari A, Roche AF, Siervogel RM (1993) Reliability of B-mode ultrasonic measurements of subcutaneous adipose tissue and intraabdominal depth: comparison with skinfold thicknesses. Int J Obes Relat Metab Disord 17:475–480
Bouchard C, Depres JP et al (1993) Exercise and obesity. Obes Res 1:133–147
Bray G (1989) Classification and evaluation of the obesities. Med Clin North Am 73:161–184
Bray G (1992) Obesity: historical development of scientific and cultural ideas. In: Björntorp P, Brodoff BN (eds) Obesity. Lippincott, Philadelphia, pp 281–293
Broussard B, Johnson A et al (1991) Prevalence of obesity in American Indians and Alaska natives. Am J Clin Nutr 53:1535S–1542 S
Brown PJ (1992) The biocultural evolution of obesity: an anthropological view. In: Björntorp P, Brodoff BN (eds) JB Lippincott Co. Philadelphia Obesity, pp 320–329
Brozeck J, Grande CF, Anderson JT, Keys A (1963) Densitometric analysis of body composition from girth measurement: revision of some quantitative assumptions. Ann NY Acad Sci 110:113–140
Cederlof R, Kaij L (1970) The effect of childbearing on body weight. Acta Psychol Scand 219(S):47–49
Chowdhury B, Kvist H, Andersson B, Björntorp P, Sjöström L (1993) CT-determined changes in adipose tissue distribution during small weight reduction in obese males. Int J Obes Relat Metab Disord 17:685–691
Colt EW, Wang J, Stallone F, Van Itallie TB, Pierson Jr RN (1981) A possible low intracellular potassium in obesity. Am J Clin Nutr 34:367–372
Crandal DL, Hausman GJ, Kral JG (1997) A review of the microcirculation of adipose tissue: anatomic, metabolic, and angiogenic perspectives. Microcirculation 4:211–231
Cushman SW, Salan LB (1973) Lipolysis and triglyceride turnover in rat adipose cells: Effects of cell size. Federation Proc 32:940–940

Deurenberg P (1996) Limitations of the bioelectrical impedance method for the assessment of body fat in severe obesity. Clin Nutr 64:449S–452 S
DiGirolamo M, Fine JB, Tagra, K, Rossmanith R (1998) Qualitative regional differences in adipose tissue growth and cellularity in male Wistar rats fed *ad libitum*. Am J Physio 274:R1460–R1467
DiGirolamo M, Newby FD, Hill JO (1989) Blood lactate levels in human obesity. Int J Obes 13:394–394
DiGirolamo M, Newby FD, Lovejoy J (1992) Lactate production in adipose tissue: a regulated function with extra-adipose implications. FASEB J 6:2405–2412
DiGirolamo M, Owens JL (1976) Water content of rat adipose tissue and isolated adipocytes in relation to cell size. Am J Physiol 231:1568–1572
Durnin JV, Womersley J (1974) Body fat assessed from total body density and its estimation from skinfold thickness: measurements on 481 men and women aged from 16 to 72 years. Br J Nutr 32:77–97
Flegal KM, Troiano R et al (1995) The influence of smoking cessation on the prevalence of overweight in the United States. N Engl J Med 333:1165–1170
Flegal KM, Carroll M, Kuczmaraki RJ, Johnson CL (1998) Overweight and obesity in the United States: prevalence and trends, 1960–1994. Int J Obes 22:39–47
Forbes EG, Brown MR et al (1988) Arm muscle plus bone area: anthropometry and CAT scan compared. Am J Clin Nutr 47:929–931
Forster LA, Bloom JE et al (1986) Reproductive history and body mass index in black and white women. Prev Med 15:685–691
Goodman-Gruen D, Barrett Connor E (1996) Sex differences in measures of body fat and body distribution in the elderly. Am J Epid 143:898–906
Gortmaker S, Dietz W (1990) Inactivity, diet, and the fattening of America. J Am Dietetic Assoc 90:1247–1255
Hill JO, Peters JC (1998) Environmental contributions to the obesity epidemic. Science 280:1371–1373
Hoogwerf BJ, Nuttall FQ (1984) . Long-term weight regulation in treated hyperthyroid and hypothyroid subjects. Am J Med 76:963–970
Hortobagyi T, Israel RG, O'Brien KF (1994) Sensitivity and specificity of the Quetelet Index to assess obesity in men and women. Euro J Clin Nut 48:369–375
International Obesity Task Force. World Health Organization (1998) Obesity: preventing and managing the global epidemic. World Health Organization, Geneva
Keys A, Fidanza F, Darvonen MJ, Kimura N, Taylor HL (1972) Indices of relative weight and obesity. J Chron Dis 25:329–343
Kissebah AH (1996) Intra-abdominal fat: is it a major factor in developing diabetes and coronary artery disease? Diab Res Clin Prac 30:25–30
Klatsky A, Armstrong M (1991) Cardiovascular risk factors among Asian Americans living in northern California. Am J Pub Health 81:1423–1428
Kono T (1985) Glucose transport in adipocytes and its regulation by insulin. Int J Obes 9 [Suppl 1]: 105–108
Kowalski TJ, Wu G, Watford M (1997) Rat adipose tissue amino acid metabolism in vivo as assessed by microdialysis and arterio-venous techniques. Am J Physio (Endo and Metab) 273:E613–E622
Kuczmarski RM, Carroll M et al (1997) Varying body mass index cutoff points to describe overweight prevalence among U.S. adults: NHANES III (1988 to 1994). Obes Res 5:542–548
Kuczmarski R, Flegal KM et al (1994) Increasing prevalence of overweight among U.S. adults: the National Health and Nutrition Examination Surveys, 1960 to 1991. J Am Med Assoc 272:205–211
Kumanyika S (1994) Obesity in minority populations: an epidemiologic assessment. Obes Res 2:166–182
Lapillonne A, Braillon PM, Delmas PD, Salle BL (1997) Dual-energy x-ray absorptiometry in early life. Horm Res 48:43–49
Laskey MA (1996) Dual-energy x-ray absorptiometry and body composition. Nutrition 12:45–51

Leibel RL, Rosenbaum M, Hirsch J (1995) Changes in energy expenditure resulting from altered body weight. N Engl J Med 332(10):621–628
Lemieux SD, Prudhomme D, Bouchard C, Tremblay A, Depres JP (1996) A single threshold value of waist girth identifies normal-weight and over-weight subjects with excess visceral adipose tissue. Am J Clin Nutr 64:685–693
Lewis C, Smith D et al (1994) Associations of body mass and body fat distribution with parity among African-American and Caucasian women: the CARDIA study. Obes Res 2:517–525
Mannisto S, Pietinen P et al (1996) Reported alcohol intake, diet, and body mass index in male smokers. Euro J Clin Nutr 50:239–245
Matson, SC, Henderson KA, McGrath GJ (1997) Physical findings and symptoms of depot medroxyprogesterone acetate use in adolescent females. J Ped Adol Gyn 10:18–23
Matsuzawa Y, Fujioka S, Tokunaga K, Seichiro T (1991) Classification of obesity with respect to morbidity. Proc Soc Exp Biol Med 200:197–201
McGarry JB and Foster DW (1979) Hormonal Control of Ketogenesis. Adv Exp Med Biol 3:79–96
Metropolitan Life Insurance Company (1983) New weight standards for men and women. Statistical Bulletin Metropolitan Life Insurance Company 64:2–9
Murray CJ and Lopez AD (1996) The global burden of disease (WHO World Bank)
Newby FC, Wilson LK, Thacker SV, DiGirolamo M (1990) Adipocyte lactate production remains elevated during refeeding after fasting. Am J Physiol 259:E865–E871
Pears J, Jung RT, Gunn A (1990) Long-term weight changes in treated hyperthyroid and hypothyroid patients. Scottish Med J 35:180–182
Pierson Jr RN, Lin KHY, Phillips RA (1974) Total body potassium in health: the effects of age, sex, height, and fat. Am J Physiol 226:206–212
Pierson RN Jr, Wang J, Thornton JC (1997) Measurement of body composition: applications in hormone research. Horm Res 48:56–62
Plourde G (1997) The role of radiologic methods in assessing body composition and related metabolic parameters. Nutr Rev 55:289–296
Popkin B (1998) The obesity epidemic is a worldwide phenomenon: trends in transitional societies. Nutr Reviews 56:106–114
Price R, Charles M, et al (1993) Obesity in Pima Indians: large increases among post-World War II birth cohorts. Am J Physiol Anthropol 92:473–479
Quetelet LAJ (1981) Antropometric pour mèsure des differentes facultés de l'homme. Brussels, Belgium: C. Muquardt: 479
Rebuffé-Scrive M (1991) Neuroregulation of adipose tissue: molecular and hormonal mechanisms. Int J Obes 15:83–86
Roch AF (1985) Body composition in youth and adults. Ross Laboratories
Schoeller DA (1995) Limitations in the assessment of dietary energy intake by self-report. Metabolism 44:18–22
Schoeller DA (1996) Update: NIH consensus conference. Bioelectrical impedance analysis for the measurement of human body composition: where do we stand and what is the next step? Nutrition 12:760–762
Seidell J (1998) Societal and personal costs of obesity. Exp Clin Endocrinol Diabetes 106 [Suppl] 2:7–9.
Seidell J, Flegal K (1997) Assessing obesity: classification and epidemiology. Br Med Bull 53:238–252
Seidell JC (1997) Time trends in obesity: an epidemiological perspective. Horm Metab Res 29:155–158
Shapiro B, Wertheimer E (1948) The synthesis of fatty acids in adipose tissue in vitro. J Biol Chem 173:725
Sims EA, Danforth E, Horton ES, Bray GA, Glennon JA, Salans LB (1973) Endocrine and metabolic effects of experimental obesity in man. Recent Prog Horm Res 29:457–496

Siri WE (1969) Body composition from fluid spaces and density: analysis of methods. Techniques for measuring body composition. Washington DC, Natl Acad Sci, 223–244

Snyder WS, Cook MJ, Nasset ES (1984) Report of the task group on reference man. Oxford, Pergamon Press

Stallone DD, Stunkard AJ, Wadden TA, Foster GD, Boorstein FJ, Arger P (1991) Weight loss and body fat distribution: a feasibility study using computerized tomography. Int J Obes 15:775–780

Stevens J, Cai J et al (1998) The effects of age on the association between body mass index and mortality. N Engl J Med 338:1–7

Tornaghi G, Raiteri R, Pozzato C, Rispoli A, Bramani M, Cipolat M, Craveri A (1994) Anthropometric or ultrasonic measurements in assessment of visceral fat? A comparative study. Int J Obes 18:771–775

Vague J, Boyer J, Jubelin J, Nicolino C, Pinto C (1969) Le rapport adipo-musculaire. Bull Acad Natl Med 153:46–54

Van der Kooy K, Seidell JC (1993) Technique for the measurement of visceral fat: a practical guide. Int J Obes Relat Metab Disord 17:187–196

Wang J, Thornton J et al (1996) Comparison for body mass index and body fat percent among Puerto Ricans, blacks, whites, and Asians living in the New York City area. Obes Res 4:377–384

Wellens RL, Roche AF, Khamis HJ, Jackson AS, Pollock ML, Siervogel RM (1996) Relationship between body mass index and body composition. Obes Res 4:35–44

Welty T, Lee E et al (1995) Cardiovascular disease risk factors among American Indians, the Strong Heart Study. Am J Epid 142:269–287.

Wickelgren I (1998) Obesity: how big a problem? Science 280:1364–1366

Willett W (1998) Is dietary fat a major determinant of body fat? Am J Clin Nutr 67 [Suppl]: 556S–562 S

Williams P (1997) Relationship of distance run per week to coronary heart disease risk factors in 8283 male runners. Arch Intern Med 157:191–198

Williamson DF, Forman, MR, Binkin NJ, Gentry EM, Remington PL, Trowbridge FL (1987) Alcohol and body weight in United States adults. Am J Public Health 77:1324

Wolf A, Colditz G (1998) Current estimates of the economic cost of obesity in the United States. Obes Res 6:97–106

Zumoff B, Strain GW, Miller LK (1990) Plasma free and non-sex hormone binding globulin-testosterone are decreased in obese man in proportion to their degree of obesity. J Clin Endocrinol Metab 70:929–931

CHAPTER 2

Health Implications of Obesity

F.X. PI-SUNYER

A. Introduction

The prevalence of overweight and obesity is increasing in the United States (PI-SUNYER 1994; FLEGAL et al. 1998) and in many other nations of the world (OBESITY 1997). From a public health viewpoint, it is important to establish the health risks of such a phenomenon. Both morbidity and mortality risk have been shown to increase with increasing weight in population studies. In addition, it has become evident that central fat distribution has a negative impact on health. This chapter describes the epidemiologic evidence that defines the risks of obesity and of fat distribution on morbidity and mortality.

B. Some Epidemiological Considerations

The duration of an epidemiologic study is an important determinant of the strength of association between excess weight and health outcome (FEINLIEB 1985). Short-term studies of 10 years or less have reported little association between excess body weight and health outcome, intermediate range studies (10–20 years) have shown a weak association, and long-term studies (greater than 20 years) generally have shown a strong association (FEINLIEB 1985; GARRISON and CASTELLI 1985; RABKIN et al 1977).

The relation between obesity and health risk also depends on the age of the group being studied. Generally, epidemiologic studies have found that obesity has more impact on disease and mortality if measured from younger as opposed to older ages (GARRISON et al. 1985; RABKIN et al. 1977; STAMLER et al. 1978; RISSANEN et al. 1989; STEVENS et al. 1998; DIEHR et al. 1998). This is partly related to the fact that early obesity is much more strongly related to diseases like hypertension and diabetes than later onset obesity, when such diseases develop as independent variables related to the process of aging itself (PI-SUNYER 1994; BARRETT-CONNOR 1985; TAYBACK et al. 1990).

Weight is a changing variable, fluctuating throughout life. In the United States, weight increases on average as people move from age 20 to age 60 (KUCZMARSKI et al. 1991). However, most epidemiologic studies take a static approach in which individuals are classified by weight at the beginning of a study and are then followed longitudinally for either disease or mortality outcomes, with no "correction" for changes in weight over time. However, gain or loss may have a powerful impact on outcomes. As a result, some recent studies have attempted to relate such weight changes to morbidity and mortality results (Chan et al. 1994; Colditz et al. 1995).

As an increasing proportion of the world's population becomes obese, health risks attributable to obesity will increase, increasing the world-wide public health burden (1998).

C. Mortality

The life and medical insurance industries were the first to investigate the relationship of body weight to increased risk of disease and, particularly, to death. In the two largest of these studies, the Build and Blood Pressure Study of 1959 (Metropolitan Life Insurance Company 1960) and the Build and Blood Pressure Study of 1979 (Society of Actuaries and Association of Life Insurance Medical Directors of America 1980), a J-shaped curve of mortality was found to occur, with persons at the low and high weights showing the greatest mortality risk. There was a significant elevation in mortality in the 1959 Build Study over mean baseline weight for obese persons, and this increased with increasing overweight to reach an excess mortality of 42% at an overweight of 30% above the 1959 Metropolitan Life Insurance median standard weight. In women, there was a similar but somewhat lower increasing mortality with increasing relative weight.

In the 1979 Build Study, the weights associated with the least mortality in men under 50 years of age were those between 5% and 15% under the ideal weights of the Metropolitan Life Tables. In men, as age increased, the optimal weights for the least mortality increased somewhat, but were still below the median standard weight of the 1959 Metropolitan Life Tables. For women, the best weights with regard to mortality were 5%–15% below the standard weights for all ages (Society of Actuaries and Association of Life Insurance Medical Directors of America 1980). For both sexes, mortality increased with increasing weight, but the association was not linear, since the rise in mortality became steeper as excess weight increased.

The primary causes of death in the 1979 study were somewhat different in men and women. In men, heart and circulatory diseases, diabetes mellitus, digestive disorders, and stroke were highest, while in women they were heart and circulatory diseases and cancer (Society of Actuaries and Association of Life Insurance Medical Directors of America 1980).

Table 1 presents the data on mortality by weight and age in the healthy people studied in the 1979 Build Study (Society of Actuaries and Association of Life Insurance Medical Directors of America 1980; McCue 1981). The optimal weights with regard to mortality were below the average weights of the population studied in ostensibly healthy men all the way to age

Table 1. Variations in mortality by weight according to age and duration of policy in the Build study 1979 (after McCue 1981)

	Mortality ratios by ages (in years)					Mortality ratios by duration of policy (in years)			
	20–29	30–39	40–49	50–59	60–69	1–5	6–10	11–15	16–22
Men									
25% underweight	102	105	112	128	135	127	119	114	105
15–25% under-weight	94	93	98	113	120	110	103	99	93
5–15% underweight	95	92	93	100	100	98	97	92	93
5% underweight–5% overweight	98	95	96	94	95	94	96	95	95
5–15% overweight	103	112	109	100	99	103	103	109	Ill
15–25% overweight	125	128	118	109	101	106	114	123	131
Women									
≥25% underweight	–	–	117	146	134	167	128	134	90
15–25% underweight	118	124	110	105	106	114	105	102	107
5–15% underweight	88	101	92	93	90	92	90	94	94
5% underweight–5% overweight	112	86	96	97	101	91	94	97	97
5–15% overweight	90	99	103	99	102	95	99	103	102
15–25% overweight	118	110	115	103	103	106	103	113	112

70 years and in ostensibly healthy women to 80 years. The most common diseases listed as causes of death in both the 1959 and the 1979 Build Studies were coronary artery disease, diabetes mellitus, digestive disease, and stroke.

There are some problems with these life insurance data. They were primarily gathered from Caucasian, upper-, and middle-income persons and as such were not representative of the population as a whole. In addition, they tended to screen out persons with conditions associated with obesity, such as hypertension, diabetes, and history of heart disease. They thus reflect a "cleaned-up" version of the population that the industry considered to be at reasonable risk for life insurance.

In an effort to circumvent these criticisms, the American Cancer Society (ACS) initiated a long-term longitudinal study in the general U.S. population (Lew 1985; Lew and Garfinkel 1979). The results of this study, conducted between 1960 and 1972 on 750 000 people, were similar to the two Build studies with regard to weight and mortality. This suggested that the data obtained from the insurance pool population could be extrapolated to the general population. In all three of the studies, the lowest mortality was found at a weight lower than the population average (Metropolitan Life Insurance Company 1960; Society of Actuaries and Association of Life Insurance Medical Directors of America 1980; McCue 1981; Lew 1985; Lew and Garfinkel 1979).

Because the ACS was specially interested in cancer, it separated smokers from non-smokers, which the insurance studies had not done. Smokers have an independent risk for mortality and they are also lower weight. Since the two Build studies included smokers, they overstated the risks of being underweight and understated the mortality risk of being overweight (Metropolitan Life Insurance Company 1960; Society of Actuaries and Association of Life Insurance Medical Directors of America 1980; McCue 1981; Lew 1985; Lew and Garfinkel 1979).

Waaler (1984) reported on a very large longitudinal population study carried out in Norway. This related weight to mortality and was one of the first in which the weight was expressed in body mass index (BMI) units. BMI is weight in kilograms divided by height in meters squared. The lowest mortality was found in the BMI range of 21–25. In the younger male age groups, increases in mortality began to rise significantly at a BMI of 27, and in females at a BMI of 28. Older cohorts (55 and above) have a wider BMI range at the upper end for minimal mortality.

The Framingham study was another longitudinal investigation of a sample of the population of a representative Massachusetts town. At 26 years of follow-up, the lowest mortality in this group occurred at weights that were 5%–15% below the average weight of the population (Garrison 1983). Other longitudinal studies, in which subjects were also followed for 34 and 26 years, showed a significant independent effect of obesity on mortality (Blair et al. 1966; Manson et al. 1995; Troiano et al. 1996). The necessity of following a population long enough if a significant relationship is to be detected between obesity and mortality has been emphasized by FEINLIEB who, referring to the Framingham data, stated that "during the first 6 years, among non-smokers, mortality and weight had an inverse relation, but as time goes on the relation becomes U-shaped with a minimum mortality at a relative weight of 100%–109%. Similar relations are found among men who smoke" (FEINLIEB 1985).

The Nurses' Health Study, which is following about 116 000 US women, has reported in its 16 year follow-up that a significant association exists between BMI and mortality (MANSON et al. 1995). These nurses, being followed by periodic questionnaires, ranged from 30 to 55 years of age at the start of the study. The lowest mortality was found at a weight at least 15% lower than the average for US women of similar age. The relative risk for mortality from heart disease in non-smoking women began to rise at a BMI of 22 and was 7.7 times higher at a BMI >32. The relative risk for mortality from all causes in non-smoking women also began to rise at a BMI of 22 and was 2.2 times greater at a BMI > 32. Thus, that overweight affects mortality in women as well as men, which had been previously reported from Framingham, has been strongly confirmed by this study. TROIANO et al. (1996), in a more recent meta-analysis of ten studies, describes a J-shaped curve of mortality for both men and women, with enhanced risk at both the low and the high end of the BMI scale.

Though it is evident that the mortality risk of obesity is greater with longer prospective periods of observation, this does not mean that older persons are not at risk. For instance, risk of death was twice as great for people with BMIs greater than or equal to the 70th percentile of weight at both 55 and 65 years of age in the Framingham study (HARRIS et al. 1988), even controlling for cholesterol, glucose, and systolic blood pressure. In a more recent study, the Longitudinal Study of Aging (ALLISON et al. 1997), lowest mortality occurred at a BMI of 27–30 for men and 30–35 for women older than 70 years. STEVENS et al. (1998) have reported quite similar data.

Although the relationship of obesity to mortality is strong, as detailed above and in many other longitudinal studies, there are a number of other studies that have not found a relationship between the two (ROSE et al. 1977; KEYS 1980; KOZAREVIC et al. 1976; BORHANI et al. 1963; WESTLUND et al. 1972; ROSENMAN et al. 1976; CHAPMAN et al. 1964; PAUL et al. 1963; NOPPA et al. 1980; DYER et al. 1975; TIBBLIN et al. 1975). There are at least two possible explanations for this. First, some of the studies have been rather short, a problem that has already been discussed. Second, in many of the studies, those very diseases that are often associated with and made worse by obesity have been eliminated or controlled for. For instance, diabetic patients have been excluded from most of the studies; hypertensive persons have either been excluded or controlled for. The rationale for doing this has been that an independent effect of obesity is being looked for. The great problem is that obesity exerts much of its effect through these conditions, and the elimination of all of these so-called co-morbid conditions greatly attenuates the adverse effect of obesity. This has been discussed by Manson et al. (1987), who have shown that in 25 major prospective studies on the relationship between obesity and longevity, each study showed at least one of three biases that led to systematic underestimation of the impact of obesity on death. These three biases were:

1. Failure to control for cigarette smoking
2. Inappropriate control for the biological effects of obesity, such as hypertension and hyperglycemia
3. Failure to control for weight loss due to subclinical disease

D. Diabetes Mellitus

Diabetes mellitus has long been known to have an association to obesity. Initial cross-sectional studies of population groups showed that the heavier populations had a greater prevalence of diabetes (West et al. 1966, 1971). In a given population, as weight increases, so does the prevalence of diabetes (Pi-Sunyer 1996). In a study conducted on the Pima Indians in Arizona, there was a steady increase in the incidence of type 2 diabetes as the body mass index moved from less than 20 up to greater than 40 (Knowler et al. 1981). Similar data have been reported in other American Indian groups (Albu et al. 1997a). In the Nurses' Health Study, where over 116,000 women have been followed prospectively over 16 years, weight was a very strong predictor for the development of diabetes (Colditz et al. 1990). The excess risk began at a BMI of 22 and rose steeply thereafter. Very similar data have been recently reported in a longitudinal study of men (Chan et al. 1994). Also, the longer the duration of the obesity, the greater the incidence of diabetes (Everhart et al. 1992).

Other variables that increase the risk of developing diabetes include aging (NIH Publication No. 95–1468 1995; National Center For Health Statistics 1993, 1995), a family history of type 2 diabetes (Knowler et al. 1981; Rewers et al. 1995) and centrally located obesity (Sparrow et al. 1986; Lundgren et al. 1989; Ohlson et al. 1985; Haffner et al. 1991). Table 2 gives prevalence rates in the United States for type 2 diabetes in relation to age and BMI (Bonham and Brock 1985). In both men and women, the great impact of increasing BMI and increasing age is clear (Colditz et al. 1990; Chan et al.

Table 2. Prevalence of adult onset diabetes: United States, 1976 ($n = 64142$) (after Bonham and Brock 1985)

Age (in years)	Men, by BMI (%)				Women, by BMI (%)			
	<24	24–25	26–27	>28	<22	22–25	26–27	>28
20–44	0.5	0.5	0.6	1.3	0.5	0.6	1.1	3.4
45–54	3.0	2.5	3.4	6.9	1.6	2.1	3.6	9.1
55–64	4.3	4.7	5.9	9.9	2.5	3.7	5.1	9.6
65+	6.7	8.0	8.2	9.4	5.6	6.4	8.9	13.2

1994; EVERHART et al. 1992; NIH PUBLICATION NO.95–1468 1995; NATIONAL CENTER FOR HEALTH STATISTICS 1993, 1995). Other countries have reported similar relationships, for example Israel (MEDALIE et al. 1974), Norway (WESTLUND et al. 1972), and Sweden (LARSSON et al. 1981). A recent study in Sweden has reported a large effect of morbid obesity on the prevalence of diabetes (SJÖSTRÖM 1992).

Weight gain in the adult years greatly increases the risk of developing diabetes. Not only does increasing weight gain increase risk, but also the risk is enhanced the higher the baseline BMI is that an individual starts from (COLDITZ et al. 1990; CHAN et al. 1994). In both cross-sectional (SPARROW et al. 1986; HAFFNER et al. 1991; DESPRÉS et al. 1989) and longitudinal studies (CHAN et al. 1994; LUNDGREN et al. 1989; OHLSON et al. 1985) fat distribution has also been found to be important in diabetes risk. The greater the amount of central or upper body or abdominal obesity, the greater the risk for diabetes.

The increased incidence of diabetes that occurs with overweight is related in part to the greater insulin resistance that develops when excess fat is deposited in the body (OLEFSKY 1981). It is well known that as insulin resistance increases, the incidence of type 2 diabetes rises (REAVEN 1988). The insulin resistance occurs primarily in two tissues, adipose tissue and muscle (RABINOWITZ et al. 1962). While it is clear that there is a down-regulation of insulin receptors in these insulin sensitive tissues, a more important factor is the development of post-receptor defects in insulin action (KOLTERMAN et al. 1980). Both glucose transport (HISSIN et al. 1982) and glucose oxidation and storage (BOGARDUS et al. 1984; SEGAL et al. 1991) are affected. There is impairment in the activation of tyrosine kinase after insulin locks to its receptor in insulin-sensitive cells (CARO et al. 1989). The net effect of the insulin resistance is an increased insulin secretion by the β cells of the pancreas and a higher prevailing blood insulin level (REISBERG et al. 1999; DEFRONZO et al. 1995; DEFRONZO 1988). In those obese persons with the appropriate genetic make-up, that is to say, the propensity to diabetes, the β cell secretion eventually begins to exhaust, the early phase of insulin secretion is lost, and carbohydrate tolerance begins to become impaired (PORTE 1991; PIMENTA et al. 1995; ERIKKSON et al. 1989). Eventually, with further exhaustion of the β cells, frank diabetes supervenes.

E. Cardiovascular Disease

Cardiovascular disease (CVD) is increased in subjects who are obese. The manifestations include coronary artery disease, sudden death, congestive heart failure, and arrhythmias. Since many obese persons also develop hypertension and dyslipidemia, sometimes it has been difficult to assign causality for the cardiovascular disease to obesity independent of these other conditions. Hypertension and dyslipidemia will be addressed first in relation to obesity and then cardiovascular disease will be discussed.

I. Hypertension

Numerous cross-sectional studies have established the association of hypertension with obesity (Paffenbarger et al. 1968; Weiss et al. 1978; Stamler et al. 1978; McMahon et al. 1984; Hypertension Detection and Follow-Up Program Cooperative Group 1977; Kannel et al. 1967). The National Health and Examination Survey (NHANES II), a cross-sectional study conducted on a representative sample of the US population from 1976–1980, reported that the prevalence of hypertension in overweight adults was 2.9 times greater than that among non-overweight adults (Van Itallie 1985). In persons aged 20–44 years, the risk of hypertension was 5.6 times greater than that in persons 45–74 years old, a figure which in turn was twice as high as that for non-overweight adults (Burton et al. 1985). (In this survey, overweight was defined as a BMI of 27.3 in males and 27.8 in females (National Center for Health Statistics 1987). This NHANES data documented a much greater association between BMI and blood pressure in the 25–54 year old age group than in the 65–75 year old group (Van Itallie 1985). A similar representative survey was carried out in Canada in the 1980s. In this study, again there was increasing prevalence of hypertension as the BMI rose (Stockwell 1985). An example of this is reproduced in Fig. 1 for diastolic hypertension in women.

The increasing prevalence of hypertension with increasing weight is clearly dissected in the reports of the Framingham study. In men, for every 10% increase in relative weight, systolic blood pressure increased by 6.5 mmHg (Kannel et al. 1969). Also, the Framingham investigators reported

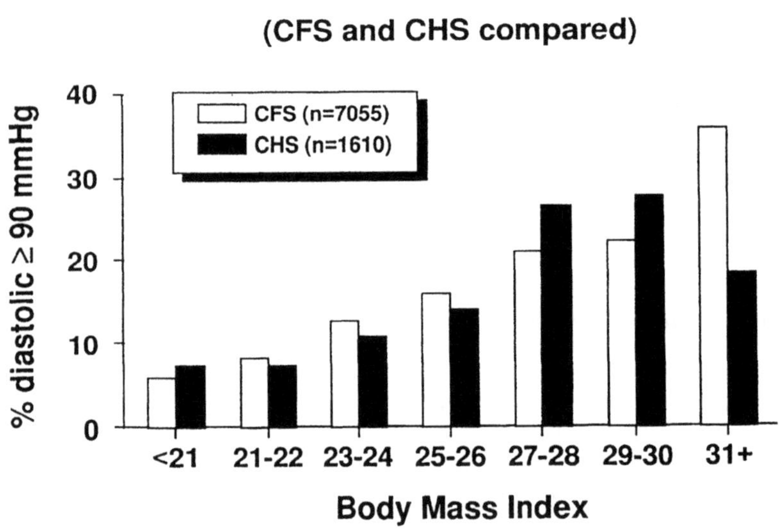

Fig. 1. BMI and diastolic hypertension in women

that persons who were only 20% overweight had an eightfold greater incidence of hypertension (BJERKEDAL 1957). In a large Swedish study in which 67000 adults were surveyed, for every 10kg increase in weight there was a 3mmHg increase in systolic and a 2mmHg increase in diastolic pressure (BJERKEDAL 1957).

In the Evans County longitudinal study, a gain in weight during a 6-year period of observation doubled the risk of hypertension (TYROLER et al. 1975). In a longer one, where college men were followed for a period of 32 years, the BMI was significantly associated with blood pressure, and weight gain significantly increased both systolic and diastolic blood pressure (GILLUM et al. 1982). In the longitudinal Nurses' Health Study, the relative risk for developing hypertension over a 16 year period went from 1.0 for a BMI < 23, to 1.67 for a BMI 23–25, to 2.8 for a BMI of 26–28, to 3.86 for a BMI of 29–31, to 5.70 for a BMI>32 (WITTEMAN et al. 1989).

In the U.S. population, blood pressure rises with age. This is not so in populations where there is not an adult body weight increase with age. Evidence for the relationship of obesity with weight can also be found in the studies which demonstrate that blood pressure drops if weight is lost (NIH 1979; DANNENBERG et al. 1988; GENUTH et al. 1974; RAMSAY et al. 1978; REISIN et al. 1978; KEMPNER et al. 1975; LANGFORD et al. 1991). This is true in young persons as well as old (WHELTON et al. 1996).

While the association between obesity and hypertension is clearly established, the pathophysiology and etiology are not. A number of mechanisms are probably responsible, with their effect varying from individual to individual. A decrease in renal filtration surface has been described in obesity, and this could lead to renal sodium retention (BRENNER et al. 1988). Also, the insulin resistance which occurs with obesity causes an elevation of both fasting and post-prandial insulin levels. Since insulin is a powerful stimulant of tubular sodium reabsorption, this also may be an important contributor (DEFRONZO et al. 1975). In addition, an enhanced activation of the sympathetic nervous system in obesity has been reported (SOWERS et al. 1982; LANDSBERG et al. 1989). This has been well described in studies on the Pima Indians (SPRAUL et al. 1994) and needs to be confirmed in other groups, since not all studies agree (YOUNG et al. 1992).

There has been increasing interest in the effect of insulin resistance on the development of hypertension in obesity. FERRANNINI (1995) reported some years ago that hypertensive persons are more likely to be insulin resistant than normotensive individuals. The exact mechanisms involved have not been fully worked out (FERRANNINI 1993). However, two possibilities relate to changes in ion transport and effects on the renin-angiotensin-aldosterone system. Alterations in sodium/hydrogen exchange, intracellular sodium/calcium levels, and intracellular sodium/magnesium levels all may be involved (TEDDE et al. 1988; JACOBS et al. 1993; RESNICK et al. 1988). Also, the activity of the renin-angiotensin-aldosterone system has been found to be elevated in obesity (ROCCHINI et al. 1986; SCAVO et al. 1968; TROVATI et al. 1989).

Hypertension leads to an increased risk of stroke. In the Framingham study, the incidence of stroke went from 22 to 30 to 49 per thousand for men younger than 50 years and from 8 to 14 to 35 per thousand for women in the same age group with relative weights of 110, 110 to 129, and 130 or more, respectively. Among persons older than 50 years, these figures went from 70 to 92 to 80 per thousand for men and from 45 to 64 to 121 per thousand for women with relative weights of 110, 110–129, and 130 or more, respectively (Hubert 1983).

II. Dyslipidemia

Many obese patients manifest dyslipidemia. The two most common abnormalities that are found in the lipid fractions are an increase in the serum triglycerides (Stamler 1979; Allbrink et al. 1962) and a decrease of HDL-cholesterol (Lamarche et al. 1993; Laakso et al. 1990). Very often, as weight increases, the levels of triglyceride rise (Lamarche et al. 1993). It is very common that as the triglycerides go up, HDL-cholesterol drops (Glueck et al. 1980; Gordon et al. 1977; Kannel et al. 1979). When total or low-density lipoprotein (LDL) cholesterol have been measured in obese persons, they have been found to be normal (Montoye et al. 1966) or mildly elevated (Assmann et al. 1992). The Framingham Study did report, however, that for every 10% increase in relative weight there was an increase in plasma cholesterol of 12 mg/dl (Kannel et al. 1979). Also, in the NHANES II data, the relative risk of hypercholesterolemia (set at 6.47 mmol/l, a definite risk) was 1.5 times greater in obese than lean persons (Van Itallie 1985). In a study carried out in volunteer blood donors of both sexes, cholesterol increased as relative weight increased, and this was more pronounced in men than in women and in those subjects less than 49 years than those greater than 49 years (Farinaro et al. 1979). Even when LDL-cholesterol does not rise too much, the quality of the LDL particles changes, so that small dense particles predominate. These are more atherogenic (Lamarche et al. 1997; Reaven et al. 1993). Since, in addition to these small LDL particles, both triglycerides and HDL-cholesterol are independent risk factors for coronary heart disease, the nature of the dyslipidemia of obesity clearly predisposes to this disease (Assman et al. 1992; Manninen et al. 1992).

III. Coronary Heart Disease

Coronary heart disease (CHD) is defined by epidemiologists as non-fatal myocardial infarction and angina pectoris (Manson et al. 1990). Whether obesity is directly a cause of CHD has been the object of much controversy. The controversy is fueled by the fact that obesity enhances other risk factors that lead to CHD, for example, high blood pressure, dyslipidemia, and glucose intolerance. But in epidemiological studies it has now been shown that obesity independently increases CHD (Feinlieb 1985). This has generally been found in longer-term longitudinal studies. For instance, both the Framingham Study and the Los Angeles Heart Study showed an independent effect of obesity on CHD in their later studies whereas they had not done so in their earlier studies with a shorter follow-up period of time (HUBERT 1983; CHAPMAN et al. 1971). Other longitudinal studies came to the same positive conclusion. These include the Honolulu Heart Program(REED et al. 1991), the US National Health and Nutrition Examination Survey I Epidemiologic Follow-Up Study (FREEDMAN et al. 1995), and the Whitehall Study (FITZGERALD et al. 1992). RABKIN et al. (1977) showed an increased risk of CHD with increasing BMI in a 26 year study of men in Manitoba, and RIMM et al. (1995) showed a similar relationship in a short study of 29 122 men.

The Nurses' Health Study is following 115 886 nurses longitudinally. These women were 30–55 years old at the start of the study and were free of known coronary heart disease, stroke, and cancer. At 16 years of follow-up, BMI was positively associated with CHD, as shown in Fig. 2 (MANSON et al. 1990). It is interesting that controlling for hypertension, diabetes mellitus, and hypercholesterolemia diminished the strength of the association but did not eliminate it.

The relation between obesity and CHD mortality has also been controversial. STALLONES (1985), in reviewing seven studies examining this relationship, found that four reported a positive association and three did not. In the Pooling Study, where a number of independent studies were pooled for greater statistical power, the relationship was positive, even though some of the individual studies were not (THE POOLING PROJECT RESEARCH GROUP 1978). More recently, there have been other studies that have shown an association between obesity and CHD mortality. The Seventh Day Adventist study has reported on a 26 year study in non-smoking men (LINSTED et al. 1991). A significant increase in mortality from both cardiovascular disease and cancer was found with increasing BMI. The protective effect associated with low body weight decreased with advancing age and disappeared by 90 years of age. Other studies have shown similar relationships between body weight and mortality from CHD, with no upswing of deaths at the lower weights. These include the Albany study (GORDON et al. 1988) and the Kaiser Permanente Study (SIDNEY et al. 1987). In women, the Nurses' Health Study showed a steeply rising rate of mortality with increasing BMI, beginning at a level of BMI of 19–21.9 (MANSON et al. 1995).

F. Respiratory

The respiratory system is negatively affected by obesity. While this is not an early manifestation of mild obesity, as BMI level increases and the duration of obesity is of longer duration, the incidence of respiratory symptoms increases. Initially, there is a mechanical effect exerted by the increased fat on the chest wall and abdomen. This reduces lung volume, increases respiratory

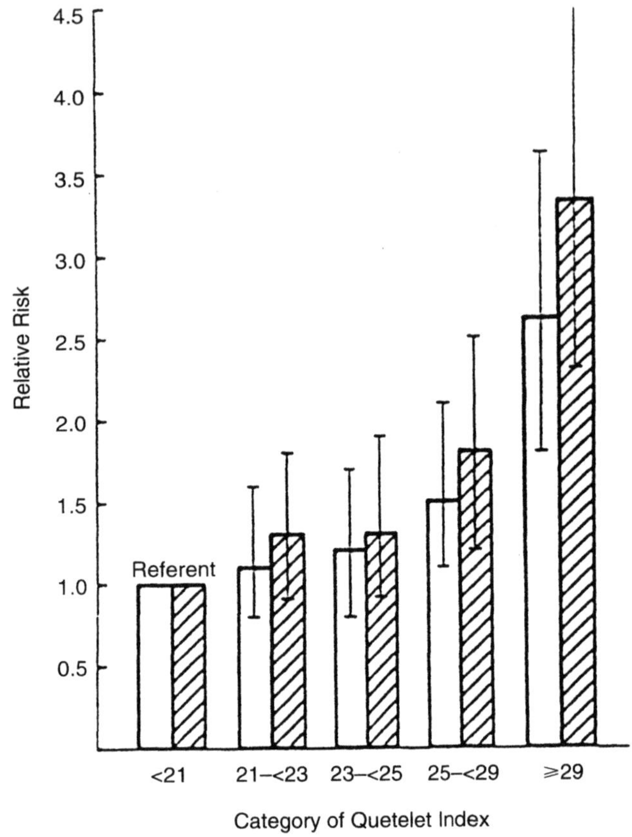

Fig. 2. Relative risks of nonfatal myocardial infarction and fatal coronary heart disease (combined), according to a category of the Quetelet index in a cohort of U.S. women 30–55 years of age in 1976. The reference category was that with an index under 21. For the other categories, *open bars* show the relative risks as adjusted for age, and *hatched bars* show the relative risks adjusted for both age and smoking. The *vertical lines* represent 95% confidence intervals

rate, and leads to decreased lung compliance (NAIMARK et al. 1960; WALTEMATH et al. 1974). Vital capacity drops, as does total lung capacity. Eventually, a ventilation-perfusion abnormality develops. This is initially characterized by hypoxia but a normal arterial pCO_2 (HOLLEY et al. 1967; DOUGLAS et al. 1972); eventually pCO_2 rises.

Sleep apnea is also common as obesity becomes more severe (SCHMIDT-NOWARA 1990; MORTONE et al. 1985; STRADLING et al. 1991; YOUNG et al. 1993). Apnea is defined as a cessation of air flow for at least 10s. It can be central (if no respiratory effort occurs), obstructive (if continued respiratory effort is noted), or mixed.

With sleep apnea, as mentioned above, oxyhemoglobin desaturation occurs and/or an arousal from sleep occurs. This eventually leads to daytime sleepiness and, more ominously, to altered cardiopulmonary function. The disorder is more common in males than females, and women generally need to be more obese for a similar level of respiratory involvement (GUILLEMINAULT et al. 1988). As apnea becomes more severe, the following complications can ensue: severe arterial hypoxemia, increased sympathetic tone, pulmonary and systemic hypertension, cardiac arrythmias, and eventually cor pulmonale (KOPELMAN et al. 1986; SHARP et al. 1980).

G. Cancer

An extraordinarily comprehensive prospective longitudinal study was carried out in the 1960s and 1970s by the American Cancer Society (ACS) to define the mortality risk from cancer. The study followed over 750000 men and women for 12 years. It found over-all mortality risk to be significantly elevated in individuals who were ≤40% overweight (1.33 for men and 1.55 for women). With regard to cancer, as shown in Table 3, obese men had higher risk for colorectal and prostate cancers and obese women for endometrial, gallbladder, cervical, ovarian, and breast cancers (LEW and GARFINKEL 1979).

In other subsequent longitudinal prospective studies, the incidence of endometrial cancer has been found to be elevated in obese women (DUNN LJ et al. 1967; MCMAHON 1974; FOLSOM et al. 1989; LAVECCHIA et al. 1982). Also, there is a positive association of breast cancer with obesity in post-menopausal women (DEWAARD 1975; PAFFENBERGER et al. 1980; LUBIN et al. 1985). In pre-menopausal women, obesity seems to be protective (CHU et al. 1991; WILLETT et al. 1985; CHUTE et al. 1991). The mechanism for this discrepancy is not clear.

Table 3. Mortality ratios for cancer sites at which incidence of overweight is greater than for average weight

	Weight Index			
Site of Cancer	110%–119%	120%–129%	130%–139%	140%+
Male				
Colon, rectum	—	—	1.53	1.73
Prostate	—	1.37	1.33	1.29
Female				
Endometrium	1.36	1.85	2.30	5.42
Uterus, unspecified	—	1.81	1.40	4.65
Cervix	—	1.51	1.42	2.39
Ovary	—	—	—	1.63
Gall Bladder	1.59	1.74	1.80	3.58
Breast	—	—	—	1.53

In men, a number of studies have found an association between obesity and colon cancer (Chute et al. 1991; Bostick et al. 1994; Lee et al. 1992). This has also been documented for women (Giovannucci et al. 1996). Obesity is also related to gallbladder cancer, more so in women than in men (Garfinkel 1986), for unknown reasons.

In the above-cited studies, it is not always clear that the association between obesity and cancer risk may not be due to a third causal factor. For example, high fat intake and high caloric intake have both been proposed as influential participants in the association. Much more nutritional epidemiology needs to be done before clear answers relating to causality can be invoked.

H. Gall Bladder Disease

Gall bladder disease is a very common health consequence of obesity. The increased risks of gall stones and cholecystitis, and the subsequent requirement for cholecystectomy have been repeatedly reported in both cross-sectional (Burnett 1971) and longitudinal (Friedman et al. 1966) studies. Risks increase with increasing weight and with advancing age (Rimm et al. 1975). In a cross-sectional study of morbidly obese individuals, the rate of gallbladder disease as defined by the presence of gall stones was found to be between 28% and 45% depending on the weight (Broomfield 1988). This is a particular problem in women, in whom the incidence of this complication is much greater than in men (Maclure et al. 1989). In women in whom obesity was defined as a BMI of 30 or greater, the Nurses' Health Study documented a yearly symptomatic gallstone incidence of more than 1% over 16 years of follow-up, and in women with a BMI greater than 40 the incidence was about 2% per year (Stampfer et al. 1992).

Why are gallstones so common in obese individuals? Two factors are particularly important. The first is that in obese persons the bile is supersaturated for cholesterol (Grundy et al. 1974), leading to the crystallization of cholesterol stones in the gallbladder. The second is that gallbladder contraction is impaired in obesity, so that the gallbladder does not empty regularly and completely, creating a pool of static bile that enhances the crystallization process (Liddle 1989; Inoue et al. 1992).

A third factor may be repeated cycles of attempts at weight loss. During a weight loss effort, when caloric intake is reduced, there is an increase in the supersaturation of the cholesterol in bile (Schreibman et al. 1974; Reuben et al. 1985; Schiffman et al. 1993). This therefore creates an increased risk during such a period. Once weight is lost and a person plateaus at a lower weight, the supersaturation decreases and the patient is at lesser risk of developing gallstones, but during the dynamic phase of weight loss the risk is increased. In those obese persons who go through numerous attempts at hypocaloric dieting, this may predispose them to an enhanced risk for gallstone formation.

I. Arthritis

Obesity enhances the risk of developing osteoarthritis. In the Framingham study, there was a strong relationship between increased weight and follow-up osteoarthritis 36 years later (Felson et al. 1988).

The association is stronger for women than for men (Felson et al. 1988). The Baltimore Longitudinal Study of Aging also showed a strong association (Hochberg et al. 1995). As obesity increases in severity, the prevalence of osteoarthritis increases (Anderson et al. 1988). In a recent study of twins, those members of a twin pair that had the higher weight had a greater amount of osteoarthritis (Cicuttini et al. 1996). In this study, it was calculated that for every kilogram increase in weight, the risk of developing osteoarthritis increases by 9–13%. In addition, it is important to point out that no evidence exists to suggest that arthritis leads to obesity (Carman et al. 1994).

J. Gout

Obesity has been shown to be associated with gout in a number of cross-sectional studies. A large study of over 70,000 women showed an increased prevalence with an increasing weight (157). Similarly, the relationship has also been found in men (Weatherall et al. 1988). In a large Canadian health survey, as the BMI increased in men from 21 to 31, the percentage with serum uric acid levels greater than 416 μmol/l increased from 7% to 31% (Canada Health Survey 1978). In women, however, the weight effect was less powerful, so that only 7% have such levels at a BMI above 31 (Canada Health Survey 1978). On the other hand, there are some data to suggest that central fat distribution is a risk factor for gout, but greater in women than in men (Seidell et al. 1985). Clearly, more studies are necessary to evaluate this relationship.

K. Abdominal Obesity

Abdominal obesity must be considered as a risk factor for health separately from total body fat (Vague 1947). The association of central or upper body obesity with abnormal glucose tolerance, insulin resistance, dyslipidemia, and hypertension (now characterized as the Metabolic Syndrome or Syndrome X) (Reaven 1988) has been described by a number of investigators (Kissebah et al. 1989; Krotkiewski et al. 1983). Epidemiological studies in both men and women, using elevated waist to hip ratios as measurement of central or upper body obesity, demonstrated that central obesity is a risk factor for cardiovascular disease,(Lapidus et al. 1984; Larsson et al. 1984), stroke (Lapidus et al. 1984), and type 2 diabetes (Després et al. 1989). Other later studies (Ohlson et al. 1985; Ducimetiere et al. 1986; Stokes et al. 1985; Donahue et al. 1987; Terry 1992; Bergström et al. 1990; Haffner et al. 1990) confirmed these.

The waist to hip ratio generally reflects the amount of abdominal subcutaneous and intra-abdominal fat in the body but to some extent it also measures gluteal muscle. Because of this, it has been suggested that utilizing waist circumference alone as a measure of central fat may be a better measure (POULIOT et al. 1994).

There has been some controversy as to whether the intra-abdominal (visceral) component or the subcutaneous component is more important in the association with the metabolic syndrome. ABATE et al. (1995) have reported that in men the subcutaneous depot is the important one. This has been confirmed by GOODPASTER et al. (1999). However, there are many other studies that implicate the visceral component (POULIOT et al. 1996; TCHERNOF et al. 1996; LEMIEUX et al. 1996a; ALBU et al. 1997b; SEIDELL et al. 1990).

It is of importance to characterize the health risks more specifically. Many studies have shown an association of central obesity with insulin resistance and hyperinsulinemia (DESPRÉS et al. 1990; DESPRÉS 1991). While the hyperinsulinemia is partly the result of B-cell compensatory increased secretion of insulin, there is a second cause, the decreased hepatic extraction of insulin (PEIRIS et al. 1986). The association has been extended from insulin resistance to type 2 diabetes (KISSEBAH et al. 1989; OHLSON et al. 1985; HAFFNER et al. 1990).

The dyslipidemia associated with central obesity has gradually been more carefully characterized. Originally, the main components were elevated triglycerides and a low HDL-cholesterol (HDL-C) (KROTKIEWSKI et al. 1983; DESPRÉS et al. 1985; KISSEBAH et al. 1982, 1985; ANDERSON et al. 1988; ALBRINK et al. 1964). Studying two subgroups of obese patients who were matched for their levels of total body fat but with either a low or high accumulation of visceral fat by CT scan, DESPRÉS et al. (1989, 1990, 1991) have shown this well. The levels of LDL-C are generally not very elevated, if at all, (HARRIS PEEPLES et al. 1989; TERRY et al. 1989). In addition, other characteristics of the dyslipidemia include a greater proportion of small, dense cholesterol ester-depleted LDL-cholesterol (LDL-C) particles (DESPRÉS 1993; DESPRÉS et al. 1993) and elevated plasma apoB concentrations (ASSMAN et al. 1992; KISSEBAH et al. 1989; TCHERNOF et al. 1996).

The importance of HDL-C as a risk factor for coronary heart disease (CHD) has been known for a long time (GORDON et al. 1977; KANNEL et al. 1979). Hypertriglyceridemia has been established as an important independent risk factor more recently (AUSTIN 1991; ASSMAN et al. 1992).

Apolipoprotein B, as the sole component of LDL, is another marker of CHD risk. In cross-sectional studies, patients with CHD have been found to have higher apoB levels (SNIDERMAN et al. 1992).

Small dense LDL-C particles are another manifestation of obesity. Two phenotypes of the pattern of LDL-C particles have been described, A and B. A has larger less dense particles and B has smaller, more dense ones. The smaller, more dense ones have been found to be more atherogenic and associated with an increased risk of CHD (AUSTIN et al. 1988, 1990). In the Quebec

Cardiovascular Study, the combination of hyperinsulinemia with these lipid abnormalities increased the risk of ischemic heart disease by more than tenfold (DESPRÉS et al. 1996; LAMARCHE et al. 1996). Thus persons with visceral obesity have a much higher risk of CHD as well as of type 2 diabetes.

The relationship of central obesity to hypertension also has been well documented (KALKHOFF et al. 1983; HARTZ et al. 1986; BLAIR et al. 1984).

A more sophisticated way to measure central fat is to do a CT scan or an MRI. With such a technique, one to three cuts 3 mm apart, using the lumbar 4–5 space as a guide, are taken. Total central fat, subcutaneous fat, and visceral fat can all be calculated (ASHWELL et al. 1984; SJÖSTRÖM et al. 1986; KVIST et al. 1988; ENZI et al. 1986). In the Quebec studies with this technique, it has been shown that individuals who have the same quantity of total body fat may have quite variable amounts of intraabdominal (visceral) fat (POULIOT et al. 1994) between CT scans and waist circumference measurement.

In the Quebec Studies, a visceral adipose tissue of 130 cm^2 or above was associated with an increased prevalence of metabolic abnormalities (LEMIEUX et al. 1996b; DESPRÉS et al. 1993). This was found to correspond closely to an abdominal girth of 1 m or above (POULIOT et al. 1994). However, these investigators also noted an accumulation of visceral fat with age (LEMIEUX et al. 1994, 1996a,b). They thus suggested a waist circumference of 90 cm for both men and women between ages of 40 and 60 as being associated with metabolic abnormalities (LEMIEUX et al. 1996a). Recently, in guidelines developed by the US National Institutes of Health, a value of 1 m (40 in) for men and 88 cm (35 in) for women have been suggested as risk factors (LEMIEUX et al. 1996a; NHLBI EXPERT PANEL et al. 1998).

Central obesity has also been shown to be associated with increased mortality. FOLSOM (1993) described this in women as have LAPIDUS et al. (1984). LARSSON et al. (1984) described it in men. TERRY (1992) also described it in male United States Army veterans.

There is also some suggestion of central obesity being related to gall bladder disease (HAFFNER et al. 1990).

Clotting abnormalities have been described in persons with increased central obesity. These include increased fibrinogen levels, increased factor VIII and factor X, and increased PAI-1 (SHIMOMURA et al. 1996). In addition, enhanced platelet aggregability has also been described in insulin resistance. As a result, it is clear that these patients are more open to risk of thrombotic events and these abnormalities may be important in the enhanced risk for myocardial infarction.

L. Conclusion

Obesity increases morbidity and mortality risk in both men and women. Abdominal fat distribution is also independently related to morbidity and mortality. Whether it is the total burden of central fat, the subcutaneous fat, or the intraabdominal visceral fat that primarily determines this risk is not clear at this time. The risks are greater the longer the duration of the obesity and the greater the degree of total fat burden and/or central fat distribution. There are definite short-term (NHLBI EXPERT PANEL et al. 1998; GOLDSTEIN 1992; PI-SUNYER 1993), and long-term (PI-SUNYER 1996) improvements in health risk factors with weight loss, so that it is reasonable to attempt to do this. Clearly, however, prevention of weight gain is the desirable goal. Such prevention would improve the quality of life and decrease health costs throughout the world.

References

Build and Blood Pressure Study (1979) Society of Actuaries and Association of Life Insurance Directors of America. (1980)
Abate N, Garg A et al (1995) Relationship of generalized and regional adiposity to insulin sensitivity in men. J Clin Invest 96:88–98
Albrink MJ, Meigs JW (1964) Interrelationship between skinfold thickness, serum lipids and blood sugar in normal men. Am J Clin Nutr 15:255–261
Albu J et al (1997a) Obesity and Diabetes. In: Bray GA, Bouchard C, James P (eds), Handbook of Obesity. Philadelphia: Marcel Decker, Inc
Albu J, Murphy L et al (1997b) Visceral fat and race-dependent health risks in obese premenopausal women. Diabetes 46:456–462
Albrink MJ, Meigs JW et al (1962) Weight gain and serum triglycerides in normal men. N Engl J Med., 484–489
Allison DB, Gallagher D et al (1997) Body mass index and all cause mortality among people age 70 and over; the Longitudinal Study of Aging. Int J Obes 21:424–431
Anderson AJ, Sobocinski KA et al (1988) Body fat distribution, plasma lipids, and lipoproteins. Arteriosclerosis 8:88–94
Anderson, JJ, Felson DT (1988) Factors associated with osteoarthritis of the knee in the first National Health and Nutrition Examination Survey (HANES I) Evidence for an association with overweight, race, and physical demands of work. Am J Med Epidemiol 128:179–189
Ashwell M, Col TJ et al (1984) Obesity; new insight into the anthropometric classification of fat distribution shown by computed tomography. Br Med J 290:1692–1694
Assman G, Schulte H (1992) Relation of high-density lipoprotein cholesterol and triglycerides to incidence of atherosclerotic coronary artery disease (the PROCAM Experience) Am J Cardiol 70:733–737
Assman G et al (1992) Obesity and hyperlipidemia: results from the perspective cardiovascular Munster (PROCAM) study. In: Björntorp P, Brodoff BN (eds) (pp 502–511) New York: JB Lippincott
Assmann G, Schulte H (1992) Role of triglycerides in coronary artery disease: lessons from the Prospective Cardiovascular Munster Study. American Journal of Cardiology 70:10H–13H
Austin MA (1991) Plasma triglyceride and coronary heart disease. [Review] [118 refs]. Arteriosclerosis Thrombosis 11:2–14
Austin MA, Breslow JL et al (1988) Low density lipoprotein subclass patterns and risk of myocardial infarction. JAMA 260:1917–1921
Austin MA, King MC et al (1990) Atherogenic lipoprotein phenotype: a proposed genetic marker for coronary heart disease. Circulation 82:495–506
Barrett-Connor E (1985) Obesity, atherosclerosis, and coronary artery disease. Ann Intern Med 103:1010–1019

Bergström RW, Newell-Morris LL et al (1990) Association of elevated fasting C-peptide level and increased intraabdominal fat distribution with the development of NIDDM in Japanese-American men. Diabetes 39:104–111
Bjerkedal T (1957) Overweight and hypertension. Acta Medica Scandinavica, 13–26
Blair D, Habicht JP et al (1984) Evidence for an increased risk of hypertension with centrally located body fat and the effect of race and sex on this risk. Am J Epidemiol 119:526–540
Blair BF, Haines LW Mortality experience according to build at higher durations. (1966). 18:35 p. Society Actuaries Trans
Bogardus C, Lillioja S et al (1984) Correlation between muscle glycogen depletion and glycogen synthase activation on in vivo insulin action in man. J Clin Invest 73:1185–1190
Bonham GS, Brock DB (1985) The relationship of diabetes with race, sex, and obesity. Am J Nutr 53:1312–1317
Borhani NO, Hechter HH et al (1963) Report of a ten-year follow-up study of the San Francisco longshoremen. J Chronic Dis 1251–1266
Bostick RM, Potter JD et al (1994) Sugar, meat, and fat intake, and nondietary risk factors for colon cancer incidence in Iowa women (United States). Cancer Causes Control 5:38–52
Brenner BM, Garcia DL et al (1988) Glomeruli and blood pressure. Less of one, more of the other? Am J Hypertens 1:335–347
Broomfield PH (1988) Effects of ursodeoxycholic acid and aspirin on the formation of lithogenic bile and gallstones during loss of weight. N Engl J Med 319:1567–1572
Burnett W (1971) The epidemiology of gall stones. Tijdschr Gastroenterol, 14:79–89
Burton BT, Foster WR et al (1985) Health implications of obesity: an NIH Consensus Development conference. Int J Obesity 9:155–170
Canada Health Survey. Health and Welfare Canada. Health and Welfare Canada. (1978). Canada Health survey.
Carman WJ, Sowers M et al (1994) Obesity as a risk factor for osteoarthritis of the hand and wrist: a prospective study. Am J Epidemiol 139:119–129
Caro JF, Dohm LG et al (1989) Cellular alterations in liver, skeletal muscle, and adipose tissue responsible for insulin resistance in obesity and type II diabetes. Diabetes/Metabolism Reviews 5:665–689
Chan JM, Rimm EB et al (1994) Obesity, fat distribution, and weight gain as risk factors for clinical diabetes in men. Diabetes Care 17:961–969
Chapman JM, Coulson AH et al (1971) The differential effect of serum cholesterol, blood pressure and weight on the incidence of myocardial infarction and angina pectoris. J Chronic Dis 23:631–645
Chapman JM, Massey FJ (1964) The interrelationship of serum cholesterol, hypertension, body weight, and risk of coronary disease: results of the first ten years' follow-up in the Los Angeles Heart Study. J Chronic Dis 933–949
Chu SY, Lee NC et al (1991) The relationship between body mass index and breast cancer among women enrolled in the Cancer and Steroid Hormone Study. J Clin Epidemiol 44:1197–1206
Chute CG, Willett WC et al (1991) A prospective study of body mass, height, and smoking on the risk of colorectal cancer in women. Cancer Causes Control 2:117–124
Cicuttini FM, Baker JR et al (1996) The association of obesity with osteoarthritis of the hand and knee in women: a twin study. J Rheumatol 23:1221–1226
Colditz GA, Willet WC et al (1995) Weight gain as a risk factor for clinical diabetes mellitus in women. Annals of Internal Medicine 481–486
Colditz GA, Willett WC et al (1990) Weight as a risk factor for clinical diabetes in women. Am J Epidemiol 132:501–513
Dannenberg AL, Garrison RJ et al (1988) Incidence of hypertension in the Framingham Study. Am J Pub Health 78:676–679

DeFronzo RA (1988) The triumvirate; β cell, muscle, liver: a collusion responsible for NIDDM. Diabetes 37:644–667

DeFronzo RA, Cooke CR et al (1975) The effect of insulin on renal handling of sodium, potassium, calcium, and phosphate in man. J Clin Invest 55:845–855

DeFronzo RA, Goodman NA et al (1995) Efficacy of metformin in patients with non-insulin-dependent diabetes mellitus. N Engl J Med 333:541–549

Deprés JP, Nadeau A et al (1989) Role of deep abdominal fat in the association between regional adipose tissue distribution and glucose tolerance in obese woman. Diabetes 38:304–309

Després JP (1991) Visceral obesity: a component of the insulin resistance–dyslipidemic syndrome. Can J Cardiol 10:17B–22B

Després JP Natl Inst Nutrit Rev. Assessing obesity: beyond the BMI. Rapport NIN. (1993). 7, 1 p

Després JP, Allard C et al (1985) Evidence for a regional component of body fatness in the association with serum lipids in men and women. Metabolism: Clinical and Experimental 34:967–973

Després JP, Ferland M et al (1989) Role of hepatic-triglyceride lipase activity in the association between intra-abdominal fat and plasma HDL cholesterol in obese women. Arteriosclerosis 9:485–492

Després JP, Lamarche B (1993) Effects of diet and physical activity on adiposity and body fat distribution: implications for the prevention of cardiovascular disease. Nutr Res Rev 6:137–159

Després JP, Lamarche B et al (1996) Hyperinsulinemia as an independent risk factor for ischemic heart disease. N Engl J Med 334:952–957

Després JP, Moorjani S et al (1990) Regional distribution of body fat, plasma lipoproteins, cardiovascular disease. Arteriosclerosis 10:497–511

Després JP, Nadeau A et al (1989) Role of deep abdominal fat in the association between regional adipose tissue distribution and glucose tolerance in obese women. Diabetes 38:304–309

Després JP, Pouliot M et al (1991) Loss of abdominal fat and metabolic response to exercise training in obese women. Am J Physiol 261: E159–E167

DeWaard F (1975) Breast cancer incidence and nutritional status with particular reference to body weight and height. Cancer Res 35:3351–3356

Diehr P, Bild DE et al (1998) Body Mass index and mortality in non-smoking older adults: the cardiovascular health study. Am J Pub Health 88:1–7

Donahue RP, Abbott RD et al (1987) Central obesity and coronary heart disease in men. Lancet 1:821–824

Douglas FG, Chong PY (1972) Influence of obesity on peripheral airways patency. J Appl Physiol 33:559–563

Ducimetiere P, Richard JL et al (1986) The pattern of subcutaneous fat distribution in middle-aged men and the risk of coronary heart disease. The Paris Prospective Study. Int J Obes 10:229–240

Dunn LJ, Bradbury JT (1967) Endocrine factors in endometrial carcinoma. Am J Obstet Gynecol 97:465–471

Dyer AR, Stamler J et al (1975) Relationship of relative weight and body mass index to 24-year mortality in the Chicago People's Gas Company Study. J Chron Disease 28:109–123

Enzi G, Gasparo M et al (1986) Subcutaneous and visceral fat distribution according to sex, age, and overweight, evaluated by computed tomography. Am J Clin Nutr 44:739–746

Erikkson J, Franssila-Kallunki A et al (1989) Early metabolic defects in persons at increased risk for non-insulin-dependent mellitus. N Engl J Med 321:337–344

Everhart JE, Pettit DJ et al (1992) Duration of obesity increases incidence of NIDDM. Diabetes 235–240

Farinaro E et al (1979) Overweight and plasma lipoprotein abnormalities in a random sample of the Neapolitan population. In: Mancini M, Lewis B, Contaldo R (eds) Medical complications of obesity. (pp 13–150). London: Academic Press
Feinlieb M (1985) Epidemiology of obesity in relation to health hazards. Ann Intern Med 103:1010–1024
Felson DT, Anderson JJ et al (1988) Obesity and knee osteoarthritis. The Framingham Study. Ann Intern Med 109:18–24
Ferrannini E (1993) The haemodynamics of obesity: a theoretical analysis. J Hypertens 10:1417–1423
Ferrannini E (1995) The phenomenon of insulin resistance: its possible relevance to hypertensive disease. In: Laragh JH, Brenner BM (eds) Hypertension: physiology, diagnosis and management. (pp 2281–2300). New York: Raven
Fitzgerald AP, Jarrett RJ (1992) Body weight and coronary heart disease mortality: an analysis in relation to age and smoking habit. 15 years follow-up data from the Whitehall Study. Int J Obes 16:119–129
Flegal KM, Carroll MD et al (1998) Overweight and obesity in the United States: prevalence and trends. Int J Obes 22:39–47
Folsom AR (1993) Body fat distribution and 5-year risk of death in older women. JAMA 269:483–487
Folsom AR, Kaye SA et al (1989) Association of incident carcinoma of the endometrium with body weight and fat distribution in older women: early findings of the Iowa Women's Health Study. Cancer Res 49:6828–6831
Freedman DA, Williamson DF et al (1995) Relation of body fat distribution to ischemic heart disease: the National Health and Nutrition Examination Survey I (NHANES I) Epidemiologic Follow-Up Study. Am J Epidemiol 142:53–63
Friedman GD, Kannel WB et al (1966) The epidemiology of gallbladder disease: observations in the Framingham Study. J Chronic Dis 19:273–292
Garfinkel L. (1986) Overweight and mortality. Cancer, 58:1826–1829
Garrison RJ, FMCWMP (1983) Cigarette smoking as a confounder of the relationship between relative weight and long-term mortality in the Framingham Study. JAMA 249:2199–2203
Garrison R, Castelli WP (1985) Weight and thirty-year mortality of men in the Framingham Study. Ann Intern Med 103:1006–1009
Genuth SM, Castro JH et al (1974) Weight reduction in obesity by out patient semi-starvation. JAMA 230:987–991
Gillum RF, Taylor HL et al (1982) Indices of obesity and blood pressure in young men followed 32 years. J Chronic Dis 35:211–219
Giovannucci E, Colditz GA et al (1996) Physical activity, obesity, and risk of colorectal adenoma in women (United States). Cancer Causes Control 7:253–263
Glueck CJ, Taylor HL et al (1980) Plasma high-density lipoprotein cholesterol: association with measurements of body mass. The Lipid Research Clinics Program Prevalence Study. Circulation 62:62–69
Goldstein DJ (1992) Beneficial health effects of modest weight loss. Int J Obes Relat Metab Disord 16:397–415
Goodpaster BH, Thaete FL et al (1999) Subcutaneous abdominal fat and thigh muscle composition predict insulin sensitivity independently of visceral fat. Diabetes 46:1579–1585
Gordon T, Castelli WP et al (1977) High density lipoprotein as a protective factor against coronary heart disease, the Framingham Study. Am J Med 62:707–714
Gordon T, Doyle JT (1988) Weight and mortality in men; the Albany Study. Int J Epidemiol 17:77–81
Gordon T, Castelli WP et al (1977) High density lipoprotein as a protective factor against coronary heart disease. The Framingham Study. Am J Med 62:707–714
Grundy SM, Duane WC et al (1974) Biliary lipid outputs in young women with cholesterol gallstones. Metabolism: Clinical and Experimental 23:67–73

Guilleminault C, Quera-Salva MA et al (1988) Women and the obstructive sleep apnea syndrome. Chest 93:104–109
Haffner S, Diehl AK et al (1990) Central adiposity with gallbladder disease in Mexican Americans. Am J Epidemiol 129:587–595
Haffner SM, Mitchell BD et al (1991) Greater influence of central distribution of adipose tissue on incidence of non-insulin-dependent diabetes in women than men. Am J Clin Nutr 53:1312–1317
Haffner SM, Stern MP et al (1990) Incidence of type II diabetes in Mexican Americans predicted by fasting insulin and glucose levels, obesity, and body-fat distribution. Diabetes 39:283–288
Harris T, Peeples L, Carpenter JW et al (1989) Alterations in low-density lipoproteins in subjects with abdominal obesity. Metabolism: Clinical and Experimental 38:1029–1036
Harris T, Cook F et al (1988) Body Mass index and mortality among non-smoking older persons. JAMA 259:1520–1524
Hartz AJ, Fischer ME et al (1986) The association of obesity with joint pain and osteoarthritis in a general population. J Chron Dis 39:311–319
Hissin PJ, Foley JE et al (1982) Mechanism of insulin-resistant glucose transport activity in the enlarged adipose cell of the aged, obese rat. J Clin Invest 70:780–790
Hochberg MC, Lethbridge-Cejku M et al (1995) The association of body weight, body fatness, and body fat distribution with osteoarthritis of the knee: data from the Baltimore Longitudinal Study of Aging. J Rheumatol 22:488–493
Holley HS, Milic-Emili J et al (1967) Regional distribution of pulmonary ventilation and perfusion in obesity. J Clin Invest 46:475–481
Hubert HB (1983) Obesity as an independent risk factor for cardiovascular disease: a 26-year follow-up of participants in the Framingham Heart Study. Circulation 67:968–977
Hypertension Detection and Follow-Up Program Cooperative Group (1977) Race, education, and prevalence of hypertension. Am J Epidemiol 106:351–361
Inoue K, Fuchigami A et al (1992) Gallbladder sludge and stone formation in relation to contractile function after gastrectomy. Ann Surg 215:19–26
Jacobs DD, Sowers JR et al (1993) Effects of weight reduction on cellular cation metabolism and vascular resistance. Hypertension 21:308–314
Kalkhoff RK, Hartz AH et al (1983) Relationship of body fat distribution to blood pressure, carbohydrate tolerance, and plasma lipids in healthy obese women. J Lab Clin Med 102:621–627
Kannel WB, Gordon T et al (1979) Obesity, lipids, and glucose intolerance. The Framingham Study. Am J Clin Nutr 32:1238–1245
Kannel WB, Brand N et al (1967) The relation of adiposity to blood pressure and development of hypertension. The Framingham Study. Ann Intern Med 67:48–49
Kannel WB et al (1979) Physiological and medical concomitants of obesity: the Framingham Study. In: Bray GA (ed) Obesity in America (pp 125–163). Washington, DC: US Dept of Health, Education, and Welfare
Kannel WB, Gordon T et al (1969) Left ventricular hypertrophy by electrocardiogram. Prevalence, incidence, and mortality in the Framingham study. Ann Intern Med 71:89–105
Kempner W, Newborg BC et al (1975) Treatment of massive obesity with rice/reduction diet program: an analysis of 106 patients with at least 45 kg weight loss. Arch Intern Med 135:1575–1583
Keys A (1980) Seven countries: a multivariate analysis of death and coronary heart disease. Cambridge, MA: Harvard University Press
Kissebah AH et al (1985) Endocrine characteristics in regional obesities: role of sex steroids. In: Vague J, Grand B et al (eds) Metabolic complications of human obesities. (pp 115–130). Amsterdam: Elsevier Science
Kissebah AH, Peiris AN (1989) Biology of regional fat distribution. Relationship to non-insulin dependent diabetes mellitus. Diabetes Metab Rev 5:83–109

Kissebah AH, Vydelingum N et al (1982) Relation of body fat distribution to metabolic complications of obesity. J Clin Endo Metab 54:254–260
Knowler WC, Pettit PJ et al (1981) Diabetes incidence in Pima Indians: contributions of obesity and parental diabetes. Am J Epidemiol 113:144–156
Kolterman OG, Insel J et al (1980) Mechanisms of insulin resistance in human obesity: evidence for receptor and postreceptor defects. J Clin Invest 65:1272–1284
Kopelman PG, Apps MC et al (1986) Nocturnal hypoxia and sleep apnoea in asymptomatic obese men. Int J Obes 10:211–217
Kozarevic D, Pirc Z et al (1976) The Yugoslavia cardiovascular disease study: II. Factors in the incidence of coronary heart disease. Am J Epidemiol 104:133–140
Krotkiewski M, Björntorp P et al (1983) Impact of obesity on metabolism in men and women. J Clin Invest 72:1150–1162
Kuczmarski RJ, Flegal KM et al (1991) Increasing prevalence of overweight among US adults. The National Health and Nutrition Examination Surveys, 1960–1991. JAMA 272:205–211
Kvist H, Chowdury B et al (1988) Total and visceral adipose tissue volumes derived from measurements with computed tomography in adult men and women: predictive equations. Am J Clin Nutr 48:1351–1361
Laakso M, Sarlund H et al (1990) Insulin resistance is associated with lipid and lipoprotein abnormalities in subjects with varying degrees of glucose tolerance. Arteriosclerosis 10:223–231
Lamarche B, Després JP et al (1993) Metabolic heterogeneity associated with high plasma triglyceride or low HDL-cholesterol levels in men. Arterioscler Thromb Vasc Biol 13:33–40
Lamarche B, Moorjani S et al (1996) Apolipoprotein A-I and B levels and the risk of ischemic heart disease during a five-year follow-up of men in the Quebec Cardiovascular Study. Circulation 94:273–278
Lamarche B, Tchernov A et al (1997) Small, dense low-density lipoprotein particles as a predictor of the risk of ischemic heart disease in men. Prospective results from the Quebec Cardiovascular Study. Circulation 69–75
Landsberg L, Krieger DL (1989) Obesity, metabolism, and the sympathetic nervous system. Am J Hypertens 2:125S–132 S
Langford HG, Davis BR et al (1991) Effect of drug and diet treatment of mild hypertension on diastolic blood pressure. The TAIM Research Group. Hypertension 17:210–217
Lapidus L, Bengtsson C et al (1984) Distribution of adipose tissue and risk of cardiovascular disease and death: a 12 year follow up of participants in the population study of women in Gothenburg, Sweden. Br Med J 289:1257–1261
Larsson B, Björntorp P et al (1981) The health consequences of moderate obesity. Int J Obes 5:97–116
Larsson B, Svärdsudd K et al (1984) Abdominal adipose tissue distribution. Obesity and risk of cardiovascular disease and death:13-year follow up of participants in the Study of Men Born in 1913. Br Med J 288:1401–1404
LaVecchia C, Franceshi S et al (1982) Oestrogens and obesity as risk factors for endometrial cancer in Italy. Int J Epidemiol 11:120–126
Lee IM, Paffenbarger RS (1992) Quetelet's index and risk of colon cancer in college alumni. J Natl Cancer Inst 84:1326–1331
Lemieux S, Després JP et al (1994) Are gender differences in cardiovascular disease risk factors explained by the level of visceral adipose tissue? 37:764
Lemieux S, Prud'homme D et al (1996a) Seven-year changes in body fat and visceral adipose tissue in women: associations with indexes of plasma glucose-insulin homeostasis. Diabetes Care 19:983–991
Lemieux S, Prud'homme D et al (1996b) A single threshold value of waist girth to identify non-obese and overweight subjects with excess visceral adipose tissue. Am J Clin Nutr 64:685–693
Lew EA (1985) Mortality and weight: insured lives and the American Cancer Society studies. Ann Intern Med 103:1024–1029

Lew EA, Garfinkel L (1979) Variations in mortality by weight among 750000 men and women. J Chronic Dis 32:563–576
Liddle RA (1989) Gallstone formation during weight-reduction dieting. Arch Intern Med 149:1750–1753
Linsted K, Tonstad S et al (1991) Body mass index and patterns of mortality among Seventh-Day Adventist men. Int J Obes 15:397–406
Lubin F, Ruder AM et al (1985) Overweight and changes in weight throughout adult life in breast cancer etiology. A case-control study. Am J Epidemiol 122:579–588
Lundgren H, Bengtsson C et al (1989) Adiposity and adipose tissue distribution in relation to incidence of diabetes in women: results from a prospective population study in Gothenburg, Sweden. Int J Obes 13:413–423
Maclure KM, Hayes KC et al (1989) Weight, diet, and the risk of symptomatic gallstones in middle-aged women. N Engl J Med 321:563–569
Manninen V, Tenkanen L et al (1992) Joint effects of serum triglyceride and LDL-cholesterol and HDL-cholesterol concentrations on coronary heart disease risk in the Helsinki Heart Study. Implications for treatment. Circulation 37–45
Manson JE, Stampfer MJ et al (1987) Body weight and longevity. A reassessment. JAMA 257:353–358
Manson JE, Colditz GA et al (1990) A prospective study of obesity and risk of coronary heart disease in women. N Engl J Med 322:882–889
Manson JE, Willett LW et al (1995) Body weight and mortality among women. N Engl J Med 333:677–685
McCue H Jr (1981) The 1979 Build and Blood Pressure Study. In: Bostrom LN, H (ed) Medical aspects of mortality statistics (pp. 182–198). Stockholm: Almquist and Wiksell International
McMahon B (1974) Risk factors for endometrial cancer. Gynecol Oncol 2:122–129
McMahon SW, Blacket RB et al (1984) Obesity, alcohol consumption and blood pressure in Australian men and women: National Heart Foundation of Australia Risk Factor Prevalence Study. J Hypertens 2:85–91
Medalie JH, Papier C et al (1974) Diabetes mellitus among 10,000 adult men. I. 5-year incidence and associated variables. Isr J Med Sci 10:681–697
Metropolitan Life Insurance Company. Build and blood pressure study 1959. Society of Actuaries. (1960). Vol I, II
Montoye HJ, Epstein FH et al (1966) Relationship between serum cholesterol and body fatness. An epidemiologic study. Am J Clin Nutr 18:397–406
Mortone PG, Dunn EV (1985) Snoring as a risk factor for disease: an epidemiological survey. Br Med J 291:630–632
Naimark A, Cherniak RM (1960) Compliance of the respiratory system and its components in health and obesity. J Appl Physiol 377–382
National Center for Health Statistics (1993) (1995) Current estimates from the National Health Interview Survey, Vital and Health Statistics. Vital and Health Statistics, 10
National Center for Health Statistics (1987) Najjar MF, Rowland M Anthropometric reference data and prevalence of overweight, United States 1976–1980. Vital and Health Stat, 11
NHLBI Expert Panel. Clinical Guidelines on the Identification, Evaluation, and Treatment of Overweight and Obesity in Adults. The Evidence Report. (1998)
NIH Publication No. 95-1468 (1995) Diabetes in America
NIH. Public Health Service: Report of the hypertension task force. DHEW Publication. (1979). 9:79
Noppa H, Bengtsson C et al (1980) Obesity in relation to morbidity and mortality from cardiovascular disease. Am J Epidemiol 111:682–692
Obesity (1997) Preventing and managing the global epidemic. Report of a WHO consultation on obesity. Geneva: World Health Organization

Ohlson LO, Larsson B et al (1985) The influence of body fat distribution on the incidence of diabetes mellitus. 13.5 years of follow up of the participants in the study of men born in 1913. Diabetes 34:1055–1058
Olefsky JM (1981) Insulin resistance and insulin action: an in vivo and in vitro perspective. Diabetes 38:148–162
Paffenbarger RS, Rhorna MC et al (1968) Chronic disease in former college students. VIII. Characteristics in youth predisposing to hypertension in later years. Am J Epidemiol 88:25–30
Paffenberger RS Jr, Kampert JB et al (1980) Characteristics that predict risk of breast cancer before and after menopause. Am J Epidemiol 112:258–268
Paul O, Lepper MH et al (1963) A longitudinal study of coronary heart disease. Circulation 20–31
Peiris AN, Mueller RA et al (1986) Splanchnic insulin metabolism in obesity. Influence of fat distribution. J Clin Invest 78:2648–2657
Pi-Sunyer FX (1993) Short-term medical benefits and adverse effects of weight loss. Ann Intern Med 722–726
Pi-Sunyer FX (1996) A review of long-term studies evaluating the efficacy of weight loss in ameliorating disorders associated with obesity. Clinical Therapeutics 18:1006–1035
Pi-Sunyer F (1994) The Fattening of America. JAMA 272:238–239
Pi-Sunyer F (1996) Weight and non-insulin dependent diabetes. Am J Clin Nutr 63:426S–429S
Pimenta W, Korytkowski M et al (1995) Pancreatic beta-cell dysfunction as the primary genetic lesion in NIDDM. JAMA 273:1855–1861
Porte D (1991) β-cells in Type II diabetes mellitus. Diabetes 40:166–180
Pouliot M, Després JP et al (1994) Waist circumference and abdominal sagittal diameter: best simple anthropometric indexes of abdominal visceral adipose tissue accumulation and related cardiovascular risk in men and women. Am J Cardiol 73:460–468
Pouliot MC, Després JP et al (1996) Visceral obesity in men: associations with glucose tolerance, plasma insulin, and lipoprotein levels. Diabetes 41:826–834
Rabinowitz D, Zierler KL (1962) Forearm metabolism in obesity and its response to intra-arterial insulin. J Clin Invest 41:2173–2181
Rabkin SW, Mathewson FA et al (1977) Relation of body weight to development of ischemic heart disease in a cohort of young North American men after a 26 year observation period. The Manitoba Study. Am J Cardiol 39:452–458
Ramsay LE, Ramsay MH et al (1978) Weight reduction in a blodd pressure clinic. Br Med J ii, 244–245
Reaven GM (1988) Banting lecture 1988. Role of insulin resistance in human disease. Diabetes 37:1595–1607
Reaven GM, Chen YDI et al (1993) Insulin resistance and hyperinsulinemia in individuals with small, dense, low density lipoprotein particles. J Clin Invest 92:141–146
Reed K, Yano, K. (1991) Predictors of arteriographically defined coronary stenosis in the Honolulu Heart Program. Am J Epidemiol 134:111–122
Reisberg RA, Boshell BR et al (1999) Insulin secretion in obesity. N Engl J Med 276:314–319
Reisin E, Abel R et al (1978) Effect of weight loss without salt restriction on the reduction of blood pressure in overweight hypertensive patients. N Engl J Med 298:1–6
Resnick LM, Gupta RK et al (1988) Intracellular free magnesium in hypertension: relation to peripheral insulin resistance. J Hypertens 6:199S–201S
Reuben A, Qureshi Y et al (1985) Effect of obesity and weight reduction on biliary cholesterol saturation and the response to chenodeoxycholic acid. Europ J Clin Invest 16:133–142
Rewers M et al (1995) Risk factors for non-insulin-dependent diabetes. In Anonymous, Diabetes in America (pp. 179–214). NIH Publication

Rimm AA, Werner LH et al (1975) Relationship of obesity and disease in 73532 weight-conscious women. Public Health Rep 90:44–54
Rimm EB, Stampfer MJ et al (1995) Body size and fat distribution as predictors of coronary heart disease among middle-aged and older US men. Am J Epidemiol 141:1117–1127
Rissanen A, Heliovaara M et al (1989) Weight and mortality in Finnish men. J Clin Epidemiol 42:781–789
Rocchini AP, Katch VL et al (1986) Role for aldosterone in blood pressure regulation of obese adolescents. Am J Cardiol 57:613–618
Rose G, Hamilton PS et al (1977) Myocardial ischemia, risk factors and death from coronary artery disease. Lancet 1:105–109
Rosenman RH, Brand RJ et al (1976) Multivariate prediction of coronary heart disease during 8.5 year follow-up in the Western Collaborative Group Study. Am J Cardiol 37:903–910
Scavo D, Borgia C et al (1968) Aspetti di funzione corticosurrenalica nell "obesita". Nota VI Il comportamento della secrezione di aldosterone e della escrezione dei suoi metaboite nel corso di alcune prove dinamiche. Fol Endocrinol 21:591–602
Schiffman ML, Sugarman HJ et al (1993) Gallstones in patients with morbid obesity. Relationship to body weight, weight loss and gallbladder bile cholesterol solubility. Int J Obese 17:153–158
Schmidt-Nowara WW (1990) Cardiovascular consequences of sleep apnea. Prog Clin Biol Res 345:630–632
Schreibman PH, Pertsemlidis D, Liu GCK, Ahrens EH (1974) Lithogenic bile: a consequence of weight reduction. J Clin Invest, 73 A(abstract)
Segal KR, Edano A et al (1991) Effect of exercise training on insulin sensitivity and glucose metabolism in lean, obese, and diabetic men. J Appl Physio 71:2402–2411
Seidell JC, Bakx JC et al (1985) Fat distribution of overweight persons in relation to morbidity and subjective health. Int J Obes 9:363–374
Seidell JC, Björntorp P et al (1990) Visceral fat accumulation in men is positively associated with insulin, glucose, and C-peptide levels, but negatively with insulin levels. Metabolism: Clinical and Experimental 39:897–901
Sharp JT, Barrocas M et al (1980) The cardiorespiratory effects of obesity. Clin Chest Med 1:103–118
Shimomura I, Funahashi T et al (1996) Enhanced expression of PAI-1 in visceral fat: possible contributor to vascular disease in obesity. Nat Med 2:800–803
Sidney S, Friedman GD et al (1987) Thinness and mortality. Am J Pub Health 77:317–322
Sjöström L, Kvist H et al (1986) Determination of total adipose tissue and body fat in women by computed tomography, 40 K, and tritium. Am J Physiol 250: E736–E745
Sjöström L (1992) Morbidity of severely obese subjects. Am J Clin Nutr 55:508S–515S
Sniderman AD, Genest J Jr (1992) The measurement of apolipoprotein B should replace the conventional lipid profile in screening for cardiovascular risk. Can J Cardiol 8:133–140
Society of Actuaries and Association of Life Insurance Medical Directors of America. (1980) Build and Blood Pressure Study 1979 (1980) Chicago. Society of Actuaries.
Sowers JR, Whitfield LA et al (1982) Role of the sympathetic nervous system in blood pressure maintenance in obesity. J Clin Endo and Metab 54:1181–1186
Sparrow D, Borkan GA et al (1986) Relationship of fat distribution to glucose tolerance. Results of computed tomography in male participants of the Normative Aging Study. Diabetes 35:411–415
Spraul M, Anderson EA et al (1994) Muscle sympathetic nerve activity in response to glucose ingestion. Impact of plasma insulin and body fat. Diabetes 43:191–196
Stallones RA (1985) Epidemiologic studies of obesity. Ann Intern Med 103:1003–1005

Stamler J (1979) Overweight, hypertension, hypercholesterolemia and coronary heart disease. In: Mancini M, Lewis B, Contaldo F (eds) Medical Complications of Obesity (pp 191–216). London: Academic Press
Stamler R, Stamler J et al (1978) Weight and blood pressure. Findings in hypertension screening of 1 million Americans. JAMA 240:1607–1610
Stampfer MJ, Maclure KM et al (1992) Risk of symptomatic gallstones in women with severe obesity. Am J Clin Nutr 55:652–658
Stevens J, Cai J et al (1998) The effect of age on the association between body-mass index and mortality. N Engl J Med 338:1–7
Stockwell H (1985) Canada Health Survey. Chron Dis Canada 6:6–13
Stokes J et al (1985) The independent contribution of various indices of obesity to the 22-year incidence of coronary heart disease: the Framingham Heart Study. In: Vague J et al (ed) Metabolic complications of human obesities (pp 49–57). New York: Elsevier Science (Biomedical Division)
Stradling JR, Crosby JH (1991) Predictors and prevalence of obstructive sleep apnoea and snoring in 1001 middle aged men. Thorax 46:85–90
Tayback M, Kumanyika S et al (1990) Body weight as a risk factor in the elderly. Arch Intern Med 150:1065–1072
Tchernof A, Lamarche B et al (1996) The dense LDL phenotype: association with plasma lipoprotein levels, visceral obesity, and hyperinsulinemia in men. Diabetes Care 9:629–637
Tedde R, Sechi LA et al (1988) In vitro action of insulin on erythrocyte sodium transport mechanisms: its possible role in the pathogenesis of arterial hypertension. Clin Exp Hypertens 10:545–549
Terry RB (1992) Waist/hip ratio, body mass index and premature cardiovascular disease mortality in US Army veterans during a twenty-three year follow-up study. Int J Obes 16:417–423
Terry RB, Wood PD et al (1989) Regional adiposity pattern in relation to lipids, lipoprotein cholesterol, and lipoprotein subfraction mass in men. J Clin Endo Metab 68: 191–199
The Pooling Project Research Group (1978) Relationship of blood pressure, serum cholesterol, smoking habit, relative weight, and ECG abnormalities to incidence of major coronary events: final report of the pooling project. J Chronic Dis 31:201–306
Tibblin HA, Wilhemsen L et al (1975) Risk factors for myocardial infarction and death due to ischemic heart disease and other causes. Am J Cardiol 35:514–522
Troiano RP, Frongillo EA Jr et al (1996) The relationship between body weight and mortality: a quantitative analysis of combined information from existing studies. Int J Obes 20:63–75
Trovati M, Massucco P et al (1989) Insulin influences the renin-angiotensin-aldosterone systems in humans. Metabolism: Clinical and Experimental 38:501–503
Tyroler HA et al (1975) Weight and hypertension; Evans County studies of blacks and whites. In: Oglesby (ed) Epidemiology and control of hypertension (pp 177–205). New York: Stratton
Vague J (1947) La differenciation sexuelle facteur determinent des formes de l'obésité. Presse Med 30:339–340
Van Itallie, TB (1985) Health implications of overweight and obesity in the United States. Ann Intern Med 103:983–988
Waaler HT (1984) Height, weight and mortality. The Norwegian experience. Acta Med Scand Suppl 679:1–56
Waltemath CL, Bergman NA (1974) Respiratory compliance in obese patients. Anesthesiology 41:84–85
Weatherall R, Shaper AG (1988) Overweight and obesity in middle-aged British men. Eur J Clin Nutr 42:221–231
Weiss YA, Safar ME et al (1978) Repeat hemodynamic determinations in borderline hypertension. Am J Med 64:382–387

West KM, Kalbfleisch JM (1966) Glucose tolerance, nutrition, and diabetes in Uruguay, Venezuela, Malaya, and East Pakistan. Diabetes 15:9–18
West KM, Kalbfleisch JM (1971) Influence of nutritional factors on prevalence of diabetes. Diabetes 20:99–108
Westlund K, Nicolaysen R (1972) Ten-year mortality and morbidity related to serum cholesterol. A follow-up of 3751 men aged 40–49. Scand J Clin Lab Invest 30:1–24
Whelton PK, Applegate WB et al (1996) Efficacy of weight loss and reduced sodium intake in the Trial of Nonpharmacologic Interventions in the Elderly (TONE). Circulation 178
Willett WC, Browne ML et al (1985) Relative weight and risk of breast cancer among premenopausal women. Am J Epidemiol 122:731–740
Witteman JCM, Willett WC et al (1989) A prospective study of nutritional factors and hypertension among US women. Circulation 1320–1327
Young JB, Macdonald IA (1992) Sympathoadrenal activity in human obesity: heterogeneity of findings since 1980. Int J Obes 16:959–967
Young T, Palta M et al (1993) The occurrence of sleep-disordered breathing among middle-aged adults. N Engl J Med 328:1230–1235

CHAPTER 3

The Molecular and Epidemiological Genetics of Obesity

Y.C. CHAGNON, L. PÉRUSSE, and C. BOUCHARD

A. Introduction

The field of the genetic and molecular basis of human obesity is currently receiving a lot of attention from both the research community and the general public. Much of the recent progress has occurred in the identification of new genes and molecules that are involved in the regulation of the energy balance, which may play a role in the obesity phenotype expression. These studies were initiated after it was observed that there is a significant heritability level for human obesity. In the first part of this chapter, we will review the recent findings on genes linked or associated with obesity phenotypes. In the second part, the genetic epidemiology studies will be summarized.

B. Molecular Genetics of Obesity

A candidate gene is a gene that could potentially play a role in determining the phenotype because of its potential involvement in its physiopathology. They include genes identified from mutations in animal models, usually rodents, and genes targeted because of their known biochemical and physiological functions in relevant biological pathways. The identification of genes that determine complex multifactorial traits such as obesity relies heavily on linkage and association studies. Table 1 presents an overview of the positive association or linkage results of obesity and body fat phenotypes with candidate genes. Here we will discuss the characteristics of genes whose products are involved in the regulation of food intake, energy expenditure, and body composition, with an emphasis on mutations that are known to cause obesity in animal models or that are related to human obesity.

Candidates for obesity also include genes and transcripts from chromosomal regions that are known to encode the genes responsible for the Mendelian syndromes that have obesity as one of their features, quantitative trait loci (QTL), identified in human-genome-wide scan (COMUZZIE et al. 1997; NORMAN et al. 1997, 1998; HANSON et al. 1998; HAGER et al. 1998; LEE et al. 1999; HIXSON et al. 1999; CHAGNON et al. 2000a) and specific chromosome searches, or QTL identified from polygenic animal models of obesity gener-

Table 1. Candidate genes which have shown evidence for the presence of association (A) or linkage (L) with BMI or body fat phenotypes (Adapted from CHAGNON et al. 1998, with permission)

Gene	Location	Analysis	Phenotype	*p* value
HSD3B1	1p13.1	A	Δskinfolds	0.04
LEPR	1p31	A	%fat	0.003
ATP1A2	1q21-q23	A	%fat	0.05
ACP1	2p25	A	BMI	0.02–0.002
		L	BMI	0.004
POMC	2p21	L[a]	leptin, fat	Lod = 2.8–4.5
APOB	2p24-p23	A	BMI, %fat	0.05–0.005
APOD	3q27-qter	A	BMI	0.006
FABP2	4q28-q31	A	BMI, fat, %fat	0.01–0.008
UCP1	4q28-q31	A	ΔBMI, Δfat, Δweight	0.05
GRL	5q31-q32	A	fat	0.007–0.003
		L	BMI	0.009
TNFα	6p21.3	A	BMI, %fat	0.02–0.01
		L[a]	%fat	0.05–0.002
LEP	7q31.3	A	Δweight, weight	0.05–0.006
		L[a]	BMI, fat, skinfolds,	
LPL	8p22	A	BMI	0.05
ADRB3	8p12-p11.2	A	BMI, fat, Δweight, weight	0.05–0.002
IGF2	11p15.5	A	BMI	0.02
SUR	11p15.1	A	obesity	0.02
		L[a]	BMI	0.003
DRD2	11q22.2-q22.3	A	weight	0.002
MC5R	18p11.2	A	BMI	0.003
		L	BMI, fat, %fat, skinfolds	0.02–0.001
MC4R	18q21.3	A	fat, %fat	0.002–0.004
INSR	19p13.3	A	obesity	0.05
LDLR	19p13.2	A	BMI, obesity	0.02–0.004
ADA	20q12-q13.11	L	BMI, fat, %fat, skinfolds	0.04–0.001
MC3R	20q12-q13.11	L	BMI, fat, %fat, skinfolds	0.04–0.001

HSD3B1, 3-beta hydroxysteroid dehydrogenase; LEPR, leptin receptor; ATP1A2, sodium potassium adenosine triphosphatase alpha-2 subunit; ACP1, acid phosphatase; POMC: proopiomelanocortin; APOB, APOD, APOA4, apolipoprotein B, D, and A4; FABP2, fatty-acid-binding protein 2; UCP1, uncoupling protein 1; GRL, glucocorticoid receptor; TNFα, tumor-necrosis factor alpha; LEP, leptin; LPL, lipoprotein lipase; ADRB3, beta-3 adrenergic receptor; SUR, sulfonylurea-receptor gene; IGF2, insulin growth factor 2; DRD2, dopamine D_2 receptor; MC3R, MC4R, MC5R, melanocortin receptors 3 to 5; INSR, insulin receptor; LDLR, low-density lipoprotein receptor; ADA: adenosine deaminase. Δ: changes.
[a] Linkages obtained using markers outside the gene.

ated by crossbreeding experiments. Until now, few of these regions have been investigated in humans with anonymous markers, usually microsatellites, for candidate genes or for linkage. Table 2 summarizes the results from such studies.

Table 2. Evidence from non-candidate genes and anonymous markers for the presence of linkage with BMI or body-fat phenotypes (Adapted from CHAGNON et al. 1998, with permission)

Marker	Location	Phenotype	*p* value
D1S193,200,476	1p32-p22	BMI, fat, skinfolds	0.02–0.009
D2S1788	2p21	leptin, fat	Lod = 2.8–4.5
D3S2432	3p24.2-p22	%fat	Lod = 2.0
GLO1	6p21.2-p21.1	skinfolds, weight	0.05–0.004
D7S680,514,530, 504,1875,495	7q	BMI, skinfolds, fat	0.02–0.0001
KEL	7q33	BMI, skinfolds	0.0001
D11S2000,2366	11q21-q22	%fat	Lod = 3.1
ESD	13q14.1-q14.2	skinfolds, % fat	0.04
P1	22q11.2-qter	weight	0.03

BF, properdin factor B; GLO1, glyoxylase I; KEL, Kell blood group; ESD, esterase D; P1, P blood group.

Table 3. Single-gene mutation in rodent models of obesity and the human gene equivalent (Adapted from CHAGNON et al. 1998, with permission)

Mutations	Genes	Chromosome	Gene product
Diabetes (*db*)	LEPR	1p31	leptin receptor
Fat (*fat*)	CPE	4q32	carboxypeptidase E
Obese (*ob*)	LEP	7q31.1	leptin
Tubby (*tub*)	TUB	11p15.5	insulin signaling protein
Agouti yellow (A^y)	ASIP	20q11.2	agouti signaling protein

I. Candidate Genes by Mutation: Single-Gene-Mutation Mouse Models

Spontaneous mutations in five different genes have been found to be responsible for obesity in mouse models. These mutations are the obese *(ob)* mutation on mouse chromosome 6, the diabetes *(db)* mutation on mouse chromosome 4, the Agouti yellow (A^y) mutation on mouse chromosome 2, the fat (*fat*) mutation on mouse chromosome 8, and the tubby (*tub*) mutation on mouse chromosome 7. All the corresponding genes have been cloned both in rodents and humans, and their gene products or possible functions identified. (Table 3).

1. The Leptin (*LEP*) Gene

The obese (*ob*) and diabetes (*db*) mutations have an almost identical phenotype when bred on the same genetic background: juvenile onset of severe obesity arising from an excess fat in all depots, excess food intake or hyperphagia, increased metabolic efficiency, hyperinsulinemia and insulin resistance, and glucose intolerance and diabetes. These mice also have an increased blood corticosterone level, hypersecretion of adrenocorticotropic hormone (ACTH), and a low level of the thyroid hormone triiodothyronin (T3). High levels of triglycerides and high-density lipoprotein cholesterol, without enhanced atherosclerotic lesions, are observed (NISHIMA et al. 1994a, b). The *Lep* gene, responsible for the obesity of the *ob* mouse, was found to encode a hormone called leptin, and the *Lepr* gene, mutant in the *db* mouse, encodes the leptin receptor.

Proleptin is a small polypeptide of 167 amino acids, encoded by two of the three exons of the gene, and includes an amino-terminal secretory signal sequence of 21 amino acids. This signal sequence is cleaved after the translocation of proleptin to the microsomes, which produce the mature peptide of 146 amino acids with an apparent molecular weight of 16 kDa. The leptin protein is highly similar among species, with 83%–84% similarity between rodents and humans (ZHANG et al. 1994; OGAWA et al. 1995). Leptin is a globular protein, similar to hemopoietic cytokines such as interleukins and granulocyte-macrophage colony stimulating factor (MADEJ et al. 1995). The protein comprises four α-helices and two short β-sheets held together by a single disulfide bond between cysteine-96 and cysteine-146 (CONSIDINE and CARO 1997).

Serum leptin is secreted by the adipose tissue and is highly correlated with fat mass, but also with other adiposity variables such as body mass index (BMI) and percent body fat (%FAT) (ROSENBAUM et al. 1996; HICKEY et al. 1996). Leptin has also been shown to be synthesized in placenta from pregnant women. There is a gender effect: all the body fat values of women are two to three times higher than those of men (see CONSIDINE and CARO 1997), probably because of the influence of sex steroids such as estrogens, progesterone, and androgens (ROSENBAUM et al. 1996). The level of leptin has been reported to be approximately fourfold greater in the serum of human obese individuals (7.5 ± 9.3 vs 31.3 ± 24.1 ng/ml) (CONSIDINE et al. 1996a). Figure 1 provides an overview of the effects of different peptides and other molecules on leptin (upper panel) and, conversely, the potential effects of leptin on different peptides (lower panel). For example, insulin or glucocorticoids stimulate the expression of leptin, whereas agonists of the β3-adrenergic receptor (ADRB3) or intracellular cyclic AMP (cAMP) lower its level (SLIEKER et al. 1996). In the former case, the stimulation of leptin will cause a decrease in food intake, as well as an increase in oxygen consumption, body temperature, and locomotion, whereas the opposite effects are observed in the latter case. On the other hand, leptin lowers the levels of the agouti-related peptide (AGRP), but stimulates the production of proopiomelanocortin (POMC) and, therefore, the production of α-melanocortin-stimulating hormone (α-MSH). α-MSH is known to be an agonist of the melanocortin receptor 4 (MC4R), whereas AGRP is known to antagonize MC4R. The resulting effect will be, in both cases, an enhanced stimulation of MC4R with a concomitant decrease in food intake and increase in energy expenditure (EE) and body temperature (T).

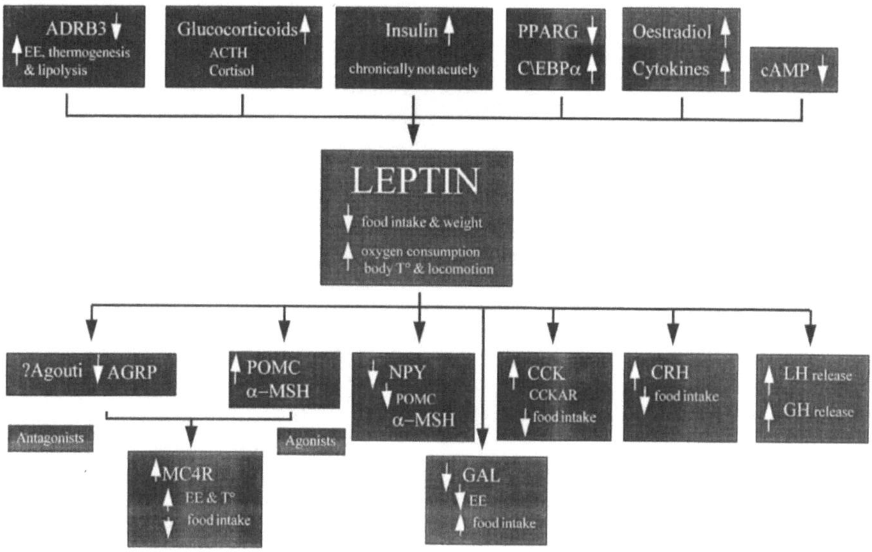

Fig. 1. Interrelations between the leptin protein and other hormones and molecules

In C57BL/6J *ob/ob* mouse, a C to T transversion in the first position of codon 105 of the leptin gene exchanges arginine for a stop codon, whereas no mRNA is detected at all in adipose tissue of the SM/Ckc-$^{+Dac}$ ob^{2J}/ob^{2J} mice, possibly because of a mutation in the promoter region of *Lep* (Zhang et al. 1994). Therefore, both strains of mice lack leptin. The human equivalent of *Lep*, designated as *LEP*, is located on chromosomes 7q31.3. Several studies tested for linkage relationships between obesity and genetic markers flanking the *LEP* gene in humans, and significant evidence of linkage with BMI, subcutaneous fat assessed from skinfolds and body fat mass was reported in some studies (Table 2; Clément et al. 1996a; Reed et al. 1996; Duggirala et al. 1996), but not in all (Norman et al. 1996; Stirling et al. 1996; Bray et al. 1996). However, since the markers used were located outside *LEP*, the positive linkages may result from an as yet unidentified gene that maps to the same chromosomal region. On the other hand, some polymorphisms in the *LEP* gene itself have been identified in the promoter region (–188 C/A) (Echwald et al. 1997a), in the untranslated exon 1 (19 A/G) (Hager et al. 1998), in exons 2 and 3 (F17L, V94M, V110M) (Considine et al. 1995; Oksanen et al. 1997), and in the 3′ UTS (9 A/G; microsatellite repeat) (Shintani et al. 1996; Oksanen et al. 1997) (Fig. 2). Significant evidence of associations between the microsatellite repeat and body weight (Shintani et al. 1996), between the V110M allelic variant and weight loss (Oskanen et al. 1997), and between 19 A/G and the blood leptin concentration (Hager et al. 1998) were reported.

Other attempts to find associations between mutations in *LEP* and human obesity have failed (Considine et al. 1995; Maffei et al. 1996; Niki et al. 1996;

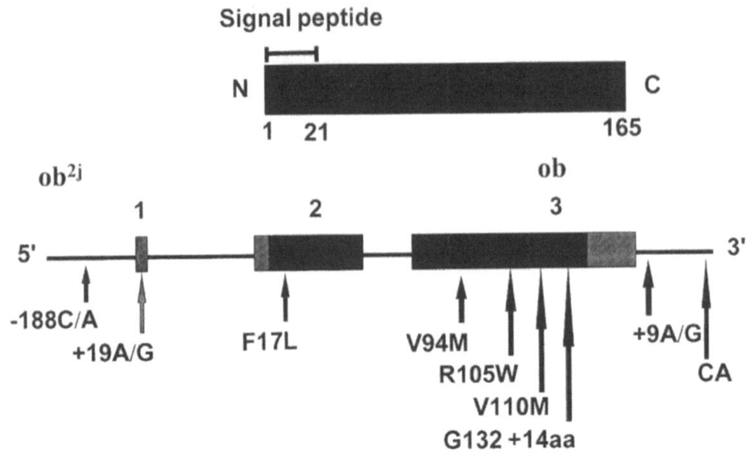

Fig. 2. Polymorphisms and mutations reported in the human leptin gene. Boxes indicate the coding translated (*black*) or untranslated (*gray*) regions. Mutations from rodents (ob^{2j} and *ob*) are indicated. Developed from ECHWALD et al. 1997a; HAGER et al. 1998; CONSIDINE et al. 1995; OKSANEN et al. 1997; SHINTANI et al. 1996; MONTAGUE et al. 1997; STROBEL et al. 1998; ZHANG et al. 1994

CARLSSON et al. 1997; ECHWALD et al. 1997a), with two exceptions. The first came from two severely obese children sibs of Pakistani origin with no detectable levels of leptin (MONTAGUE et al. 1997). These children were found to be homozygous for a frame-shift mutation in codon 133 of *LEP* resulting in the insertion of 14 aberrant amino acids followed by a stop codon (MONTAGUE et al. 1997). This mutation was shown to produce a phenotype very similar to the one resulting from the R105X mutation in *ob/ob* mice with early onset of severe obesity and hyperphagia. The second was, in homozygous form, in two female and one male Turkish sibs, a C to T transition in the first base of codon 105, identical to the mutation leading to the appearance of a premature stop codon in *ob/ob* mouse, and which resulted in the substitution of an arginine for a tryptophan (R105W) (STROBEL et al. 1998). These subjects were hyperphagic and showed impaired processing of the leptin protein at the origin of morbid obesity for two of them, the third one being moderately obese, with hypogonadism from a hypothalamic-pituitary-hormone insufficiency (STROBEL et al. 1998).

2. The Leptin-Receptor (*LEPR*) Gene

The rodent leptin-receptor *Lepr* gene encodes a protein of 1164 amino acids with a hydrophobic 22 amino acids secretory signal sequence, a predicted 23 amino acids transmembrane domain, and a cytoplasmic domain of variable length and constitution (TARTAGLIA et al. 1995; CHEN et al. 1996; LEE et al. 1996). Five differentially spliced transcripts, four long isoforms, identical up to the K889 amino acid, which include the transmembrane domain and differing intracellular domain of the protein, and one short and soluble form, identical up to the H796 amino acid, without these domains, have been observed (LEE et al. 1996). *Lepr* has many of the features of a cytokine receptor, with two WSXSW motives in the extracellular domain, janus kinase (JAK) interaction motives, and a signal transducer and activator of transcription (STAT) motive in the intracellular domain of the longest form of the receptor, which give *Lepr* the signaling capabilities of an interleukin-6-type cytokine receptor (BAUMANN et al. 1996). In situ hybridization studies have detected the extracellular domain of *Lepr* in arcuate, paraventricular, ventromedial, and dorsomedial nuclei of the hypothalamus and in the choroid plexus and leptomeninges in the rodent brain (MERCER et al. 1996; SCHWARTZ et al. 1996a), whereas the long isoform was detected mainly in the hypothalamus (MERCER et al. 1996). Northern blot analysis has also shown strong signal for the presence of the extracellular domain of *Lepr* in lung and kidney, and weaker signals in liver and skeletal muscle (TARTAGLIA et al. 1995). It is predicted that the longest form of *Lepr*, which includes the JAK and STAT motives, is the signaling receptor, whereas the shorter form, without the transmembrane and intracellular domains, could act as a transporter of leptin across the blood–CSF barrier in a saturable system (BANKS et al. 1996). In fact, in humans, the ratio of CSF to serum leptin decreases with increasing BMI, despite the positive correlation between leptin in the CSF and BMI (SCHWARTZ et al. 1996b; CARO et al. 1996).

Lepr is mutant in the *db* and the db^{2j} mice, and in the Zucker (*fa*) and Koletsky (fa^{k}) rats (CHUA et al. 1996a,b; PHILLIPS et al. 1996; IIDA et al. 1996; WU-PENG et al. 1997; TAKAYA et al. 1996; WHITE et al. 1997; CHUA et al. 1997). In the *db* mouse, a point mutation creates a splice-donor site in the 3′-UTS of the penultimate exon 17, specific to the short-form transcript, resulting in the insertion of 106 nucleotides from exon 17 into the intracellular coding region of the longest (signaling) isoform of *Lepr*. The resulting frame shift changes the long isoform for the short nonfunctional form of the receptor (CHEN et al. 1996; LEE et al. 1996). In the *fa* rat, the mutation in *Lepr* results in a single amino acid substitution Q269P, reducing the intracellular transport capacity of *Lepr* (CHUA et al. 1996b), possibly by interfering in its dimerization (PHILIPPS et al. 1996) or by effectively activating *Lepr* (WHITE et al. 1997). Finally, in the db^{2j} mouse and fa^{k} rat, a nonsense mutation, respectively in codons 625 and 763 that specify a tryptophan and a tyrosine residue, introduces a stop codon in the extracellular domain of *Lepr*, thereby producing a truncated nonfunctional receptor (TAKAYA et al. 1996; WU-PENG et al. 1997). Both rat *fa* and fa^{k} mutations result in deficient plasma-to-CSF transport of leptin (WU-PENG et al. 1997).

The human equivalent of *Lepr*, named *LEPR*, is located at 1p31. The structures of the *LEPR* gene and protein are very similar to those of the rodents, with a similarity of 78% and 71% for the extracellular and intracellular domains, respectively, between mouse and human proteins (TARTAGLIA et al. 1995; CHEN et al. 1996), and an overall similarity of 76% to the rat pro-tein (PHILIPPS et al. 1996). Multiple transcripts were also detected in humans (CIOFFI et al. 1996); these transcripts were localized to the hypothalamus (CONSIDINE et al. 1996b) and the choroid plexus epithelium, the ependymal lining, and neurons of the hypothalamic nuclei (arcuate, suprachiasmatic, mammillary, paraventricular, dorsomedial, supraoptic, and posterior), nucleus basalis of Meynert, inferior olivary nucleus, and cerebellar Purkinje cells (COUCE et al. 1997). The immunoreactivity of *LEPR* in the human brain was the same for 17 lean, 14 obese, and 4 NIDDM subjects (COUCE et al. 1997). No evidence of linkage between three markers surrounding *LEPR* and BMI or body fat in 217 Pima Indian families, a population with a high prevalence of obesity and diabetes, was observed (NORMAN et al. 1996), as in obese Finnish sibs (OHMAN et al. 1999). Suggestive linkages between three markers in the 1p32-p22 chromosomal region and adiposity phenotypes were reported in the Québec Family Study (CHAGNON et al. 1997a).

LEPR is encoded by 20 exons, of which the first two are untranslated. Some 20 different polymorphisms in *LEPR*, with nine allelic variations in exon coding sequences of the gene, six of them resulting in amino acid changes, as well as six nucleotide changes, two microsatellite repeats, and restriction site polymorphisms in different introns, an alternative spliced transcript, a missplicing mutation for exon 18, and a 45 bp insertion/deletion in the 3′-UTS, have been reported (CHUNG et al. 1996; CHUNG et al. 1997; THOMPSON et al. 1997; GOTODA et al. 1997; ECHWALD et al. 1997b; MATSUOKA et al. 1997; FRANCKE et al. 1997; ROLLAND et al. 1998; CLÉMENT et al. 1998; see also CHAGNON et al. 1999). Based on a study of 10 obese and 10 control Pima Indians, relatively weak evidence for an association between percent body fat and three different allelic variations in introns 16 and 19, and in exon 20 of *LEPR*, was at first reported (THOMPSON et al. 1997). In a second study, a weak association with BMI was observed in a British male population with another variant in exon 14 (K656N), but only in leaner (BMI $\leq 28\,kg/m^2$) subjects (GOTODA et al. 1997). More recently, in the Québec Family Study, positive linkages and weak associations between Q223R and a CTTT microsatellite repeat in intron 16 of LEPR and adiposity phenotypes were found (CHAGNON et al. 1999), and strong associations in the HERITAGE family study (CHAGNON et al. 2000b). On the other hand, a splice-mutation that resulted in skipping of exon 18, and a truncated leptin receptor lacking both the transmembrane and the intracellular portion of *LEPR*, was reported for a consanguineous family of Kabilian origin (CLÉMENT et al. 1998). In this family of nine children, three sisters were shown to be homozygotes for the mutation, which produced an early-onset morbid obesity, which develops in the first months of life, with hyperphagia, and hypogonadotropic hypogonadism, but, in contrast to *db* mouse, normal glycemia and insulinemia, and normal triglyceridemia and cholesterolemia. These show the direct role of *LEPR* in the control of adiposity and body composition in humans, and good evidence that DNA sequence variation in the *LEPR* gene could contribute to the more common forms of human obesity.

3. The Agouti Signaling-Protein (*ASIP*) Gene

The agouti yellow (A^y) mouse has an obesity phenotype similar to that of the *ob* and *db* mice, except that it entails the adult onset of a more moderate obesity with normal levels of corticosterone (HERBERG and COLEMAN 1977). As in other rodent obesity models, with the exception of the *ob* mouse, which does not produce leptin at all, the A^y mouse has elevated levels of leptin (MAFFEI et al. 1995), which reflects the increase in adipose tissue mass. A^y is also resistant to leptin administered peripherally or intracerebroventricularly (HALAAS et al. 1997). The A^y mRNA, which produces a 131 amino acid protein, was shown to be expressed exclusively in the skin of neonatal mice (BULTMAN et al. 1992). It was shown that the ectopic expression of agouti produces the obesity phenotype (KLEBIG et al. 1995), and that the agouti-specific expression in adipose tissue in mouse was permissive for obesity (MYNATT et al. 1997). It has been hypothesized that the effect of agouti could be mediated by the intracellular Ca^{2+} levels ($[Ca^{2+}]_i$) in skeletal muscle whereby an increase in $[Ca^{2+}]_i$ could lead to insulin resistance, hyperinsulinemia, and obesity (ZEMEL et al. 1995; KIM et al. 1996). Other mechanisms that have been proposed to explain the obesity phenotype in these mice include a depressed basal lipolytic rate in adipocytes, resulting from altered intracellular cAMP levels, or a decreased adrenergic tone, which may trigger hyperphagia and increased efficiency of food utilization, both mechanisms arising from antagonizing α-MSH activation of melanocortin receptors (KLEBIG et al. 1995).

Five melanocortin receptors (MC1R to MC5R) have been identified so far. These receptors exhibit various binding properties with their ligands. Agouti shifts the MSH functional curve of these receptors (LU et al. 1994; KIM et al. 1997), without affecting maximal activation; this suggests that agouti acts as a competitive antagonist, inhibiting agonist binding. MC3R and MC5R are not antagonized by agouti in rodents, but human HEK-293 cells transfected with the *MC3R* gene showed an increase of $[Ca^{2+}]_i$ in the presence of agouti. MC4R is clearly antagonized by human agouti, and the intracerebroventricular administration of an agonist of MC4R, as well as MC3R, inhibits feeding in four mouse models of hyperphagia, including *agouti*; therefore these receptors exert a tonic inhibitory effect on feeding behavior (FAN et al. 1997). Moreover, knockout mice for *MC4R* depict several aspects of the *agouti* obesity syndrome, including hyperphagia, hyperinsulinemia, and late-onset obesity (HUSZAR et al. 1997).

The human homolog of agouti, renamed agouti-signaling protein encoded by the *ASP* (WILSON et al. 1995) or *ASIP* gene, has been reported to be expressed in adipose tissue and testis (KWON et al. 1994), in heart, ovary, and, at a lower level, in liver and kidney (WILSON et al. 1995). The agouti protein is 132 amino acids long, with a peptide signal of 22 amino acids, a central basic region of 28 amino acids, and a conserved cysteine-rich region in the COOH-terminal part of the protein, including 11 cysteines. The *ASIP* gene has been located to 20q11.2. No mutation in *ASIP* has been found yet, and no significant linkage or association between markers in the proximity of *ASIP* and obesity has been found in two studies (Xu et al. 1995; Norman et al. 1996). On the other hand, positive results were observed between three markers, slightly telomeric to *ASIP*, and percent body fat, in the Québec Family Study (Lembertas et al. 1997), with obesity status in Americans (Lee et al 1999), and with 24 hours respiratory quotient in Pima Indians (Norman et al. 1998). In a recent human-genome-wide search for obesity genes in a Mexican-American population, the *POMC* gene was proposed as a candidate gene for the strong linkage between a marker located on 2p21 and leptin and fat mass level (Comuzzie et al. 1997). *POMC* is produced by neurons in the hypothalamic arcuate nucleus (Kiss et al. 1984), and one of its posttranslational products is α-MSH, the ligand of the MCRs. Central *POMC* and leptin pathways are also reported to have independent and additive effects on murine obesity (Boston et al. 1997). A early-onset obesity with adrenal insufficiency were observed in a compound heterozygote for two mutations in exon 3 of POMC (G7013T, C7133Δ) which interfere with synthesis of ACTH and α-MSH, whereas an homozygote for a mutation in exon 2 (C3804A) showed no POMC translation (Krude et al. 1998). Polymorphisms in the human *MC3R*, *MC4R*, and *MC5R* genes were tested for linkage and association with obesity in the Québec Family Study. There was evidence of linkage between adiposity variables and *MC3R* (Lembertas et al. 1997) as well as *MC5R*, but not with *MC4R* (Chagnon et al. 1997b). On the other hand, associations were observed for both *MC4R* and *MC5R* and adiposity, but only in women (Chagnon et al. 1997b). Three mutations (ΔCTCT at codon 211; insertion GATT at codon 246; Tyr35X) within MC4R has been observed. In all cases, a truncated non-functional protein was produced resulting in obesity (Yeo et al. 1998; Vaisse et al. 1998; Hinney et al. 1999).

A gene coding for an agouti-related protein (AGRP) was cloned recently, after a search in the expressed-sequence tag (ESTs) bank for a characteristic pattern of cysteine spacing in the COOH-terminal region of agouti (Ollmann et al. 1997; Shutter et al. 1997). This pattern is homologous to ion-channel blockers in spider and snail toxins (Ichida et al. 1993; Figueiredo et al. 1995), and is reminiscent of the agouti effect on increasing intracellular calcium. The mouse *Agrp* gene is expressed in the arcuate nucleus of the hypothalamus and in the adrenal gland; it antagonizes *Mc3r* and *Mc4r*, and has an 8- to 10-fold increased expression in *ob/ob* mice (Ollmann et al. 1997; Shutter et al. 1997). Moreover, transgenic mice overexpressing either mouse or human genes, the latter located at 16q22, developed an obesity phenotype, but without the skin pigmentation observed in *Agouti* mouse; this indicates that the effect of AGRP is not mediated by the Mc1r protein (Ollmann et al. 1997; Graham et al. 1997). The AGRP protein is very similar to the ASIP protein, with an identical length of 132 amino acids, the cysteine-rich region with cysteines 9 out of 11 times, but without the central basic region of ASIP (SHUTTER et al. 1997). There have so far been no linkage and mutation studies reported for the *AGRP* gene in humans.

4. The Fat (*FAT-CPE*) Gene

The *fat* mouse is characterized by early-onset hyperproinsulinemia (about 4 weeks) with hyperglycemia but a lack of insulin resistance. This is followed by a later-onset obesity than in *ob* and *db* mice, which develops slowly between 6 and 8 weeks of age. A multi-endocrine deficiency, probably involving some degree of ACTH deficiency, is suspected in the *fat* mouse, presumably resulting in decreased gluconeogenesis and/or lowered insulin counterregulation. A Ser202Pro mutation in the carboxypeptidase E (*cpe*) gene, which codes for an enzyme involved in the processing of prohormones such as POMC and pro-insulin, results in the non-secretion and the absence of maturation of *cpe*; this mutation is thought to be the cause of the obesity phenotype in these mice (NAGGERT et al. 1995).

In humans, there is no evidence of linkage between markers surrounding *CPE* at 4q28-q31 and percent body fat in Pima Indians (NORMAN et al. 1996). However, a mutation in the gene of another prohormone-processing enzyme, the prohormone convertase 1 or PC1, was found in one obese patient (JACKSON et al. 1997). The proband was a woman with postprandial hypoglycemia episodes, with strong obesity during childhood, hypogonadotropic hypogonadism, and corticotropic insufficiency. Pro-insulin, des-65,66 pro-insulin, and POMC were elevated; this suggested that a prohormone-processing deficiency was at the origin of this pathology. The two alleles of *PC1* were mutated in this woman, to result in the absence of PC1 activity. The first allele was mutated in the splicing donor site of intron 5, with a protein truncated in the middle of the catalytic domain resulting, whereas the second allele had a Gly483Arg substitution, which resulted in the absence of maturation and in the retention of the proPC1 in the endoplasmic reticulum (JACKSON et al. 1997). The link between this inactivation of PC1 and obesity still needs to be clarified, but it probably involves energy-balance control by neuropeptides.

5. The Tubby (*TUB*) Gene

The *tubby* mutation produces a late-onset, as with *fat*, moderate type of obesity. The function of the protein encoded by *Tub* has not yet been established, but based on homology with other genes, it has been suggested that tubby could be related to apoptosis or death programming of the cells (NOBEN-TRAUTH et al. 1996). A more recent study implicates *Tub* in intracellular signaling by insulin (KAPELLER et al. 1999). The *tub* mouse has a G to T transversion in the splicing donor site of intron 12, resulting in the addition of 24 aberrant amino acids and the lack of the 44 carboxyl amino acids (NOBEN-TRAUTH et al. 1996). No mutation in the human equivalent of the gene has been reported yet. However, negative linkage results have been observed in Pima Indians, between five markers located in the corresponding human chromosomal region at 11p15.1 and BMI and percent body fat (NORMAN et al. 1996) and in Finnish obese sib pairs (OHMAN et al. 1999).

II. Functional Candidate Genes

Candidate genes can also be defined on the basis of their roles in relevant biochemical and physiological functions. By definition, a whole series of such candidate genes may be the target of study. For instance, it is likely that there are candidate genes for body mass, body composition, body fat distribution, level of abdominal visceral fat, appetite and satiety signals, resting metabolic rate, diet-induced thermogenesis, physical activity level, nutrient partitioning, comorbidities of obesity, and others. Several functional candidate genes have been tested for association or linkage with obesity phenotypes, but there have been more negative than positive findings (see CHAGNON et al. 1998 for a review). A summary of the positive findings obtained with such candidate genes is given in Table 1. In this section, we will describe some of the candidate-gene results pertaining to energy intake, energy expenditure, and body fat in more detail.

1. Energy Intake

Several hormones, neurotransmitters, and peptides have been shown to be implicated in the control of eating behavior, either by stimulating or inhibiting food intake (see LEIBOWITZ and HOEBEL 1998 for a review) (Fig. 3). In the brain, two centers of appetite control have been uncovered by selective destruction in experiments conducted on rodents. The lateral hypothalamus (LH) was identified as a feeding center because its destruction caused experimental animals to stop eating and to starve to death. On the other hand, the ventromedial hypothalamus (VMH) was dubbed as a satiety center, since its destruction turned animals into chronic overeaters.

a) Feeding Center

Different appetite-stimulating peptides are observed in LH, such as neuropeptide Y (NPY), galanin (GAL), melanin-concentrating hormone (MCH) and the hypocretin/orexin peptides which trigger eating in the brain, probably through cell-surface orphan receptors. One of these peptides, NPY, has a strong stimulatory effect on food intake, especially carbohydrate intake, and on the utilization of carbohydrate for lipogenesis (LEIBOWITZ and HOEBEL 1998). NPY is a 36 amino acids peptide with a C-terminal amide which has been highly conserved during evolution, as there is only one amino acid difference between the sequence of the pig and that of humans, the rabbit, the rat, and the mouse, where the latter four have identical sequences (LARHAMMAR et al. 1993). In *ob/ob*, NPY is overexpressed, but if these mice

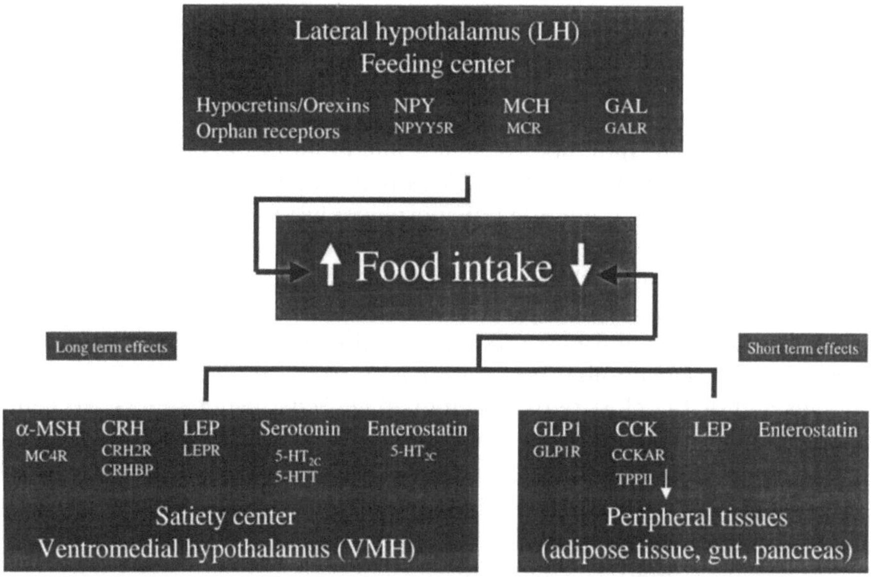

Fig. 3. Localization and effect of different peptides on food intake

are made deficient in NPY, they become less obese because of reduced food intake, and are less severely affected by *diabetes*; this suggests that NPY is partly a modulator of leptin deficiency (Erickson et al. 1996). The *NPY* gene, located at 11p15.1, has no known polymorphism, and no linkage between neighboring markers and obesity in humans has been reported (Roche et al. 1997), except with the lean part of body composition (Chagnon et al. 2000a). Various types of NPY receptors have been identified in humans, two of these receptors, *NPY1R* and *NPY5R*, map to the same locus at 4q31.3-q32 but in opposite orientations. *NPY1R* and *NPY5R* were recently investigated for genetic polymorphisms in their coding regions among overweight and obese subjects (BMI $\geq 27\,kg/m^2$; Roche et al. 1997), and for *NPY5R* differences among obese, underweight, and anorexia nervosa subjects (Rosenkranz et al. 1998). No polymorphism was found for *NPY1R*, whereas a Glu to Ala change at amino acid 4 in the first intracellular domain of NPY5R was found in one subject, and a nucleotide substitution (1278 G/A) at codon 426, which does not change the glycine at this position, was observed at similar frequencies (11%–14%) among the three groups of subjects (Rosenkranz et al. 1998).

Galanin (GAL) is a 29 amino acid peptide, observed widely in animals including fish, reptiles, amphibians, birds, insects, and mammals where it is ubiquitously expressed throughout the central and peripheral nervous system. In addition to stimulating food intake and reducing energy expenditure, GAL is also involved in the control of learning, memory, pain, sexual behavior, gastrointestinal motility, heart rate, gastric- and pancreatic exocrine secretions, and in pancreatic-, hypothalamic- and pituitary endocrine secretions. In rat,

GAL mRNA production and release rise markedly in animals that consume a high amount of fat (45% of the diet), with a concomitant gain in weight (LEIBOWITZ et al. 1998). Also, in obesity-prone as opposed to resistant rats, the expression of GAL in the brain is greater, especially in the paraventricular nucleus and median eminence. No significant difference in plasma galanin concentrations was found for 30 obese, 35 normal-weight and 11 anorectic women (INVITTI et al. 1995). No genetic polymorphism has yet been reported for the *GAL* gene. On the other hand, a GAL receptor (*GALR*), located at 18q23, was shown to belong to the superfamily of receptors, with seven transmembrane domains coupled to G-proteins (HABERT-ORTOLI et al. 1994; NICHOLL et al. 1995; BURGEVIN et al. 1995). GALR activation results in inhibition of adenylyl cyclase and phospholipase C, an opening of the potassium/ATP-sensitive channels and/or the closure of calcium/voltage-sensitive channels. No genetic studies on the *GALR* gene have been reported so far.

The melanin-concentrating hormone (MCH) is a 19 amino acid peptide expressed in the lateral hypothalamus and the zona incerta. The role of MCH in mammals is uncertain, but it has been associated with obesity since a differential display analysis of the mRNA expressed in the hypothalamus of *ob/ob* vs *ob/+* mice found a 50%–80% increased expression of *Mch* mRNA in *ob/ob* animals (QU et al. 1996). Fasting raised the level of MCH, and its injection in the rat brain increased eating in a way similar to that brought about by the orexigenic effect of NPY. The *Mch* gene also encodes two other peptides, the first one being a 13 amino acid peptide designated neuropeptide E1, which has the same epitope as the corticotropin-releasing hormone (CRH), and the second is a 19 amino acid peptide, neuropeptide GE, whose sequence is homologous to that of the human-growth-hormone-releasing factor. MCH antagonizes the action of α-MSH on melanocortin receptors, and has been shown to be the natural ligand of a orphan G-protein-coupled receptor, the somatostatin-like receptor (SLC1; SAITO et al. 1999; SHIMOMURA et al. 1999). SLC1 gene is located on chromosome 1q32.1 (JUNG et al. 1997), whereas the human *MCH* gene is located at 12q23.1 and has not been isolated yet.

Two related peptides, the hypocretins 1 and 2 (GAUTVIK et al. 1996; DE LECEA et al. 1998), also named orexins A and B (SAKURAI et al. 1998), are expressed in the hypothalamus of the rat where they trigger eating by the animals (SAKURAI et al. 1998). These peptides were shown to be ligands of cell-surface orphan receptors (SAKURAI et al. 1998). Injection of hypocretins/orexins in rat brain causes the animal to eat three to six times more than the control animals, whereas starvation upregulates their expression. The analysis of these two peptides and their genes (SAKURAI et al. 1999) is at an early stage.

b) Satiety Center

The leptin has become the most studied satiety hormone since the cloning of its gene from the *ob* mouse. Its effects and interactions with other molecules are depicted in Fig. 1.

The corticotropin-releasing hormone (CRH) is the primary hypothalamic releasing factor. Its relevance to obesity can be inferred from data showing that adrenolectomy, which increases CRH synthesis and secretion, reverses all known forms of obesity. CRH is an inhibitor of food intake and accelerates the metabolic rate. Intracerebral administration of leptin increases the expression of CRH. Urocortin (UCN), a homolog of CRH, has similar effects. The anorexic effects of CRH and UCN are mediated by the second (CRH2R) of the two CRH receptors. A CRH-binding protein (CRH-BP) can inactivate both CRH and UCN, while a ligand that is able to dissociate CRH or UCN from CRH-BP replenishes brain levels with free CRH and UCN, and suppresses body weight gain significantly. The corresponding genes in humans are located at 8q13 and on chromosome 2 for CRH and UCN respectively, and polymorphisms within these genes have not been studied yet in relation to obesity.

Serotonin (5-HT) is involved in a broad range of biological, physiological, and behavioral functions such as eating, motor activity, mood, sleep, sex drive, thermoregulation, cardiovascular, and respiratory activity. Several lines of evidence implicate the serotonergic system in body weight regulation and more specifically in eating behavior (see Hinney et al. 1997). Drugs that either directly or indirectly increase postsynaptic serotonergic stimulation decrease hunger and the consumption of food, whereas decreased serotonergic transmission results in hyperphagia. Serotonin reuptake inhibitors decrease the meal size and rate of eating in obese and lean subjects. Various receptors for 5-HT are known and some of these, the 5-HT2 receptors, have been shown to be involved in the control of both food intake and energy expenditure. More particularly, knockout mice for the 5-HT2C subtype, expressed predominantly in the medial hypothalamus, and for which the gene is located at Xq24, have a significantly elevated (13%) body weight compared to wild-type siblings (Tecott et al. 1995). In humans, a Cys23Ser polymorphism in the *5-HT2C* gene did not occur in different allele frequencies in 241 obese (BMI ≥ 97th percentile), 80 normal-weight (BMI 5th–85th percentile) and 92 underweight (BMI ≤ 15th percentile) subjects, and did not show linkage in 83 families with an obese proband (Lentes et al. 1997). On the other hand, a polymorphism located about 1 kb upstream of the coding region of the gene encoding the serotonin-transport (5-HTT) protein was associated with a higher concentration of *5-HTT* mRNA in cells that were homozygous for the long form than in those of the other two genotypes (Lesch et al. 1996). As previously found for the 5-HT2C receptor, the polymorphism in *5-HTT* did not occur in any different allele frequencies in 385 obese, 112 underweight, and 96 anorexia nervosa subjects, and no linkage was observed in 98 families with obese probands (Hinney et al. 1997). Another peptide, enterostatin, which is the pentapeptide-activation fragment of procolipase, was shown, in rats, to selectively inhibit intake of dietary fat via κ-opioidergic and serotonergic components, through peripheral and central mechanisms (Ookuma et al. 1997; Erlanson-Albertsson and York 1997). No genetic study of enterostatin has been reported thus far.

Some peptides produced from peripheral tissues also have short-term effects on food intake. Two of these peptides are leptin and enterostatin. Other peptides have exclusively short-term effects. Among these peptides are cholecystokinin (CCK) and glucagon-like peptide 1 (GLP1). Administration of CCK inhibits food intake and decreases meal size in rats through a satiating effect. In vivo, CCK is released from the upper small intestine following stimulation by ingested foods. The satiating effect of CCK is mediated by a CCKA receptor (CCKAR) located in the pyloric sphincter and on vagal afferent terminals in the duodenum and liver. Spontaneous knockout of the *CCKAR* gene in Otsuka Long Evans rats produces increased meal size and obesity. On the other hand, CCK is selectively inhibited by the tripeptidyl peptidase II enzyme (TPPII) (ROSE et al. 1996), which has been shown to be expressed in CCK-responsive neurons. The genes of *CCK* and *TPPII* were located respectively at 3p22-p21.3 and 13q32-q33, but no specific genetic study for these genes in humans has been reported so far.

Finally, GLP1, an intestinal peptide derived from proglucagon, is known for its insulinotropic and antidiabetic effects (HOLZ et al. 1995). GLP1 and its receptor (GLP1R) are found in the hypothalamus, and the administration of GLP1 into central cerebral ventricules, not peripherally, powerfully inhibits feeding in fasted rats (TURTON et al. 1996). Exendin, an antagonist of GLP1R, doubled food intake in satiated rats and raised the feeding response to NPY. GLP1 reduces short- but not long-term food intake or body weight in lean and obese rats (DONAHEY et al. 1998), similarly to the CCK effects. The chromosomal location of the human genes for *GLP1* and *GLP1R* are 2q23 and 17q25 respectively, and no genetic study of these two genes has been reported yet.

2. Energy Expenditure

Several candidate genes have the potential to influence energy expenditure and thus contribute to the development of obesity. Uncoupling proteins are membrane mitochondrial proton transporters whose functions are to "uncouple" mitochondrial respiration from the production of ATP, thereby generating heat. Uncoupling protein 1 (UCP1) is expressed exclusively in brown adipose tissue, while other forms of the protein, called UCP2 and UCP3, are expressed in a wider range of tissue, including white adipose and muscle tissue for UCP2 (FLEURY et al. 1997; GIMENO et al. 1997), and almost exclusive expression in muscle tissue for UCP3 (LISS et al. 1998). In a 12-year follow-up study, OPPERT et al. (1994) found that carriers of a Bcl I RFLP variant of the *UCP1* gene were more prevalent in high-fat gainers than in low-fat gainers. An other studies has also shown a relation between UCP1 and weight loss (FUMERON et al. 1996). More recently, a strong linkage between markers encompassing the chromosomal region of the *UCP2* and *UCP3* genes on chromosome 11q23 and resting metabolic rate was uncovered (BOUCHARD et al. 1997). UCP2 and UCP3 were also associated to metabolic rate and BMI (WALDER et al. 1998; CASSELL et al. 1999; ARGYROPOULOS et al. 1998).

Adrenergic receptors play an important role in the mobilization of lipids stored in the adipocytes through the regulation of catecholamine-induced lipolysis, and thus represent candidate genes for energy balance. Genetic variation in the $\alpha2$-, $\beta2$-, and $\beta3$-adrenergic receptors were investigated for their potential role in obesity. The most frequently studied polymorphism of the adrenergic receptors is a tryptophan to arginine substitution in codon 64 (Trp64Arg) in the first intracellular loop of the $\beta3$-adrenergic receptor (*ADRB3*). Results have been controversial: weak associations were reported for increased susceptibility to weight gain in a French group of morbidly obese subjects (CLÉMENT et al. 1995) and with a higher waist to hip circumference ratio and early onset of NIDDM in Finnish women (WIDEN et al. 1995), whereas negative results were obtained in the majority of the numerous other studies (see CHAGNON et al. 1997c). On the other hand, in a more recent analysis of the Trp64Arg variation, using a locus with a strong linkage with leptin and fat mass (COMUZZIE et al. 1997) to select a subgroup of subjects identical by descent, a more constant association with different adiposity variables such as BMI, percent body fat, and skinfold thickness was reported (MITCHELL et al. 1998). There is also an additive effect of the *UCP1* mutation, described by OPPERT et al. (1994), and the Trp64Arg mutation in *ADRB3* in morbid obesity (CLÉMENT et al. 1996b).

There have been few other reports on polymorphims in other adrenergic receptors. Polymorphisms of the $\alpha2$- and $\beta2$-adrenergic receptor genes were tested for association with indicators of body fat and fat distribution in the Québec Family Study with negative results, except for an association in women between a polymorphism in the $\alpha2$-adrenergic receptor and the tendency to accumulate more fat in the truncal-abdominal area compared to the extremities (OPPERT et al. 1995). More recently, two polymorphisms in codons 16 (Gly16Arg) and 27 (Glu27Gln) of the $\beta2$-adrenergic receptor gene were tested for association with obesity in 140 Swedish women widely divergent in body fat mass, and the Gln27Glu polymorphism was found to be markedly associated with obesity defined as a BMI value above $27\,kg/m^2$ (LARGE et al. 1997). The fat mass of women homozygous for the polymorphism ($n = 22$) was on average 20 kg higher than those without it ($n = 49$). On the other hand, the Gly16Arg polymorphism was found to be associated with an increased agonist affinity for the receptor, but not with obesity (LARGE et al. 1997).

3. Body Fat

Besides its role as a lipid-storage organ, the adipocytes produce various factors that might serve as feedback signals to regulate fat-cell size and adipose-tissue metabolism (FRIED and RUSSEL 1998). Among these factors are cytokines, such as the tumor-necrosis factor α (TNF-α), whose expression is increased with obesity in rodents and in humans. The mRNA level of TNF-α was shown to increase after a gain in adipose tissue, and to decrease with weight loss, whereas an inverse relationship between LPL activity and TNF-α expression was also observed (KERN et al. 1995; HOTAMISLIGIL et al. 1995). The expression of TNF-α in adipose tissue was also found to be correlated with the degree of hyperinsulinemia and glucose-disposal rate during an euglycemic clamp. These findings suggest that TNF-α may play a role in the development of insulin resistance and obesity. A study conducted in Pima Indians suggested a linkage between body fat and a marker near the TNF-α gene as well as significant evidence for an association with the BMI (NORMAN et al. 1995). More recently, a polymorphism in the promoter of the TNF-α gene was found to be associated with increased percent body fat and serum leptin levels and decreased insulin sensitivity (FERNANDEZ-REAL et al. 1997).

Glucocorticoids and sex steroids are known to be important correlates of body-fat distribution, and some of the genes involved in the metabolism of these hormones are considered to be strong candidate genes for fat distribution. Glucocorticoids promote visceral fat accumulation, and omental fat cells have a higher glucocorticoid-binding capacity than subcutaneous fat cells have. A linkage between obesity (BMI > $27\,kg/m^2$) and a Bcl I RFLP at the glucocorticoid receptor (*GRL*) gene locus was reported (CLÉMENT et al. 1996c), whereas an association with abdominal visceral fat level was observed (BUEMANN et al. 1997). Previously, a partial antisense RNA knockout mouse of the *GRL* gene was related to obesity (PÉPIN et al. 1992). The β3-hydroxysteroid dehydrogenase (HSD3B1) enzyme plays a key role in the biosynthesis of steroid hormones such as progesterone, glucocorticoids, mineralocorticoids, androgens, and estrogens. The 12-year change in abdominal skinfold was associated, in women of the Québec Family Study, with a Bgl II RFLP in the *HSD3B1* gene (VOHL et al. 1994).

Insulin-like growth factors (IGF) mediate the action of growth hormone under the control of insulin, and therefore play a role in body mass development throughout the lifespan. In the spontaneously diabetic Goto–Kakizaki (GK) rat strain, a locus for impaired glucose tolerance and adiposity was identified near the insulin-like growth factor II (*IGF2*) gene (GAUGUIER et al. 1996). Recently, a polymorphism in the 3′-untranslated region of the *IGF2* gene was shown to be associated with BMI in 1474 middle-aged men (O'DELL et al. 1997), as a microsatellite DNA marker located in the 5′ region of IGF1 was associated with body fat and fat free mass (SUN et al. 1999). Morbidly obese patients, who were homozygotes for a mutation in the leptin receptor, had, unexpectedly, lower levels of IGF1 and IGF-binding-3 protein (CLÉMENT et al. 1998). The insulin-receptor substrate-1 (IRS-1) is the main substrate for the insulin- and IGF-1 receptors. A Gly972Arg mutation in *IRS-1* showed a significant interaction with obesity that was associated with a 50% reduction in insulin sensitivity (CLAUSEN et al. 1995). Moreover, the Gly972Arg mutation may cause a predisposition to NIDDM in the presence of excess body weight (SIGAL et al. 1996).

The development of obesity requires the continuous differentiation of new adipocytes to meet the demands of fat accretion over time. The peroxisome proliferator activated receptor γ (PPARγ) is a dominant activator of fat-cell differentiation from pre-adipocytes into adipocytes. Two different *PPARγ* splice variants, *PPARγ1* and PPARγ2, expressed abundantly and mainly in adipose tissue, were cloned both in mouse and humans. An increase of 40% in the *PPARγ2* mRNA, but not *PPARγ1*, was found in obesity; this increase dropped, whereas mean PPARγ2 mRNA expression fell by 25% with a 10% weight loss (VIDAL-PUIG et al. 1997).

C. Genetic Epidemiology of Human Obesity

Chronic diseases such as obesity result from the complex interactions between genetic and environmental factors. Genetic epidemiology studies of obesity have been useful in quantifying the extent of familial aggregation, in estimating the heritability of obesity phenotypes, in characterizing the transmission of the underlying genes through segregation analysis, and in determining whether shared genetic and/or environmental factors contribute to the covariation among obesity and its co-morbidities. A detailed review of the genetic epidemiology studies of human obesity can be found elsewhere (BOUCHARD et al. 1998). This section will briefly review only the most recent findings.

I. Familial Aggregation

Although it is well known that obesity aggregates in families, the risk of becoming obese when a first-degree relative is overweight or obese is not properly understood. This risk can be quantified with a statistical value called the lambda coefficient (λ_S) which is defined as the ratio of the risk of being obese when a biological relative is obese compared to the risk in the population at large, i.e., the prevalence of obesity (RISCH 1990). Estimates of this familial risk based on the body mass index were recently reported (ALLISON et al. 1996; LEE et al. 1997). Age- and gender-standardized risk ratios obtained from 2349 first degree relatives of 840 obese probands and from 5851 participants of the National Health and Nutrition Examination Survey III (NHANES III) (see Fig. 4) reveal that the prevalence of obesity (BMI $\geq$ 30 kg/m^2) is about 2 times higher in families of obese individuals than in the population at large (LEE et al. 1997). The risk increases with the severity of obesity in the proband as well as with the BMI threshold used to define the obesity. Thus, the risk of extreme obesity (BMI $\geq$ 45 kg/m^2) is about eight times higher in families of extremely obese (BMI $\geq$ 45 kg/m^2) subjects. These results clearly illustrate that obesity aggregates in families and that the level of familial aggregation tends to increase with the severity of the condition.

II. Heritability Studies

Despite the large number of heritability studies of obesity, there is no consensus among researchers on the importance of genetic factors in the familial

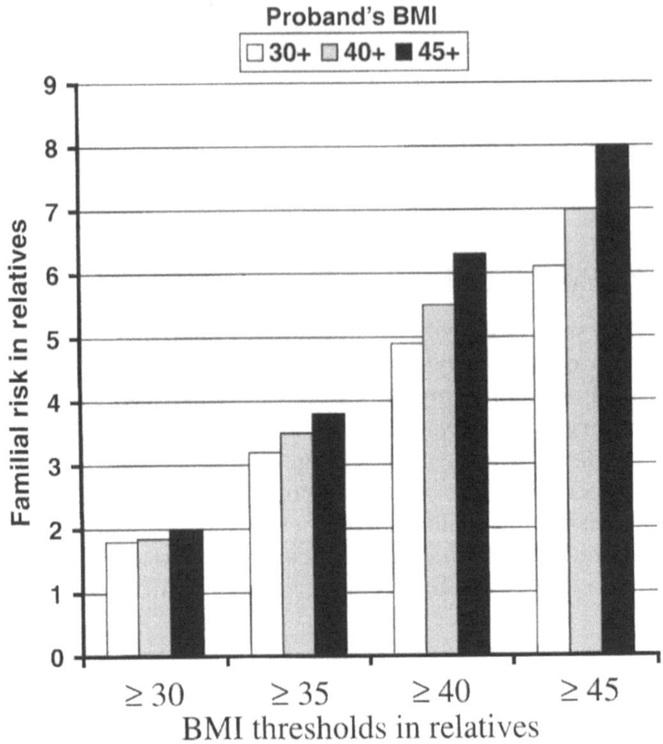

Fig. 4. Familial risk of obesity in relatives of obese probands. Values are age- and sex-standardized risk ratios of the prevalence rate of obesity in relatives of probands in relation to the population prevalence. Adapted from LEE et al. 1997

aggregation of obesity. Heritability estimates ranging from almost zero to values as high as 90% have been reported for the body mass index (STUNKARD et al. 1986; BOUCHARD et al. 1988). The use of different designs (twin, family, and adoption studies) and the often small sample sizes could explain such a wide variation in the estimates of heritability. The heritability found from twin studies (50%–80%) is highest, it is intermediate with nuclear family data (30%–50%), and lowest when derived from adoption data (10%–30%). When based on several types of relatives used jointly in the same design, the heritability estimates typically cluster around 25%–40% of the age- and gender-adjusted phenotype variance.

The genetics of fat distribution has also been investigated in family studies, and the evidence available, which is mainly derived from skinfold measurements, reveals that, after adjusting for total amount of body fat, about 30%–50% of the variance in the distribution of subcutaneous fat is accounted for by genetic factors (BOUCHARD et al. 1993; BOUCHARD et al. 1998 for a review). A few studies have estimated the heritability of central-abdominal fat and abdominal-visceral fat (AVF) levels. In a study based on 25 MZ and 18 DZ female twin pairs, the heritability of central-abdominal fat, measured by dual-energy X-ray absorptiometry (DEXA), was found to reach 73% of the age- and total-body-fat-adjusted phenotypic variance (CAREY et al. 1996). The first heritability estimates of the AVF depot based on direct measurements were recently reported. In one study, AVF area was measured by computed tomography between L4 and L5 vertebrae in 366 adult subjects from the Québec Family Study (PÉRUSSE et al. 1996). After adjustment for total body fat, the results were compatible with a genetic effect accounting for 56% of the variance in AVF. In another study based on 437 subjects participating in the HERITAGE Family Study, a heritability estimate of 48% was reported for AVF after adjustment for total body fat (RICE et al. 1997a). These results suggest that the amount of visceral fat, independent of total body fat mass, is strongly determined by genetic factors.

Circulating levels of leptin are increased in most obese individuals; this reflects their higher fat mass, but also suggests that they could be resistant to the action of leptin. The heritability of plasma leptin levels was recently estimated in two different populations. In one study based on a sample of 361 individuals from 118 African-American families, a heritability estimate of 39% was reported for age-adjusted leptin levels (ROTIMI et al. 1997). In another study, the heritability of leptin was estimated to be 63% in Mexican-American families (COMUZZIE et al. 1997). These results indicate that about 40%–60% of the variance in blood leptin levels is potentially attributable to genetic factors.

III. Segregation Analysis

The heritability of obesity phenotypes results from polygenic effects, i.e., from the effects of several genes, each having a small influence on the phenotype. In addition to polygenic effects, a phenotype could also be influenced by the segregation of a single locus that has a large impact on the phenotype, a so-called major-gene effect. In segregation analysis, the phenotype is assumed to be influenced by the independent and additive contributions of a major-gene effect, a multifactorial background due to polygenes, and a unique environmental component (residual). Several studies have tested the hypothesis of the segregation of a major gene for BMI or height-adjusted body mass, body-fat mass, fat distribution, or abdominal visceral fat. In general, segregation studies of BMI, or some other measure of height-adjusted weight, provided evidence for the segregation of a recessive locus with a frequency of about 0.2 that accounts for 35%–45% of the variance. For fat mass and/or percent body fat, three studies reported evidence for the segregation of a recessive locus (RICE et al. 1993; COMUZZIE et al. 1995; RICE et al. 1997b). The most recent of these studies, performed on subjects from the HERITAGE Family Study, showed that 64% of the variance in age-adjusted fat mass could be accounted for by the segregation of a major gene, with an estimated frequency of 0.2 for the high-fat allele in this population (RICE et al. 1997b).

The hypothesis of the segregation of a major gene for AVF was recently examined in the Québec Family Study (BOUCHARD et al. 1996) and HERITAGE Family Study (RICE et al. 1997b) cohorts. Very similar results were observed in both studies, with evidence for a putative recessive locus accounting for 51% and 57% of the variance of AVF, with a multifactorial component accounting for an additional 21% and 17% of the variance. However, in both studies, the evidence for the contribution of a major gene was considerably reduced when AVF was analyzed with statistical control over differences in total body fat, thus suggesting that the major gene may influence fat mass rather than AVF amount per se. In summary, results from segregation studies consistently indicate that body fat content is influenced by the segregation of an autosomal recessive locus; this explains up to about 65% of the variance in addition to the multifactorial component which accounts for 20%–40% of the variance. Unfortunately, the molecular genetic studies have not yet uncovered any gene that could be a strong candidate for these major gene effects.

IV. Genetic Covariation Between Obesity and Co-Morbidities

Few attempts have been made to determine whether or not, and to what extent, the associations between obesity and its co-morbidities share common genetic and/or environmental etiologies. Using data from the Québec Family Study, we examined the shared genetic basis between measures of body fat and various co-morbidities of obesity. In one of these studies, we investigated the familial clustering of body fat and fat distribution with fasting insulin (RICE et al. 1996). The results of this study are summarized in Fig. 5. This figure illustrates the influence of specific genetic (G1 and G2) and environmental (E1 and E2) factors on body fat and glucose and insulin levels. The figure also indicates that

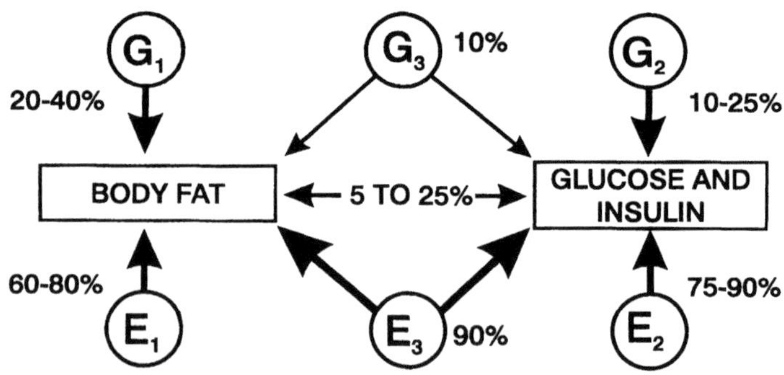

Fig. 5. Overview of the genetic (*G*) and environmental (*E*) effects in the covariation between body fat and fasting glucose and insulin levels in the Québec Family Study. See text for details. (From RICE et al. 1996a)

shared genetic (G3) and environmental (E3) factors contribute to the common variance of 5%–25% between indicators of body fat and insulin within individuals. These results suggest that there is a shared genetic basis for the two traits, with a bivariate heritability estimate of about 10% (RICE et al. 1996).

In another study, we investigated the cross-trait familial resemblance between measures of body fat and fat distribution with resting blood pressure (RICE et al. 1994). The results indicated that upper-body fat, as assessed by the trunk to extremity skinfolds ratio, was the strongest interindividual obesity correlate of blood pressure, particularly diastolic blood pressure. A bivariate heritability estimate of 33% was found between upper body fat and diastolic blood pressure (RICE et al. 1994). More recently, we examined the cross-trait familial resemblance between body fat and blood lipids (PÉRUSSE et al. 1997). The results provided evidence for a significant familial resemblance in the covariation between body fat and blood lipids, but this cross-trait familial resemblance was largely attributable to sibling cross-trait correlations, while parent–offspring and spouse cross-trait correlations were not significant. Although the hypothesis of a shared genetic basis for body fat and blood lipids cannot be excluded from this pattern of cross-trait correlations, the results suggest that shared familial environment rather than shared genes are responsible for the covariation observed between body fat and blood lipids. This finding corroborates the results of another study, in which common environmental factors rather than shared genes were found to explain the covariation of blood lipid and adiposity measures (MAHANEY et al. 1995). Significant genetic correlations were also reported between insulin levels, measured in the fasting state and two hours following a glucose load, and several traits associated with the metabolic syndrome, including BMI, HDL-cholesterol, waist-to-hip ratio, and subscapular-to-triceps skinfolds ratio (MITCHELL et al. 1996), and with various measures of HDL fractions and size of HDL particles (RAINWATER et al. 1997).

Other studies based on twin data investigated the genetic basis of the relationships between obesity and its co-morbidities. Based on data from 2508 adult MZ and DZ twins, CARMELLI et al. (1994) reported higher concordance rates in MZ twins compared to DZ twins for the clustering of hypertension, diabetes, and obesity and suggested, based on a path analysis model, that a common latent factor influenced by genetic (59%) and environmental (41%) effects was involved in this clustering.

The genetic and environmental etiologies of five risk-factor traits related to obesity were investigated in a sample of 289 elderly (52–86 years) MZ and DZ twins, of whom 140 pairs had been reared apart (HONG et al. 1997a). The results indicated that the cross-trait correlations between BMI and other metabolic phenotypes, such as triglycerides and HDL-cholesterol levels, systolic blood pressure, and insulin resistance, were higher in MZ than in DZ twins, suggesting a shared genetic basis in the covariation between BMI and features of the metabolic syndrome. The authors estimated that 52% of the genetic variance of BMI was shared with the other phenotypes (HONG et al.

1997a). These results suggest the presence of genetic mechanisms underlying the association between obesity and its most common co-morbidities.

A twin study investigating the relationship between insulin levels and BMI in 165 female MZ twin pairs, showed that, after removal of genetic influences by analysis of the intrapair differences, BMI remained significantly correlated with insulin, suggesting that the association between these two traits is partly independent of genetic factors (Mayer et al. 1996). In another study, based on 97 pairs of MZ twins, significant correlations were reported between intrapair differences in fasting insulin and intrapair differences in fasting glucose, triglycerides, and BMI, suggesting that nongenetic factors also contribute to the covariation between these risk-factor phenotypes (Hong et al. 1997b). Finally, a study based on 140 MZ and 96 DZ pairs of nondiabetic women twins revealed a significant genetic effect for each of three uncorrelated factors derived from a principal-component analysis of 10 risk-factor phenotypes (Edwards et al. 1997). These three factors were body mass/fat distribution, insulin/glucose, and blood lipids and their heritabilities were 61%, 87%, and 25%, respectively.

D. Summary

The limited information available on the magnitude of the familial risk of obesity reveals that there are children and adults with a significantly higher predisposition to become obese as they grow older. The most significant advances in the last few years have been the cloning of the genes that are responsible for the obesity and co-morbidities in the single-gene rodent models of obesity. A new biochemical pathway, with a large spectrum of action, has been uncovered from the characterization of leptin. The identification, in humans, of mutations in two of these genes, LEP and LEPR, and in the prohormone-convertase-1, the proopiomelanocortin, and the melanocortin receptor 4 gene, represents the first direct evidences of the involvement of genes in human obesity. The results of the two genome-wide scan efforts that have been published so far are disappointing in that they have yielded relatively few new candidate chromosomal regions or candidate genes for the genetic basis of adiposity and obesity in humans. Other scans on different human populations, and the cloning of the genes responsible for the quantitative trait loci observed in multigenic animal models of obesity will uncover additional and more promising candidate genes that will contribute to the ongoing efforts to identify the genetic and molecular basis of the common forms of human obesity.

References

Argyropoulos G, Brown AM, Willi SM et al (1998) Effects of mutations in the human uncoupling protein 3 gene on the respiratory quotient and fat oxidation in severe obesity and type 2 diabetes. J Clin Invest 102:1345–1351

Allison DB, Faith MS, Nathan JS (1996) Risch's lambda values for human obesity. Int J Obes 20:990–999
Banks WA, Kastin AJ, Huang W, Jaspan JB, Maness LM (1996) Leptin enters the brain by a saturable system independent of insulin. Peptides 17:305–311
Baumann H, Morella KK, White DW et al (1996) The full-length leptin receptor has signaling capabilities of interleukin 6-type cytokine receptors Proc Nat Acad Sci 93:8374–8378
Boston BA, Blaydon KM, Varnerin J, Cone RD (1997) Independent and additive effects of central POMC and leptin pathways on murine obesity. Science 278: 1641–1644
Bouchard C, Pérusse L, Leblanc C, Tremblay A, Thériault G (1988) Inheritance of the amount and distribution of human body fat. Int J Obes 12:205–215
Bouchard C, Rice T, Lemieux S, Després JP, Pérusse L, Rao DC (1996) Major gene for abdominal visceral fat area in the Quebec Family study. Int J Obes 20:420–427
Bouchard C, Pérusse L, Chagnon YC, Warden C, Ricquier D (1997) Linkage between markers in the vicinity of the uncoupling protein 2 gene and resting metabolic rate in humans. Hum Mol Genet 6:1887–1889
Bouchard C, Pérusse L, Rice T, Rao DC (1998) The genetics of obesity. In: Bray GA, Bouchard C, James WPT (eds) Handbook of obesity. Dekker, New York, pp 157–190
Bouchard C, Després JP, Mauriège P (1993) Genetic and nongenetic determinants of regional fat distribution. Endocr Rev 14:72–93
Bray MS, Boerwinkle E, Hanis CL (1996) Ob gene not linked to human obesity in Mexican American affected sib pairs from Starr County, Texas. Hum Genet 98:590–595
Buemann B, Vohl MC, Chagnon M et al (1997) Abdominal visceral fat is associated with a BclI restriction fragment length polymorphism at the glucocorticoid receptor gene locus. Obes Res 5:186–192
Bultman SJ, Michaud EJ, Woychik RP (1992) Molecular characterization of the mouse agouti locus. Cell 71:1195–1204
Burgevin MC, Loquet I, Quarteronet D, Habert-Ortoli E (1995) Cloning, pharmacological characterization, and anatomical distribution of a rat cDNA encoding for a galanin receptor. J Mol Neurosci 6:33–41
Carey DGP, Nguyen TV, Campbell LV, Chisholm DJ, Kelly P (1996) Genetic influences on central abdominal fat: a twin study. Int J Obes 20:722–726
Carlsson B, Lindell K, Gabrielsson B et al (1997) Obese (ob) gene defects are rare in human obesity. Obes Res 5:30–35
Carmelli D, Cardon LR, Fabsitz R (1994) Clustering of hypertension, diabetes, and obesity in adult male twins: same genes or same environments? Am J Hum Genet 55:566–573
Caro JF, Kolaczynski JW, Nyce MR et al (1996) Decreased cerebrospinal-fluid/serum leptin ratio in obesity: a possible mechanism for leptin resistance. Lancet 348:159–161
Cassell PG, Neverova M, Janmohamed S et al (1999) An uncoupling protein 2 gene variant is associated with a raised body mass index but not Type II diabetes. Diabetologia 42:688–692
Chagnon YC, Pérusse L, Lamothe M et al (1997a) Suggestive linkages between markers on human 1p32-p22 and body fat and insulin levels in the Quebec Family Study. Obes Res 5:115–121
Chagnon YC, Chen WJ, Pérusse L et al (1997b) Linkage and association studies between the melanocortin receptors 4 and 5 genes and obesity-related phenotypes in the Québec Family Study. Mol Med 3:663–673
Chagnon YC, Pérusse L, Bouchard C (1997c) Familial aggregation of obesity, candidate genes and quantitative trait loci. Curr Opin Lipidol 8:205–211
Chagnon YC, Pérusse L, Bouchard C (1998) The human obesity gene map: the 1997 update. Obes Res 6:76–92

Chagnon YC, Chung WK, Pérusse L, Chagnon M, Leibel RL, Bouchard C (1999) Linkages and associations between the leptin receptor (LEPR) gene and human body composition in the Québec Family Study (QFS). Int J Obes 23:278–286
Chagnon YC, Wilmore JH, Borecki IB et al (2000b) Linkages and associations between the leptin receptor (LEPR) gene and adiposity in the HERITAGE Family Study. J Clin Endo Metab 85:29–34
Chagnon YC, Borecki IB, Pérusse L et al (2000a) Genes related to the fat free body mass: results from a genome-wide search in the Québec Family Study. Metabolism 49:1–6
Chen H, Charlat O, Tartaglia LA et al (1996) Evidence that the diabetes gene encodes the leptin receptor: Identification of a mutation in the leptin receptor gene in db/db mice. Cell 84:491–495
Chua Jr. SC, Koutras IK, Han L, Liu S-M, Kay J, Young SJ, Chung WK, Leibel RL (1997) Fine structure of the murine leptin receptor gene: Splice site suppression is required to form two alternatively spliced transcripts. Genomics 45:264–270
Chua Jr. SC, Chung WK, Wu-Peng XS et al (1996a) Phenotypes of mouse diabetes and rat fatty due to mutations in the OB (leptin) receptor. Science 271:994–996
Chua SC, White DW, Wu-Peng XS et al (1996b) Phenotype of fatty due to Gln269Pro mutation in the leptin receptor (Lepr). Diabetes 45:1141–1143
Chung WK, Power-Kehoe L, Chua M, Lee R, Leibel RL (1996) Genomic structure of the human OB receptor and the identification of two novel intronic microsatellites. Genome Res 6:1192–1199
Chung WK, Power-Kehoe L, Chua M et al (1997) Exonic and intronic variation in the leptin receptor (OBR) of obese humans. Diabetes 46:1509–1511
Cioffi JA, Shafer AW, Zupancic TJ et al (1996) Novel B219/OB receptor isoforms: possible role of leptin in hematopoiesis and reproduction. Nat-Med 2:585–589
Clausen J, Hansen T, Bjorbaek C et al (1995) Insulin resistance: interactions between obesity and a common variant of insulin substrate-1. Lancet 346:397–402
Clément K, Vaisse C, Manning BSJ et al (1995) Genetic variation in the $\beta 3$-adrenergic receptor and an increased capacity to gain weight in patients with morbid obesity. N Engl J Med 333:352–354
Clément K, Garner C, Hager J et al (1996a) Indication for linkage of the human OB gene region with extreme obesity. Diabetes 45:687–690
Clément K, Ruiz J, Cassard-Doulcier AM et al (1996b) Additive effect of A→G (–3826) variant of the uncoupling protein gene and the Trp64Arg mutation of the beta 3-adrenergic receptor gene on weight gain in morbid obesity. Int J Obes 20: 1062–1066
Clément K, Philipi A, Jury C et al (1996c) Candidate gene approach of familial morbid obesity: linkage analysis of the glucocorticoid receptor gene. Int J Obes 20:507–512
Clément K, Vaisse C, Lahlou N et al (1998) A mutation in the human leptin receptor gene causes obesity and pituitary dysfunction. Nature 392:398–401
Comuzzie AG, Blangero J, Mahaney MC et al (1995) Major gene with sex-specific effects influences fat mass in Mexican Americans. Genet. Epidemiol. 12:475–488
Comuzzie AG, Hixson JE, Almasy L et al (1997) A major quantitative trait locus determining serum leptin levels and fat is located on human chromosome 2. Nat Genet 15:273–276
Considine RV, Considine EL, Williams CJ et al (1995) Evidence against either a premature stop codon or the absence of obese gene mRNA in human obesity. J Clin Invest 95:2986–2988
Considine RV, Sinha MK, Heiman ML et al (1996a) Serum immunoreactive-leptin concentrations in normal-weight and obese humans. N Engl J Med 334:292–295
Considine RV, Considine LE, Williams CJ, Hyde TM, Caro JF (1996b) The hypothalamic leptin receptor in human: identification of incidental sequence polymorphisms and absence of the db/db mouse and fa/fa rat mutations. Diabetes 45:992–994
Considine RV, Caro JF (1997) Leptin and the regulation of body weight. Int J Cell Biol 29:1255–1272

Couce ME, Burguera B, Parisi JE, Jensen MD, Lloyd RV (1997) Localization of leptin receptor in the human brain. Neuroendocrinol 66:145–150
De Lecea L, Kilduff TS, Peyron C et al (1998) The hypocretins hypothalamus specific peptides with neuroexcitatory activity. Proc Nat Acad Sci USA 95:322–327
Donahey JC, van-Dijk G, Woods SC, Seeley RJ (1998) Intraventricular GLP-1 reduces short- but not long-term food intake or body weight in lean and obese rats. Brain Res 779:75–83
Duggirala R, Stern MP, Mitchell BD et al (1996) Quantitative variation in obesity-related traits and insulin precursors linked to the OB gene region on human chromosome 7. Am J Hum Genet 59:694–703
Echwald SM, Rasmussen SB, Sorensen TIA et al (1997a) Identification of two novel missense mutations in the human OB gene. Int J Obes 21:321–326
Echwald SM, Sorensen TD, Sorensen TIA et al (1997b) Amino acid variants in the human leptin receptor: Lack of association to juvenile onset of obesity. Biochem Biophys Res Comm 233:248–252
Edwards KL, Newman B, Mayer E, Selby JV, Krauss RM, Austin MA (1997) Heritability of the insulin resistance syndrome in women twins. Genet Epidemiol 14:241–253
Erlanson-Albertsson C, York D (1997) Enterostatin – a peptide regulating fat intake. Obes Res 5:360–372
Erickson JC, Hollopeter G, Palmiter RD (1996) Attenuation of the obesity syndrome of ob/ob mice by the loss of neuropeptide Y. Science 274:1704–1706
Fan W, Boston BA, Kesterson RA, Hruby VJ, Cone RD (1997) Role of melanocortinergic neurons in feeding and the agouti obesity syndrome. Nature 385:165–168
Fernandez-Real JM, Gutierrez C, Ricart W et al (1997) The TNF-α gene Nco I polymorphism influences the relationship among insulin resistance, percent body fat, and increased serum leptin levels. Diabetes 46:1468–1472
Figueiredo SG, Garcia ME, Valentim AC, Cordeiro MN, Diniz CR, Richardson M (1995) Purification and amino acid sequence of the insecticidal neurotoxin Tx4 (6–1) from the venom of the "armed" spider Phoneutria nigriventer (Keys). Toxicon 33:83–93
Fleury C, Neverova M, Collins S et al (1997) Uncoupling protein-2: a novel gene linked to obesity and hyperinsulinemia. Nat Genet 15:269–272
Francke S, Clément K, Dina C et al (1997) Genetic studies of the leptin receptor gene in morbidly obese French Caucasian families. Hum Genet 100:491–496
Fried SK, Russell CD (1998) Diverse roles of adipose tissue in the regulation of systematic metabolism and energy balance. In: Bray GA, Bouchard C, James WPT (eds) Handbook of Obesity. Dekker, New York, pp397–413
Fumeron F, Durack-Bown I, Betoulle D et al (1996) Polymorphisms of uncoupling protein (UCP) and beta 3 adrenoreceptor genes in obese people submitted to a low calorie diet. Int J Obes 20:1051–1054
Gauguier D, Froguel P, Parent V et al (1996) Chromosomal mapping of genetic loci associated with non-insulin dependent diabetes in the GK rat. Nat Genet 12:38–43
Gautvik KM, De Lecea L, Gautvik VT et al (1996) Overview of the most prevalent hypothalamus-specific mRNAs, as identified by directional tag PCR subtraction. Proc Nat Acad Sci USA 93:8733–8738
Gimeno RE, Dembski M, Weng X et al (1997) Cloning and characterization of an uncoupling protein homolog. A potential molecular mediator of human thermogenesis. Diabetes 46:900–906
Gotoda T, Manning BS, Goldstone AP et al (1997) Leptin receptor gene variation and obesity: lack of association in a white British male population. Hum Mol Genet 6:869–876
Graham M, Shutter JR, Sarmiento U, Sarosi I, Stark KL (1997) Overexpression of AGRT leads to obesity in transgenic mice. Nat Genet 17:273–274

Habert-Ortoli E, Amiranoff B, Loquet I, Laburthe M, Mayaux JF (1994) Molecular cloning of a functional human galanin receptor. Proc Natl Acad Sci USA 91:9780–9783

Hager J, Clément K, Francke S et al (1998) A polymorphism in the 5′ untranslated region of the human ob gene is associated with low leptin levels. Int J Obes 22:200–205

Hager J, Dina C, Francke S et al (1998) A genome-wide scan for human obesity genes reveals a major susceptibility locus on chromosome 10. Nat Genet 20:304–308

Halaas JL, Boozer C, Blair-West J, Fidahusein N, Denton DA, Friedman JM (1997) Physiological response to long-term peripheral and central leptin infusion in lean and obese mice. Proc Natl Acad Sci USA 94:8878–8883

Hanson RL, Ehm MG, Pettitt DJ et al (1998) An autosomal genomic scan for loci linked to Type II diabetes mellitus and body-mass index in Pima Indians. Am J Hum Genet 63:1130–1138

Herberg L, Coleman DL (1977) Laboratory animals exhibiting obesity and diabetes syndromes. Metabolism 26:59–99

Hickey MS, Israel RG, Gardiner SN et al (1996) Gender differences in serum leptin levels in humans. Bioch Mol Med 59:1–6

Hinney A, Barth N, Ziegler A et al (1997) Serotonin transporter gene-linked polymorphic region: allele distributions in relationship to body weight and in anorexia nervosa. Pharmacology Lett 61:295–303

Hinney A, Schmidt A, Nottebom K et al (1999) Several mutations in the melanocortin-4 receptor gene including a nonsense and a frameshift mutation associated with dominantly inherited obesity in humans. J Clin Endocrinol Metab 84:1483–1486

Hixson J, Almasy L, Cole S et al (1999) Normal variation in leptin levels is associated with polymorphisms in the proopiomelanocortin gene, POMC. J Clin Endocrinol Metab 84:3187–3191

Holz GG 4th, Leech CA, Habener JF (1995) Activation of a cAMP-regulated Ca(2+)-signaling pathway in pancreatic beta-cells by the insulinotropic hormone glucagon-like peptide-1. J Biol Chem 270:17749–17757

Hong Y, Pedersen NL, Brismar K, de Faire U (1997a) Genetic and environmental architecture of the features of the insulin-resistance syndrome. Am J Hum Genet 60:143–152

Hong Y, Brismar K, Hall K, Pedersen NI, deFaire U (1997b) Associations between insulin-like growth factor-I (IGF-I), IGF-binding protein-1, insulin and other metabolic measures after controlling for genetic influences: results from middle-aged and elderly monozygotic twins. J Endocrinol 153:251–257

Hotamisligil GS, Arner P, Caro JF, Atkinson RL, Spiegelman BM (1995) Increased adipose tissue expression of tumor necrosis factor-α in human obesity and insulin resistance. J Clin Invest 95:2409–2415

Huszar D, Lynch CA, Fairchild-Huntress V et al (1997) Targeted disruption of the melanocortin-4 receptor results in obesity in mice. Cell 88:131–141

Ichida S, Wada T, Sekiguchi M, Kishino H, Okazaki Y, Akimoto T (1993) Characteristics of specific 125I-omega-conotoxin GVIA binding in rat whole brain. Neurochem Res 18:1137–1144

Iida M, Murakami T, Ishida K, Mizuno A, Kuwajima M, Shima K (1996) Phenotype-linked amino acid alteration in leptin receptor cDNA from Zucker Fatty (fa/fa) rat. Bioch Biophys Res Comm 222:19–26

Invitti C, Brunani A, Pasqualinotto L et al (1995) Plasma galanin concentrations in obese, normal weight and anorectic women. Int J Obes 19:347–349

Jackson RS, Creemers JWM, Ohagi S et al (1997) Obesity and impaired prohormone processing associated with mutations in the human prohormone convertase 1 gene. Nat Genet 16:303–306

Jung BP, Nguyen T, Kolakowski LF Jr et al (1997) Discovery of a novel human G protein-coupled receptor gene (GPR25) located on chromosome 1. Biochem Biophys Res Commun 230:69–72

Kapeller R, Moriarty A, Strauss A et al (1999) Tyrosine phosphorylation of tub and its association with Src homology 2 domain-containing proteins implicate tub in intracellular signaling by insulin. J Biol Chem 274:24980–24986
Kern PA, Saghizadeh M, Ong JM, Bosch R J, Deem R, Simsolo RB (1995) The expression of tumor necrosis factor in human adipose tissue. J Clin Invest 95:2111–2119
Kim JH, Mynatt RL, Moore JW, Woychik RP, Moustaid N, Zemel MB (1996) The effects of calcium channel blockade on agouti-induced obesity. FASEB J 10:1646–1652
Kim JH, Kiefer LL, Woychik RP et al (1997) Agouti regulation of intracellular calcium: role of melanocortin receptors. Am J Physiol 272 (Endocrinol Metab 35):E379–E384
Kiss JZ, Cassell MD, Palkovits M (1984) Analysis of the ACTH/beta-End/alpha-MSH-immunoreactive afferent input to the hypothalamic paraventricular nucleus of rat. Brain Res 324:91–99
Klebig ML, Wilkinson JE, Geisler JG, Woychik RP (1995) Ectopic expression of the agouti gene in transgenic mice causes obesity, features of type II diabetes, and yellow fur. Proc Natl Acad Sci USA 92:4728–4732
Krude H, Biebermann H, Luck W, Horn R, Brabant G, Gruters A (1998) Severe early-onset obesity, adrenal insufficiency and red hair pigmentation caused by POMC mutations in humans. Nat Genet 19:155–157
Kwon HY, Bultman SJ, Löffler C et al (1994) Molecular structure and chromosomal mapping of the human homolog of the agouti gene. Proc Natl Acad Sci USA 91:9760–9764
Large V, Hellström L, Reunisdottir S et al (1997) Human beta-2 adrenoreceptor gene polymorphism are highly frequent in obesity and associate with altered adipocyte beta-2 adrenoreceptor function. J Clin Invest 100:3005–3013
Larhammar D, Blomqvist AG, Söderberg C (1993) Evolution of neuropeptide Y and its related peptides. Comp Biochem Physiol 106C:747–752
Lee JH, Reed DR, Price RA (1997) Familial risk ratios for extreme obesity: implications for mapping human obesity genes. Int J Obes 21:935–940
Lee G-H, Proenca R, Montez JM et al (1996) Abnormal splicing of the leptin receptor in diabetic mice. Nature 379:632–635
Lee JH, Reed DR, Li WD et al (1999) Genome scan for human obesity and linkage to markers in 20q13. Am J Hum Genet 64:196–209
Leibowitz SF, Akabayashi A, Alexander JT, Wang J (1998) Gonadal steroids and hypothalamic galanin and neuropeptide Y: role in eating behavior and body weight control in female rats. Endocrinology 139:1771–1780
Leibowitz SF, Hoebel BG (1998) Behavioral neuroscience of obesity. In: Bray GA, Bouchard C, James WPT (eds) Handbook of Obesity. Dekker Inc, New York, pp 313–358
Lembertas A, Pérusse L, Chagnon YC et al (1997) Identification of an obesity quantitative trait locus on mouse chromosome 2 and evidence of linkage to body fat and insulin on the human homologous region 20q. J Clin Invest 100:1240–1247
Lentes K-U, Hinney A, Ziegler A et al (1997) Evaluation of CYS23SER mutation within the human 5-HT2c receptor gene: no evidence for an association of the mutant allele with obesity or underweight in children, adolescents and young adults. Pharmacology Lett 61:9–16
Lesch KP, Bengel D, Heils A et al (1996) Association of anxiety-related traits with a polymorphism in the serotonin transporter gene regulatory region. Science 274:1527–1531
Liu Q, Bai C, Chen F, Wang R, MacDonald T, Gu M, Zhang Q, Morsy MA, Caskey CT (1998) Uncoupling protein-3: a muscle-specific gene upregulated by leptin in ob/ob mice. Gene 207:1–7
Lu D, Willard D, Patel IR et al (1994) Agouti protein is an antagonist of the melanocyte-stimulating hormone receptor. Nature 371:799–802

Madej T, Boguski MS, Bryant SH (1995) Threading analysis suggests that the obese gene product may be a helical cytokine. FEBS Lett 373:13–18

Maffei M, Halaas J, Ravussin E, et-al (1995) Leptin levels in human and rodent: measurement of plasma leptin and ob RNA in obese and weight-reduced subjects. Nat Med 1:1155–1161

Maffei M, Stoffel M, Barone M et al (1996) Absence of mutations in the human OB gene in obese/diabetic subjects. Diabetes 45:679–682

Mahaney MC, Blangero J, Comuzzie AG, VandeBerg JL, Stern MP, MacCluer JW (1995) Plasma HDL cholesterol, triglycerides, and adiposity. A quantitative genetic test of the conjoint trait hypothesis in the San Antonio Family Heart study. Circulation 92:3240–3248

Matsuoka N, Ogawa Y, Hosoda K et al (1997) Human leptin receptor gene in obese Japanese subjects: evidence against either obesity-causing mutations or association of sequence variants with obesity. Diabetologia 40:1204–1210

Mayer EJ, Newman B, Austin MA et al (1996) Genetic and environmental influences on insulin levels and the insulin resistance syndrome: an analysis of women twins. Am J Epidemiol 143:323–332

Mercer JG, Hoggard N, Williams LM, Lawrence CB, Hannah LT, Trayhurn P (1996) Localization of leptin receptor mRNA and the long form splice variant (Ob-Rb) in mouse hypothalamus and adjacent brain regions by in situ hybridization. FEBS Lett 387:113–116

Mitchell BD, Kammerer CM, Mahaney MC et al (1996) Genetic analysis of the IRS. Pleiotropic effects of genes influencing insulin levels on lipoprotein and obesity measures. Arterioscler. Thromb Vasc Biol 16:281–288

Mitchell BD, Blangero J, Comuzzie AG et al (1998) A paired sibling analysis of the beta-3 adrenergic receptor and obesity in Mexican Americans. J Clin Invest 101:584–587

Montague CT, Farooqi IS, Whitehead JP et al (1997) Congenital leptin deficiency is associated with severe early-onset obesity in humans. Nature 387:903–908

Mynatt RL, Miltenberger RJ, Klebig ML et al (1997) Combined effects of insulin treatment and adipose tissue-specific agouti expression on the development of obesity. Proc Nat Acad Sci USA 94:919–922

Naggert JK, Fricker LD, Varlamov O et al (1995) Hyperproinsulinaemia in obese fat/fat mice associated with a carboxypeptidase E mutation which reduces enzyme activity. Nat Genet 10:135–142

Nicholl J, Kofler B, Sutherland GR, Shine J, Iismaa TP (1995) Assignment of the gene encoding human galanin receptor (GALNR) to 18q23 by in situ hybridization. Genomics 30:629–630

Niki T, Mori H, Tamori Y et al (1996) Human Obese gene: Molecular screening in Japanese and Asian Indian NIDDM patients associated with obesity. Diabetes 45:675–678

Nishina PM, Lowe S, Wang J, Paigen B (1994a) Characterization of plasma lipids in genetically obese mice: the mutants obese, diabetes, fat, tubby, and lethal yellow. Metabolism 43:549–553

Nishina PM, Naggert JK, Verstuyft J, Paigen B (1994b) Atherosclerosis in genetically obese mice: the mutants obese, diabetes, fat, tubby, and lethal yellow. Metabolism 43:554–558

Noben-Trauth K, Naggert JK, North MA, Nishina PM (1996) A candidate gene for the mouse mutation tubby. Nature 380:534–538

Norman RA, Bogardus C, Ravussin E (1995) Linkage between obesity and a marker near the tumor necrosis-a locus in Pima Indians. J Clin Invest 96:158–162

Norman RA, Leibel RL, Chung WK et al (1996) Absence of linkage of obesity and energy metabolism to markers flanking homologous of rodent obesity genes in Pima Indians. Diabetes 45:1229–1232

Norman RA, Thompson DB, Foroud T et al (1997) Genomewide search for genes influencing percent body fat in Pima Indians: suggestive linkage at chromosome 11q21–q22. Am J Hum Genet 60:166–173

Norman R, Tataranni P, Pratley R et al (1998) Autosomal genomic scan for loci linked to obesity and energy metabolism in Pima Indians. Am J Hum Genet 62:659–668
Ogawa Y, Masuzaki H, Isse N et al (1995) Molecular cloning of rat obese cDNA and augmented gene expression in genetically obese Zucker Fatty (fa/fa) rats. J Clin Invest 96:1647–1652
Ohman M, Oksanen L, Kainulainen K et al (1999) Testing of human homologues of murine obesity genes as candidate regions in Finnish obese sib pairs. Eur J Hum Genet 7:117–124
Oksanen L, Kainulainen K, Heiman M, Mustajoki P, Kauppinen-Makelin R, Kontula K (1997) Novel polymorphism of the human ob gene promoter in lean and morbidly obese subjects. Int J Obes 21:489–494
Ollmann MM, Wilson BD, Yang YK et al (1997) Antagonism of central melanocortin receptors in vitro and in vivo by agouti-related protein. Science 278:135–138
Ookuma K, Barton C, York DA, Bray GA (1997) Effect of enterostatin and kappa-opioids on macronutrient selection and consumption. Peptides 18:785–91
Oppert JM, Tourville J, Chagnon M et al (1995) DNA polymorphisms in a2- and β2-adrenoceptor genes and regional fat distribution in humans: association and linkage studies. Obes Res 3:249–255
Oppert JM, Vohl MC, Chagnon M et al (1994) DNA polymorphism in the uncoupling protein (UCP) gene and human body fat. Int J Obes 18:526–531
O'Dell SD, Miller GJ, Cooper JA et al (1997) ApaI polymorphism in insulin-like growth factor II (IGF2) gene and weight in middle-aged males. Int J Obes 21:822–825
Pépin MC, Pothier F, Barden N (1992) Impaired type II glucocorticoid receptor function in mice bearing antisense RNA transgene. Nature 355:725–728
Pérusse L, Després J-P, Lemieux S, Rice T, Rao DC, Bouchard C (1996) Familial aggregation of abdominal visceral fat level: results from the Quebec family study. Metabolism 45:378–382
Pérusse L, Rice T, Després JP, Bouchard C, Rao DC (1997) Cross-trait familial resemblance for body fat and blood lipids: familial correlations in the Quebec Family Study. Arteriosclerosis Thrombosis 17:3270–3277
Phillips MS, Liu Q, Hammond HA et al (1996) Leptin receptor missense mutation in the fatty Zucker rat. Nat Genet 13:18–19
Qu D, Ludwig DS, Gammeltoft S et al (1996) A role for melanin-concentrating hormone in the central regulation of feeding behavior. Nature 380:243–247
Rainwater DL, Mitchell BD, Mahaney MC, Haffner SM (1997) Genetic relationship between measures of HDL phenotypes and insulin concentrations. Arterioscler Thromb Vasc Biol 17:3414–3419
Reed D, Ding Y, Xu W, Cather C, Green ED, Price RA (1996) Extreme obesity may be linked to markers flanking the human OB gene. Diabetes 45:691–694
Rice T, Borecki IB, Bouchard C, Rao DC (1993) Segregation analysis of fat mass and other body composition measures derived from underwater weighing. Am J Hum Genet 52:967–973
Rice T, Province M, Pérusse L, Bouchard C, Rao DC (1994) Cross-trait familial resemblance for body fat and blood pressure: familial correlations in the Québec family study. Am J Hum Genet 55:1019–1029
Rice T, Nadeau A, Pérusse L, Bouchard C, Rao DC (1996) Familial correlations in the Quebec Family Study: cross-trait familial resemblance for body fat with plasma glucose and insulin. Diabetologia 39:1357–1364
Rice T, Després J-P, Daw EW et al (1997a) Familial resemblance for abdominal visceral fat: the HERITAGE family study. Int J Obes 21:1024–1031
Rice T, Després J-P, Pérusse L et al (1997b) Segregation analysis of abdominal visceral fat: the HERITAGE family study. Obes Res 5:417–424
Risch N (1990) Linkage strategies for genetically complex traits. I. Multilocus models. Am J Hum Genet 46:222–228
Roche C, Boutin P, Dina C et al (1997) Genetic studies of neuropeptide Y and neuropeptide Y receptors Y1 and Y5 regions in morbid obesity. Diabetologia 40: 671–675

Rolland V, Clément K, Dugail I et al (1998) Leptin receptor gene in a large cohort of massively obese subjects: No identification of the fa/fa rat mutation. Detection of an intronic variant with no association with obesity. Obes Res 6:122–127
Rose C, Vargas F, Facchinetti P et al (1996) Characterization and inhibition of a cholecystokinin-inactivating serine peptidase. Nature 380:403–409
Rosenbaum M, Nicolson M, Hirsch J et al (1996) Effects of gender, body composition, and menopause on plasma concentration of leptin. J Clin End Metab 81:3424–3427
Rosenkranz K, Hinney A, Ziegler A et al (1998) Screening for mutations in the neuropeptide Y Y5 receptor gene in cohorts belonging to different weight extremes. Int J Obes 22:157–163
Rotimi C, Luke A, Li Z, Compton J, Bowsher R, Cooper R (1997) Heritability of plasma leptin in a population sample of African-American families. Genet. Epidemiol 14:255–263
Saito Y, Nothacker HP, Wang Z, Lin SH, Leslie F, Civelli O (1999) Molecular characterization of the melanin-concentrating-hormone receptor. Nature 400:265–269
Sakurai T, Amemiya A, Ishii M et al (1998) Orexins and orexin receptors: a family of hypothalamic neuropeptides and G protein-coupled receptors that regulate feeding behavior. Cell 92:573–585
Sakurai T, Moriguchi T, Furuya K, Kajiwara N, Nakamura T, Yanagisawa M, Goto K (1999) Structure and function of human prepro-orexin gene. J Biol Chem 274:17771–17776
Schwartz MW, Seely RJ, Campfield LA, Burn P, Baskin DJ (1996a) Identification of targets for leptin action in rat hypothalamus. J Clin Invest 98:1101–1106
Schwartz MW, Peskind E, Raskind M, Boyko EJ, Porte D Jr (1996b) Cerebrospinal fluid leptin levels: relationship to plasma levels and to adiposity in humans. NatureMed 2:589–593
Shimomura Y, Mori M, Sugo T et al (1999) Isolation and identification of melanin-concentrating hormone as the endogenous ligand of the SLC-1 receptor. Biochem Biophys. Res Commun 261:622–626
Shintani M, Ikegami H, Yamato E et al (1996) A novel microsatellite polymorphism in the human OB gene: A highly polymorphic marker for linkage analysis. Diabetologia 39:1398–1401
Shutter JR, Graham M, Kinsey AC, Scully S, Lüthy R, Stark KL (1997) Hypothalamic expression of ART, a novel gene related to agouti, is up-regulated in obese and diabetic mutant mice. Genes & Development 11:593–602
Slieker LJ, Sloop KW, Surface PL et al (1996) Regulation of expression of ob mRNA and protein by glucocorticoids and cAMP. J Biol Chem 271:5301–5304
Sigal RJ, Doria A, Warram JH, Krolewski AS (1996) Codon 972 polymorphism in the insulin receptor substrate-1 gene, obesity, and risk of noninsulin-dependent diabetes mellitus. J Clin Endocrinol Metab 81:1657–1659
Stirling B, Cox NJ, Bell GI, Hanis CL, Spielman RS, Concannon P (1996) Identification of microsatellite markers near the human ob gene and linkage studies in NIDDM-affected sib pairs. Diabetes 44:999–1001
Strobel A, Issad T, Camoin L, Ozata M, Strosberg DA (1998) A leptin missense mutation associated with hypogonadism and morbid obesity. Nat Genet 18:213–215
Stunkard AJ, Foch TT, Hrubec Z (1986) A Twin Study of Human Obesity. JAMA 256:51–54
Sun G, Gagnon J, Chagnon YC et al (1999) Association and linkage between an insulin-like growth factor-1 gene polymorphism and fat free mass in the HERITAGE Family Study. Int J Obes 23:929–935
Takaya K, Ogawa Y, Hiraoka J et al (1996) Nonsense mutation of leptin receptor in the obese spontaneously hypertensive Koletsky rat. Nat Genet 14:130–131
Tartaglia LA, Dembski M, Weng X et al (1995) Identification and expression cloning of a leptin receptor, OB-R. Cell 83:1263–1271
Tecott LH, Sun LM, Akana SF et al (1995) Eating disorder and epilepsy in mice lacking 5-HT_{2c} serotonin receptors. Nature 374:542–546

Thompson DB, Ravussin E, Bennett PH, Bogardus C (1997) Structure and sequence variation at the human leptin receptor gene in lean and obese Pima Indians. Hum Mol Genet 6:675–679
Turton MD, O'Shea D, Gunn I et al (1996) A role for glucagon-like peptide-1 in the central regulation of feeding. Nature 379:69–72
Vaisse C, Clément K, Guy-Grand B, Froguel P (1998) A frameshift mutation in human MC4R is associated with a dominant form of obesity. Nat Genet 20:113–114
Vidal-Puig AJ, Considine RV, Jimenez-Linan M et al (1997) Peroxisome proliferator-activated receptor gene expression in human tissues. Effects of obesity, weight loss, and regulation by insulin and glucocorticoids. J Clin Invest 99:2416–2422
Vohl MC, Dionne FT, Pérusse L, Dériaz O, Chagnon M, Bouchard C (1994) Relation between BglII polymorphism in 3β-hydroxysteroid dehydrogenase gene and adipose-tissue distribution in human. Obes Res 2:444–449
Walder K, Norman R, Hanson R et al (1998) Association between uncoupling protein polymorphisms (UCP2-UCP3) and energy metabolism obesity in Pima Indians. Hum Mol Genet 7:1431–1435
White DW, Wang Y, Chua SC et al (1997) Constitutive and impaired signaling of leptin receptors containing the Gln Pro extracellular domain *fatty* mutation. Proc Natl Acad Sci USA 94:10657–10662
Widén E, Lehto M, Kanninen T, Walston J, Shuldiner AR, Groop LC (1995) Association of a polymorphism in the β3-adrenergic-receptor gene with features of the insulin resistance syndrome in Finns. N Engl J Med 333:348–351
Wilson BD, Ollmann MM, Kang L, Stoffel M, Bell GI, Barsh GS (1995) Structure and function of ASP, the human homolog of the mouse agouti gene. Hum Mol Genet 4:223–230
Wu-Peng SX, Chua SC Jr, Okada N, Liu S-M, Nicolson M, Leibel RL (1997) Phenotype of the obese Koletsky (f) rat due to Tyr763Stop mutation in the extracellular domain of the leptin receptor (Lepr). Evidence for deficient plasma-to-CSF transport of leptin in both the Zucker and Koletsky obese rat. Diabetes 46:513–518
Xu W, Reed DR, Ding Y, Price RA (1995) Absence of linkage between human obesity and the mouse agouti homologous region (20q11.2) or other markers spanning chromosome 20q. Obes Res 3:559–562
Yeo GSH, Farooqi IS, Aminian S, Halsall DJ, Stanhope RG (1998) A frameshift mutation in MC4R associated with dominantly inherited human obesity. Nat Genet 20:111–112
Zemel MB, Kim JH, Woychik RP et al (1995) Agouti regulation of intracellular calcium: Role in the insulin resistance of viable yellow mice. Proc Natl Acad Sci USA 92:4733–4737
Zhang Y, Proenca R, Maffel M, Barone M, Leopold L, Friedman JM (1994) Positional cloning of the mouse obese gene and its human homologue. Science 372:425–432

CHAPTER 4

The Influence of Obesity on the Development of Non-Insulin-Dependent Diabetes Mellitus

J.W. KOLACZYNSKI and B.J. GOLDSTEIN

The majority of patients diagnosed with NIDDM or type 2 diabetes have been obese for many years and are obese at the time of diagnosis. Moreover, quite commonly, weight gain continues throughout the years of living with the disease and, unquestionably, contributes through different pathogenetic mechanisms to the development of the numerous co-morbid conditions which add to the complexity of the management, health care costs, and worsening of disease outcomes. In contrast, weight loss is often associated with significant improvement of the metabolic control and, on occasion, even leads to a complete reversal of the diabetic state. This review will summarize some of the mechanisms involved in the development of type 2 diabetes in obese and overweight individuals. A broad scheme highlighting the pathophysiology of this process is shown in Fig. 1.

A. Epidemiological and Experimental Perspective

Several large epidemiological studies in men and women document that the risk of NIDDM rises exponentially with increasing body weight (COLDITZ et al. 1995; CHAN et al. 1994; PERRY et al. 1995). In both men and women the increased risk of NIDDM was already evident in those subjects who, by BMI criteria, fell below the cutoff dividing lean from obese (COLDITZ et al. 1995; CHAN et al. 1994; PERRY et al. 1995). An interesting caveat to these observations is that the escalating epidemic of obesity in the United States over two decades ending in 1990, was not matched by increase in the incidence rate of NIDDM. There were 23.7 known NIDDM cases per 1000 US residents in 1978 and 25.3 per 1000 in 1990. In the subsequent three years a "catch-up" growth in prevalence of diabetes was noted, reaching 29.0, 29.5, and 30.7 in 1990, 1991, and 1993, respectively (KENNY et al. 1995). A lag period between development of obesity and diabetic state is certainly a possibility (BENNETT 1990; LE STUNFF and BOUGNERES 1995) but one can also argue that this may also represent an improvement in the diagnosis of the disease at the asymptomatic stage.

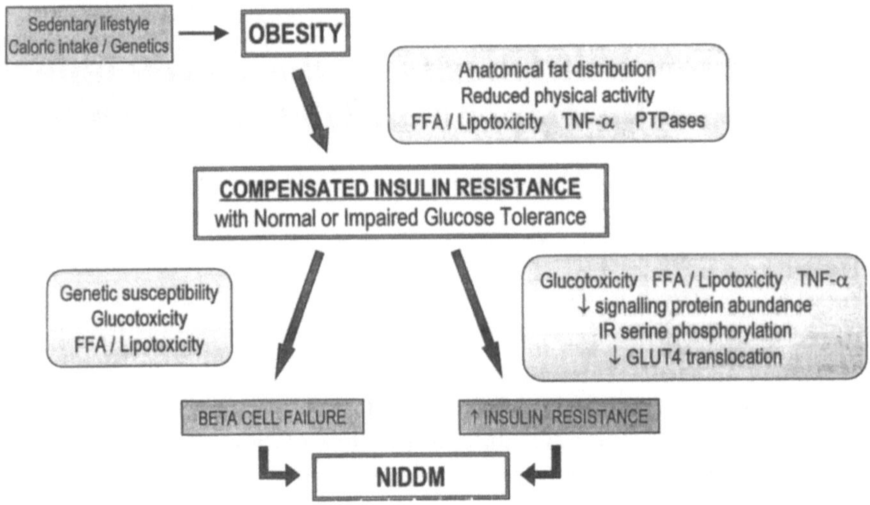

Fig. 1. Factors postulated to play important roles in the pathogenesis of insulin resistance and type 2 diabetes in obese human subjects. Most obese individuals are insulin resistant, and have normal or impaired glucose tolerance because of compensatory hyperinsulinemia from pancreatic β-cell hypersecretion. Additional factors as shown come into play and lead to worsening or persistent insulin resistance coupled with failure of the pancreatic β-cells to sustain effective insulin output, leading to the hyperglycemia characteristic of overt diabetes

I. Experimental Weight Gain and Insulin Action

The studies in a group of prisoners (SIMS et al. 1973) documented that experimental weight gain causes hyperinsulinemia and resistance of the enlarged adipocytes to insulin action on lipolysis and glucose oxidation, respectively. Overfeeding of twins for 100 days with excess of 84,000 calories, under a rooming-in setting, caused a variable weight gain and development of hyperinsulinemia indicative of development of insulin resistance (BOUCHARD et al. 1990). In both experimental settings, physical activity during the periods of overfeeding was limited (SIMS et al. 1973; BOUCHARD et al. 1990). In addition, weight gain brought about by overfeeding requires provision of calories that exceeded the threshold of satiety (SIMS et al. 1973). Once the period of overfeeding was terminated, the majority of the study subjects did return to their initial weight and eating habits (SIMS et al. 1973). We have made similar observations in recent studies (KOLACZYNSKI et al. 1996a; MARCO et al. 1998). An interesting finding was that, in spite of gain in adipose tissue mass due to overfeeding, maintenance of unrestrained physical activity does not result in worsening of insulin action in the healthy individuals (MARCO et al. 1998).

II. Overweight, Physical Activity, and the Risk of NIDDM

A protective effect of physical activity has been confirmed by several large prospective studies (HELMRICH et al. 1991; MANSON et al. 1992; MANSON et al. 1991; GURWITZ et al. 1994; BURCHFIED et al. 1995). In all of them the overweight individuals who exercised had a significantly reduced incidence of NIDDM compared to their sedentary counterparts. The studies have also confirmed that even a modest level of physical activity confers protection against NIDDM. As calculated by HELMRICH et al., a 500 kcal/week increment in energy expenditure is associated with a 6% decline in the age-adjusted risk of NIDDM (HELMRICH et al. 1991).

The above epidemiological and experimental data clearly indicate that weight gain or established obesity in association with sedentary lifestyle contribute most to the metabolic sequel that leads to the diabetic state. One can assume that both obesity and physical inactivity contribute to development of NIDDM through either a common pathophysiological mechanism or that the different pathways produce an additive or synergistic effect. In practice, the contribution of each of the conditions is difficult to separate, as obesity and inactivity usually coexist. At any rate, the 6% decline in the age-adjusted risk of NIDDM associated with exercise-induced increase in daily energy expenditure appears to be preserved even when body weight does not change (HELMRICH et al. 1991). It is thus the positive energy balance rather than weight (fat mass) per se that appears to be the most metabolically detrimental.

III. Effect of Increased Physical Activity on Energy Balance in Obesity

Short-term studies in obese men (PI-SUNYER 1985) and women (WOO et al. 1982) indicate that with free access to food, increased physical activity does not produce changes in caloric intake. In contrast, under similar experimental conditions, the lean, habitually physically active individuals appear to match their energy intake to the energy expenditure (WOO and PI-SUNYER 1985). As pointed out by FLATT, one of the explanations may be that the effect of exercise in obese individuals can lower the average 24-h respiratory quotient (RQ; FLATT 1995). Individuals predisposed to weight gain have higher RQ's (ZURLO et al. 1990) (i.e., they are "carbohydrate oxidizers") and weight continues to increase until either the expanding fat mass overrides this primary phenomenon with increased FFA availability for the oxidation, or the average daily RQ equals the food quotient (FQ; FLATT 1995). Thus, exercise in individuals predisposed to weight gain appears to be a peculiar substitute for an expansion of the adipose tissue mass. Therefore, perhaps, in the obese, previously sedentary people, the exercise-induced lowering of the average 24-h RQ may be, at least in the short-term, paradoxically be perceived as a state of satiety.

IV. Physical Activity and Insulin Sensitivity in Obese Individuals

Physical activity is associated with improvement in insulin sensitivity. Except for extreme situations, such as long-term exhausting exercise (e.g., after running a marathon; Tuominen et al. 1996), or eccentric type of exercise (e.g., running downhill; Kirwan et al. 1992), both endurance training, and resistance training are associated with the post-exercise improvement of insulin action. Even a single bout of acute exercise enhances insulin sensitivity in insulin-resistant obese individuals and NIDDM patients (Devlin et al. 1987; Burstein et al. 1990; Perseghin et al. 1997). The effect is short-lived and lasts from several hours up to couple of days (Burstein et al. 1985). Chronic exercise training may (Segal et al. 1991) or may not (Oshida et al. 1989) be associated with improvement in physical fitness (as measured by the increase in VO_{2max}), but whether it adds to the improvement in insulin sensitivity (a carry-over effect) in obese patients with or without NIDDM, as observed for a single bout of exercise, is still a debatable issue (Perseghin et al. 1997; Segal et al. 1991). At any rate, physical fitness allows for longer bouts of exercising and thus helps by reducing the excessive storage of fat.

B. Fat Distribution and Risk of Diabetes

The surrogate clinical markers of increased abdominal fat accumulation are waist circumference, waist–hip ratio, and the sagittal diameter. The direct methods applicable to humans are based on direct visualization of the intraabdominal fat depot with CT and MRI. In general, truncal, and especially intraabdominal accumulation of body fat is associated with higher plasma glucose, and insulin levels indicative of insulin resistance, and thus, confers a greater risk for development of NIDDM (Vague 1956; Kissebah et al. 1982; Krotkiewski et al. 1983; Ohlson et al. 1985). The interplay between upper body fat accumulation and insulin resistance is, in fact, complex and not completely understood. The most commonly cited pathogenetic sequel hinges on the high lipid turnover rate in intraperitoneal fat with increased sensitivity of these fat depots to catecholamine-induced lipolysis (Ostman et al. 1979; Rebuffe-Scrive et al. 1989). The resulting increased portal and systemic concentration of fatty acids is believed to have an adverse effect on hepatic metabolism, muscle insulin sensitivity, and insulin secretion (Bjorntorp 1990; Boden 1996). The studies that incorporated MRI to measure intraabdominal fat depots indicate that the actual situation may be far more complex. For example, a recent study conducted in lean and obese men indicated that the subcutaneous truncal fat appears to be equally, if not more, important compared to intraperitoneal fat for the development of both hepatic and peripheral insulin resistance (Abate et al. 1995). In contrast, in non-obese premenopausal females, it is the total fat mass which is inversely related to the degree of insulin sensitivity, while body fat distribution has no additional effect

(BONORA et al. 1992). In obese premenopausal females, however, while total fat mass had no relationship with insulin-mediated glucose metabolism, the upper body and particularly visceral fat accumulation was inversely associated with insulin-mediated glucose metabolism (BONORA et al. 1992). The mechanisms for the observed differences between males and females and between different degrees of adiposity in females are unclear.

C. Transition from an Obese to a Diabetic State – Nutritional and Humoral Mediators

Given that development of NIDDM usually requires the coexistence of the inherited defect both in pancreatic insulin secretory dynamics and insulin resistance, obesity is a condition in which these abnormalities may become clinically more apparent. Since obesity is synonymous with adipose excess, logic would imply that fat has an adverse effect. Owing to progress in recent years, adipose tissue is now regarded as not only a redundant depot of the expendable fuels, but also as hormonally active tissue, connected by a feedback system with the brain. In addition, there are numerous local factors acting in the tissue, all of which can modulate tissue growth and metabolism. The role of some of these factors will be discussed below.

I. Free Fatty Acids (FFA)

1. Effect of FFA on Glucose Utilization in Skeletal Muscle

The concept that elevated blood levels of FFA play a key role in the development of insulin resistance in obesity and NIDDM is more than 30 years old (RANDLE et al. 1963). According to this concept, the increased levels of the circulating FFA would compete with glucose as the physiological substrate in skeletal muscle. The resulting increase in FFA oxidation produces an increased acetyl CoA:CoA-SH ratio, and an increased citrate concentration. The increased concentration of acetyl CoA, through activation of pyruvate dehydrogenase kinase, will lead to phosphorylation, and thus inactivation, of pyruvate dehydrogenase, the critical enzyme controlling entry of pyruvate into oxidative metabolism. In addition, the increase in cytoplasmic citrate concentration inhibits phosphofructokinase 1 and thus glycolysis itself. The resulting increase in glucose-6-phosphate inhibits hexokinase and finally glucose uptake (RANDLE et al. 1963). For many years, this hypothesis generated major criticism until it was realized that the crucial factor influencing full operability of the glucose–fatty acid cycle is time. As elegantly shown by BODEN et al. in humans in vivo, the pathogenetic scheme unfolds in three major steps (BODEN et al. 1996). The earliest demonstrable defect, widely confirmed by others (WOLFE et al. 1989; LEE et al. 1990; BONADONNA et al. 1989), is the inhibition of insulin-stimulated carbohydrate oxidation by FFA without change in glucose uptake (Boden 1996). On the other hand, maintenance of FFA elevation for longer than 4h is associated with concentration-dependent reduction in insulin-stimulated glucose uptake in normal men (Boden et al. 1991) and patients with NIDDM (Boden and Chen 1995). The resulting reduction in glycogen synthesis was initially the most compatible with an FFA-induced defect in glucose transport or phosphorylation, since insulin-stimulated glucose uptake, synthesis, and glycolysis were inhibited to the same extent, while glycogen synthase activity remained unchanged (Boden and Chen 1995). When FFA concentration remain over 500μM for longer than 4h, however, the reduction in insulin-stimulated glucose uptake is associated with an increase in muscle glucose-6-phosphate concentrations and decrease in glycogen synthase activity (Boden et al. 1994).

One remaining concern is the physiological or pathophysiological relevance of the conditions used in the above experimental setting (i.e., concurrent hyperinsulinemia and FFA elevation). Under free living conditions, meal-associated rise in insulin in lean and obese individuals cause similar absolute and relative declines in plasma FFA concentrations (Golay et al. 1986). In addition, during oral glucose challenge in lean, obese, and obese NIDDM patients, similar suppression of FFA has been observed (Groop et al. 1991a). Under normal feeding conditions, the only time when FFA are allowed to rise (and they do) are the brief preprandial periods and the nocturnal period (sleep) (Golay et al. 1986). During these periods, the contribution of muscle to glucose disposal is far less important, but it is certainly possible that rising fatty acids may have a priming effect on insulin resistance which becomes so apparent in the postprandial state. In this discussion one can also not ignore the fact that following mixed ("physiological") meal intake, there is a postprandial elevation in circulating triglycerides which can carry FFA, through the local breakdown by lipoprotein lipase, to the sites of their deposition in adipose tissue and skeletal muscle (Nguyen et al. 1996). Thus, in normal physiology, glucose–fatty acid competition for oxidative and non-oxidative disposal and storage is a "daily routine". There is only a short window (about 2h) when the fuel competition does not affect the insulin-stimulated glucose uptake. Thus, insufficient or dysfunctional insulin secretion, as seen in glucose intolerance and type 2 diabetes, will cause prolongation of the postprandial glucose elevation (i.e., exceeding the 2-h window) with subsequent further deterioration of insulin action in muscle. In this context, the reduced lipoprotein lipase activity in skeletal muscle in insulin-resistant states including diabetes and obesity (Nguyen et al. 1996. Richelsen et al. 1993) may be viewed, at least in part, as a defense mechanism (limitation of the local FFA flux).

Although the above considerations serve well to explain the physiological importance of the available experimental data, the validation of the significance of the glucose–FFA cycle with experiments utilizing limb (forearm, leg) balance techniques in humans have yielded conflicting results (see Sect. C.I.4).

2. FFA and Fat Storage in Skeletal Muscle

As shown by KELLY et al., FFA may impair insulin-stimulated muscle-glucose uptake only if FFA uptake by skeletal muscle is increased (KELLY et al. 1993). After entering the muscle cell, FFA can be either oxidized or stored as triglycerides. The increased storage of fat in adipose tissue causes the tissue hypertrophy or hyperplasia. There is no evidence that this same rule applies to the muscle, as the primary role of muscle is not fuel storage but the use of fuel, to generate energy for contraction. One can thus assume that, analogously to glycogen metabolism, the skeletal muscle cell ability to store triglycerides is limited and finite. Therefore, with the coexistent glycogen and fat deposition without their concomitant respective oxidative and nonoxidative loss (i.e., low-turnover state) the transport of fatty acids and glucose will have to decline (a hypothetical competition "for storage space of unused fuels"). The Randle hypothesis does not take this factor into account, nor any of the alternative hypotheses proposed thus far. At any rate, there is growing evidence that the increased accumulation of triglycerides in the muscle cells may significantly contribute to the development of insulin resistance. The concept that human skeletal muscle may contain an intracellular lipid pool is not new (DAGENAIS et al. 1976).

PHILLIPS et al. found a negative correlation between insulin-induced activation of glycogen synthase and skeletal muscle lipid content in middle-aged non-diabetic women (PHILLIPS et al. 1996). Similar findings were reported by PAN et al. for obese and non-obese male Pima Indians (PAN et al. 1997). Multiple regression analysis showed that skeletal muscle triglyceride content is inversely related to insulin resistance, independent of obesity and body fat distribution (PAN et al. 1997). In this study, the intracellular muscle fat and upper body fat distribution were the strongest independent contributors to insulin resistance (PAN et al. 1997).

The detrimental effects of fatty acids on glucose metabolism in the muscle can also be the result of lipotoxicity (term first used by UNGER, UNGER 1995) or be the consequence of the exposure of the cell to fatty acids, and not be related to their effect on the substrate competition for storage space and their oxidation. Fatty acids can reduce glucose transport/phosphorylation by the accumulation of the products of the hexosamine–phosphate pathway (HAWKINS et al. 1996), a quantitatively relatively minor pathway of glucose and amino acid metabolism. This pathway is also implicated in the development of glucose toxicity (loss of insulin sensitivity due to hyperglycemia) (MARSHALL et al. 1991). The resulting accumulation of glucosamine pathway metabolites (aminosugars), including *N*-acetylglucosamine, *N*-acetylglucosamine-6-phosphate, and others, has a powerful effect on the ability of insulin to stimulate GLUT4 translocation to the plasma membrane with subsequent decline in insulin-stimulated glucose transport. (MARSHALL et al. 1991; BARON et al. 1995). Fatty acids can also interfere with GLUT4 gene expression (LONG and PEKALA 1996), but insulin resistance in human obesity and

NIDDM is not associated with quantitative defects in GLUT4 in the muscle (GARVEY et al. 1992).

3. Effects of FFA on Glucose Production in the Liver

Increased exposure of the liver to fatty acids, especially those generated by intraabdominal fat depots, stimulates the intrahepatic fat oxidation, initiating a sequence of events resulting in upregulation of gluconeogenesis (WILLIAMSON et al. 1966). Specifically, the accumulation of acetyl-CoA would increase activity of pyruvate carboxykinase, that is, the first committed step in gluconeogenesis. Stimulation of gluconeogenesis itself is insufficient to increase hepatic glucose output in the presence of adequate pancreatic insulin-secretory reserve and intact hepatic autoregulation (CLORE et al. 1991). In the presence of fatty acid elevation, the dose–response curve of insulin inhibition of hepatic glucose production, however, is shifted to the right, indicating a decline in hepatic insulin sensitivity (REBRIN et al. 1995). Disabling the insulin-secretory compensation by insulin clamping (inhibition of endogenous insulin secretion by somatostatin with maintenance of basal insulinemia by infusion of exogenous insulin) has been shown, in healthy volunteers, to unmask the stimulatory effect of fatty acids on hepatic glucose production (BODEN and JADALI 1991). One of the extreme concepts pertinent to the above observations is the Single Gateway Hypothesis, which states that insulin controls hepatic glucose output indirectly via suppression of adipocyte lipolysis. The major problems challenging the validity of the hypothesis is that, first, insulin inhibits lipolysis more potently than it suppresses the hepatic glucose production (NURJHAN et al. 1986; GROOP et al. 1991). Secondly, although a reduction in the sensitivity of insulin resulting in the suppression of the endogenous glucose production can consistently be demonstrated for obese subjects, the sensitivity of the antilipolytic effect of insulin (when normalized by the fat mass) remains unchanged (GROOP et al. 1991; GROOP et al. 1992).

4. Is the Liver a Major Site of Fatty Acid Oxidation in Obesity with and without Diabetes?

The numerous studies where indirect calorimetry was utilized to assess fuel partitioning for oxidation in postabsorptive state, document that the respiratory quotient is reduced in obesity and obese patients with NIDDM, indicating increased oxidation of fat (FELBER et al. 1981; FELBER et al. 1987; GOILAY et al. 1988). Although increased glucose oxidation seems to mark predisposition to the obese state (ZURLO et al. 1990), the increased lipid oxidation becomes evident once obesity develops, and appears to be independent of the duration of the obesity (ZURLO et al. 1990). At the skeletal muscle level, however, as can be assessed with the application of leg-balance technology, a somewhat different picture emerges. Specifically, as studied by Colberg and colleagues, in premenopausal women with upper body fat distribution, the postabsorptive uptake and oxidation of fatty acids is diminished (COLBERG et al. 1995). In addition, a positive relationship is found between postabsorptive rates of FFA oxidation across the leg and rates of insulin-stimulated glucose storage (COLBERG et al. 1995). Consistent with this observation is the finding that in healthy lean individuals, in postabsorptive conditions, uptake and oxidation of FFA in skeletal muscle is relatively high, compared to insulin-resistant obese controls (BALTZAN et al. 1962; DAGENAIS et al. 1976). In type 2 diabetes, the fasting hyperglycemia, in addition to the defect associated with insulin resistance, is an important factor that inhibits skeletal muscle FFA uptake and oxidation in the postabsorptive state (KELLY and SIMONEAU 1994). After a mixed meal ingestion, the fractional extraction (but not the total uptake) of FFA by leg muscle continues to be reduced, and as observed in the postabsorptive state, the defect in lipid oxidation persists (KELLY and SIMONEAU 1994). One can therefore deduce that if the insulin resistance of obesity and insulin resistance/hyperglycemia of type 2 diabetes cause defective FFA extraction and oxidation across the skeletal muscle, there must be a compensatory increase in FFA extraction and oxidation in the liver. As argued by JEQUIER, the increased lipid oxidation in the liver would explain the observed increase in total body lipid oxidation (JEQUIER 1996). In addition, gluconeogenesis, which is increased in the presence of elevated FFA, lowers the systemic respiratory quotient, which may further contribute to the overestimation of the total body lipid oxidation in the postabsorptive state (JEQUIER 1996).

5. Effects of FFA on Insulin Secretion

Hyperinsulinemia has been a consistently observed feature of human obesity since insulin assays became available (KARAM et al. 1963; PERLEY and KIPNIS 1966). Hyperinsulinemia is mainly the consequence of increased insulin secretion and partly due to reduced insulin clearance from the circulation (POLONSKY et al. 1988). In vivo and in vitro data from rodent models of genetic diabetic and non-diabetic obesity (UNGER 1995) indicate that acutely increased FFA levels stimulate both basal insulin secretion and insulin secretory response to glucose. The partial dependence of basal insulin secretion on acute alterations in FFA levels has recently been demonstrated in humans (BODEN et al. 1998; BODEN et al. 1999). In contrast, the long-term effect of FFA is dependent on the β-cell vulnerability to the chronic FFA overload, of which increased accumulation of intracellular triglycerides is a feature. The rise in intracellular fat precedes glucose incompetence (i.e., reduction in insulin-stimulated glucose secretion) and fasting hyperglycemia in the obese rodents predestined to develop diabetes (UNGER 1995). Whether or not a similar sequence occurs during the transition from normoglycemia to type 2 diabetes in human obesity is still unknown. Prolonged exposure to elevated concentrations of FFA in the presence of moderate hyperglycemia (~8.8 mM hyperglycemic clamp combined with infusion of Intralipid with heparin) in normal volunteers was associated with increased insulin secretory rate throughout the 48-h study (BODEN et al. 1995). In contrast, in normal subjects, the sole exposure to elevated FFA alone for 24h produced a biphasic effect on acute insulin response to intravenous boluses of glucose, with an increase followed by a decrease, observed at 6h and at the end of the fat infusion, respectively (PAOLISSO et al. 1995). The discrepancies in the outcomes of the above-cited studies may be more apparent than real. The concept, at least partially, that consolidates these two important observations is that hyperglycemia augments glycolytic flux, with the subsequent rise in one of the Krebs-cycle intermediates, which inhibits fatty acyl-CoA entry (activated energy-rich intracellular form of FFA) for beta-oxidation (PRENTKI et al. 1995). The resulting cytosolic accumulation of fatty acyl-CoA leads to a stimulation of insulin secretion (PRENTKI et al. 1995). On the other hand, under normoglycemic conditions, the increased fatty acid entry will, en masse, be channeled for beta-oxidation with subsequent inhibition of the glycolytic flux of glucose, its oxidation, and therefore a loss in its competence as a secretagogue (PRENTKI et al. 1995). The concept, again, does not take into consideration that prolonged β-cell exposure to FFA overload may, in susceptible individuals, ultimately cause increased accumulation of the intracellular triglycerides and the sequence of events described in obese rodents developing diabetes (UNGER 1995).

6. Obesity, Fatty Acids, and Type 2 Diabetes – Is There a Link?

Since obesity represents a disequilibrium between storage and utilization of fat, one can definitely search for the causal association implied by this basic principle. If the expansion of fat depots is brought about by the parallel increase in the fat cellularity, or fat redistribution from the depots characterized by high FFA turnover (such as intraabdominal ones) to sites with less lipolytic activity (such as femoral or gluteal fat), such a scenario is likely to produce no metabolic consequences. If insulin resistance preexists as a primary defect, it may affect fat exchange between high-turnover and low-turnover fat depots due to coexistent hyperinsulinemia. Thus, either the primary defect in fat hyperplasia, or fat redistribution due to primary insulin resistance will ultimately lead to increase in lipolysis by the enlarged adipocytes with the resulting rise in FFA and worsening of hyperinsulinemia. The elevated FFA will in turn be channeled to the liver for oxidation, stimulation of gluconeogenesis, storage, and production of triglyceride-rich lipoproteins (mainly VLDL). Whether or not the FFA:glucose competition in the skeletal muscle as delineated by RANDLE and further substantiated by BODEN and colleagues exists, is still a matter of controversy. The compromising scenario would be the proposed competition of intramuscular fat and glycogen for the storage space which can affect uptake of the respective substrates through numerous feedback systems such as stimulation of glucosamine pathway, reduction in LPL activity, reduction in GLUT4 translocation. Since the compensatory rise in insulin levels and subsequently in glucose levels will favor glucose as a substrate, it would explain the preferential use of this fuel for oxidation by the skeletal muscle in the obese individuals with and without type 2 diabetes. From this scenario one can easily understand the beneficial effect of therapeutic interventions such as reduction in caloric intake, or increase in physical activity which both contribute to the increase in the utilization of the stored energy. Although attractive, this hypothesis is still largely unproven. In addition, it ignores the influence of other factors, discussed below, which may play an important role.

II. Adipose Tissue as an Endocrine Organ

1. Tumor Necrosis Factor (TNF) α

As in various animal models of obesity (HOTAMISLIGIL et al. 1993; HOFMANN et al. 1994), TNF-α messenger RNA and protein expression is elevated in adipose tissue of obese humans (HOTAMISLIGIL et al. 1995; KERN et al. 1995). In contrast, weight loss is associated with downregulation of the TNF-α expression (HOTAMISLIGIL et al. 1995; KERN et al. 1995). The molecular basis of TNF-α-induced insulin resistance has been partially worked out and linked to inhibition of insulin-receptor tyrosine kinase activity (HOTAMISLIGIL et al. 1996). Specifically, TNF-α induces serine (instead of tyrosine) phosphorylation of insulin-receptor substrate-1 (IRS-1); the abnormally phosphorylated IRS-1, in turn, serves as an inhibitor of the insulin receptor tyrosine kinase (HOTAMISLIGIL et al. 1996). TNF-α released to the systemic circulation (endocrine effect) or generated locally (paracrine/autocrine effect) (SAGHIZADEH et al. 1996) could also produce the same molecular sequel in the skeletal muscle. These data have stimulated an interest in targeting TNF-α, as a way to modulate insulin resistance associated with obesity and type 2 diabetes, and was substantiated in a clinical trial with anti-TNF-α neutralizing antibody (OFEI et al. 1996). Surprisingly, however, obese type 2 diabetes patients showed no improvement in insulin sensitivity during the 4-week trial period (OFEI et al. 1996).

2. Endogenous Ligands of the PPAR-γ Nuclear Receptor and Adipose Tissue Cellularity

A dramatic improvement in our understanding of adipocyte development and insulin action occurred when members of the new class of the drugs used to enhance insulin sensitivity, the thiazolidinediones were shown to be potent ligands for receptors involved in the molecular processes of adipogenesis. In this particular context, the common molecular link was a member of the peroxisome-proliferator-activated receptor (PPAR) subfamily of nuclear hormone receptors called PPAR-γ (TONTONOZ et al. 1994). The expression of PPAR-γ, especially one of its splice variants, PPAR-γ2, is highly adipose-tissue selective (VIDAL-PUIG et al. 1997). Further work established that PPAR-γ plays a crucial role in the process of adipogenesis, and that activation of that factor in any fibroblastic cell expressing it (fibroblasts, preadipocytes, myoblasts) can convert it into fully differentiated adipocyte (TONTONOZ et al. 1994a; SPIEGELMAN 1998). The search for natural high-affinity ligands of this nuclear receptor has so far identified prostanoid 15-deoxy$\Delta^{12,14}$PGJ2 as one of them (FORMAN et al. 1995; KLIEWER et al. 1995). The low-affinity ligands are several polyunsaturated fatty acids, including linoleic acid (KLIEWER et al. 1997). The importance of these natural ligands in the differentiation of adipocytes in vivo has not been established yet. At any rate, stimulation of adipogenesis by recruitment of fibroblast-like fat cells precursors (preadipocytes), will promote a hyperplastic mode of adipose tissue expansion. The consequence of a shift towards higher fat cellularity, even if fat mass does not change, is improvement of insulin action to inhibit lipolysis (and therefore FFA flux to the general circulation), with stimulation of glucose uptake and fat storage as a result (SPIEGELMAN 1998).

The evidence that PPAR-γ is the major target for the thazolidinediones is very strong (SPIEGELMAN 1998). There is also a growing evidence, although so far limited to observations in rodents, that an improvement in insulin sensitivity brought about by treatment with thazolidinediones can be explained by their stimulating effect on adipose tissue cellularity (HALLAKOU et al. 1997).

From the above data it may be concluded that the detrimental metabolic effect of obesity is not related to the size of fat mass per se but to the fat (i.e., triglyceride) mass per adipose cell. Thus, any defect in the proportional growth of fat cell number to the fat storage is expected to lead to severe metabolic consequences, one manifestation of which is elevation in FFA. An extreme example of this scenario is severe insulin resistance associated with lipoatrophy (BEYLOT et al. 1988). One can also make a connection between low birth weight due to underdevelopment of fat tissue and the development of metabolic syndrome in adult life. Again, the concept is not new, but it resurfaced with the new data discussed above.

3. Leptin

The discovery of the ob gene product leptin (ZHANG et al. 1994) and its role in body fat regulation through activation of the hypothalamic receptors (PELLEYMOUNTER et al. 1995), has taught us a very important lesson, namely, that the largest body depot of fuel (i.e., fat in adipose tissue) is controlled by the brain through a specific feedback system. Since leptin is discussed in detail elsewhere in this volume, we will limit our discussion to several conceptual issues. First, can we draw a similar analogy, namely that the other major fuel depots, such as intrahepatic and intramuscular glycogen and fat, are under feedback regulation by the brain through yet undiscovered humoral or neural factors? Secondly, do these depots communicate and through what mechanism? As discussed above in the context of the role of FFA in the pathogenesis of insulin resistance, we can explain a number of pathogenetic mechanisms based on the substrate (fuel) competition. However, it is not known whether this a primary mechanism, or if perhaps another regulatory effect contributes through circulating mediators acting directly or by signaling across the blood–brain barrier. Leptin is one of the most recent examples of this effect; this demonstrates that such a concept is feasible and may well involve a number of new signaling hormones. Various forms of leptin receptors (TARTAGLIA 1997) are present in numerous organs including adipose tissue (KUTOH et al. 1998), pancreatic islets (KULKARNI et al. 1997), skeletal muscle (MUOIO et al. 1997), gonads (KARLSSON et al. 1997), and adrenals (CAO et al. 1997), suggesting a direct modulating effect of leptin.

One interesting aspect of leptin physiology is that experiments in humans demonstrated a duality of leptin regulation: it is fat-mass dependent and independent. Acute caloric deprivation, such as short-term starvation in healthy individuals, produces a profound downregulation of the circulating leptin levels that is out of proportion with the concomitant (negligible) loss of fat stores (KOLACZYNSKI et al. 1996; BODEN et al. 1996). Under these conditions, infusion of a small amount of glucose, sufficient to inhibit ketogenesis (up to 2.0 mg/kg/min), would prevent the food deprivation-induced downregulation in the circulating leptin (KOLACZYNSKI et al. 1996; BODEN et al. 1996). Ketogenesis in the liver during starvation is the marker of the depletion of the intrahepatic glycogen stores (OWEN et al. 1974). In addition, comparable rates of glucose infusion has been associated with no net utilization of hepatic glycogen in humans (TUONIAN et al. 1996). Moreover, strenuous physical exercise that uses up the stored fuels in the muscle is also associated with downregulation in circulating leptin levels (TUOMINEN et al. 1997). Thus, all these data provide circumstantial evidence that there may be communication between intrahepatic-, skeletal muscle-, and adipose tissue fuel depots, marked by fat-mass independent changes in the circulating leptin levels.

D. Skeletal Muscle Anatomy in Obesity and Type 2 Diabetes

The discussion on the role of obesity in the development of diabetes has so far focused mostly on the effect of fat-derived signals. The picture is far more complex. The questions that deserve brief discussion are: is there an anatomical basis for metabolic derangements in obesity leading to diabetes other than peculiar forms of fat distributions? Is there any functional defect affecting substrate delivery? These questions are especially pertinent to the skeletal muscle which in the postprandial state has to absorb the bulk of the delivered glucose. The components of an effective system to accomplish that goal are: (a) adequate blood flow, (b) effective transfer of insulin and glucose across the endothelial wall of the muscle capillaries, and (c) sufficient mass of the insulin-sensitive muscle units.

I. Skeletal Muscle Blood Flow

Insulin stimulates, probably through a central mechanism, the sympathetic nervous system activity (ANDERSON et al. 1991), a mechanism which is thought to cause vasoconstriction and has been implicated in the pathomechanisms of hypertension commonly seen in obese and non-obese insulin-resistant individuals (REAVEN et al. 1996). In normal humans, the sympathetic activation by exogenous hyperinsulinemia is observable after a short time lag of about 1–2h (ANDERSON et al. 1991; VOLLENWEIDER et al. 1994). This effect can be drastically reduced by dexamethasone (SCHERRER et al. 1993), strongly suggesting involvement of the hypothalamo–pituitary–adrenal axis. Obese, hyperinsulinemic, insulin-resistant individuals without evidence of autonomic neuropathy (normal response to Valsalva maneuver and immersion in ice-water) have elevated baseline sympathetic nerve activity that does not increase further in response to short-term insulin infusion, indicating that the sympathoexcitation has already been maximal (VOLLENWEIDER et al. 1994). In contrast, in obese Pima Indians, insulin resistance and hyperinsulinemia is associated with reduced baseline sympathetic neural activity and preserved sympathetic responsiveness to insulin infusions (SPRAUL et al. 1993; SPRAUL et al. 1996). The reasons for the observed discrepancies between Caucasian obese patients and the obese Pima Indians is unclear, but has been implicated to be associated with increased predisposition to weight gain in the latter (SPRAUL et al. 1993). Finally, the sustained sympathetic activation has been hypothesized to be responsible for the changes in the skeletal muscle capillary density (rarefaction) and the muscle fiber composition (see Sects. D.II and D.III).

On the other hand, insulin acting locally, probably via a nitric oxide-mediated mechanism (SCHERRER et al. 1994; STEINBERG et al. 1994), causes vasodilation and increased blood flow in the skeletal muscle (NUUTILA et al. 1996). This mechanism has been shown to be defective in obese individuals with and without diabetes compared to normal controls, when studied under comparable levels of hyperinsulinemia (LAAKSO et al. 1990; LAAKSO et al. 1992). The defect in insulin-induced vasodilation may, to some extent, be compensated by higher postprandial levels of insulin in obese subjects (EGAN and STEPNIAKOWSKI 1994). Thus, it appears that, as long as the insulin-secretory capacity is preserved in insulin resistance associated with obesity, a similar postprandial increase in the skeletal blood flow will be observed. Interestingly, as with sympathoexcitation, the direct vasodilatory effect of insulin in skeletal muscle occurs after an approximately 2-h lag. One can therefore question the physiological importance of these neural and rheological phenomena in normal postprandial conditions in which the 2-h period is sufficient to complete the food-derived glucose in the skeletal muscle. It must be kept in mind, however, that in obese and non-obese insulin resistant individuals with glucose intolerance or type 2 diabetes, the postprandial glucose elevation lasts longer, and that glucose uptake is in large part brought about by the mass action of hyperglycemia, which may be more dependent on the status of the vascular tone and the resulting blood flow. The persistent hyperglycemia of overt diabetes, by causing activation of protein kinase C (HENRION et al. 1994), nitric oxide quenching by glycosylated proteins, and by oxidation damage (TING et al. 1996), may cause a decline in nitric oxide generation and/or its vasodilatory action, thus limiting the utility of vasodilation as the important compensatory mechanism.

II. Skeletal Muscle Capillary Density and Endothelial Barrier

Insulin resistance is characterized by the reduced capillary density in the skeletal muscle (LILLIOJA et al. 1987). A similar degree of density reduction has been observed in men and women with type 2 diabetes and non-diabetic individuals matched for gender, age, degree of obesity, and body fat distribution (MARIN et al. 1994). The reduced capillary density elongates the diffusion distance from capillary to muscle cells and may therefore affect glucose transfer. The quantitative contribution of this finding to the observed reduction in insulin-stimulated glucose utilization remains unknown.

The diffusion of insulin across the vascular endothelium can also be a potential step that limits insulin action. Insulin crosses the endothelium to reach the interstitial space by a receptor-mediated mechanism (KING and JOHNSON 1985). The efficiency of insulin transfer can be assessed by determining the circulating plasma-lymph gradient. A recent study, in which such an approach was used, documented that transport of insulin from plasma to lymph was normal in obese, non-diabetic insulin-resistant subjects (CASTILLO et al. 1994). In addition, although a remarkable delay in insulin action has been documented for obese individuals with and without type 2 diabetes, it was not due to delay in activation in insulin-receptor tyrosine kinase (NOLAN et al. 1997). From these data one can conclude that transendothelial transport of insulin in obesity and NIDDM does not contribute to the defects in insulin action observed in these conditions (NOLAN et al. 1997).

III. Skeletal Muscle Fiber Characteristics in Obesity and Type 2 Diabetes

Skeletal muscle is composed of slow-twitch, oxidative, and insulin-sensitive type 1 fibers, and fast-twitch, glycolytic, insulin-resistant type 2 fibers. The proportion of each fiber type appears to be genetically determined, but it may well be environmentally (such as sedentary lifestyle) modified (LILLIOJA et al. 1987; KRIKETOS et al. 1996), or be the consequence of hyperinsulinemia, or increased sympathetic tone (ZEMAN et al. 1988). Several reports have documented that the proportion of type 2 fibers is increased in insulin-resistant states, including upper-body obesity and type 2 diabetes (LILLIOJA et al. 1987; KRIKETOS et al. 1996; KROTKIEWSKI et al. 1990). In contrast, the study of GARVEY et al., which included lean healthy subjects, obese individuals with normal glucose tolerance and IGT, as well as lean and obese patients with type 2 diabetes, failed to detect any differences in muscle-fiber composition between the groups (GARVEY et al. 1992).

Thus, all the studies document that, although some morphological and rheological differences may exist in the skeletal muscle of obese insulin-resistant patients with and without type 2 diabetes, their contribution to the observed reduction in insulin-stimulated glucose utilization is still not clearly established and deserves further investigation.

E. The Influence of Obesity on Insulin Signaling at a Molecular Level

As discussed above, multiple defects in insulin action in obesity and type 2 diabetes have been described. The molecular nature of these abnormalities is currently under intensive investigation (for recent reviews, see WHITE and KAHN 1994; DEFRONZO 1998; KAHN 1998; KRUSZYNSKA and OLEFSKY 1996), and some of these defects have been substantiated. The identified defects include a decrease in insulin binding due to receptor downregulation as well as defects in signaling by the insulin receptor kinase that occur in association with alterations in the phosphorylation state of post-receptor substrates (CARO et al, 1989; HURRELL et al. 1989; TAKAYAMA et al. 1988; GOODYEAR et al. 1995; OLEFSKY et al. 1995; RONDINONE et al. 1997). Some of the abnormalities can be inherited (and be permanent) conferring a genetic susceptibility to the development of the obese state and type 2 diabetes. On the other hand, the acquired defects can surface in established obesity and diabetes, and may, at least in part, be reversible. For example, in obesity with and without diabetes, downregulation of the insulin receptor may be an acquired abnormality since it can revert to near-normal expression in adipose tissue and muscle after weight reduction (FREIDENBERG et al. 1988; FRIEDMAN et al. 1992). Below we will discuss some of the defects in insulin-receptor-mediated tyrosine phosphorylation of specific substrates participating in the postreceptor signal propagation, and, finally, discuss the pathogenetic role of the cell molecular machinery participating in signal termination.

I. Insulin-Receptor-Mediated Phosphorylation Cascade ("On Signal")

Studies of adipocytes in obese patients have documented that both autophosphorylation and the activity towards exogenous peptide substrates of the insulin receptor tyrosine kinase are essentially normal in obesity (FREIDENBERG et al. 1988; SINHA et al. 1987; FREIDENBERG et al. 1987). Similarly, in human hepatocytes, and skeletal muscle, the tyrosine kinase activity expressed per amount of receptor protein is normal (CARO et al. 1986; GOODYEAR et al. 1995; NOLAN et al. 1994). Downstream transmission of the insulin receptor signal through its substrate proteins, however, may be a site of abnormal function in obesity. To this end, both insulin receptor and IRS-1 protein levels in obese subjects are significantly decreased compared to those levels in lean subjects (GOODYEAR et al. 1995); this occurs despite the fact that in obese subjects the phosphorylation of IRS-1 by the insulin receptor tyrosine kinase, and by the same token the kinase activity, is normal in the adipocytes (THIES et al. 1990) and muscle strips (GOODYEAR et al. 1995). The reduced overall abundance of tyrosine phosphorylated IRS-1, and the resulting loss of the cumulative signal strength (i.e., a subthreshold activation) may account for subsequent limitation of the downstream signal propagation.

One of the major signaling proteins for the metabolic effects of insulin, which is activated by docking with tyrosine-phosphorylated IRS-1, is phosphatidylinositol (PI) 3′-kinase. Activation of PI 3′-kinase elicits the serine phosphorylation of cellular proteins and increases the abundance of phosphoinositides phosphorylated at the D-3 position which activates downstream cellular serine/threonine kinases and it appears to be necessary for insulin-stimulated translocation of glucose transporters (KELLY et al. 1992; BACKER et al. 1992). In skeletal muscle strips from obese individuals, the levels of the 85 kDa regulatory/docking subunit of PI 3′-kinase are ~35% lower than in lean subjects (GOODYEAR et al. 1995). Conversely, the insulin-stimulated PI-3 kinase activity associated with IRS-1 was reduced 3.5-fold in obese subjects compared to the level of stimulation in control individuals, partly due to decreases in PI 3′kinase enzyme protein content (GOODYEAR et al. 1995a). Complementary to these findings are in vivo studies, utilizing NMR technology, documenting that one of the major defects in glucose disposal in skeletal muscle involves an early step in either glucose transport or glucose phosphorylation (PETERSEN et al. 1998), suggesting that a defect in glucose-transporter translocation may underlie the block in glucose disposal characteristic of the obese state. In addition, it can be demonstrated that there is a reduction in insulin-sensitive glucose transporter, GLUT4, mRNA, and protein in the adipose tissue in human obesity with and without diabetes (GARVEY et al. 1991; GIACCHETTI et al. 1994; GARVEY et al. 1988). Interestingly, this abnormality appears to be limited to the fat cell, since the levels of GLUT4 mRNA and protein in skeletal muscle from obese subjects is normal (GARVEY et al. 1992; KAHN et al. 1992).

II. Potential Role of Protein-Tyrosine Phosphatases (PTPases) – "Off Signal"

Increased expression of specific PTPase enzymes that negatively regulate the insulin action pathway may be involved in the pathogenesis of cellular insulin resistance in obesity and NIDDM (GOLDSTEIN 1996). PTPases, and in particular the transmembrane PTPase LAR and the intracellular enzyme PTP1B, can act as negative regulators of the insulin-signaling pathway by balancing the steady-state of tyrosine phosphorylation of the insulin receptor and its substrate proteins (for a recent review, see GOLDSTEIN et al. 1998). Also, studies showed that vanadate and pervanadate compounds, which may act as PTPase inhibitors in animals as well as in isolated cells, can enhance insulin action and ameliorate diabetes in a variety of animal models associated with obesity (SHECHTER 1990; MEYEROVITCH et al. 1991; POSNER et al. 1994; SHISHEVA et al. 1994). Clinical trials in humans have also recently provided evidence that vanadate or related compounds may potentially be effective as anti-diabetic agents (GOLDFINE et al. 1995; COHEN et al. 1995).

1. Skeletal Muscle of Obese Subjects

In obese, insulin-resistant Pima Native Americans, basal PTPase activity in skeletal muscle was shown to be elevated by 33% (MCGUIRE et al. 1991). In contrast, in a different group of obese, non-diabetic subjects, KUSARI et al. (1994) observed a decrease in skeletal muscle PTPase activity of a similar magnitude. In our own studies (AHMAD et al 1997), PTPase activity, with the insulin receptor as substrate in subcellular fractions from non-diabetic obese subjects, was significantly increased to 140%–170% of the level in lean controls and was strongly linearly correlated with the BMI. Quantification by immunoblot analysis demonstrated that the abundance of LAR and PTP1B phosphatases had the greatest increase (~3.0-fold) in the particulate fraction from obese, non-diabetic subjects. However, of the four major enzymes examined, only immunodepletion of LAR from the particulate fraction resulted in near normalization of the PTPase activity in obese subjects compared to the activity in control subjects.

2. Adipose Tissue of Obese Subjects

In our recent studies (AHMAD et al. 1995), obese subjects had a 1.7-fold increase in PTPase activity in subcutaneous abdominal fat that was strikingly correlated with BMI. This increase in tissue PTPase activity was associated with a prominent 2.0-fold rise in the abundance of LAR. Furthermore, immunodepletion of LAR protein (but not of PTP1B or SHP-2) resulted in normalization of the PTPase activity towards the insulin receptor (AHMAD et al. 1995). Interestingly, in obese subjects, a 10% loss of initial body weight was associated with a reduction in PTPase activity in subcutaneous adipose, and a decrease in the abundance of LAR and PTP1B by 21% and 27% (AHMAD et al. 1997a). Weight loss was also associated with a 26% decrease in fasting insulin levels (AHMAD et al. 1997a). The magnitude of insulin decline correlated with the reduction in the abundance of both LAR and PTP1B.

3. Studies on Skeletal Muscle of Subjects with Type 2 Diabetes

Interestingly, in subjects with overt type 2 diabetes, the available data suggests that tissue PTPase activities may actually be decreased. KUSARI et al. (KUSARI et al. 1994) found that skeletal muscle particulate fraction PTPase activity against RCM lysozyme was reduced by ~20% in subjects with NIDDM. This was associated with an average decrease of 38% in the abundance of PTP1B

Table 1. Alterations in tissue PTPase activity and the abundance of candidate PTPase enzymes in adipose and skeletal muscle tissue of human subjects with obesity and type 2 diabetes

	Obese non-diabetic	Obese NIDDM
Tissue PTPase activity		
Adipose tissue	↑	Not determined
Skeletal muscle	↑	↓
Abundance of PTPases		
Adipose tissue	↑ LAR	Not determined
Skeletal muscle	↑ LAR, PTP1B	↓ LAR, PTP1B

See text for literature references.

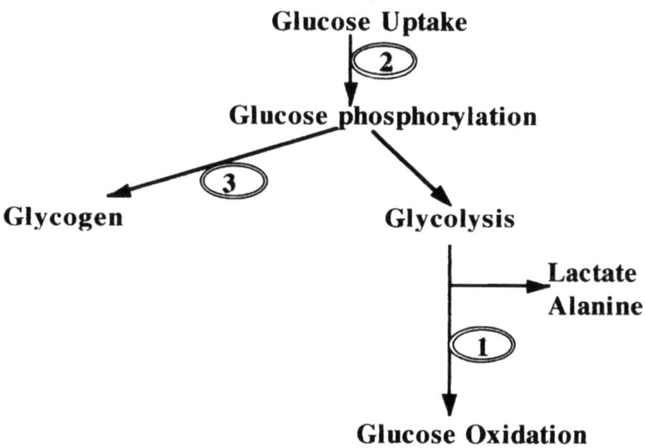

Fig. 2. Sequence of defects in insulin-stimulated glucose utilization produced by FFA (adapted from BODEN 1996). The earliest phenomenon [1] is inhibition in glucose oxidation without decline in glucose uptake. The second defect (detectable after a lag of 3–4 h) is reduction in glucose uptake and phosphorylation to G-6-P [2]. In the final step [3] the reduction in glucose uptake is associated with decline in glycogen synthase activity and increased accumulation of G-6-P

protein abundance in the diabetic subjects, although the abundance of other PTPase homologs was not assessed. WORM et al. (WORM et al. 1996) also noted a marked decrease in soluble PTPase activity in skeletal muscle biopsies from patients with NIDDM, down to 42% of the activity in the control subjects. We recently found that PTPase activity towards the insulin receptor in the cytosol and particulate fractions of skeletal muscle from obese subjects with NIDDM was significantly decreased, to 39% of the level in controls, and that the mass of several specific enzymes was decreased accordingly (AHMAD 1997). The greatest change occurred with LAR and PTP1B, which were decreased to 44% and 37% of the abundance in the control individuals, respectively.

The above data suggest that PTPase activity may be upregulated by hyperinsulinemia (associated with insulin resistance of obesity) and downregulated by hyperglycemia in the established diabetes (Table 1). Clearly, further work needs to be done with a larger number of subjects in various states associated with insulin resistance, to provide additional evidence in support of the potential pathophysiological relationships between PTPase activities and tissue insulin resistance and whether or not they may be amenable to therapeutic intervention.

F. Summary and Conclusions

The connection between obesity and type 2 diabetes is clearly established epidemiologically, and the majority of patients with diabetes are obese. The pathogenetic sequel leading from obesity to the diabetic state involves numerous behavioral, humoral, chemical, and molecular mechanisms, all of which have received even attention in the preceding sections. Some of these mechanisms are compounded in Fig. 2. All the described disturbances have been or can be potential targets for therapeutic interventions, but they are still unlikely to substitute primary prevention aimed at the preservation of a healthy lifestyle, leanness, normal body composition, and physical activity, especially in individuals born with the "thrifty" genes and a familiar predisposition.

References

Abate N, Garg A, Peshock RM, Stray-Gundersen J, Grundy SM (1995) Relationship of generalized and regional adiposity to insulin sensitivity in men. J Clin Invest 96:88–98

Ahmad F, Considine RV, Goldstein BJ (1995) Increased abundance of the receptor-type protein-tyrosine phosphatase LAR accounts for the elevated insulin receptor dephosphorylating activity in adipose tissue of obese human subjects. J Clin Invest 95:2806–2812

Ahmad F, Azevedo JL, Cortright R, Dohm GL, Goldstein BJ (1997) Alterations in skeletal muscle protein-tyrosine phosphatase activity and expression in insulin-resistant human obesity and diabetes. J Clin Invest 100:449–458

Ahmad F, Considine RV, Bauer TL, Ohannesian JP, Marco CC, Goldstein BJ (1997a) Improved sensitivity to insulin in obese subjects following weight loss is accompanied by reduced protein-tyrosine phosphatases in adipose tissue. Metabolism 46:1140–1145

Anderson EA, Hoffman RP, Balon TW, Sinkey CA, Mark AL (1991) Hyperinsulinemia produces both sympathetic neural activation and vasodilation in normal humans. J Clin Invest 88:1282–1290

Backer JM, Schroeder GG, Kahn CR, Myers MG, Wilden PA, Cahill DA, White MF (1992) Insulin stimulation of phosphatidylinositol 3-kinase activity maps to insulin receptor regions required for endogenous substrate phosphorylation. J Biol Chem 267:1367–1374

Baltzan MA, Andres R, Cader G, Zierler KL (1962) Heterogeneity of forearm metabolism with special reference to free fatty acids. J Clin Invest 41:116–125

Baron AD, Zhu J, Weldon H, Maianu L, Garvey WT (1995) Glucosamine induces insulin resistance in vivo by affecting GLUT4 translocation in skeletal muscle. J Clin Invest 96:2792–2801
Bennett PH (1990) Epidemiology of diabetes mellitus. In: Rifkin H, Porte D Jr (eds) Ellenberg and Rifkin's diabetes mellitus. New York, Elsevier, 363–77
Beylot M, Sautot G, Laville M, Cohen R (1988) Metabolic studies on lipoatrophic diabetes: mechanism of hyperglycemia and evidence of resistance to insulin of lipid metabolism. Diabetes Metab 14:20–24
Bjorntorp P (1990) "Portal" adipose tissue as a generator of risk factors for cardiovascular disease and diabetes. Arteriosclerosis 10:4903–496
Boden G (1996) Role of fatty acids in the pathogenesis of insulin resistance and NIDDM. Diabetes 45:3–10
Boden G, Chen X (1995a) Effects of fat on glucose uptake and utilization in patients with non-insulin-dependent diabetes. J Clin Invest 96:1261–1268
Boden G, Chen X (1999) Effects of fatty acids and ketone bodies on basal insulin secretion in type 2 diabetes. Diabetes 48:577–583
Boden G, Jadali F (1991) Effects of lipid on basal carbohydrate metabolism in normal men. Diabetes 40:686–692
Boden G, Jadali F, White J, Liang Y, Mozoli M, Chen X, Coleman E, Smith C (1991) Effects of fat on insulin-stimulated carbohydrate metabolism in normal men. J Clin Invest 88:960–966
Boden G, Chen X, Ruiz J, White JV, Rossetti L (1994) Mechanisms of fatty acid-induced inhibition of glucose uptake. J Clin Invest 93:2438–2446
Boden G, Chen X, Rosner J, Barton M (1995) Effects of a 48-h fat infusion on insulin secretion and glucose utilization. Diabetes 44:1239–1242
Boden G, Chen X, Mozzoli M, Ryan I (1996) Effect of fasting on serum leptin in normal human subjects. J Clin Endocrinol Metab 81:3424–3427
Boden G, Chen X, Iqbal N (1998) Acute lowering of plasma fatty acids lowers basal insulin secretion in diabetic and nondiabetic subjects. Diabetes 47:1609–1612
Bonadonna RC, Zych K, Boni C, Ferrannini E, DeFronzo RA (1989) Time dependence of the interaction between lipid and glucose in humans. Am J Physiol 257:E49–E56
Bonora E, Del Prato S, Bonadonna RC, Gulli G, Solini A, Shank ML, Ghiatas AA, Lancaster JL, Kilcoyne RF, Alyassin AM, De Fronzo R (1992) Total body fat content and fat topography are associated differently with in vivo glucose metabolism in nonobese and obese nondiabetic women. Diabetes 41:1151–1159
Bouchard C, Tremblay A, Despres JP, Nadeau A, Lupien PJ, Therriault G, Dussault J, Moorjani S, Pinault S, Fournier G (1990) The response to long-term overfeeding in identical twins N Engl J Med 322:1477–1482
Burchfied CM, Sharp DS, Curb JD, Rodriguez BL, Hwang LJ, Marcus EB, Yano K (1995) Physical activity and incidence of diabetes: the Honolulu Heart Program. Am J Epidemiol 141:360–368
Burstein R, Polychronakos C, Toews CJ, MacDougal JD, Guyda HJ, Posner BI (1985) Acute reversal of the enhanced insulin action in trained athletes. Association with insulin receptor changes. Diabetes 34:756–760
Burstein R, Epstein Y, Shapiro Y, Charuzi I, Karnieli E (1990) Effect of an acute bout of exercise on glucose disposal in human obesity. J Appl Physiol 69:299–304
Cao GY, Considine RV, Lynn RB (1997) Leptin receptors in the adrenal medulla of the rat. Am J Physiol 273(2 Pt 1):E448–E4452
Caro JF, Ittoop O, Pories WJ, Meelheim D, Flickinger EG, Thomas F, Jenquin M, Silverman JF, Khazanie PG, Sinha MK (1986): Studies on the mechanism of insulin resistance in the liver from humans with noninsulin-dependent diabetes. Insulin action and binding in isolated hepatocytes, insulin receptor structure, and kinase activity. J Clin Invest 78:249–258
Caro JF, Dohm GL, Pories WJ, Sinha MK (1989) Cellular alterations in liver, skeletal muscle, and adipose tissue responsible for insulin resistance in obesity and type II diabetes. Diabetes-Metab Revs 5:665–689

Castillo C, Bogartus C, Bergman R, Thuillez P, Lillioja S (1994) Interstitial insulin concentrations determine glucose uptake rates but not insulin resistance in lean and obese men. J Clin Invest 93:10–16
Chan JM, Rimm EB, Colditz GA, Stampfer MJ, Willett WC (1994) Obesity, fat distribution, and weight gain as risk factors for clinical diabetes in men. Diabetes Care 17:961–969
Clore JN, Glickman PS, Nestler JE, Blackard WG (1991) In vivo evidence for hepatic autoregulation during FFA-stimulated gluconeogenesis in normal humans. Am J Physiol 261:E425–E429
Cohen N, Halberstam M, Shlimovich P, Chang CJ, Shamoon H, Rossetti L (1995) Oral vanadyl sulfate improves hepatic and peripheral insulin sensitivity in patients with non-insulin-dependent diabetes mellitus. J Clin Invest 95:2501–2509
Colberg SR, Simoneau J-A, Thaete FL, DE Kelley (1995) Skeletal muscle utilization of free fatty acids in women with visceral obesity. J Clin Invest 95:1846–1853
Colditz GA, Willett WC, Rotnitzky A, Manson JE (1995) Weight gain as a risk factor for clinical diabetes mellitus in women. Ann Intern Med 122:481–486
Dagenais GR, Tancredi RG, Zierler KL (1976) Free fatty acid oxidation by forearm muscle at rest, and evidence for an intramuscular lipid pool in the human forearm. J Clin Invest 58:421–431
DeFronzo RA (1998) Pathogenesis of type 2 diabetes: metabolic and molecular implications for identifying diabetes genes. Diabetes Revs 5:177–269
Devlin JT, Hirshman M, Horton ED, Horton ES (1987) Enhanced peripheral and splanchnic insulin sensitivity in NIDDM men after single bout of exercise. Diabetes 36:434–439
Egan BM, Stepniakowski K (1994) Compensatory hyperinsulinemia and the forearm vasodilator response during an oral glucose tolerance test in obese hypertensives. J Hypertens 12:1061–1067
Felber JP, Meyer HU, Curchod B, Iselin HU, Rousselle J, Maeder E, Pahud P, Jequier E (1981) Glucose storage and oxidation in different degrees of obesity measured by continuous indirect calorimetry. Diabetologia 20:39–44
Felber JP, Ferrannini E, Golay A, Meyer HU, Theibaud D, Curchod B, Maeder E, Jequier E, DeFronzo RA (1987) Role of lipid oxidation in the pathogenesis of insulin resistance of obesity and type II diabetes. Diabetes 36:1341–1350
Felber JP, Golay A, Jequier E, Curchod B, Temler E, DeFronzo RA, Ferranini E (1988) The metabolic consequences of long-term human obesity. Int J Obesity 12:377–389
Flatt JP (1995) McCollum Award Lecture, 1995: Diet, lifestyle, and weight maintenance. Am J Clin Nutr 62:820–836
Forman BM, Tontonoz P, Chen J, Brun RP, Spiegelman BM, Evans RM (1995) 15-deoxy$\Delta^{12,14}$ prostaglandin J2 is a ligand for the adipocyte determination factor PPAR-γ. Cell 83:803–812
Freidenberg GR, Henry RR, Klein HH, Reichart DR, and Olefsky JM (1987) Decreased kinase activity of insulin receptors from adipocytes of non-insulin-dependent diabetic subjects. J Clin Invest 79:240–250
Freidenberg GR, Reichart D, Olefsky JM, Henry RR (1988) Reversibility of defective adipocyte insulin receptor kinase activity in non-insulin-dependent diabetes mellitus. Effect of weight loss. J Clin Invest 82:1398–1406
Friedman JE, Dohm GL, Leggettfrazier N, Elton CW, Tapscott EB, Pories WP, Caro JF (1992) Restoration of insulin responsiveness in skeletal muscle of morbidly obese patients after weight loss – effect on muscle glucose transport and glucose transporter GLUT4. J Clin Invest 89:701–705
Garvey WT, Maianu L, Huecksteadt TP, Birnbaum MJ, Molina JM, Ciaraldi TP (1991) Pretranslational suppression of a glucose transporter protein causes insulin resistance in adipocytes from patients with non-insulin-dependent diabetes mellitus and obesity. J Clin Invest 87:1072–1081
Garvey WT, Maianu L, Hancock JA, Golichowski AM, Baron A (1992) Gene expression and GLUT4 in skeletal muscle from insulin-resistant patients with obesity, IGT, GDM, and NIDDM. Diabetes 41:465–475

Giacchetti G, Faloia E, Taccaliti A, Morosini PP, Arnaldi G, Mantero F, Accili D, De Pirro, R (1994) Decreased expression of insulin-sensitive glucose transporter mRNA (GLUT-4) in adipose tissue of non-insulin-dependent diabetic and obese patients: evaluation by a simplified quantitative PCR assay. J Endocrinol Invest 17:709–715

Golay A, Swislowski ALM, Chen YDI, Jaspan JB, Reaven GM (1986) Effect of obesity on ambient plasma glucose, free fatty acid, insulin, growth hormone, and glucagon concentrations. J Clin Endocrinol Metab 63:481–484

Golay A, Felber JP, Jequier E, DeFronzo RA, Ferrannini E (1988) Metabolic basis of obesity and non-insulin dependent diabetes mellitus. Diabetes Metab Rev 4:727–747

Goldfine AB, Simonson DC, Folli F, Patti ME, Kahn CR (1995) Metabolic effects of sodium metavanadate in humans with insulin- dependent and noninsulin-dependent diabetes mellitus in vivo and in vitro studies. J Clin Endo Metab 80: 3311–3320

Goldstein BJ (1996) Protein-tyrosine phosphatases and the regulation of insulin action. In: Diabetes Mellitus: A Fundamental and Clinical Text, edited by D. LeRoith, J. M. Olefsky, and S. I. Taylor, Philadelphia: Lippincott, pp 174–186

Goldstein BJ, Li PM, Ding W, Ahmad F, Zhang WR (1998) Regulation of insulin action by protein-tyrosine phosphatases. Vitamins and Hormones, vol 54, pp 67–96

Goodyear LJ, Giorgino F, Sherman LA, Carey J, Smith RJ, Dohm GL (1995) Insulin receptor phosphorylation, insulin receptor substrate-1 phosphorylation, and phosphatidylinositol 3-kinase activity are decreased in intact skeletal muscle strips from obese subjects. J Clin Invest 95:2195–2204

Groop LC, Saloranta C, Shank M, Bonadonna RC, Ferrannini E, De Fronzo RA (1991) The role of free fatty acid metabolism in the pathogenesis of insulin resistance in obesity and non-insulin-dependent diabetes mellitus. J Clin Endocrin Metab 72:96–107

Groop LC, Bonnadonna RC, Simonson DC, Petrides AS, Shank M, DeFronzo RA (1992) Effect of insulin on oxidative and nonoxidative pathways of free fatty acid metabolism in human obesity. Am J Physiol 263 (Endocrinol Metab. 26):E79–E84

Gurwitz JH, Field TS, Glynn RJ, Manson JE, Avorn J, Taylor JO, Hennekens CH (1994) Risk factors for non-insulin-dependent diabetes mellitus requiring treatment in the elderly. J Am Geriatric Soc 42:1235–1240

Hallakou S, Doare L, Foufelle F, Kergoat M, Guerre-Millo M, Berthault MF, Dugail I, Morin J, Auwerx J, Ferre P (1997) Pioglitazone induces in vivo adipocyte differentiation in the obese Zucker fa/fa rat. Diabetes 46:1393–1399

Hawkins S, Barzilai N, Liu R, Chen W, Rosetti L (1996) Increased FFA and hyperglycemia result in a similar increase in carbon flux into the glucosamine pathway and peripheral insulin resistance. Diabetes 45 [Suppl 2]:11 A

Helmrich SP, Ragland DR, Leung RW, Paffenbarger RS Jr (1991) Physical activity and reduced occurrence of non-insulin dependent diabetes mellitus. N Engl J Med 325:147–152

Henrion D, Laher L (1994) Insulin potentiates norepinephrine-induced vascular tone by activation of protein kinase C and tyrosine kinase. Can J Physiol Pharmacol 72:849–854

Hofmann C, Lorentz K, Braithwaite SS, Colca JR, Palazuk BJJR, Hotamisligil GS, Spiegelman BM (1994) Altered gene expression for tumor necrosis factor-α and its receptors during drug and dietary modulation of insulin resistance. Endocrinology 134:264–270

Hotamisligil GS, Shargill NS, Spiegelman BM (1993) Adipose expression of tumor necrosis factor-alpha; direct role in obesity-linked insulin resistance. Science 259:87–91

Hotamisligil GS, Arner P, Caro JF, Atkinson RL, Spiegelman BM (1995) Increased adipose expression of tumor necrosis factor-α in human obesity and insulin resistance. J Clin Invest 95:2409–2415

Hotamisligil GH, Peraldi P, Budavari A, Ellis R, White MF, Spiegielman BM (1996) IRS-1 mediated inhibition of insulin receptor tyrosine kinase activity in TNF-α and obesity-induced insulin resistance. Science 271:665–670
Hurrell DG, Pedersen O, Kahn CR (1989) Alterations in the hepatic insulin receptor kinase in genetic and acquired obesity in rats. Clin Exp Pharmacol Physiol 125: 2454–2462
Jequier E (1996) A metabolic perspective on the interaction between obesity and diabetes. Current Concepts in Endocrinology and Diabetes 3:10–15.
Kahn BB, Pedersen O (1992) Tissue-specific regulation of glucose transporters in different forms of obesity. Proc Soc Exp Biol Med 200:214–217
Kahn BB (1998) Type 2 diabetes – when insulin secretion fails to compensate for insulin resistance. Cell 92:593–596
Karam JH, Grodsky GM, Forsham PH (1963) Excessive insulin response to glucose in obese subjects as measured by immunochemical assay. Diabetes 12:197–204
Karlsson C, Lindell K, Svensson E, Bergh C, Lind P, Billig H, Carlsson LM, Carlsson B (1997) Expression of functional leptin receptors in human ovary. J Clin Endocrinol Metab 82:4144–4148
Kelly DE, Simoneau JA (1994) Impaired free fatty acid utilization by skeletal muscle in non-insulin dependent diabetes mellitus. J Clin Invest 94:2349–2356
Kelly DE, Mokan M, Simoneau JA, Mandarino LJ (1993) Interaction between glucose and fatty acid metabolism in human skeletal muscle. J Clin Invest 92:91–98
Kelly KL, Ruderman NB, Chen KS (1992) Phosphatidylinositol-3-kinase in isolated rat adipocytes – Activation by insulin and subcellular distribution. J Biol Chem 267: 3423–3428
Kenny SJ, Aubert RE, Geiss LS (1995) Prevalence and incidence of non-insulin dependent diabetes mellitus. In: Diabetes in America. Harris MI (ed) Washington DC: US Government Printing Office, pp 37–46
Kern PA, Saghizadeg M, Ong JM, Bosch RJ, Deem R, Simsolo RB (1995) The expression of tumor necrosis factor in adipose tissue: regulation by obesity, weight loss, and relationship to lipoprotein lipase. J Clin Invest 95:2111–2119
King GL, Johnson SM (1985) Receptor-mediated transport of insulin across endothelial cells. Science 227:1583–1586
Kirwan JP, Hickner RC, Yarasheski KE, Kohrt WM, Wiethop BV, Holloszy JO (1992) Eccentric exercise induces transient insulin resistance in healthy individuals. J Appl Physiol 72:2197–2202
Kissebah AH, Vydelingum N, Murray R, Evans DJ, Hartz AJ, Kallhoff RK, Adams PW (1982) Relation of body fat distribution to metabolic complications of obesity. J Clin Endocrinol Metab 54:254–260
Kliewer SA, Lenhard JM, Willson TM, Patel I, Morris DC, Lehmann JM (1995) A prostaglandin J2 metabolite binds peroxisome proliferator-activated receptor gamma and promotes adipocyte differentiation. Cell 83:813–819
Kliewer SA, Sundseth SS, Jones SA, Brown PJ, Wisley GB, Koble CS, Devchand P, Wahli W, Willson TM, Lenhard JM, Lehmann JM (1997) Fatty acids and eicosanoids regulate gene expression through direct interactions with peroxisome proliferator-activated receptors alpha and gamma. Proc Natl Acad Sci USA 94:4318–4323
Kolaczynski JW, Considine RV, Ohannesian J, Marco C, Opentanova I, Nyce MR, Myint M, Caro JF (1996) Responses of leptin to short-term fasting and refeeding in humans: a link with ketogenesis but not ketones themselves. Diabetes 45: 1511–1515
Kolaczynski JW, Ohannesian J, Considine RV, Marco C, Caro JF (1996) Response of leptin to short-term and prolonged overfeeding in humans. J Clin Endo Metab 81:4162–4165
Kriketos AD, Pan DA, Lillioja S, Cooney GJ, Baur LA, Milner MR, Sutton JR, Jenkins AB, Bogartus C, Storlien LH (1996) Interrelationships between muscle morphology, insulin action and adiposity. Am J Physiol 270(6 Pt 2):R1332–R1339

Krotkiewski M, Bjorntorp P, Sjostrom L, Smoth U (1983) Impact of obesity on metabolism in men and women: importance of regional adipose tissue distribution. J Clin Invest 72:1150–1162
Krotkiewski M, Seidell JC, Bjorntorp P (1990) Glucose tolerance and hyperinsulinemia in obese women: role of adipose tissue distribution, muscle fibre characteristics and androgens. J Intern Med 228:385–392
Kruszynska YT, Olefsky JM (1996) Cellular and molecular mechanisms of non-insulin dependent diabetes mellitus. J Invest Med 44:413–428
Kulkarni RN, Wang ZL, Wang RM, Hurley JD, Smith DM, Ghatei MA, Withers DJ, Gardiner JV, Bailey CJ, Bloom SR (1997) Leptin rapidly suppresses insulin release from insulinoma cells, rat and human islets and, in vivo, in mice. J Clin Invest 100:2729–2736
Kusari J, Kenner KA, Suh KI, Hill DE, Henry RR (1994) Skeletal muscle protein tyrosine phosphatase activity and tyrosine phosphatase 1B protein content are associated with insulin action and resistance. J Clin Invest 93:1156–1162
Kutoh E, Boss O, Levasseur F, Giacobino JP (1998) Quantification of the full length leptin receptor (OB-Rb) in human brown and white adipose tissue. Life Sci 62:445–451.
Laakso M, Edelman SV, Brechtel G, Baron AD (1990) Decreased effect of insulin to stimulate skeletal muscle blood flow in obese men. J Clin Invest 85:1844–1852
Laakso M, Edelman SV, Brechtel G, Baron AD (1992) Impaired insulin-mediated skeletal muscle blood flow in patients with NIDDM. Diabetes 41:1076–83
Le Stunff C, Bougneres P (1994) Early changes in postprandial insulin secretion, not in insulin sensitivity, characterize juvenile obesity. Diabetes 43:696–702
Lee HU, Le HK, Koh CS, Min HK (1990) Artificial induction of intravascular lipolysis by lipid–heparin infusions leads to insulin resistance in man. Diabetologia 31:285–290
Lillioja S, Young AA, Cutler CL, Ivy JL, Abbott WG, Zawadzki JK, Yki-Jarvinen H, Chriostin L, Secomb TW, Bogartus C (1987) Skeletal muscle capillary density and fiber type are possible determinants of in vivo insulin resistance in man. J Clin Invest 80:415–24.
Long SD, Pekala PH (1996) Lipid mediators of insulin resistance: ceramide signalling downregulates GLUT4 gene transcription in 3T3-L1 adipocytes. Biochemical Journal. 319:179–84
Manson JE, Rimm EB, Stampfer MJ, Willett WC, Hennekens CH (1991) Physical activity and incidence of non-insulin dependent diabetes mellitus in women Lancet 338:774–778
Manson JE, Nathan DM, Krolewski AS, Stampfer MJ, Willett WC, Hennekens CH (1992) A prospective study of exercise and incidence of diabetes among US male physicians. JAMA 268:63–67
Marco CC, Ohannesian JP, Najm PS, Caro JF, Kolaczynski JW (1998) Small weight gain is not associated with development of insulin resistance in healthy young individuals. Diabetes 47 [Suppl 1]:A80, (Abstract)
Marin P, Andersson B, Krotkiewski M, Bjontorp P (1994) Muscle fiber composition and capillary density in women and men with NIDDM. Diabetes Care 17:382–386
Marshall S, Bacote V, Traxinger RR (1991) Discovery of a metabolic pathway mediating glucose-induced desensitization of glucose transport system: role of hexosamine biosynthesis in the induction of insulin resistance. J Biol Chem 266:4706–4712
McGuire MC, Fields RM, Nyomba BL, Raz I, Bogardus C, Tonks NK, Sommercorn J (1991) Abnormal regulation of protein tyrosine phosphatase activities in skeletal muscle of insulin-resistant humans. Diabetes 40:939–942
Meyerovitch J, Rothenberg PL, ShechterY, Bonner-Weir S, Kahn CR (1991) Vanadate normalizes hyperglycemia in two mouse models of non-insulin dependent diabetes mellitus. J Clin Invest 87:1286–1294

Muoio DM, Dohn GL, Fiedorek FT Jr, Tapscott EB, Coleman RA (1997) Leptin directly alters lipid partitioning in skeletal muscle. Diabetes 46:1360–1363
Nguyen TT, Mijares AH, Johnson CM, Jensen MD (1996) Postprandial leg and splanchnic fatty acid metabolism in nonobese men and women. Am J Physiol 271(6 Pt 1):E965–E972
Nolan JJ, Freidenberg GR, Henry R, Reichart D, Olefsky JM (1994) Role of human skeletal muscle insulin receptor kinase in the in vivo insulin resistance of noninsulin-dependent diabetes mellitus and obesity. J Clin Endocrinol Metab 78: 471–477
Nolan JJ, Ludvik B, Baloga J, Reichart D, Olefsky JM (1997) Mechanisms of the kinetic defect in insulin action in obesity and NIDDM. Diabetes 46:994–1000
Nurjhan N, Campbell PJ, Kennedy FP, Miles JM, Gerich JE (1986) Insulin dose–response characteristics for suppression of glycerol release and conversion to glucose in humans. Diabetes 35:1326–30
Nuutila P, Raitakari M, Laine H (1996) Role of blood flow in regulating insulin-stimulated glucose uptake in humans. J Clin Invest 97:1741–1747
Ofei F, Hurel S, Newkirk J, Sopwith M, Taylor R (1996) Effects of an engineered human anti-TNF-α antibody (CDP571) on insulin sensitivity and glycemic control in patients with NIDDM. Diabetes 45:881–885
Ohlson LO, Larsson B, Svardsudd K, Welin L, Eriksson H, Wilhelmsen L, Bjontorp P, Tibbin G (1985) The influence of body fat distribution on the incidence of diabetes mellitus. 13.5 years of follow-up of the participants in the study of men born in 1913. Diabetes 34:1055–1058
Olefsky JM, Nolan JJ (1995) Insulin resistance and non-insulin-dependent diabetes mellitus: cellular and molecular mechanisms. Am J Clin Nutr 61:S980–S986
Oshida Y, Yamanouchi K, Hayamizu S, Sato Y (1989) Long-term mild jogging increases insulin action despite no influence on body mass index or VO_{2max}. J Appl Physiol 66:2206–2210
Ostman J, Arner P, Engfeldt P, Kager L (1979) Regional differences in the control of lipolysis in human adipose tissue. Metabolism Clin and Exp 28:1198–1205
Owen OE, Reichard GA Jr, Boden G, Schuman C (1974) Comparative measurements of glucose, beta-hydroxybutyrate, acetoacetate, and insulin in blood and cerebrospinal fluid during starvation. Metabolism 23:714–718
Pan DA, Lillioja S, Kriketos AD, Milner MR, Baur LA, Bogartus C, Jenkins AB, Storlien LH (1997) Skeletal muscle triglyceride levels are inversely related to insulin action. Diabetes 46:983–988
Paolisso G, Gambardella A, Amato L, Tortoriello R, D'Amore A, Varricchio M, D'Onoforio F (1995) Opposite effects of short- and long-term fatty acid infusion on insulin secretion in healthy subjects. Diabetologia 38:1295–1299
Pelleymounter MA, Cullen MJ, Baker MB, Hecht R, Winters D, Boone T, Collins F (1995) Effect of the obese gene product on body weight regulation in ob/ob mice. Science 269:540–543
Perley M, Kipnis DM (1966) Plasma insulin responses to glucose and tolbutamide of normal weight and obese diabetic and non-diabetic subjects. Diabetes 15:867–874
Perry IJ, Wannamethee SG, Walker MK, Thompson AG, Whincup PH, Shaper AG (1995) Prospective study of risk factors for development of non-insulin dependent diabetes in middle-aged British men. BMJ 3:560–564
Perseghin G, Price TB, Petersen KF, Roden M, Cline GW, Gerow K, Rothman DL, Shulman G (1997) Increased glucose transport phosphorylation and muscle glycogen synthesis after exercise training in insulin resistant subjects. N Engl J Med 335:1357–1362
Petersen KF, Hendler R, Price T, Perseghin G, Rothman DL, Held N, Amatruda JM, Shulman GI (1998) C-13/P-31 NMR studies on the mechanism of insulin resistance in obesity. Diabetes 47:381–386
Phillips DI, Caddy S, Ilic V, Fielding BA, Frayn KN, Borthwick AC, Taylor R (1996) Intramuscular triglyceride and muscle insulin sensitivity: evidence for a relationship in nondiabetic subjects. Metabol Clin & Exp 45:947–950

Pi-Sunyer FX (1985) Effects of exercise on food intake. In: Hirsch J, Van Itallie TB (eds) Recent Advances in Obesity Research IV, pp 368–373. London, John Libbey
Pollare T, Vessby B, Lithell H (1991) Lipoprotein lipase activity in skeletal muscle is related to insulin sensitivity. Arterioscler Thromb 11:1192–1203
Polonsky KS, Given BD, Hirsch L, Shapiro ET, Tillil H, Beebe C, Galloway JA, Frank BH, Karrison T, Van Cauter E (1988) Quantitative study of insulin secretion and clearance in normal and obese subjects. J Clin Invest 81:435–441
Posner BI, Faure R, Burgess JW, Bevan AP, Lachance D, Zhangsun G, Fantus IG, Ng JB, Hall DA, Lum BS, Shaver A (1994) Peroxovanadium compounds – A new class of potent phosphotyrosine phosphatase inhibitors which are insulin mimetics. J Biol Chem 269:4596–4604
Prentki M, Corkey BE (1995) Are the beta cell signalling molecules malonyl-CoA and cytosolic long-chain acyl-CoA implicated in multiple tissue defects of obesity and NIDDM? Diabetes 44:863–870
Randle PJ, Garland PB, Hales CN, Newsholme EA (1963) The glucose fatty-acid cycle: its role in insulin sensitivity and metabolic disturbances of diabetes mellitus. Lancet I:785–789
Reaven GM, Lithell H, Landsberg L (1996) Hypertension and associated metabolic abnormalities – the role of insulin resistance and sympathoadrenal system. N Engl J Med 334:374–381
Rebrin K, Steil GM, Getty L, Bergman RN (1995) Free fatty acids as a link in the regulation of hepatic glucose output by peripheral insulin. Diabetes 44:1038–1045
Rebuffe-Scrive M, Andersson B, Olbe L, Bjorntorp P (1989) Metabolism of adipose tissue in intraabdominal depots of non-obese men and women. Metabolism 38:453–458
Richelsen B, Pedersen SB, Moller-Pedersen T, Schmit O, Moller N, Borglum JD (1993) Lipoprotein lipase activity in muscle tissue influenced by fatness, fat distribution and insulin in obese females. Eur J Clin Invest 23:226–233
Rondinone CM, Wang LM, Lonnroth P, Wesslau C, Pierce JH, Smith U (1997) Insulin receptor substrate (IRS)-1 is reduced and IRS-2 is the main docking protein for phosphatidylinositol 3-kinase in adipocytes from subjects with non-insulin-dependent diabetes mellitus. Proc Natl Acad Sci USA 94(8):4171–4175
Saghizadeh M, Ong JM, Garvey WT, Henry RR, Kern PA (1996) The expression of TNF-α by human muscle; relationship to insulin resistance. J Clin Invest 97:1111–1116
Segal KR, Edano A, Abalos A, Albu J, Blando L, Tomas MB, Pi-Sunyer FX (1991) Effect of exercise training on insulin sensitivity and glucose metabolism in lean, obese and diabetic men. J Appl Physiol 71:2402–2411
Shechter Y (1990) Insulin-mimetic effects of vanadate: possible implications for future treatment of diabetes. Diabetes 39:1–5
Scherrer U, Vollenweider P, Randlin D, Jequier E, Nicod P, Tappy L (1993) Suppression of insulin induced sympathetic activation and vasodilation by dexamethasone in humans. Circulation 88:388–394
Scherrer U, Randin D, Vollenweider P, Lollenweider L, Nicod P (1994) Nitric oxide release accounts for insulin's vascular effects in humans. J Clin Invest 94:2511–2515
Shisheva A, Ikonomov O, Shechter Y (1994) The protein tyrosine phosphatase inhibitor, pervanadate, is a powerful antidiabetic agent in streptozotocin-treated diabetic rats. Clin Exp Pharm Physiol 134:507–510
Sims EA, Danforth E, Horton ES, Bray GA, Glennon JA, Salans LB (1973) Endocrine and metabolic effects of experimental obesity in men. Recent Prog Horm Res 29:457–496
Sinha MK, Pories WJ, Flickinger EG, Meelheim D, Caro JF (1987) Insulin-receptor kinase activity of adipose tissue from morbidly obese humans with and without NIDDM. Diabetes 36:620–625
Spiegelman BM (1998) PPAR-γ adipogenic regulator and thazolidinedione receptor. Diabetes 47:507–514

Spraul M, Ravussin E, Fontvieille AM, Rising R, Larson DE, Anderson EA (1993) Reduced sympathetic nerve activity: a potential mechanism predisposing to body weight gain. J Clin Invest 92:1730–1735
Spraul M, Ravussin E, Baron AD (1996) Lack of relationship between muscle sympathetic nerve activity and skeletal muscle vasodilation in response to insulin infusion. Diabetologia 39:91–96
Steinberg HO, Brechtel G, Johnson A, Baron AD (1994) Insulin-mediated skeletal muscle vasodilation is nitric oxide dependent: a novel action of insulin to increase nitric oxide release. J Clin Invest 94:1172–1179
Takayama S, Kahn CR, Kubo K, Foley JE (1988) Alterations in insulin receptor autophosphorylation in insulin resistance: Correlation with altered sensitivity to glucose transport and anti-lipolysis to insulin. J Clin Endo Metab 66:992–990
Tartaglia LA (1997) The leptin receptor. J Biol Chem 272:6093–6096
Thies RS, Molina JM, Ciaraldi TP, Freidenberg GR, Olefsky JM (1990) Insulin-receptor autophosphorylation and endogenous substrate phosphorylation in human adipocytes from control, obese, and NIDDM subjects. Diabetes 39(2): 250–259
Ting HH, Timimi FK, Boles KS, Creager SJ, Ganz P, Creager MA (1996) Vitamin C improves endothelium-dependent vasodilation in patients with non-insulin-dependent diabetes mellitus. Journal of Clinical Investigation 97:22–28
Tontonoz P, Hu E, Graves RA, Budavari AB, Spiegelman BM (1994) mPPAR-γ2: tissue-specific regulator of an adipocyte enhancer. Genes Dev 8:1224–1234
Tontonoz P, Hu E, Spiegelman BM (1994) Stimulation of adipogenesis in fibroblasts by PPAR-γ2, a lipid-activated transcription factor. Cell 79:1147–1156
Tounian P, Schneiter P, Henry S, Tappy L (1996) Effects of infused glucose on glycogen metabolism in healthy humans. Clin Physiol 16:403–416
Tuominen JA, Ebeling P, Bourney R, Koranyi L, Lamminen A, Rapola J, Sane T, Vuorinen-Markkola H, Koivisto VA (1996) Postmarathon paradox: insulin resistance in the face of glycogen depletion. Am J Physiol 270:E336–E343
Tuominen JA, Ebeling P, Laquier FW, Heiman ML, Stephens T, Koivisto VA (1997) Serum leptin levels and fuel homeostasis in healthy man. Europ J Clin Invest 27:206–211
Unger RH (1995) Lipotoxicity in the pathogenesis of obesity-dependent NIDDM: genetic and clinical implications. Diabetes 44:863–870
Vague J (1956) The degree of masculine differentiation of obesities: a factor determining predisposition to diabetes, atherosclerosis, gout, and uric calculus disease. Am J Clin Nutr 4:20–34
Vidal-Puig AJ, Considine RV, Jimenez-Linan M, Werman A, Pories WJ, Caro JF, Flier JS (1997) Peroxisome-proliferator-activated receptor gene expression in human tissue: effects of obesity, weight loss, and regulation by insulin and glucocorticoids. J Clin Invest 99:2416–2422
Vollenweider P, Randlin D, Tappy L (1994) Impaired insulin-induced sympathetic neural activation and vasodilation in skeletal muscle in obese humans. J Clin Invest 93:2365–2371
White MF, Kahn CR (1994) The insulin signalling system. J Biol Chem 269:1–4
Williamson JR, Kreisberg RA, Felts PW (1966) Mechanism for the stimulation of gluconeogenesis of fatty acids in perfused rat liver. Proc Natl Acad Sci USA 56:247–254
Wolfe B, Klein MS, Peters EJ, Schmidt BF, Wolfe RR (1989) Effect of elevated free fatty acids on glucose oxidation in normal humans. Metab Clin Exp 37:323–329
Woo R, Garrow JS, Pi-Sunyer FX (1982) Effect of exercise on spontaneous calorie intake in obesity. Am J Clin Nutr 36:470
Woo R, Pi-Sunyer FX (1985) Effect of increased physical activity on voluntary intake in lean women. Metabolism 34:836

Worm D, Vinten J, Staehr P, Henriksen JE, Handberg A, Beck-Nielsen H (1996) Altered basal and insulin-stimulated phosphotyrosine phosphatase (PTPase) activity in skeletal muscle fractions from NIDDM patients compared with control subjects. Diabetologia 39:1208–1214

Zeman RJ, Ludemann R, Easton TG, Etlinger JD (1988) Slow to fast alterations in skeletal muscle fibers caused by clenbuterol, a $\beta 2$-receptor agonist. Am J Physiol 254:E726–E732

Zhang Y, Proenca R, Maffei M, Barone M, Leopold L, Friedman JM (1994) Positional cloning of the mouse obese gene and its human homologue. Nature 377:527–529

Zurlo F, Lillioja S, Espositio-Del Puente A, Nyomba BL, Raz I, Saad MF, Swinburn BA, Knowler WC, Bogardus C, Ravussin E (1990) Low ratio of fat to carbohydrate oxidation as a predictor of weight gain: study of 24-h RQ. Am J Physiol 259:E650–E657

CHAPTER 5

Fat Metabolism in Obesity, with Special Emphasis on the Insulin Resistance and Elevated Lipid Mobilization in Visceral Obesity

P. BJÖRNTORP

A. Introduction

Adipose tissue is the main tissue for storage of lipid energy. A normal subject carries about 10–15 kg fat, corresponding to the energy needs of about a month. Obviously this large energy depot transports large amounts of triglyceride in and out of its specialized cells for storage, the adipocytes. Input of fat is accomplished by hydrolysis of circulating triglycerides, while storage triglycerides also need to be hydrolyzed before being mobilized. The main products of hydrolysis are glycerol and free fatty acids (FFA) a fraction of which is re-esterified for further storage. The net outflux of FFA seems to be dependent on the free pool of adipocyte FFA in relation to blood circulation, transporting the FFA to their various destinations.

As will be described in the following, there is now considerable evidence of a heterogeneity of white adipose tissue. These differences seem to be more of a quantitative than qualitative nature. For example in the rat the turnover of lipid is clearly higher in mesenteric adipocytes than in other examined regions (LI and BJÖRNTORP 1995). When precursor cells to adipocytes are isolated from each of these regions and cultured to full maturity under identical conditions the differences in lipid uptake, and mobilization as well as density of endocrine receptors become equal (LI et al. 1993). These results indicate that factors in the microenvironment rather than inherent characteristics of the adipocyte genome are different in different regions.

In this review the main emphasis will be placed on lipid mobilization and its regulation in obesity. There are two primary reasons for this. Mobilized FFA are currently the focus of interest in the pathogenesis of insulin resistance and its associated diseases (REAVEN 1995). Another reason is simply the limitations of space.

B. Basic Mechanisms

I. Lipid Mobilization

Triglyceride hydrolysis in the adipocyte occurs through catalysis by a hormone sensitive lipase, which, when activated by phosphorylation, cleaves triglycerides into FFA and glycerol. This is occurs step-wise, and when the monoacylglycerol stage is reached an additional lipase will remove the remaining fatty acid with a high velocity. Diacylglycerols are apparently accumulated to some degree, but are assumed to be of insignificant quantitative importance (ARNER and ÖSTMAN 1967). Therefore, the release of free glycerol is frequently used as a measurement of lipolysis, particularly since free glycerol is not an efficient substrate for fatty acid reesterification.

The hormone sensitive lipase is phosphorylated by a cyclic AMP-regulated protein kinase A. The concentration of cyclic AMP is dependent on a balance between an adenyl cyclase and a phosphodiesterase. This cascade is regulated by several hormonal receptors at the cell surface. In man those of the adrenergic systems are most important, and are both stimulatory and inhibitory β- and α-adrenergic receptors which signal to the lipolytic cascade via G_s and G_i proteins. Insulin is mainly active by powerful inhibition of this system at the levels of the phosphodiesterase and lipase phosphorylation (for more detailed review see LÖNNROTH and SMITH 1992).

The steroid hormone superfamily exerts powerful regulatory input which affect various steps of the lipolysis "trigger" system, described above, by "permissive," usually stimulating actions. This occurs via genomic interactions of the complex formed by the hormone and its receptor, which interact with the appropriate DNA sequence to induce transcriptional effects and subsequent protein synthesis for the parts of the lipolysis machinery, from the "triggering" receptors to the lipase.

The lipid accumulating system catches circulating triglyceride particles, mainly chylomicra, and very low density lipoproteins in capillary-bound lipopolysaccharides, where triglycerides are hydrolyzed by lipoprotein lipase, produced by the adjacent adipocytes. The resulting FFA are believed to be transferred to these adipocytes by lateral transfusion via cell membranes and subsequently reesterified to storage triglycerides. The glycerol backbone is delivered mainly from the glycolytic pathway and is therefore dependent on glucose uptake and insulin, which also stimulate lipoprotein lipase activity (for further details, see review in AILHAUD 1990).

Both the lipid mobilization and accumulation systems are highly dependent on the microenvironment in adipose tissue. Sympathetic drive seems to be the main stimulatory factor for lipolysis in human adipose tissue. Blood flow is an important regulator for lipolysis, because when FFA concentration builds up in adipocytes, it inhibits lipolysis by a feed-back mechanism. Blood flow is also important for lipid uptake, simply by delivering substrate particles to the triglyceride accumulating system (for a detailed review see LÖNNROTH and SMITH 1992; AILHAUD 1990).

Adipose tissue metabolism is highly species dependent. This review will only deal with human data, except where only animal data are available.

II. Adipose Tissue Storage of Triglycerides

The net result of the mass of an adipose tissue is of course the end result of the balance between input and output of triglycerides, particularly since fat normally occupies about 90% of the mass. The triglycerides are collected in adipocytes, and the capacity of an adipose tissue to store fat is dependent on the size and number of the adipocytes. Available evidence now suggests that, at least at adult age, the size variation of the adipocytes is the more flexible component, at least up to what seems to be a maximal size, where new fat cells seem to be formed, perhaps via cell to cell interactions, which might well be influenced by external factors (AILHAUD et al. 1972).

C. FFA Metabolism in Obesity

The concentrations of circulating FFA and free glycerol are elevated in obesity, and are apparently proportional to adipose tissue mass (BJÖRNTORP and ÖSTMAN 1971). In the steady state the concentrations of both are usually proportional to their turnover rate and assumed to be the net result of adipose tissue lipid mobilization (BJÖRNTORP and ÖSTMAN 1971). Larger fat cells mobilize more FFA than smaller ones (JACOBSSON and SMITH 1972), but since there are more small fat cells than enlarged ones per unit of adipose tissue mass, this seems to even up to similar total lipid mobilization.

There is clearly a difference in lipid mobilization in subgroups of obesity. With abdominal, central, visceral obesity FFA turnover is elevated in comparison with peripheral, gluteo-femoral obesity at the same level of elevated total body fat mass (BJÖRNTORP 1991). This is most likely due to differences in triglyceride turnover in different adipose tissues. The rank order seems to be intraabdominal, visceral > subcutaneous abdominal > subcutaneous gluteo-femoral adipose tissues. When these tissues are enlarged individually the net result of systemic FFA concentration will be dependent on which depot is enlarged. Consequently, when intraabdominal and abdominal subcutaneous tissues are enlarged, a higher FFA concentration will be the result than when gluteo-femoral tissues are enlarged. The former is a typical male and the latter a typical female characteristic (BJÖRNTORP 1991).

The reason for this inherent regional difference in adipocyte metabolism seems to be explained by the variation in density of specific hormonal receptors on or in adipocytes. This seems to be valid for both adrenergic hormones, which trigger lipolysis directly, and steroid hormones, which regulate lipolysis by "permissive" actions, usually via genomic interactions and protein synthesis. Furthermore, blood flow and innervation vary markedly and are of fundamental importance for adipocyte functions (BJÖRNTORP 1996). There is also a variation in the insulin receptor density in different adipose tissue depots (BOLINDER et al. 1983).

The question of why FFA mobilization is elevated in obesity, particularly in abdominal obesity, is of major interest because there is convincing evidence that this is followed by generation of several risk predictors of prevalent disease, such as type I diabetes mellitus. Abdominal obesity is a powerful predictor of not only diabetes but also cardiovascular disease and stroke (BJÖRNTORP 1991).

Could the elevated FFA concentration in obesity simply be a consequence of an elevated fat mass? In this alternate explanation it is thought that obesity with too many fat cells, where each adipocyte is contributing with a normal amount of FFA, will show an increased FFA mobilization just as a mass effect. The highest FFA concentrations are, however, found in centralized obesity where adipocytes in central regions are usually hypertrophied. This is also a condition with severe insulin resistance and elevated circulating insulin concentrations (BJÖRNTORP 1991). Insulin is a powerful inhibitor of fat mobilization and would be expected to control lipolysis in this condition, if effective. It is therefore unlikely that a mass effect is a primary cause of increased FFA mobilization in obesity, and that some other mechanism provides the pathogenetic background. Such an aberration might, however, be amplified by adipose tissue enlargement.

Could an increased stimulation of the lipolytic system explain the elevated FFA in obesity? Arguments for this possibility might be sought in the status of the systems regulating lipolysis. The major trigger of lipolysis in humans seems to be the sympathetic nervous system. Previous studies have indicated that sympathetic nervous activity is elevated in obesity in general (BJÖRNTORP et al. 1977). Recent studies suggest that in central obesity not only is such activity increased in the steady state, but also after challenges such as lunch and laboratory stress tests (LJUNG et al., 2000).

The secretion of steroid hormones, which provide the basis for the lipolytic actions of the catecholamines from the sympathetic nervous system by permissive actions, are also abnormal in obesity. Particularly in central obesity, cortisol production seems to be elevated due to a hypersensitivity of the hypothalamo-pituitary-adrenal axis (BJÖRNTORP 1993). Cortisol in the presence of growth hormone (GH) has long been known to increase lipolysis, an effect prevented by inhibitors of protein synthesis (FAIN et al. 1965). In human adipose tissue under fully controlled conditions in culture, cortisol in the presence of insulin shows an inhibition of lipolysis. However, when GH is added, a brisk stimulation of lipolysis occurs even in the presence of insulin (OTTOSSON et al. 1999). In patients with Cushing's syndrome lipolysis after norepinephrine stimulation is lower than normal, particularly in the abdominal subcutaneous region (REBUFFÉ-SCRIVE et al. 1988a). Administration of 15 mg prednisolone daily for a week to normal, slightly obese women was, however, followed by elevated lipolysis in abdominal subcutaneous adipocytes (REBUFFÉ-SCRIVE et al. 1988b). In both conditions, hyperinsulinemia prevailed. In the latter experiment no change was seen in FFA concentrations before and during an oral glucose tolerance test. Elevation of FFA after glucocorticoid administration has, however, clearly been shown previously in other studies (DREILING et al. 1962; DIVERTIE et al. 1991).

The situation is thus complex. At the cellular level cortisol needs the presence of GH to exert a lipolytic effect, and this seems to occur even in the presence of insulin (OTTOSSON et al. 1999). Apparently this effect varies in different regions of adipose tissue, but the net effect in systemic circulation seems in general to be an elevated FFA concentration. We thus have a system which is dependent on cortisol, GH, and insulin and which varies in different adipose tissue regions. These regions might be characterized by different responsiveness due to variations in density of the specific hormonal receptors. Visceral adipocytes seem to have a high density of the glucocorticoid receptors (REBUFFÉ-SCRIVE et al. 1985), and a low density of insulin receptors (BOLINDER et al. 1983). In addition, the net outflow of FFA from adipose tissue is regulated by the intracellular reesterification process, which in turn is dependent on continuous influx of glucose. Cortisol inhibits glucose transport in adipose tissue (CARTER-SU and OKAMOTO 1987). Blood flow in adipose tissue may also change due to these endocrine events.

The antilipolytic effect of insulin is a powerful regulator of net outflux of FFA from adipose tissue. In tissue culture experiments with human adipose tissue cortisol apparently diminishes the antilipolytic effect of insulin (CIGOLINI and SMITH 1979).

In an attempt to summarize it seems likely that the effects of glucocorticoid exposure in vivo on lipid mobilization in adipose tissue is stimulatory. The mechanism of this at the cellular level is not known and is apparently an integrated action between glucocorticoids, GH, insulin, blood flow, and innervation. In addition, these effects seem to vary in different regions of adipose tissue.

Steroid hormones other than cortisol also exert permissive effects on adrenergic triggering of lipolysis. Testosterone (T) has marked such effects. At the cellular level a specific androgen receptor is responsible for lipolytic and other signals. This receptor is upregulated by androgens (DE PERGOLA et al. 1990), and mediates an increase of the density of lipolytic adrenergic receptors, and stimulates several steps in the following lipolytic cascade (XU et al. 1990, 1991). Androgens thus exert powerful permissive effects on adipose tissue lipolysis. These actions are also like those of cortisol, varying quantitatively in different regions. Strangely enough the net effects of testosterone on systemic circulating FFA concentrations do not seem to have been studied in controlled experiments.

Estrogens and progesterone most likely also affect adipose tissue lipolysis with a regional variation, but the mechanisms have not been possible to disclose in the apparent functional absence of specific hormonal receptors in human adipose tissue (REBUFFÉ-SCRIVE et al. 1986; BRÖNNEGÅRD et al. 1994).

Several of the mentioned steroid hormones show abnormal secretion in obesity, particularly in abdominal obesity. This is of interest because of the clearly elevated FFA levels of this condition. In the following an attempt will be made to discuss the background to elevated systemic FFA in abdominal obesity, based on the metabolic and endocrine perturbations described in that condition.

In abdominal obesity a hypersensitive HPA axis seems to be an important feature (BJÖRNTORP 1993). This results in periodically elevated cortisol secretion. Furthermore, perhaps as a consequence of the activation of the hypothalamo-pituitary-adrenal axis, sex steroid and growth hormone secretions are low (BJÖRNTORP 1993). This would result in a situation with a low permissive background for lipolysis stimulation. Central obesity is a condition with severe insulin resistance and hyperinsulinemia which would be expected to control lipolysis efficiently. Still, there is clearly an elevated FFA turnover (KISSEBAH et al. 1982). A remaining possibility thus seems to be that the widespread insulin resistance among various organs also affects adipose tissue, making the antilipolytic effect of insulin inefficient.

The results of studies in vivo on the insulin inhibition of FFA concentration in circulation have consistently shown a diminished effect in obesity, suggesting an insulin resistance of lipid mobilization (GROOP et al. 1991, 1992; BONADONNA et al. 1990; HOWARD et al. 1984; CHEN et al. 1987; ZANCANARO et al. 1990; FELBER et al. 1988). The question then is where this insulin resistance is localized.

In vitro data suggest a normal (ARNER et al. 1981; LÖNNROTH et al. 1983) or diminished (KASHIWAGI et al. 1984) antilipolytic effect of insulin in subjects with obesity. These data are, however, not entirely conclusive because the subjects studied were not free of diabetes or impaired glucose tolerance, the region of adipose tissue studied varied, and obese subjects were not subdivided according to the regional enlargement of adipose tissues. The ideal study should address non-diabetic, insulin-resistant, visceral obesity, the condition where clearly elevated FFA is combined with insulin resistance and hyperinsulinemia, and where consequently resistance to the antilipolytic effect of insulin would be expected. The antilipolytic effect of insulin, measured in vitro, shows strong negative correlations with systemic insulin sensitivity measured with the euglycemic clamp ($r \approx 0.8$) (MÅRIN et al. 1992). This was particularly the case with adipocytes from visceral fat depots ($r \approx 0.9$). Such adipocytes also showed a lower sensitivity of the antilipolytic effect of insulin than those from subcutaneous depots. Such strong statistical correlations do not tell anything about cause-effect relationships but they indicate closely related phenomena. Speculatively one may suggest that general, systemic insulin resistance, as measured with the clamp, also involves the antilipolytic mechanisms in adipose tissue. The resulting excess of FFA mobilization might be responsible for systemic insulin resistance by primarily muscular interactions, a Randle effect (RANDLE et al. 1963). The insulin resistance generated by glucocorticoids have been suggested to be mediated via FFA (GUILLAUME-GENTIL et al. 1993).

The observation that the antilipolytic effect is less sensitive in visceral fat cells may have bearing on the pronounced elevation of FFA concentrations with associated insulin resistance in central, visceral obesity. Although the visceral fat depots are normally comparably small in comparisons with total adipose tissue, with central obesity they are, by definition, enlarged. They would then be expected to expose the liver for high concentrations of FFA. This in turn has known consequences for synthesis of very low density lipoproteins, and gluconeogenesis which become elevated, and may also decrease hepatic insulin clearance, resulting in the generation of well established risk factors for prevalent disease (BJÖRNTORP 1990). Although a large fraction of portal FFA is captured by the liver it may well be that FFA delivered from visceral fat depots may contribute to systemic FFA with a significant fraction in visceral obesity (BJÖRNTORP 1994).

Taken together these considerations may indicate that the antilipolytic effect of insulin is an important regulator of portal and systemic FFA concentrations. Since this effect seems less sensitive in visceral than other fat cells, visceral obesity will be characterized by elevated FFA in spite of hyperinsulinemia, and explain the apparent insulin resistance of adipose tissue in terms of diminished inhibition of systemic FFA concentrations. An additional factor might be the elevated cortisol secretion in visceral obesity (BJÖRNTORP 1993), which would amplify such a phenomenon. Cortisol has been shown to decrease the sensitivity of the antilipolytic effect of insulin (CIGOLINI and SMITH 1979), an effect which might be more pronounced in visceral than other adipocytes due to the high density of the glucocorticoid receptor in visceral fat cells (REBUFFÉ-SCRIVE et al. 1985). In fact, available data would be compatible with the following chain of events. Elevated cortisol secretion is affecting an enlarged visceral fat mass with adipocytes with a high density of glucocorticoid receptors. This will result in a blunted antilipolytic effect of insulin and an elevation of lipid mobilization with outflux of increased concentrations of FFA in both portal and systemic circulation.

Visceral obesity is characterized by marked insulin resistance in combination with elevated FFA concentrations. Which of these phenomena comes first? Several possibilities may be discussed with each of the three players, visceral obesity, insulin resistance, and elevated FFA concentrations as primary actors.

Visceral obesity may be associated with elevated FFA mobilization due to the inherent low sensitivity of an enlarged visceral depot to insulin on inhibition of fat mobilization. This may well be amplified by cortisol. Circulating FFA may then generate "risk factors" via hepatic mechanisms as well as induce systemic insulin resistance.

Another alternative is that elevated FFA are a primary pathogenetic player and may be followed by insulin resistance, but it is hard to understand elevated FFA in a chain of events with visceral obesity as a secondary consequence.

Systemic insulin resistance as a primary event is unlikely to be followed by elevated FFA, in fact the opposite would be expected due to the hyperinsulinemia. Furthermore, there does not seem to be a plausible mechanism whereby insulin resistance could cause visceral obesity.

This seems to leave us with the first alternate hypothesis as the best explanation. This is also supported by experimental evidence in terms of measurements of the antilipolytic effects of insulin in visceral adipocytes, and its strong relationships to an expected consequence of elevated FFA concentrations, namely insulin resistance (MÅRIN et al. 1992). In addition, the elevated cortisol secretion in visceral obesity (BJÖRNTORP 1993), high density of glucocorticoid receptors in visceral adipocytes (REBUFFÉ-SCRIVE et al. 1985), as well as the effect of cortisol to diminish the antilipolytic action of insulin (CIGOLINI and SMITH 1979) add to the support of this mechanistic chain of events.

Another question then is to try to understand the mechanism whereby visceral accumulation of depot fat is occurring. This has been discussed in detail previously (BJÖRNTORP 1993). In brief, cortisol probably plays a major role here by specific stimulation of lipoprotein lipase activity and diminution of lipolysis. It may be speculated that these lipid-accumulating mechanisms override the insensitivity of the antilipolytic effect of insulin, which would be expected to be followed by a diminution of visceral fat mass. In other words, the cortisol effects on the lipid-accumulating side balance the end result to a net accumulation of lipid in spite of a leak of FFA due to a diminished control by insulin. Together this would result in a rapid turnover of visceral adipose triglycerides, and this has been observed (LI and BJÖRNTORP 1995; MÅRIN et al. 1995).

In addition to effects of cortisol, the low levels of sex steroid and growth hormones in abdominal obesity seem to amplify visceral accumulation of triglycerides, because these hormones act mainly to prevent enlargement of visceral fat depots via effects opposite those of cortisol on lipoprotein lipase and lipolysis. Unfortunately very little is known of the potential effects of these hormones on the antilipolytic action of insulin.

D. Genetic Influence

Visceral obesity (BOUCHARD et al. 1993) and insulin resistance (GROOP et al. 1996) both have important genetic components. This also holds for cortisol secretion (MEIKLE et al. 1988) and lipolysis regulation in adipose tissue (ARNER 1995). The syndrome of abdominal obesity with elevated FFA and insulin resistance may thus originate from several genetic susceptibilities. The current state of information does not seem to be sufficient to allow research to focus on a limited number of candidate genes. If the analysis of the situation given above turns out to be correct, then the genes controlling the antilipolytic effect of insulin will be of interest. Unfortunately, the mechanistic details of this process are currently not sufficiently well known.

References

Ailhaud G, Grimaldi P, Négrel R (1972) Cellular and molecular aspects of adipose tissue development. Annu Rev Nutr 12:207–233

Ailhaud G (1990) Cellular and secreted lipoprotein lipase revisited. Clin Biochem 23:343–347

Arner P, Bolinder J, Engfeldt P, Östman J (1981) The antilipolytic effect of insulin in human adipose tissue in obesity, diabetes mellitus, hyperinsulinemia, and starvation. Metabolism 30:753–760

Arner P, Östman J (1967) Mono- and diacylglycerols in human adipose tissue. Biochim Biophys Acta 369:209–211

Arner P (1995) The ß-adrenergic receptor – a cause and cure of obesity. New Enlg J Med 333:382–383

Björntorp P, Holm G, Jacobsson B, Schiller-de Jounge K, Lundberg P-A, Sjöström L, Smith U, Sullivan L (1977) Physical training in human hyperplastic obesity. IV. Effects on the hormonal status. Metabolism 26:319–328

Björntorp P, Östman J (1971) Human adipose tissue. Dynamics and regulation. Adv Metab Dis 5:277–327

Björntorp P (1990) "Portal" adipose tissue as a generator of risk factors for cardiovascular disease and diabetes. Arteriosclerosis 10:493–496

Björntorp P (1994) Fatty acids, hyperinsulinemia, and insulin resistance: which comes first? Current Opinion Lipidology 5:166–174

Björntorp P (1991) Metabolic implications of body fat distribution. Diabetes Care 14:1132–1143

Björntorp P (1996) The regulation of adipose tissue distribution in humans. Int J Obesity 20:291–302

Björntorp P (1993) Visceral obesity: A "Civilization syndrome". Obes Res 1:206–222

Bolinder J, Kager L, Östman J, Arner P (1983) Differences at the receptor and postreceptor levels between human omental and subcutaneous adipose tissue in the action of insulin on lipolysis. Diabetes 32:117–122

Bonadonna RC, Groop LC, Kraemer N, Ferrannini E, DelPrato S, DeFronzo RA (1990) Obesity and insulin resistance in humans: A dose-response study. Metabolism 39:452–459

Bouchard C, Despres JP, Mauriege P (1993) Genetic and non-genetic determinants of regional fat distribution. Endocr Rev 14:72–93

Brönnegård M, Ottosson M, Böös J, Marcus C, Björntorp P (1994) Lack of evidence for estrogen and progesterone receptor in human adipose tissue. J Steroid Biochem Mol Biol 51:275–81

Carter-Su C, Okamoto K (1987) Effect of insulin and glucocorticoids on glucose transporters in rat adipocytes. Am J Physiol 252:E441–E453

Chen Y-DI, Golay A, Swislocki ALM, Reaven GM (1987) Resistance to insulin suppression of plasma free fatty acid concentrations and insulin stimulation of glucose uptake in noninsulin-dependent diabetes mellitus. J Clin Endocrinol Metab 64:17–21

Cigolini M, Smith U (1979) Human adipose tissue in culture. VIII. Studies on the insulin-antagonistic effect of glucocorticoids. Metabolism 28:502–510

De Pergola G, Xu X, Yang S, Giorgino R, Björntorp P (1990) Up-regulation of androgen receptor binding in male rat fat pad adipose precursor cells exposed to testosterone: study in a whole cell assay system. J Steroid Biochem Molec 37:553–558

Divertie GD, Jensen MD, Miles JM (1991) Stimulation of lipolysis in humans by physiological hypercortisolemia. Diabetes 40:1228–1232

Dreiling DA, Bierman EL, Debons AF, Elstach P, Schwartz CL (1962) Effect of ACTH, hydrocortisone and glucagon on plasma nonesterified fatty acid concentration (NEFA) in normal subjects and in patients with liver disease. Metabolism 11:571–578

Fain JN, Kovacev VP, Scow RO (1965) Effect of growth hormone and dexamethasone on lipolysis and metabolism in isolated fat cells of the rat. J Biol Chem 240:3522–3527
Felber JP, Golay A, Felley C, Jecquier E (1988) Regulation of glucose storage in obesity and diabetes: Metabolic Aspects. Diabetes Metab Rev 4:691–700
Groop L, Forsblom C, Lehtovirta M, Tuomi T, Karanko S, Nissén M, Ehrnström B-O, Forsén B, Isomaa B, Snickars B, Taskinen M-R (1996) Metabolic consequences of a family history of NIDDM (The Botnia Study). Evidence of sex-specific parental effects. Diabetes 45:1585–1593
Groop LC, Bonadonna RC, Simonson DC, Petrides AS, Shank M, DeFronzo RA (1992) Effect of insulin on oxidative and non-oxidative pathways of free fatty acid metabolism in human obesity. Am J Physiol 263:E79–E84
Groop LC, Saloranta C, Shank M, Bonadonna RC, Ferrannini E, DeFronzo RA (1991) The role of free fatty acid metabolism in the pathogenesis of insulin resistance in obesity and non insulin-dependent diabetes mellitus. J Clin Endocrinol Metab 72:96–107
Guillaume-Gentil C, Assimacopoulos-Jeannet F, Jeaurenaud B (1993) Involvement of non-esterified fatty acid oxidation in glucocorticoid-induced peripheral insulin resistance in rats. Diabetologia 36:899–906
Howard B, Klimies J, Vasquez B, Brady D, Nagulesparan M, Unger RH (1984) The antilipolytic action of insulin in obese subjects with resistance to its glucoregulatory action. J Clin Endocrinol Metab 58:544–551
Jacobsson B, Smith U (1972) Effect of cell size on lipolysis and antilipolytic action of insulin in human fat cells. J Lipid Res 13:651–654
Kashiwagi A, Bogardus C, Lillioja S (1984) In vitro insensitivity of glucose transport and antilipolysis to insulin due to receptor and postreceptor abnormalities in obese Pima Indians with normal glucose tolerance. Metabolism 33:772–777
Kissebah AH, Vydelingum N, Murray R, Evans OJ, Hartz AJ, Kalkhoff RV, Adams PW (1982) Relation of body fat distribution to metabolic complications of obesity. J Clin Endocrinol Metab 54:254–260
Li M, Björntorp P (1995) Effects of testosterone and triglyceride uptake and mobilization in different adipose tissues in male rats in vivo. Obes Res 3:113–119
Li M, Yang S, Björntorp P (1993) Metabolism of different adipose tissue in vivo in the rat. Obes Res 1:459–468
Ljung T, Holm G, Friberg P, Andersson B, Mårin P, Bengtsson B-Å, Svensson J, Dallman M, McEwen B, Björntorp P (2000) The activity of the hypothalamic-pituitary-adrenal axis and the sympathetic nervous system in men with elevated waist/hip circumference ratio. Obes Res, in print
Lönnroth P, Digirolamo M, Krotkiewski M, Smith U (1983) Insulin binding and responsiveness in fat cells from patients with reduced glucose tolerance and type II diabetes. Diabetes 32:748–754
Lönnroth P, Smith U (1992) Intermediary metabolism with an emphasis on lipid metabolism, adipose tissue, and fat cell metabolism. In: Björntorp P, Brodoff BN (eds) "Obesity". Lippincott, New York, pp 3–14
Mårin P, Andersson B, Ottosson M, Olbe L, Chowdhury B, Kvist H, Holm G, Sjöström L, Björntorp P (1992) The morphology and metabolism of intraabdominal adipose tissue in men. Metabolism 41:11, 1242–1248
Mårin P, Gustafsson C, Odén B, Björntorp P (1995) Assimilation and mobilization of triglycerides in subcutaneous abdominal and femoral adipose tissue in vivo in men: Effects of androgens. J Clin Endocrinol Metab 80:239–243
Meikle AW, Stringham JD, Woodward MG, Bishop DT (1988) Heritability of variation of plasma cortisol levels. Metabolism 37:514–517
Ottosson M, Lönnroth P, Björntorp P, Edén S (1999). Differential effects of cortisol and growth hormone on lipolysis in human adipose tissue. Submitted for publication

Randle PJ, Garland PB, Hales CN, Newsholme EA (1963) The glucose-fatty acid cycle. Its role in insulin sensitivity and the metabolic disturbances of diabetes mellitus. Lancet 1:785–789

Reaven GH (1995) Pathophysiology of insulin resistance in human disease. Physiol Rev 75:473–486

Rebuffé-Scrive M, Eldh J, Hafström L-O, Björntorp P (1986) Metabolism of mammary, abdominal and femoral adipocytes in women before and after menopause. Metabolism 35:9, 792–797

Rebuffé-Scrive M, Krotkiewski M, Elfverson J, Björntorp P (1988a) Muscle and adipose tissue morphology and metabolism in Cushing's syndrome. J Clin Endocrinol Metab 67:1122–1128

Rebuffé-Scrive M, Lönnroth P, Andersson B, Smith U, Björntorp P (1988b) Effects of short-term prednisolone administration on the metabolism of human subcutaneous adipose tissue. J Obesity Weight Regul 7:22–33

Rebuffé-Scrive M, Lundholm K, Björntorp P (1985) Glucocorticoid hormone binding to human adipose tissue. Eur J Clin Invest 15:267–271

Xu X, De Pergola G, Björntorp P (1991) Testosterone increases lipolysis and the number of beta-adrenoceptors in male rat adipocytes. Endocrinology 128:379–382

Xu X, de Pergola G, Björntorp P (1990) The effects of androgens on the regulation of lipolysis in adipose precursor cells. Endocrinology 126:1229–1234

Zancanaro C, Cigolini M, Bonora E, Moghetti P, Querena M, Muggeo M (1990) Glucose and insulin suppression of plasma free fatty acids in obese subjects with normal glucose tolerance or mild, newly diagnosed type 2 (non-insulin-dependent) diabetes. J Endocrinol Invest 13:55–59

CHAPTER 6

The Regulation of Body Weight: Set-Points and Obesity

M.D. HIRVONEN and R.E. KEESEY

A. Introduction

It is widely recognized that physiological variables such as body core temperature, body water, and blood glucose are maintained within fairly tight tolerances. A century and a half ago, French physician Claude Bernard marveled at the stability of the "milieu interieur," and the ability of the body both to monitor and control its internal state with apparent precision. The term "homeostasis" was later introduced by CANNON (1932) to denote the condition of physiological stability that characterizes our internal environment. The general viewpoint, viz. that a set of monitored variables and control processes exist for the purpose of maintaining a stable internal environment, provided an empirically testable model, which formed the basis for the scientific investigation of regulatory physiology in the twentieth century.

I. Set-Points

Assuming that a variable is subject to regulatory control necessarily implies that a particular set level of that variable be specified and maintained. The term "set-point" originated with engineers of the 1950s who designed environmental control systems. In this context, a set-point is defined as an independent and adjustable signal which serves as a standard value against which a feedback signal from the controlled system can be compared. Thus, a set-point establishes the desired value of a controlled variable, allowing deviations from the set value to be monitored and quantified. If the feedback signal exceeds preset limits, appropriate effector mechanisms are evoked so as to restore the regulated variable to its preset level.

These concepts of controlled processes and set-points were readily adapted to the study of physiological variables by regulatory biologists. For example, the literature reveals a widespread application of these ideas to the study of body temperature regulation. The argument will be made here that set-point regulators and control systems also provide fruitful models for understanding how body weight and body energy are stably maintained.

II. Body Weight and Energy

While this chapter focuses on the regulation of body weight, it should be noted that body weight, per se, is probably NOT a single regulated variable. Rather, changes in body weight reflect underlying changes in body energy stores in the form of proteins (muscles and organs), fats (adipose tissue depots), and carbohydrates (glycogen stored in muscle and liver). Protein serves a primary structural and functional role, making up the major portion of cellular membranes, muscles, and organs. While protein can be utilized as a source of fuel, the loss of body protein is necessarily associated with a loss of function. Small changes in body protein are associated with changes in body weight, but the majority of weight change can reasonably be attributed to other factors. On the other hand, the primary function of carbohydrate and fat stores is to supply energy as needed, providing a continuous flow of substrate for energy-requiring physiological processes. Glycogen is an easily accessible source of energy for short-term expenditures. But stores of glycogen are limited, comprising only about 1% of total body weight. Therefore, significant changes in body mass cannot be attributed to alterations in energy stored in the form of carbohydrate. Triglycerides released from adipose tissue are mobilized more slowly, but can sustain longer periods of expenditure. Adipose tissue makes up about 20% of body weight in the average person, making it a much better candidate variable for investigating changes in body weight. It is estimated that the fat stores of a person of average weight are adequate to meet his or her energy needs for a period of approximately 2 months.

III. Body Energy Set-Point

Maintenance of a stable body energy content requires energy intake to be equal to energy expenditure. Changes in body energy stores, therefore, occur when a disparity develops between energy intake and energy expenditure. Thus, from a body energy set-point perspective, one can view total body energy as a regulated variable, with energy intake and energy expenditure serving as the primary effector mechanisms through whose control energy balance at a whole-body level is achieved and sustained.

IV. Supporting a Set-Point Argument

Practically speaking, in order to make the argument that body weight is regulated by a set-point mechanism, it must be shown that certain conditions are met. First, the regulated variable in question should show some degree of stability. Second, there must be an active defense of the level of the variable the control system is set to maintain. That is, when the regulated variable is perturbed (i.e., displaced from set-point), appropriate effector mechanisms should be evoked. And the magnitude of response of those effector mechanisms should be proportional to the degree of displacement from set-point.

1. Stability

In the case of body weight, examination of a variety of species suggests that stability is clearly the norm. Animals provided with nutritionally-adequate, balanced diets maintain body weights at levels that are stable and appropriate for age, gender, and species. That is not to say that body weights never change. On the contrary, the balance of body energy intake, expenditure, and storage is a dynamic, ongoing process, and body weights of individuals are influenced by genetic, environmental, and developmental factors. In Fig. 1, the body weights of the control rats (open circles) shows the typical growth pattern of a group of adult male rats measured across 6 months. Note that the body weights of these animals increase with time, but in a predictable manner. In this example, stability is implied by the predictable pattern of growth. That is, the body weights do not vary dramatically from day to day, or even week to week. Rather, the rats gain weight steadily, at a rate predictable in accordance with age, gender and strain.

Similarly, if you measured the daily body weight of a person for a period of several months, the variation in body weight is small, generally estimated to be less than 2%. Given the abundance and ready availability of a wide variety of foodstuffs in today's world, this degree of stability is remarkable. It should, however, be noted that the body weight set-point is not necessarily a

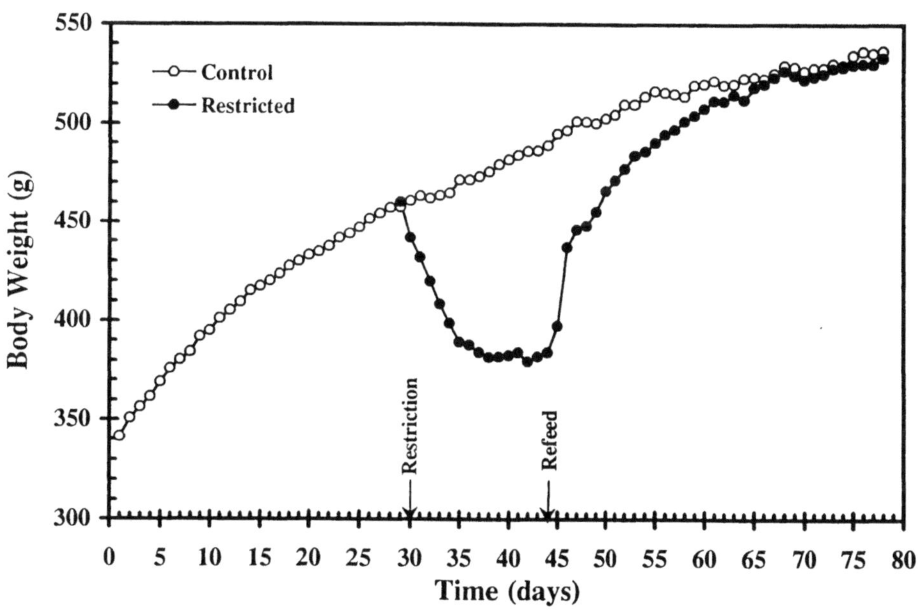

Fig. 1. The daily body weights of control rats (*open circles*, n = 5) and restricted rats (*closed circles*, n = 5). Fourteen days of caloric restriction resulted in a loss of approximately 20% of pre-restriction body mass, which the rats regained once free feeding was restored (adapted from MITCHEL and KEESEY 1978)

condition over which individuals appear to have voluntary control. People may not maintain the body weight that they would personally choose, although their body weights are still quite stable. This stability strongly suggests an underlying regulatory process, although stability alone is not sufficient to establish the existence of an underlying set-point mechanism.

2. Active Defense

The essential evidence needed to support a set-point mechanism is that of resistance to displacement from the level of stable body weight maintenance. As an example of the active defense of body weight set-point, Fig. 1 shows the body weight curves of two groups of rats, one group (closed circles) was calorically restricted for 14 days, producing a decline in body weight of approximately 20%, before being returned to ad libitum feeding. Note that the decline of body weight was rapid during the first 6 days of caloric restriction, but the rate of loss diminished across the next 8 days. Since the rats consumed the same amount of food each day for the entire 14-day restriction period, the diminishing rate of loss suggests that physiological adaptations occurred which progressively limited the loss of body weight. Indeed, the daily energy expenditure of the restricted rats was found to be significantly depressed, thus rendering them more efficient utilizers of energy compared to non-restricted rats. Further evidence of this increased efficiency can be inferred from their rate of weight gain upon re-feeding. The rate of gain is greater initially, due to the rats ingesting larger amounts of food while still retaining their increased efficiency of energy utilization. As the body weights of previously restricted rats approached the level of non-restricted rats, both their total daily calories ingested and their rate of energy expenditure begin to approximate the levels of non-restricted rats. Thus, the weight adjustments depicted in Fig. 1 provide an example of the proportional response of regulatory mechanisms evoked by a displacement from body weight set-point.

As suggested in the previous example, weight loss induced by caloric restriction has been shown to be accompanied by a reduction in energy expenditure. Some decline in energy expenditure is to be expected with weight reduction, since less tissue should require less energy for its maintenance. However the declines in energy expenditure seen following restriction are much larger than would be predicted on the basis of tissue losses. For example, in rats a decline in body mass of 14% caused a decline in daily energy expenditure of more than 24% (CORBETT et al. 1985). During food restriction, these alterations in energy expenditure serve to blunt further weight loss. Even when ad libitum feeding conditions are reinstated, energy expenditure remains below expected levels until body weight is restored to pre-deprivation levels. In addition to changes in energy expenditure, alterations in energy intake are also observed when body weight is displaced from set-point. For example, food intake is concurrently increased when free feeding is reinstated following periods of food restriction, and remains above normal levels until body weight is restored to the appropriate level. Changes in both energy expenditure and energy intake are also evoked when body weight is elevated acutely. In this case, daily energy expenditure is elevated to an extent proportionately greater than expected from the added tissue, while food intake is dramatically reduced. These adjustments in energy expenditure and intake then persist until body weight is returned to normal levels. Thus, coordinated alterations in both energy expenditure and energy intake are means by which weight perturbations are resisted, and by which body weight is restored following its displacement from set-point.

B. The Dieter's Dilemma

The following example offers a set-point interpretation of the effects of dieting on energy expenditure and body weight. Imagine two individuals, closely matched for body weight, body fat, caloric intake, age, gender, etc. One of these individuals begins a caloric restriction program (i.e., a diet) designed to reduce body weight, while the other eats normal amounts and maintains a stable body weight. The body weight of the calorically-restricted person declines relatively rapidly at first. But, as weight declines, energy requirements decline as well, until the individual's daily energy expenditure matches energy intake. With the resulting balance between energy intake and expenditure, body weight stabilizes at a reduced level. Thus, even if the individual stays on the diet, body weight will not decline any further. Discouraged by the "failure" of the diet, the individual then resumes eating normal meals. Initially, the regain of body weight is rapid, because the caloric intake is now substantially greater than energy needed at the reduced body weight. In fact, with only normal levels of intake, weight gain will continue (albeit at decreasing rates) until energy intake and energy expenditure again come back into balance at the individual's pre-restriction or set-point body weight. The surprising result of this example is that, if the total calories ingested and expended by the dieter across the loss and regain cycle were calculated, the sum would be considerably less than for the non-dieter. Yet, in spite of this reduced energy flux, the dieter ends up weighing exactly the same as the non-dieter at the end of the weight loss and regain cycle.

The decline in daily energy expenditure which can occur with diet-induced reductions in body weight likewise accounts for another commonly reported paradox. Recognizing that individuals can differ in their body weight set-points, consider what happens when a person spontaneously maintaining a body weight of 200 lbs decides to diet and lower his weight to that of a friend who weighs 180 lbs. Though he is successful in lowering his weight to exactly that of his friend, he finds that they differ in other respects. Most significantly, his daily energy expenditure is substantially less than that of his friend and the only way he can remain at 180 lbs is to eat significantly less. Thus, the metabolic adaptations which we seem to have evolved to stabilize body weight at the set-point level can produce the paradoxical condition whereby a dieter must eat considerably less to maintain a reduced body weight than would a non-dieter spontaneously maintaining exactly the same weight.

C. Changing the Set-Point

The previous discussion of adaptive alterations in energy intake and expenditure following perturbation of body weight is strong evidence in favor of the idea that an individual's body weight is maintained at some preferred value, or set-point. Furthermore it implies that compensation for short-term changes in body weight will occur automatically, and that body weight will tend to return to the set-point level. How then is it possible to achieve a lasting change in body weight?

I. Upregulation of the Set-Point

We have all experienced or observed body weight increases of individuals that appear to be relatively permanent. How do these apparent upward changes in the set-point for body weight occur? It is not easy to answer to this question, but it appears that consideration of the influence of diet on fat cell morphology may constitute a basis for one explanation. As noted earlier, energy expenditure is sharply elevated when body weight rises above the set-point. Thus, weight gains resulting from exposure to diets that promote overconsumption are initially resisted by higher rates of metabolism. Likewise, if soon returned to a regular diet the body weight gains are quickly reversed and the initial level of body weight is restored. However, it has been reported that, with long-term exposure to weight-promoting high fat diets, this metabolic resistance to weight gain declines. The energy expenditure of rats exposed to a high fat diet for six months was, for example, higher than lean controls, but appropriate for the larger body mass these rats now maintained. And when returned to a regular diet, these rats failed to return to their former weight but instead remained at the elevated body weight induced by high fat feeding. Thus, it appears that sustained maintenance of elevated body weights can lead to adaptations which cause an irreversible elevation in the body weight set-point. These observations stimulate one to ask what changes prolonged exposure to high fat diets induce so as to lessen the initial resistance to weight gain and support the maintenance of a higher level?

II. The Fat Cell Size and Number: A Hypothesis

An explanation of this diet-induced elevation of body weight set-point following large and sustained weight gains may reside in our understanding of adipose cell physiology. The size of body fat stores depends both on the number and size of fat cells. HIRSCH et al. (1989) has suggested that fat cells have a limited storage capacity, and that fat is stored in existing fat cells only until they reach a certain size. Once this threshold size is reached, new fat cells are either recruited from a population of pre-adipocyte cells, or new adipocytes are produced. This recruitment of new cells may be the key to understanding the differential effects of short vs long term body fat increases on set-point. Reversible (smaller and short-term) increases in body weight are apparently the result of increasing the size of existing fat cells. When the conditions that induce fat gain are removed, the fat cells simply reduce their average size and body weight returns to pre-gain levels. Irreversible (larger and sustained) increases in body weight are likely associated with increases in cell number. The recruitment of new adipose cells into the active cell pool appears to be a unidirectional process. Once activated, adipocytes apparently cannot return to the pre-adipocyte stage, suggesting that, while new fat cells can be acquired, they cannot be lost. Thus, body fat (and body weight) increases can be temporary if fat is added to existing adipose cells. But if that increase is sufficiently large and prolonged, new fat cells form and the change becomes permanent. At that point, reducing body weight by dieting can reduce the size of fat cells, but not the fat cell number. Reducing adipose cell size then triggers both behavioral (increased energy intake) and physiological (reduced energy expenditure) adaptations which tend to promote the restoration of fat cells to their optimal size, and maintenance of body fat at the new level.

The fat cell number and size hypothesis suggests that body weight regulatory mechanisms are organized so as to monitor and maintain fat cell size, not fat cell number. That is why two individuals can both be regulating body weight normally and effectively, yet maintain very different levels of body weight and body fat. An obese individual is typically supporting a much larger number of adipocytes, although the average size of those adipocytes need be no different from those of a lean individual.

D. Assessing Set-Points in Individuals

Knowing that the displacement of body weight from set-point alters energy expenditure and energy intake is of little clinical value if one does not have a means of specifying where an individual's body weight stands relative to his or her set-point. Measuring body weight is easy, but determining set-point is a bit more difficult. Assessing set-point is possible, however, and a method for doing so is proposed below.

I. Determining Daily Energy Needs

When a free-feeding organism is in energy balance at a stable body weight, its energy needs are predictable from the mass of tissues it maintains. Max Kleiber, a noted animal nutritionist, examined the relationship between body mass and daily resting energy expenditure for a variety of species, ranging in size from small to large birds and mammals. He observed that resting energy expenditure increased at a rate 3/4 that of body mass. This relationship was expressed mathematically as $REE_{kcal/d} = kBW_{kg}^{(0.75)}$, where REE is resting energy expenditure in kilocalories/day, BW is body weight in kilograms, and k is a constant (the average value of k ≅ 69 across the species studied by KLEIBER (1975)).

1. Using the Resting Energy Expenditure and Body Weight to Assess Set-Points

The extension of this expenditure-mass principle to different sized members within a given species is also possible. Figure 2 illustrates a simple linear relationship between resting energy expenditure and body mass in 56 free feeding, weight stable, male rats of the same age and strain. In this group of rats, body weight has considerable predictive power, accounting for ~61% of the variation in resting energy expenditure. Furthermore, the resulting regres-

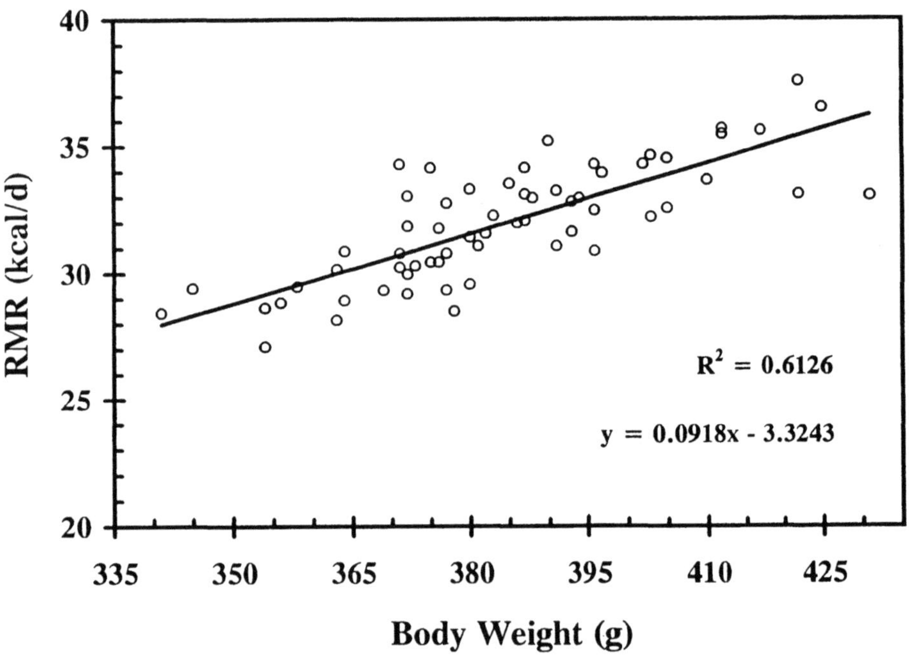

Fig. 2. A simple linear regression model of the relationship between resting energy expenditure (REE) and body weight in 56 free-feeding adult, male, Sprague-Dawley, rats. Note that body weight is a good predictor of energy requirements, accounting for approximately 61% of the variance in resting energy expenditure

sion equation (Y = 0.092X – 3.32) provides a means of estimating the daily resting energy expenditure of any rat in this population, given its body weight.

However, as discussed earlier, displacement of body weight from set-point alters the basic mass-expenditure relationship for any individual from this group. Only when a rat is at its set-point, can its resting energy expenditure be predicted accurately from linear relationship depicted in Fig. 2. Reducing the body weight of any rat below its spontaneously maintained level reduces energy expenditure more than would be predicted from the change in body mass. The implication is that an animal that is below its set-point will necessarily display a resting energy expenditure significantly lower than an animal of the same weight that has not been displaced from set-point. Conversely, an animal that is above its set-point should display a significantly elevated resting energy expenditure, compared to animals of similar weight that are not displaced from set-point. Only a rat that was maintaining body mass at set-point would be expected to show a resting energy expenditure that fell within the range appropriate for that body weight, as predicted by the linear relationship shown in Fig. 2.

To illustrate the point discussed above, consider the following observations (R.E. Keesey and M.D. Hirvonen, unpublished). The same 56 rats whose body weights and energy expenditures are depicted in Fig. 3 were subsequently divided into 7 groups of 8 animals each, matched for body weight. The groups were then subjected to a food restriction protocol, receiving 100%, 90%, 85%, 80%, 70%, 60%, or 50% of their average daily food intake for a period of 1 week. Body weight and resting energy expenditure were assessed following the restriction period. Figure 3 shows both (1) the observed change in resting energy expenditure as a function of the change in body weight for these 8 groups of rats, and (2) the change predicted on the basis of body weight from the relationship in Fig. 2. It can be seen (Fig. 3) that the greater the restriction-induced deviation from set-point (i.e., the weight maintained spontaneously while eating ad libitum and in energy balance), the greater the decline in resting energy expenditure below the value expected from the mass-expenditure relationship which characterizes this population (see Fig. 2).

There is only a small range of deviation from normal weight where the predicted resting energy expenditure and observed resting energy expenditure coincide. Supporting the notion that when an organism is at set-point, its energy expenditure is predictable as a function of the mass of tissue it maintains. This concept can be exploited further to determine whether an organism is above, below, or very close to its set-point body mass. Only when at its body weight set-point will its expenditure coincide with the expected value. Thus, by determining the basic relationship between resting energy expenditure and body weight within a given species, one can assess whether an individual from that population is at, above, or below its individual set-point.

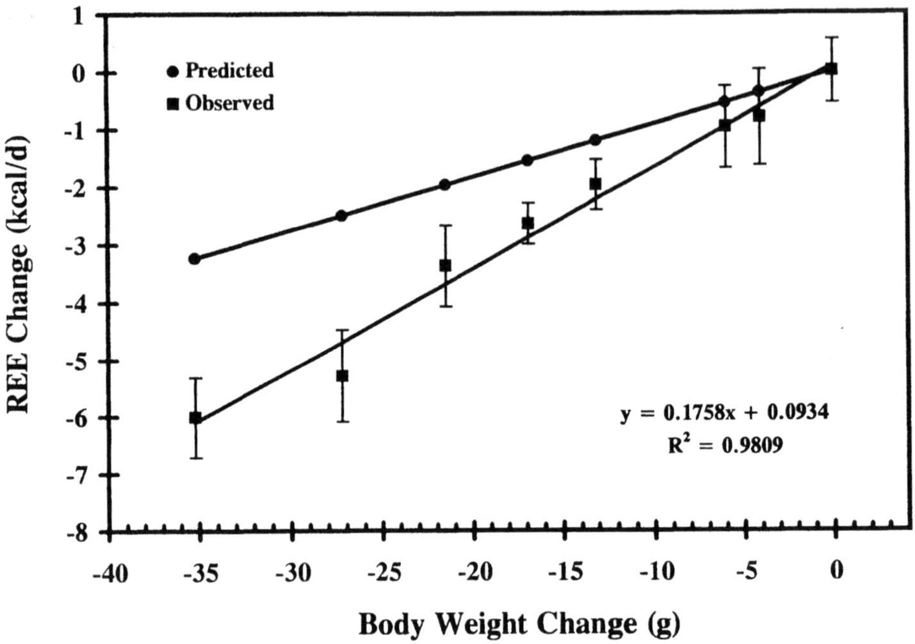

Fig. 3. The changes in resting energy expenditure (REE) predicted by the loss of body weight in calorically-restricted rats (using the regression equation derived in Fig. 2) compared to the observed resting energy expenditure. Note the proportionally larger deviation of observed from predicted resting energy expenditure associated with increasing losses of body weight

E. Set-Point Analysis of a Pharmacological Effect on Body Weight

A variety of pharmacological agents are known to have effects on body weight. The application of set-point theory to account for the effects of pharmacological agents on body weight is perhaps best illustrated by specific example. To that end, we will consider the effects of one such agent, nicotine, on body weight.

I. Effects of Nicotine on Body Weight

There is considerable evidence that smokers weigh less than nonsmokers, and that ex-smokers commonly gain weight soon after quitting. In controlled studies in animals, the acute administration of nicotine leads to short-term increases in energy expenditure and decreases in food intake and body weight. With chronic administration of nicotine, however, body weight stabilizes (at a reduced level compared to control rats), and both the rats' energy expenditure and food intake are then the same as those of rats spontaneously maintaining similar body weights. Increasing the dose of nicotine reinstates the acute depression of food intake and elevation of energy expenditure, causing further loss of body weight. But even at the higher dose, body weight subsequently stabilizes (at an even lower level) and food intake and energy expenditure return to levels appropriate for the body weight then maintained.

1. Tolerance Interpretation

On the basis of these observations, it could be surmised that a primary or direct effect of nicotine is to stimulate energy expenditure and suppress food intake, thereby causing body weight to decline. It might also be assumed that, after chronic exposure to nicotine, pharmacological tolerance causes these primary effects to abate, and body weight losses to cease. Consistent with this interpretation is the additional observation that increasing the dose of nicotine can reinstate the primary food intake and energy expenditure effects, although tolerance will again develop at any (nonlethal) dose level. This explanation is simple, concise, and adequately accounts for the given data. However, there is an alternative to this "tolerance" explanation, which deserves consideration.

2. Set-Point Interpretation

The effects of nicotine administration on energy expenditure, food intake, and body weight fit nicely into a set-point framework. In this scenario, it is assumed that the primary effect of nicotine is to lower the body weight set-point, while changes in food intake and energy expenditure are seen as representing the secondary (effector) mechanisms by which body weight is reduced to its new (lower) set-point. If indeed nicotine does lower the set-point for body weight, a series of predictions concerning changes in food intake, energy expenditure, and body weight can be deduced. First, the administration of nicotine would induce a disparity between the current body weight and the new (lower) set-point. This displacement of set-point would trigger an appropriate response from effector mechanisms (i.e., a reduction in food intake and an increase in energy expenditure) resulting in the loss of body weight. Eventually, however, when body weight is again consonant with set-point, the alterations in food intake and energy expenditure would abate. Body weight would then stabilize at the new set-point, and food intake and energy expenditure would be at the level of other rats spontaneously maintaining this body weight. Increasing the dose of nicotine would be expected to reduce the set-point further, creating another difference between set-point and body weight. As a result, appropriate effector adjustments (food intake and energy expenditure) would again be evoked, leading to a further loss of body weight. Food intake and energy expenditure would again come back into balance when body weight reached the new lower set-point.

3. Choosing Between Tolerance and Set-Point Interpretations

Both theoretical approaches described above provide a plausible explanation of the effects of nicotine administration on body weight, food intake, and energy expenditure. In order to distinguish between these two competing viewpoints, it is necessary to devise and implement experimental conditions under which each theory produces unique predictions. What, for example, would be the result if the body weight of rats was reduced by caloric restriction prior to the administration of nicotine?

a) Predictions Based on the Tolerance Interpretation

The "tolerance" hypothesis assumes that the depression of food intake and elevation of energy expenditure are primary effects of nicotine administration, with weight change occurring as a secondary consequence of these effects. Certainly, tolerance would not develop prior to exposure to a specific pharmacological agent, so tolerance should develop only after exposure to nicotine for some finite length of time. Since there is no obvious reason why changing body weight should alter the development of tolerance, the time frame of tolerance development might be expected to be very similar to that seen in animals of normal body weight. Therefore, the tolerance hypothesis would appear to predict that reducing body weight prior to administration of nicotine would not materially change the effects on food intake and energy expenditure from that seen in rats not subjected to weight loss. In other words, the apparent prediction is that food intake will be depressed and energy expenditure elevated in both normal weight and weight-reduced rats, until there had been sufficient exposure to nicotine for tolerance to develop. While tolerance develops, further weight losses would also be expected to occur in both normal weight rats and rats already weight-reduced.

b) Predictions Based on the Set-Point Interpretation

The "set-point" hypothesis assumes that the primary effects of nicotine are to depress the body weight set-point, and that adjustments in food intake and energy expenditure are the secondary mechanisms by which body weight is then reduced to the lower set-point. Since keeping body weight commensurate with set-point is primary, altering body weight prior to nicotine administration should significantly alter the initial food intake and energy expenditure response to nicotine. Lowering body weight prior to nicotine administration should markedly reduce the disparity between the new set-point induced by nicotine and the existing weight and therefore blunt the usual effects of nicotine on food intake and energy expenditure. Were body weight reduced to a level that exactly coincided with the nicotine-induced depression of set-point, there should be no detectable suppression of food intake or elevation of energy expenditure upon nicotine administration in reduced weight rats. Were body weight lowered below the set-point induced by nicotine, there should actually be a reversal of the effector response (i.e., food intake should actually increase, and energy expenditure decrease) following nicotine administration. In each case the coordinated change in food intake and energy expenditure evoked upon nicotine administration should serve simply to move body weight toward the new set-point.

4. Results of Lowering Body Weight Prior to Nicotine Administration

The results of an experiment in which body weight was lowered prior to nicotine administration were reported by SCHWID et al. (1992). The depression of food intake and elevation of energy expenditure observed in rats given nicotine at normal body weights was found to be diminished or even eliminated in rats with reduced body weights. These findings tend to support the interpretation that nicotine acts to reduce the set-point for body weight, and questions the "tolerance" interpretation of its effects on energy intake and expenditure.

SCHWID et al. (1992) also reported that increasing body weight prior to nicotine administration prolonged the period of depressed food intake and elevated energy expenditure, although both of these parameters eventually normalized when body weight was eventually reduced. Notably, rats with initially elevated, normal, and reduced weights all stabilized at the same reduced body weight following nicotine treatment. When nicotine administration ceased, all rats returned to the level of body weight displayed by non-treated control rats. These results suggest that nicotine indeed lowers the body weight set-point, but that the effects of nicotine on body weight set-point are reversible.

5. Further Evidence of Set-Point Shifts in Nicotine Treated Rats

As discussed earlier, making a set-point argument requires demonstrating not only stability of the putative regulated variable, but the active defense of that level. After 14 days of nicotine treatment, both control and nicotine-treated rats maintained stable body weights, although the treated rats maintained body weight at a level ~8% below nontreated rats (SCHWID et al. 1992). Defense of this lowered level implies that nicotine- treated rats should resist displacement from their reduced body weight by appropriate effector actions, in this case by adjustments both in food intake and energy expenditure. SCHWID et al. (1992) addressed this issue of active defense in their report. Half of control and nicotine-treated rats were calorically-restricted to produce a further 8% decline in their respective body weights. The energy expenditure of the restricted groups, both control and nicotine, was reduced significantly relative to non-restricted control and nicotine-treated rats, respectively. By design, the weight of the restricted control rats was reduced to the same level that the non-restricted but nicotine-treated rats spontaneously maintained. Thus, while exactly the same weight as the nicotine-treated rats, the energy expenditure of restricted control rats was significantly lower. The nicotine-treated rats likewise showed a significantly reduced energy expenditure with caloric restriction, but only after their body weight had been lowered below the reduced level they maintained spontaneously. In other words, the control and nicotine rats differed in the body weight each defended, yet both displayed an active defense of their spontaneously-maintained body weights.

An active defense was also evident in the energy intake of restricted control and nicotine-treated rats. When calorically-restricted rats were allowed to feed freely again, each displayed elevated levels of food intake relative to non-restricted rats. Both control and nicotine-treated rats restored body weight to pre-restriction levels in less than 2 weeks, and initially elevated food intakes declined progressively to normal levels across the same period. This pattern of altered food intake associated with the restoration of body weight to the pre-restriction levels in both control and nicotine-treated rats suggests an active defense of body weight set-points by these rats.

Thus, it appears that the evidence favors a set-point interpretation of nicotine's effects on body weight. Nicotine-treated rats maintain stable but reduced body weights, relative to control rats. Lowering body weight prior to nicotine administration diminishes or eliminates the depression in food intake and elevation of energy expenditure displayed by rats treated with nicotine at normal body weights. Nicotine-treated rats also displayed coordinated adjustments in food intake and energy expenditure which counter efforts to displace them from the reduced weight levels they spontaneously maintained. Thus, rats treated with nicotine appear to maintain and defend a lower set-point and apparently do so in the same manner and as effectively as control rats defend a normal body weight.

F. Set-Point Signaling Mechanism(s)

Though it appears that a set-point for body energy exists, the means by which deviations from set-point might be detected and/or signaled are as yet poorly understood. It is apparent, however, that maintaining a set-point would require some form of feedback signal that varies proportionally with body fat content, or perhaps total body energy. The recent discovery of several biologically-active molecules, levels of which appear to co-vary with body energy, has raised the possibility that these may be key components in the signaling of body energy levels to the set-point control system.

I. Leptin

Leptin is a 16-kilodalton protein produced by the obesity (*ob*) gene of mice. Mice that have defective copies of the *ob* gene (*ob/ob* mice) become grossly obese, and leptin injections restore their body weight to normal levels. Soon after its discovery, it was shown that leptin is produced by adipocytes, and the larger the adipose store the greater the leptin production. This led researchers to suggest that leptin functions as a "hormonal lipostat," in other words, the body fat signaling molecule. This suggestion fits well with the previous proposal that adipocyte size is a key determining factor in the regulatory control of body energy stores. Rapid weight gain would tend to be manifested as larger adipocytes and a greater adipose mass, thus increasing leptin production and promoting appropriate effector mechanism responses to limit weight gain. Conversely, rapid weight loss would reduce fat cell size, leading to lower circulating leptin and evoke mechanisms appropriate for limiting weight loss.

Shortly after the discovery of leptin, investigators identified the leptin receptor and showed that a receptor mutation was a primary factor in several other animal models of obesity, including the *db* (or diabetic) mouse, the corpulent rat, and the Zucker fatty (*fa/fa*) rat. In each of these animal models, obesity appears to result from an inability to respond appropriately to circulating leptin, rather than a lack of leptin itself. Further evidence for a possible role of leptin in obesity is the finding that leptin receptors are found in the arcuate and paraventricular nuclei of the hypothalamus, areas thought to be intimately involved in the control of energy intake and expenditure.

One problem with the leptin set-point signal theory is that there are no indications that human obesity is associated with a defective *ob* gene or receptor. In addition, other research has shown that obese humans have high circulating leptin levels, not low levels like the *ob/ob* mice. Rather, it appears that obese people produce adequate amounts of leptin, and that leptin receptors are present, but somehow the leptin fails to trigger an appropriate effector response. CARO et al. (1996) found that the spinal fluid of obese individuals contains only slightly more leptin than lean individuals, in spite of fivefold higher blood leptin levels. Thus, it seems likely that obesity in humans results from either a peripheral insensitivity to circulating leptin or a lack of leptin access to receptors in the brain.

It is significant that, in the evolutionary development of mammals, it has been starvation rather than obesity that has posed the more serious threat to the survival of the species. Thus, it has been suggested (AHIMA et al. 1996; FLIER and MARATOS-FLIER 1998) that another role for leptin may be to signal declining energy reserves (rather than simply reflecting static levels of fat stores). Falling leptin levels appear to trigger adaptive responses that are characteristic of a starving organism, and these responses are reversed by the administration of leptin.

II. Leptin, Neuropeptide-Y, and Melanocortin

It seems apparent that, while leptin is produced peripherally, its critical site of action is likely the brain. Two other bioactive molecules, neuropeptide-Y and melanocortin, may act in concert with leptin in the regulation of body energy. NPY has been shown to be a potent stimulator of appetite. Production of neuropeptide-Y is elevated in *ob/ob* mice, and injection of neuropeptide-Y centrally induces many of the symptoms associated with the *ob/ob* syndrome. Also, leptin receptors have been identified in the arcuate nucleus, a major site of central neuropeptide-Y production. It has thus been suggested (SCHWARTZ et al. 1996; ERICKSON et al. 1996) that neuropeptide-Y and leptin may act in concert in the control of energy intake, with leptin serving to hold neuropeptide-Y levels in check. Falling leptin levels would therefore be expected to lead to increased neuropeptide-Y production, and stimulate increased energy intake. Melanocortin receptors have also been identified in the arcuate nucleus, and peptides that bind to these receptors have been shown to suppress appetite. This leads to speculation that melanocortin might interact with leptin to prevent weight gain. There remains, however, a great deal of research to be done to clarify the roles of leptin, neuropeptide-Y and melanocortin in the control of body weight and body fat content. However, such research holds the promise of uncovering ways of intervening in the set-point signaling system so as to modify or alter the level at which body weight is actively maintained.

G. Why High Set-Points?

In general, evolutionary pressures select for factors that promote survival and reproductive success. Yet the prevalence of obesity, a condition that would be considered detrimental to survival, is currently on the rise. To understand why evolutionary pressures would favor selection of genes that today appear to promote obesity, one must compare the environment in which humans evolved to the one in which we currently exist. Seasonal variability of foodstuffs was a major factor in the survival of our ancestors. Clearly, the ability to store substantial amounts of energy rapidly when food was readily available was an asset in an environment where periodic scarcity of food was the norm. Thus, the genetic predisposition toward rapid assimilation of energy very likely provided a distinct advantage when a period of plenty was often followed by a period of scarcity. When food was scarce, and energy intake not sufficient to meet daily energy needs, the ability to lower resting energy expenditure adaptively would also convey an advantage. By lowering energy needs, the amount of stored energy needed to supplement energy intake would be minimized. This ability extends survival time, and increases the opportunity to reproduce. Another factor that likely contributed to the prevalence of obesity in today's society was lifespan. That is, the lifespan of our ancestors was very limited compared to today. Given an expected lifespan of 30 years or less, obesity and related diseases that occur in adults over the age of 30 would hardly be expected to exert significant influence on the selection process.

Thus, in an environment where food is often scarce, evolution would favor those who were able both to store energy efficiently during times of plenty, and to reduce energy needs when faced with a scarcity of food. These attributes would tend to promote survival and future reproductive success in our short-lived ancestors. Unfortunately, those same survival traits tend to favor obesity in the current population of long-lived, relatively sedentary humans living in an environment where a wide variety of highly palatable foods are continuously available.

An example of the influence of natural selection on obesity is evident in the work of RAVUSSIN et al. (1990) with the Pima Indians. The Pima culture developed in arid regions of Arizona, where, historically, food supplies were often limited. RAVUSSIN et al. (1990) has reported that obesity in Pima Indians is associated with a "low metabolic rate." One interpretation of this finding is that evolving in this harsh environment has left modern Pimas with what has been referred to as a "thrifty" metabolism. Low metabolic rate, and thus reduced energy needs, was a distinct advantage for the ancestors of modern Pimas, undoubtedly enhancing their ability to store more energy when food was available, and thus to survive and reproduce successfully. Today, however, the same thrifty metabolism in combination with an abundant food supply is likely a major contributor to the high level of obesity among their descendents.

H. Set-Point Perspective on Obesity

In the preceding pages, the idea has been advanced that set-points models are instructive in understanding how energy intake and energy expenditure are controlled in the process of regulating body weight and adiposity at specific levels. It is further shown that the condition of obesity can be viewed as one of regulation at an elevated set-point. The general lack of success in treating obesity by dietary means is consistent with such a view. Efforts by the obese to lose weight by dieting are resisted both by an enhanced appetite and a reduced rate of energy expenditure (LEIBEL et al. 1995). Thus, even when obese individuals are able to achieve some measure of weight loss, the associated metabolic adjustments favor its eventual recovery.

Given what we now know about the system for regulating body energy, in particular the dilemma it poses for the treatment of obesity by dietary means, one is drawn to conclude that a more promising approach to achieving significant and sustainable weight loss in the obese would be one based upon a strategy of adjusting either the regulatory set-point or the feedback signal to which the regulatory system responds. Clearly, were a means discovered for reducing set-point and/or the feedback signal for body energy in obese individuals, the very same physiological processes that otherwise act to resist displacement from an elevated adiposity level would instead work to facilitate fat loss and the eventual maintenance of a lowered level of body fat.

The goal of finding ways to modify set-points is not as remote as it may seem at first glance. In laboratory experiments it has been shown that lesions of the lateral hypothalamus (Keesey and Corbett 1989) or the administration of certain toxins (Peterson et al. 1984) produce a chronic reduction in the body weight set-point. Earlier in this chapter, data were presented indicating that nicotine's effects on body weight are apparently achieved by a resetting of the level at which the treated rats regulate body weight. While such procedures and/or agents are clearly not viable solutions to treating obese individuals, they do indicate the feasibility of an approach predicated upon resetting the weight level maintained by the regulatory system.

Indeed, there are indications that certain drugs used in the treatment of obesity (e.g., fenfluramine) may work by reducing the body weight set-point (Stunkard 1982). This possibly takes on additional significance in light of recent findings (Stock 1997) that sibutramine, a noradrenaline and 5-hydroxytryptamine reuptake inhibitor, causes weight loss in laboratory rodents by concurrently inhibiting food intake and enhancing energy expenditure. While further research is needed before concluding that sibutramine is a bona-fide set-point reducing drug, these initial results are consistent with such an interpretation. In any event, approaches which have a resetting of the system for energy regulation as their specific goal would seem to hold the greatest promise for the development of effective means for the treatment of the major medical problem obesity presents.

References

Ahima RS, Prabakaran D, Mantzoros C, Qu D, Lowell B, Maratos-Flier E, Flier JS (1996) Role of leptin in the neuroendocrine response to fasting. Nature 382:250–252

Cannon WB (1932) The wisdom of the body. New York, WW Norton

Caro JF, Sinha MK, Kolaczynski JS, Zhang PL, Considine RV (1996) Leptin: The tale of an obesity gene. Diabetes 45:1455–1462

Corbett SW, Wilterdink EJ, Keesey RE (1985) Resting oxygen consumption in over- and underfed rats with lateral hypothalamic lesions. Physiol Behav 35:971–977

Erickson JC, Hollopeter G, Palmiter RD (1996) Attenuation of the obesity syndrome of ob/ob mice by the loss of neuropeptide Y. Science 274:1704–1707

Flier JS, Maratos-Flier E (1998) Obesity and the hypothalamus: Novel peptides for new pathways. Cell 92:437–440

Hirsch J, Fried SK, Edens NK, Leibel RL (1989) The fat cell. Med Clin North Am 73(1):83–96

Keesey RE, Corbett SW (1989) Body weight regulation, thermogenesis, and the lateral hypothalamus. In: Lardy H, Stratman F (eds) Hormones, thermogenesis, and obesity. New York, Elsevier, 219–232

Kleiber M (1975) The fire of life: an introduction to animal energetics. New York: Robert E. Krieger Co

Leibel RL, Rosenbaum M, Hirsch J (1995) Changes in energy expenditure resulting from altered body weight. N Engl J Med 332:621–628

Mitchel JS, Keesey RE (1977) Defense of a lowered weight maintenance level by lateral hypothalamically lesioned rats: Evidence from a restriction-refeeding regimen. Physiol Behav 18:1121–1125

Peterson RE, Seefeld MD, Christian BS, Potter CL, Kelling CK, Keesey RE (1984) The wasting syndrome in 2,3,7,8-tetrachlorodibenzo-p-dioxin toxicity: basic features and their interpretation. In: Poland A, Kimbrough R (eds) Banbury report 18: biological mechanisms of dioxin action. Cold Spring Harbor Laboratory, 291–307

Ravussin E, Zurlo R, Ferraro R, Bogardus C (1990) Energy expenditure in man: Determinants and risk factors for body weight gain. In: Oomura Y et al (eds) Progress in obesity research. London, John Libbey, LTD
Schwartz MW, Seeley RJ, Campfield LA, Burn P, Baskin DG (1996) Identification of targets of leptin action in the rat hypothalamus. J Clin Invest 98(5):1101–1106
Schwid SR, Hirvonen MD, Keesey RE (1992) Nicotine effects on body weight: a regulatory perspective. Am J of Clin Nutr 55:878–884
Stock MJ (1997) Sibutramine: a review of the pharmacology of a novel anti-obesity agent. Int J Obes 21 [Suppl 1]:S25–S29
Stunkard AJ (1982) Anorectic agents lower a body weight set-point. Life Sci 30:2043–2055

Section II
Pharmacological and Other Treatments

CHAPTER 7
Serotonin Drugs and the Treatment of Obesity

J.C.G. HALFORD and J.E. BLUNDELL

A. Introduction

The monoamine neurotransmitter, serotonin (5-HT, 5-hydroxytryptamine) was first linked to the control of food intake, and of feeding behavior, more than 20 years ago. Early studies showed that increasing CNS 5-HT levels by using 5-HT precursors such as tryptophan and 5-hydroxytryptophan (5-HTP) produced a significant reduction in the food intake of laboratory animals. Increasing 5-HT activity by other mechanisms such as directly administering 5-HT into the CNS, blocking synaptic 5-HT breakdown or directly stimulating (agonising) 5-HT receptors also produced this hypophagic response (BRAY and YORK 1972; JESPERSON and SCHEEL-KRUGER 1973; BARRETT and MCSHERRY 1975; PINDER 1975; GARATTINI and SAMANI, 1976). Moreover, it was noted that neurotoxic lessioning of 5-HT neurons (with 5,7-dihyroxytrytamine, 5,7-DHT), or preventing 5-HT synthesis (using para-chlorophenylanine, pCPA) which depletes neuronal 5-HT, not only prevented 5-HT induced hypophagia but increased food intake (MACKENZIE et al. 1979).

Even with the relatively non-specific pharmacological tools available to early researchers, it became apparent that CNS 5-HT was involved in controlling food intake. This evidence was reviewed in the first volume of the International Journal of Obesity (BLUNDELL 1977). Blundell proposed that the 5-HT system not only had an inhibitory role in feeding, but was also a key satiety factor (part of a natural energy intake control mechanism). As 5-HT reduced food intake and subsequently reduced body mass in animals, 5-HT manipulations could potentially be used to reduce human food intake and body weight. Thus, understanding the role of 5-HT in the regulation food intake could have implications for the study of the etiology and treatment of obesity.

B. 5-HT Synthesis, Storage, Release, and Re-Uptake

Neuronal 5-HT is synthesized from the essential amino acid tryptophan. In the cytoplasm of the cell body dietary l-tryptophan is hydoxylated to 5-hydroxytryptophan (5-HTP) by the enzyme tryptophan hydroxylase. Since this enzyme is normally not saturated with substrate, the availability of the essential amino acid tryptophan in the diet is generally regarded as the rate-limiting step in serotonin biosynthesis. Increased neuronal firing of the 5-HT neuron may stimulate tryptophan hydroxylase activity, to make available more 5-HTP. The 5-HTP is then transported down the axon to the nerve terminal, where the final step of 5-HT synthesis generally occurs. At the terminal 5-HTP is rapidly decarboxylated by the enzyme L-amino acid decarboxylase to produce 5-HT. This rapid conversion generally means there are normally relatively low concentrations of 5-HTP in 5-HT neurons unless this metabolic step is experimentally blocked. Most of the 5-HT produced is stored in pre-synaptic vesicles (taken up via a vesicle membrane transport). 5-HT is released into the synaptic cleft in response to the arriving action potential (calcium dependent). Synaptic 5-HT continues to stimulate pre- and post-synaptic receptors until it is either reabsorbed into the pre-synaptic neuron for re-use (sodium dependent), or is converted to 5-hydoxyindole acetic acid (5-HIAA) by monoamine oxidase (MAO).

C. 5-HT Receptors

Even the most comprehensive review of 5-HT receptors at the time of writing will most probably be out of date within 18 months. Advances in the discovery and identification of novel 5-HT receptors has recently taken place. The functional significance of many of these new 5-HT receptors remains to be determined. GADDUM and PICARELLI (1957) first divided 5-HT receptors into D and M (which now correspond to 5-HT_2 and 5-HT_3). PEROUTKA and SNYDER (1978) identified two distinct 5-HT receptors, 5-HT_1 and 5-HT_2. This gave three 5-HT receptors – 5-HT_1, 5-HT_2, and 5-HT_3 (BRADLEY NOMENCLATURE COMMITTEE 1986). Since this time cloning and radioligand techniques have allowed the further subdivision of 5-HT receptor into 14 distinct subtypes (see HOYER and MARTIN (1997) for review). These subtypes are 5-HT_{1A}, 5-$HT_{1B(r\ \mathrm{and}\ h)}$, 5-$HT_{1D}$,[1] 5-$ht_{1E}$,[2] 5-$ht_{1F}$, 5-$HT_{2A}$, 5-$HT_{2B}$, 5-$HT_{2C}$,[3] 5-$HT_3$, 5-$HT_4$, 5-$ht_{5\alpha}$, 5-$ht_{5\beta}$, 5-$HT_6$, and 5-$HT_7$. At the present time the 5-HT receptors most directly implicated in feeding control are 5-HT_{1A}, 5-HT_{1B}, and 5-HT_{2C}. Post-synaptic 5-HT_{1B} and the 5-HT_{2C} receptors are generally believed to be involved in the 5-HT satiety system (Sect. D.III). 5-HT_{1A} receptors are believed to function as autoreceptors and so their activation is generally believed to reverse 5-HT induce hypophagia possibly by limiting both the synthesis and release of 5-HT (Sect. D.III). It is possible that some of the more recently discovered 5-HT receptors (5-ht_{1E}, 5-ht_{1F}, 5-$ht_{5\alpha}$, 5-$ht_{5\beta}$, 5-HT_6, and 5-HT_7) could have a role in the control of food intake. This research will require

[1] r and h, rodent and human.
[2] ht, functional significance remains to be determined.
[3] Previously termed 5-HT_{1C}.

the development of specific pharmacological tools to selectively stimulate or block these receptors.

D. 5-HT and Food Intake (Animal Studies)

5-HT induced suppression of food intake has classically been investigated using general 5-HT manipulations (i.e. drugs which promote the release of 5-HT and/or prevent its re-uptake). These are then opposed by selective antagonists of various 5-HT receptor subtypes to ascertain the precise mode of drug-induced hypophagia. Such antagonists include metergoline (5-$HT_{1/2}$), methysergide (5-$HT_{1/2C}$), mianserin (5-HT_2), ritanserin (5-HT_2), ketanserin (5-HT_2), (±)cyanopindol (5-$HT_{1\,A/1B}$), SB 200646 (5-$HT_{2B/2C}$), and GR 127935 (5-$HT_{1B/1D}$). More selective antagonists are now becoming available. It is worth noting that such antagonist blocking studies can only provided information on the mechanism but not the nature of drug induced hypophagia (i.e. whether it is consistent with the natural operation of satiety).

I. Fenfluramine

Some of the earliest experiments studying the role of 5-HT in food intake used the 5-HT activating drug fenfluramine (5-HT releaser and re-uptake inhibitor). When considering the action of fenfluramine it is important to draw the distinction between more recent experiments which use D-fenfluramine, the most 5-HT specific fenfluramine isomer, and earlier studies which use DL-fenfluramine, a mixture of both isomers containing the less potent and less 5-HT specific L-fenfluramine. The use of selective 5-HT receptor antagonists of various 5-HT_1 and 5-HT_2 receptor subtypes has indicated D-fenfluramine-induced hypophagia is mediated by 5-HT_{1B} receptors (CLIFTON 1994 for review). This was based on the evidence of a number of studies which blocked D-fenfluramine actions with antagonists of 5-$HT_{1\,A/1B}$ receptors, but failed to block D-fenfluramine action with antagonists of 5-$HT_{2\,A/2C}$ receptors (NEILL and COOPER 1989; SAMANIN et al. 1989; NEILL et al. 1990). However, not all experimental evidence confirms this central role of 5-HT_{1B} receptors. More recently, DOURISH (1995) has successfully blocked D-fenfluramine induced hypophagia with a selective 5-$HT_{2B/2C}$ antagonist SB 200646 A. In addition, a selective 5-$HT_{1B/1D}$ antagonist, GR 127935, failed to block DL-fenfluramine action on food intake (DOURISH 1995).

The precise pharmacological mechanism of D-fenfluramine induced hypophagia still needs to be determined and may involve direct agonist action at some receptors (Curzon et al. 1997). If D-fenfluramine acts solely by promoting the release, and preventing re-uptake of 5-HT, when neuronal 5-HT is depleted (using pCPA) D-fenfluramine-induced hypophagia should be attenuated. However, GIBSON et al. (1993) demonstrated D-fenfluramine still potently reduced food intake in the absence of 5-HT. An in vivo dialysis study by the same group confirmed that a rise in 5-HT concentration (produced by release and/or by re-uptake inhibition) was not needed to produce D-fenfluramine hypophagia (OLUYOMI et al. 1994). RAITERI et al. (1995) also found no relationship between D-fenfluramine induced 5-HT release and its hypophagic effects. Therefore, it is possible that D-fenfluramine, and its active metabolite D-norfenfluramine, directly agonise 5-HT feeding receptors. CURZON et al. (1997) suggest that D-fenfluramine may induce hypophagia by directly acting on 5-HT_{2C} and other unspecified 5-HT receptors.

II. Fluoxetine

The selective serotoninergic re-uptake inhibitor (SSRI) fluoxetine produces a reliable reduction in food intake which is not easily blocked by 5-HT antagonists (WONG et al. 1988; GRIGNASCHI and SAMANIN 1992; LIGHTOWLER et al. 1996). CACCIA et al. (1992) noted that brain concentrations of fluoxetine, and its active metabolite norfluoxetine, required to reduce food intake were several times higher than required to block 5-HT re-uptake. The dose required to produced hypophagia was equivalent to that which blocks the re-uptake of dopamine and nor-adrenaline, and promotes dopamine release (CACCIA et al. 1992). This evidence suggests fluoxetine-induced hypophagia is not solely dependent on its 5-HT activity but on general monoamine activity. However, in contrast to this proposition fluoxetine induced hypophagia has been partially, and fully blocked by the $5\text{-HT}_{1/2}$ antagonist metergoline (LEE and CLIFTON 1992; HALFORD and BLUNDELL 1996a – respectively). Both these studies used lower doses of fluoxetine than previous blocking studies. Using differing behavioral analysis techniques, both studies also demonstrated that metergoline did not reverse all fluoxetine induced changes to feeding behavior (LEE and CLIFTON 1992: HALFORD and BLUNDELL 1996a). At higher doses fluoxetine may block food intake by non-5-HT mechanisms which induce changes in feeding behavior not related to satiety (HALFORD and BLUNDELL 1996a). Pharmacological evidence suggests that fluoxetine may act as a direct 5-HT_{2C} receptor antagonist (LIGHTOWLER et al. 1994). If this is so it would suggest 5-HT_{1B} receptors mediated the fluoxetine induced hypophagia observed by LEE and CLIFTON (1992) and HALFORD and BLUNDELL (1996a).

Blocking the synthesis of 5-HT (using pCPA) does not attenuate fluoxetine induced hypophagia (LIGHTOWLER et al. 1996). This study suggests that like D-fenfluramine, fluoxetine-induced hypophagia, at the doses used, does not appear to be related to increases in synaptic 5-HT. In the same study LIGHTOWLER et al. (1996) also failed to block fluoxetine induced hypophagia with metergoline ($5\text{-HT}_{1/2}$ antagonist) therefore fluoxetine induced anorexia did not appear dependent on the direct agonism 5-HT_{1B} and 5-HT_{2C} feeding receptors. CURZON et al. (1997) suggested that fluoxetine reduces food intake independently of direct action of 5-HT. Again, the precise mechanism by which fluoxetine reduces food intake still needs to be determined. The fact the fluoxetine is pharmacologically active as a 5-HT re-uptake inhibitor at a dose much lower than that needed to induced hypophagia demonstrates this point (CACCIA et al. 1992).

In contrast to fluoxetine, the hypophagia induced by another SSRI sertraline (KOE et al. 1983) appears to be more reliably blocked by 5-HT antagonists acting on 5-HT_{1B} and 5-HT_{2C} receptors, and by 5,7-DHT lesions (LUCKI et al. 1988). This initial study supports the 5-HT hypothesis. Whether the mechanism of action is solely re-uptake inhibition or whether direct activation of post-synaptic 5-HT feeding receptors may contribute to sertraline's effects remain to be determined. Further pre-clinical data on the nature of sertraline-induced hypophagia, comparable to that we currently possess for D-fenfluramine and fluoxetine, are not as yet available.

III. Selective 5-HT Agonists

Direct agonists of specific 5-HT receptors have allowed further research into the role of 5-HT receptors in the control food intake. However, most selective 5-HT agonists utilized in food intake studies have been active at two or more differing 5-HT receptors. For example, mCPP and TFMPP have affinity for both 5-HT_{1B} and 5-HT_{2C} receptors, RU-24969 has an affinity for both 5-HT_{1A} and 5-HT_{1B} receptors, and MK-212 and DOI have affinity for all 5-HT_2 receptors. However, all these drugs directly activate the 5-HT_{1B} and/or 5-HT_{2C} receptors linked to satiety. Agonists of single 5-HT receptors are only now becoming available. For example the highly selective 5-HT_{1B} agonist CP-94,253 (KOE et al. 1992) has been shown to reduce food intake potently in the rat (KOE et al. 1992; HALFORD and BLUNDELL 1996b). RU-24969 ($5\text{-HT}_{1A/1B}$ antagonist) also potently reduces food intake (KENNETT et al. 1987). As action on the 5-HT_{1A} auto-receptor should not induce hypophagia, and may even produce hyperphagia, RU-24646-induced hypophagia should be mediated by the 5-HT_{1B} receptor (KENNETT and CURZON 1988a; CURZON 1990). These studies suggest direct agonism of the 5-HT_{1B} receptor alone is sufficient to induce hypophagia.

The hypophagia induced by TFMPP and mCPP ($5\text{-HT}_{1B/2C}$ agonists) is blocked by selective 5-HT_2 receptors antagonists, but the expression of this effect may depend on the activation of 5-HT_{1B} receptors (KENNETT and CURZON 1988a,b, 1991; CURZON 1990). The $5\text{-HT}_{2C/2B}$ antagonist SB 200646 A, and the novel 5-HT_{2C} receptor antagonist SB 242084 have both been shown to block mCPP-induced anorexia, suggesting mCPP activation of the 5-HT_{2C} receptor alone is sufficient to induce hypophagia (KENNETT et al. 1994, 1997). Selective $5\text{-HT}_{2A/2B/2C}$ agonists such as MK-212, DOI, and DOM also potently reduce food intake (CLINESCHIMDT et al. 1978; KING et al. 1989; AULAKH et al. 1994, 1995). However, antagonist blocking and cross tolerance studies suggest that only MK-212 induced hypophagia is mediated by the 5-HT_{2C} receptor (KING et al. 1989; AULAKH et al. 1994, 1995). It is also worth noting that transgenic mice, which lack functional 5-HT_{2C} receptors show both hyperphagia and obesity (TECOTT et al. 1995).

E. 5-HT and Feeding Behavior (Animal Studies)

It is important to distinguish between drugs which reduce food by acting on the natural satiety mechanisms and those which reduce food intake by inducing nausea, sedation, hyperactivity, or malaise. In both animal and human studies the structure of feeding behavior has been used to diagnose the nature of drug-induced hypophagia. Early pre-clinical studies comparing fenfluramine with amphetamine showed both drugs produced similar effects of food intake but distinct effects on feeding behavior (BLUNDELL and LATHAM 1979). One of the most detailed means of studying the feeding behavior is the Behavioral Satiety Sequence (BSS) (ANTIN et al. 1975; REVIEW Halford et al. 1998). The BSS is a specific stochastic sequence of behavior which rats display after a meal (eating → activity → grooming → resting). Pre-feeding, which enhances the natural process of satiety, advances the elements but maintains the structure of the BSS. Thus, hypophagic drugs which preserve the BSS are enhancing satiety. Hypophagia due to sedation, nausea, hyperactivity etc. disrupts the structure of the BSS (HALFORD and BLUNDELL 1998; HALFORD et al. 1998).

Many studies have shown that 5-HT releasing and re-uptake inhibiting drugs produce changes in feeding behavior, as measured by the BSS or other behavioral assays, which are consistent with the operation of satiety. Fenfluramine and D-fenfluramine has been shown to adjust feeding behavior and the BSS in a manner consistent with the operation of satiety (BLUNDELL and LATHAM 1978, 1980; BLUNDELL and MCARTHUR 1981; HALFORD and BLUNDELL 1993). Fluoxetine, sertraline and other SSRIs have been shown to produce similar effects (CLIFTON et al. 1989; HALFORD and BLUNDELL 1993; SIMANSKY and VAIDYA 1990; MCGUIRK et al. 1992). The novel anti-obesity compound sibutramine (5-HT and noradrenaline re-uptake inhibitor – SNRI) also enhances the BSS (HALFORD et al. 1995). Of the drugs which are selective agonists of 5-HT receptors the 5-HT_{1B} agonist CP –94253 and the 5-$HT_{1B/2C}$ agonists mCPP and TFMPP produce changes in feeding behavior consistent with the operation of satiety (HALFORD and BLUNDELL 1996b; SIMANSKY and VAIDYA 1990; KITCHENER and DOURISH 1994). Other agonists such as RU-24969 (5-$HT_{1A/1B}$), MK-212 (5-HT_2) and DOI (5-HT_2) disrupt feeding behavior by inducing either hyper-locomotion or sedation (KITCHENER and DOURISH 1994; HALFORD and BLUNDELL 1996b; SIMANSKY and VAIDYA 1990). Direct agonism of 5-HT receptors, other than 5-HT_{1B} and 5-HT_{2C}, may lead to disruption of feeding behavior (DOI and MK-212 disruption due to 5-HT_{2A} action). Future antagonist blocking studies may determine the pharmacological mechanisms of this disruption and its contribution to the hypophagia produced by these drugs.

F. Hypothalamic 5-HT

Early studies of the role of hypothalamic 5-HT found that serotoninergic agents injected into the medial hypothalamus potently reduced food intake

(LEIBOWITZ 1980). A key site appears to be the para ventricular nucleus (PVN) (LEIBOWITZ et al.1981; LEIBOWITZ and SHOR-POSNER 1986). 5-HT, fluoxetine, and D-norfenfluramine injected into the PVN produced a potent selective reduction in carbohydrate intake (LEIBOWITZ et al. 1989, 1990, 1992). The 5-HT antagonist metergoline produced the opposite effects (increased feeding and specific increases in carbohydrate consumption). It is not clear whether changes in 5-HT within the PVN (HUSTON et al. 1988; MACOR et al. 1990) or adjacent areas mediate 5-HT induced hypophagia (FLETCHER and COSCINA 1993; COSCINA et al. 1994; CURRIE and COSCINA 1996). Neuropeptide Y (NPY) is also associated with the PVN. NPY is a potent stimulator of food intake, particularly carbohydrate intake, and appear to oppose the action of 5-HT. 5-HT drugs (fluoxetine, mCPP) block increases in feeding induced by NPY (GRIGNASCHI et al. 1995; DRYDEN et al. 1996a,b). Blocking 5-HT synthesis (pCPA), antagonizing other 5-HT receptors, or stimulating the 5-HT_{1A} autoreceptor results in increases in the level of NPY in the PVN (DRYDEN et al. 1995, 1996b,c). This has led to the suggestion that 5-HT, and 5-HT drugs, may at least partly inhibit food intake by inhibition NPY synthesis (DRYDEN and WILLIAMS 1996).

G. 5-HT Diet Composition and Food Choice

Studies of hypothalamic 5-HT suggest that it is involved in systems regulating macronutrient intake. A key issue in the treatment of obesity is the possibility of altering food choice. Consumption of energy dense foods leads to positive energy balance and weight gain. As fat contributes the most to energy density (double the kcals per gram than carbohydrate and protein) consumption of a high fat diet predisposes to weight gain (POPPITT and PRENTICE 1996; MACDIARAMID et al. 1997). Evidence also exists that excess consumption of carbohydrates also leads to significant energy storage (HORTON et al. 1995). Therefore the development of an anti-obesity agent which specifically reduces the intake of foods that promote positive energy balance (high in fat and refined sugars) would be an ideal tool in a weight control treatment regime.

It has been proposed that the 5-HT mechanism functions as a sensor the of plasma amino acid ratio (FERNSTORM and WURTMAN 1973; WURTMAN and FERNSTORM 1974). This was based on the observation of a relationship between the T:LNAA (tryptophan: large neutral amino acids) plasma ratio and the relative proportions of dietary carbohydrate and protein consumed. Increased carbohydrate intake raised the T:LNAA ratio (TEFF et al. 1989). The resulting increase in available tryptophan increased 5-HT synthesis and release. Therefore 5-HT function was directly linked to the nutritional status of the body, and consequently altered subsequent feeding behavior to maintain energy balance. The relative strength of this system has been questioned (BLUNDELL and HILL 1987). However, along with a 5-HT-vagal link with the peripheral satiety factor CCK (sensitive to fat intake), the T:LNAA ratio is one theoretical way in which macronutrient consumption could modulate the direction of feeding behavior.

Evidence indicates carbohydrate and protein action on the T: LNAA ratio is altered in patients with anorexia nervosa (PIRKE et al. 1986) and obesity (ASHLEY et al. 1985). Glucose pre-loading does not correct this abnormality. BREWERTON (1995) reviewed evidence of general 5-HT dys-regulation in subjects with eating and weight disorders (specifically those suffering from Bulimia Nervosa). BREWERTON noted some clinical features of eating disorders, such as feeding disturbance, depression and impulsivity, all indicated abnormal or reduced 5-HT function. Moreover, psychobiological stressors such as binge eating (see later) and/or dieting 'may perturb and interact with a vulnerable 5-HT system.' Additionally, JIMERSON et al. (1997) suggested individual differences in 5-HT functioning may be a dispositional risk factor for abnormal eating behavior. Thus, aberrations of normal nutritional status could have a potent effect on 5-HT functioning, or vice versa. Indeed 5-HT levels, especially in women (ANDERSON et al. 1990; WALSH et al. 1995), do appear to be sensitive to dieting. The prolactin response to tryptophan is elevated in women following 3 weeks adherence to a low calorie diet (LCD) (1000 kcal/day) (GOODWIN et al. 1987). The LCD also leads to a fall plasma tryptophan, decreasing the T: LNAA ratio (GOODWIN et al. 1990). Prolactin response to D-fenfluramine or mCPP (5-HT_{2C} agonist) in healthy female subjects showed that moderate dieting also causes 5-HT_{2C} receptor supersensitivity (WALSH et al. 1995; COWEN et al. 1996). This evidence suggests a sensitive connection between nutritional status and 5-HT metabolism.

I. Diet Selection in Animal Studies

The relationship between 5-HT and nutrient intake is conceptualized as a system in which the post-absorptive action of carbohydrate acts on CNS 5-HT synthesis via tryptophan uptake into brain. This system has provided a basis for numerous studies on the action of acute and chronic doses of 5-HT drugs on macronutrient selection in animals. In these studies animals are presented with a choice of differing diets varying in macronutrient composition (diet selection paradigms). These studies are fraught with methodological problems which in turn result in a lack of definitive results (BLUNDELL 1993). In two diet choice paradigms the proportion of protein and carbohydrate are varied, but fat levels are held constant, precluding the study of 5-HT effects on fat intake. Some studies showed selective reduction in carbohydrate intake (LUO and LI 1990). Other studies showed 5-HT drugs reduced the intake of diets except those high in carbohydrate (LAWTON and BLUNDELL 1992, 1993). Two diet choice selective paradigms demonstrate 5-HT suppression of intake is not dependent on the food containing large amounts of carbohydrate. Additionally, low carbohydrate diets do not prevent 5-HT induced hypophagia. Three choice diet selection paradigms consist of diets containing largely one macronutrient, or alternatively pure macronutrients, allowing a clearer identification of 5-HT macronutrient specific effects (KANARAK 1987). The injection of 5-HT into the PVN reduces both carbohydrate and fat (LEIBOWITZ and SHOR-POSNER 1986).

Some chronic 5-HT studies use the phenomenon of dietary induced obesity (SCLAFANI 1984) based on procedures such as cafeteria feeding. In the cafeteria diet paradigm animals are offered a choice between a standard laboratory chow control or the laboratory chow together with one or two highly palatable composite foods (ROGERS and BLUNDELL 1984). Long-term exposure to the cafeteria choice leads to dietary induced obesity. An alternative paradigm involves laboratory chow supplements with an other highly preferred foods (fat or sugar). BLUNDELL and HILL (1988) noted that supplemented diets increased daily intake, meal size, and induce obesity. In this paradigm rats chronically treated with D-fenfluramine and offered a fat supplemented diet do not display the dietary induced obesity seen in the non-drug control animals. D-Fenfluramine reduced body weight and the consumption of both supplemented and standard diets (BLUNDELL and HILL 1984; PRATS et al. 1989). High fat diets (50%–60% fat) can also be used to induce obesity in rats (compared with a 4%–5% low fat control). In Osborne-Mendel rats, which normally feed readily on high fat diets and gain weight, chronic treatment with D-fenfluramine reduced food intake and prevented weight gain with no sign of drug tolerance (FISLER et al. 1993). Therefore, exposure to high fat diets does not weaken the hypophagic effects of D-fenfluramine (BLUNDELL and HILL 1989). D-Fenfluramine reduces the consumption of high fat diets which could be of therapeutic value in obesity treatment if demonstrated in humans (BLUNDELL et al. 1995).

II. Food Choice in Humans

As with animal studies, human research is affected by methodological issues (BLUNDELL and LAWTON 1995). Human studies also initially tested protein and carbohydrate differences, in which fat intake (which was high in both options) was held constant (WURTMAN and WURTMAN 1977a,b). 5-HT manipulation reduced carbohydrate intake to a greater degree than protein intake but any reduction in either choice involved an obligatory reduction in fat intake (WURTMAN et al. 1985). Therefore, 5-HT drugs could be considered to reduce meals high in carbohydrate and fat selectively. In free feeding studies, the consumption of high protein, not high carbohydrate snacks were reduced by 5-HT manipulation (WURTMAN et al. 1982). Again the high protein snacks had a high fat content. It should be noted that D-fenfluramine's most dramatic effect on human feeding behavior is a reduction in snacking. This also involves an obligatory, reduction in fat intake.

Do 5-HT drugs directly and selectively reduce fat intake? Free selection designs incorporating foods differing in all macronutrients have been employed to investigate this. BLUNDELL and HILL (1986) found that D-fenfluramine hypophagia in humans was not macronutrient specific. In another short term study, fluoxetine did not give rise to a selective macronutrient effect (LAWTON et al. 1995). However, other studies found D-fenfluramine, fluoxetine, and the novel 5-$HT_{1B/1D}$ agonist sumatriptan, selectively reduced fat intake (GOODALL et al. 1993; MCGUIRK and SILVERSTONE 1990). Additionally, a clinical study by LAFRENIERE et al. (1993) found that chronic administration of D-fenfluramine selectively reduced fat intake. At the end of 3 months the energy intake of the D-fenfluramine group was 16% lower than the placebo control, a 13% reduction in energy from meal intake and a 23% reduction in energy from snacks. The energy reduction was characterized by a decrease in the percent energy as fat from 34% to 30%, a reduction of about 25% in total fat consumption. Similarly, PRENTICE et al. (1997) noted D-fenfluramine produced the greatest effect on energy balance in a high fat diet. GREEN et al. (1997) noted that D-fenfluramine induced the largest reduction in daily energy intake in obese females offered high fat sweet snacks. As obese women show a preference for sweet high fat foods (DREWNOWSKI et al. 1992), this may have important implications for treating obesity.

H. 5-HT and Appetite Motivation in Humans

In animals, activation of post synaptic 5-HT receptors (r5-HT_{1B} and 5-HT_{2C}) have been shown to reduce potently food intake in a manner consistent with satiety. In humans, 5-HT drug action on food intake, feeding behavior (meals and snacks), and on subjective ratings of appetite such as hunger also produce consistent results. Most human studies have used racemic fenfluramine (D and L isomers), D-fenfluramine and fluoxetine.

I. Hunger

Subjective feelings of hunger are conceptualized as the conscious expression of appetite which is believed to drive feeding behavior. Therefore the action of 5-HT drugs on subjective ratings of hunger and fullness is great importance. D-Fenfluramine has been shown to reduce hunger both before and after a test meal (BLUNDELL and HILL 1986; GOODALL and SILVERSTONE 1988). In the obese, both D-fenfluramine and a glucose preload produced similar reductions in hunger possibly acting on the same underlying mechanisms of satiety (BLUNDELL and HILL 1988). Tracking daily hunger ratings across the day, D-fenfluramine maintained suppression of post-prandial hunger (HILL and BLUNDELL 1990) indicative of enhanced post meal satiety. Therefore, the action of 5-HT drugs would to be to intensify the satiating power (measured by subjective ratings of hunger, fullness, and prospective consumption) of specific foods (BLUNDELL and HILL 1989). Following chronic administration D-fenfluramine increased post meal satiety, and also reduced hunger at specific points during the diurnal cycle (HILL et al. 1995). D-Fenfluramine also reduced the frequency and potency of urges to eat. In obese females D-fenfluramine reduced the willingness to initiate (reduced hunger), and maintain (increase within-meal satiation) eating episodes (HILL et al. 1995).

II. Food Intake

Again most data on 5-HT drug effects on food intake is based on racemic fenfluramine, D-fenfluramine, and fluoxetine (believed to act pre-synaptically to release 5-HT and block its re-uptake but see Sect. D.I). We have previously noted the effect of these drugs on the intake of foods with specific macronutrient content (Sect. G.II). There is also evidence for the effect of 5-HT precursors tryptophan (BLUNDELL and HILL 1988) and 5-HTP (CANGIANO et al. 1992), and of the selective 5-HT_{2C} agonist mCPP (COWEN et al. 1996) and 5-$HT_{1B/1D}$ agonist sumatriptan (BOELES et al. 1997). Additionally, selective 5-HT antagonists (metergoline and ritanserin) have been used to determine the pharmacological mechanism responsible for 5-HT drug induced hypophagia in humans (GOODALL and SILVERSTONE 1988; GOODALL et al. 1993). For D-fenfluramine the extent of observed energy reduction in various studies ranges from 11% to 40%. This variance is dependent on a variety of clinical or experimental differences between studies. However, significant reductions in food intake are apparent in D-fenfluramine treatments varying in length from a day to six months.

5-HT drugs restrain the motivation to eat and food consumption is maintained for considerable periods of time. After a 12-month trial of D-fenfluramine, withdrawal led to an immediate rise in daily energy consumed (GUY-GRAND et al. 1990). D-Fenfluramine had maintained a strong influence on food intake for the entire 12 months, although weight loss stopped after 6 months. D-Fenfluramine reduced weight (and reported hunger) to a point of physiological resistance at which equilibrium between hunger urges and drug anorectic activity had been reached. The rebound in hunger after drug withdrawal (at 12 months) demonstrates a lack of tolerance to the hypophagic effects of D-fenfluramine. D-Fenfluramine, despite the fact that most reduction in food intake was in the first 6 months, maintained an energy intake reduction of 6% to 10% over the course of the year.

Another demonstration of the potency of 5-HT drugs to inhibit human food intake is their use in combination with or after a very low calorie diet (VLCD). The VLCD reduced weight in the obese by 14 kg in eight weeks (FINER et al. 1992) which normally creates a strong disposition to resume overeating and gain the weight lost. However, patients given D-fenfluramine continued to lose weight. D-Fenfluramine overcame the physiological and psychological drive to eat following substantial weight reduction. However, in a similar study, sertraline (SSRI) failed to block weight regain after termination of a VLCD (WADDEN et al. 1995).

Direct agonism of 5-HT receptors also potently reduces food intake. mCPP (5-$HT_{1B/2C}$ agonist), which was shown to reduce food intake in the rat via 5-HT_{2C} receptors, also reduces food intake in humans (WALSH et al. 1994), and enhances plasma prolactin concentrations following dieting (COWEN et al. 1996). As dieting affects plasma tryptophan and 5-HT sensitivity in the brain (ANDERSON et al. 1990; WALSH et al. 1994) it has been argued that the 5-HT_{2C} receptor may mediate appetite control in humans (COWEN et al. 1996). Ritanserin, the 5-HT_2 antagonist, blocks D-fenfluramine-induced hypophagia in humans which supports the pivotal role of 5-HT_{2C} in human appetite control (GOODALL et al. 1993). Up to recently no involvement of 5-HT_{1B} receptors in mCPP-induced anorexia was considered as there was no equivalent to the rodent 5-HT_{1B} receptor in the human. However, the human 5-$HT_{1D\beta}$ has been reclassified as a human 5-HT_{1B} receptor (HOYER and MARTIN 1997). Recently, a novel 5-$HT_{1B/1D}$ agonist, sumatriptan has been found to decrease food intake, and specifically high fat foods in healthy women (BOELES et al. 1997). The development of specific 5-HT_{2C}, human 5-HT_{1B} and human 5-$HT_{1B/2C}$ agonists would be an important strategy for the advancement of drug treatment in obesity. Indeed the combined action at 5-HT_{2C} and 5-HT_{1B} receptors should produce a potent suppression of food intake and hunger.

I. Binge Eating in the Obese

STUNKARD (1959) first noted binge eating was a distinct factor in the obese. Estimates of binge eating behavior in the obese vary depending on the criteria used. SPITZER et al. (1992) found 30.1% of obese subjects displayed binge eating behavior without purging, compared with the prevalence of 2% in normal weight subjects. Bingeing was more likely to occur in women than men. Bingeing behavior has been explained in terms of 5-HT dys-regulation (BREWERTON 1995; JIMERSON et al. 1997). Therefore drugs which increase CNS 5-HT levels and sustain satiety, should also counter the effects of 5-HT dysregulation and may prevent the disinhibition which leads to a binge eating episodes. D-Fenfluramine has been shown to normalize eating patterns of obese women suffering binge eating (STUNKARD et al. 1996). However, the benefits of fluoxetine were no different in obese women with or without bingeing behavior (GREENO and WING 1996). The lack of available data on the effects of 5-HT drugs on binge eating behavior in the obese makes it difficult to assess whether or not 5-HT drugs can specifically normalize aberrant eating behavior, and if so by what mechanism. However, estimation of the prevalence of binge eating in the obese means this behavior may be a key target for future pharmacological intervention.

J. 5-HT Genetics and Obesity

Twin and family studies of the heritability of adiposity in humans have demonstrated a genetic link to the development of obesity (BOUCHARD 1989, 1991; SØRESON et al. 1989; ALLISON et al. 1996). Differences in the key genes alter the functioning of mechanisms controlling food intake and energy balance could be an inherited predisposing factor to becoming obese. Distinct, human populations such as the Pima Indians of North America appear to have a strong genetic disposition to become obese when they come in contact with energy dense high fat foodstuffs (RAVUSSIN et al. 1994). Additionally, the breeding of genetically obese strains of animals (such as the *ob/ob* or *db/db* mice) demonstrate the potent effects of single genes on body mass and the metabolism of energy. Such animals when placed in an environment with a plentiful supply of palatable food quick become vastly overweight. Consequently, with the advancement of techniques able to isolate specific genes, much research in locating genes responsible for causing obesity has been done. An example of this was the discovery of a gene responsible for the production of the adipose tissue feedback hormone termed leptin or *ob* protein (CAMPFIELD et al. 1995; HALAAS et al. 1995; PELLEYMOUNTER et al. 1995).

TECOTT et al. (1995) demonstrated that transgenic mice lacking functional 5-HT_{2C} receptors overate and quickly became obese. Thus, a genetic alteration of the 5-HT system could produce obesity. However, no difference in the expression of the mutant alleles of a gene controlling the 5-HT_{2C} receptor were observed between obese or normal weight adolescent humans (LENTUS et al. 1997).

Mutations in this gene did not appear to have any role in the etiology of human obesity. Its seems unlikely that the human obese phenotype is controlled solely by one gene. The occurrence of and variation in the obese phenotype within the human population would suggest instead that the genetic contribution to human obesity is polymorphic. As with the Pima Indians, the occurrence of obesity is also dependent on environment factors, specifically food quality, and availability. It is possible that differences in some of the genes controlling various aspects of the 5-HT system may contribute with others factors, to the development of obesity. These genes may control presynaptic neuronal 5-HT metabolism, or some aspect of the functioning of the one or more of the specific 5-HT feeding receptors (5-HT_{2C}, 5-HT_{1B} or 5-HT_{1A}).

K. Serotonin and the Risk Factors for Overeating

Many humans, possibly a majority, demonstrate a vulnerability to gain weight. One reason for this is the apparent ease with which certain people develop a positive energy balance. The risk factors which provoke the occurrence of a positive energy balance reside within the biological system, in the environment (food supply itself), and at the interface between food and the biological system. Some of the most potent risk factors which generate unwanted overconsumption (energy intake rising above energy expenditure) have been described elsewhere (BLUNDELL and MACDIARMID 1997).

Research which has accumulated over the course of more than a decade indicates that serotoninergic drugs significantly reduce all of the major risk

Table 1. Serotoninergic drugs reduce the risk factors for overeating

Risk Factor	Effect of drug	Reference
Urge to eat	Strength of urge reduced	HILL et al. (1995)
Inter-meal eating motivation	Suppression of desire to eat between meals	BLUNDELL and HILL (1987)
Intensity of hunger	Decreased	GOODALL and SILVERSTONE (1988)
Snacking frequency	Reduced	HILL and BLUNDELL (1986)
High fat foods (acute)	Reduced	WURTMAN et al. (1985)
Dietary fat (long-term)	Selective reduction (25%)	LAFRENIERE et al. (1993)
Weak post-ingestive satiety	Increased satiating action of food	BLUNDELL and HILL (1989)
Carbohydrate craving	Reduction of carbohydrate (and fat) during "craving phases"	BREZINZKI et al. (1990)
Low mood and loss of control	Improved mood and increased feeling of control of eating	HILL et al. (1995)
Potent hedonic food qualities (high palatability and high energy density)	Suppressed intake of food with strong sensory attractiveness	WURTMAN et al. (1987)

factors for overeating (see Table 1). It follows that if 5-HT drugs can antagonize the risk factors then one consequence will be a reduced level of energy intake. The evidence shows that 5-HT drugs have the capacity to inhibit consumption even in the face of potent risk factors such as strong urges to eat, high levels of hunger, the presence of high-fat foods, and the presence of high-palatability foods with powerful hedonic qualities. Some urges to eat and weak satiety signals can be considered as psychobiological markers of overconsumption. Factors concerning the eating pattern (large meals, binges, or increases tendency to snack) can be considered as behavioral markers, and factors in the food supply such as high fat, high-palatability foods (stimulating sensory-nutrient combinations) can be considered environmental markers of overeating. It is therefore important that the action of drugs, mediated via serotoninergic processes, are effective against the psychobiological, behavioral, and environmental markers of over-consumption.

L. Overview

Pre-clinical rodent studies implicate 5-HT_{1B} and 5-HT_{2C} receptors in mediating 5-HT drug-induced reduction in food intake. Behavioral studies indicate this effect is consistent with the operation of post meal satiety. A consistent pattern of reduction in food intake and hunger induced by 5-HT drugs in humans is also seen, although its mechanism is less understood. Some 5-HT drugs can restrain appetite and food for a period of one year, or after a very low calorie diet demonstrating the potency of 5-HT action. Recent studies with selective 5-HT agonists have implicated human 5-HT_{1B} and 5-HT_{2C} receptors in the actions of 5-HT drugs.

Animal studies, despite methodological problems, indicate that certain 5-HT drugs such as D-fenfluramine can inhibit the consumption of high fat diets. Human studies also demonstrate D-fenfluramine and other 5-HT drugs suppress the eating of highly palatable fat foods, and may induce the selective avoidance of dietary fat, useful in targeting snacking behavior. It remains unclear whether 5-HT drugs can specifically reduce bingeing behavior which is a characteristic of certain but not all obese subjects. Despite the recent withdrawal of two major 5-HT drugs from the anti-obesity market, 5-HT drugs have useful role in the treatment of obesity by adjusting the biological mechanism involved in appetite to control the positive energy balance and weight gain. The development of safe and effective 5-HT selective receptor agonists (5-HT_{1B} and/or 5-HT_{2C}) may prove effective and precise pharmacological therapy for obesity.

Many people are vulnerable to weight gain because of a permissive physiological system in the presence of a potent and stimulating food supply in a provocative environment. This combination appears to lead easily to a positive energy balance. Certain risk factors for over-consumption have been identified. Evidence, both experimental and clinical, indicates that 5-HT drugs such as D-fenfluramine can effectively reduce the potency of these risk factors (urges to eat energy dense foods, high levels of dietary fat, hunger, large meals and frequent snacks, strong sensory attractiveness of foods, and loss of control and relatively weak satiety signals), and thereby lessen their impact on energy intake. Obese people find it extremely difficult to prevent themselves drifting into positive energy balance through general over-consumption (termed passive) or through active food seeking (such as binge eating). 5-HT drugs are clearly effective in inhibiting the intake of some of the most palatable and sensorily attractive foods available. Serotoninergic drugs can provide biological assistance to allow obese people to control those risk factors for weight gain operating via over-consumption.

References

Allison DB, Kaprio J, Korkeila M, Koskenvuo M, Neale MC, Hayakawa K (1996) The heritability of body mass index among an international sample of monozygotic twins reared apart. Int J Obesity 20:501–506

Anderson IM, Parry-Billins M, Newsholme EA, Fairburn CG, Cowen PJ (1990) Dieting reduced plasma tryptophan and alters brain 5-HT functioning in women. Psychological Medicine 20:785–791

Antin J, Gibbs J, Holt J Young RC, Smith GP (1975) Cholecystokinin elicits the complete behavioral sequence of satiety in rats. J Comp Physiol Psych 89:784–760
Ashley DVM, Fleury MO, Golay A, Maeder E, Leathwood PD (1985) Evidence for diminished brain 5-HT biosynthesis in obese diabetic humans. Am J Clin Nutr 42:1240–1245
Aulukh CS, Mazzola-Pomietto P, Hilihan-Giblin BA, Murphy DL (1995) Lack of cross tolerance for hypophagia induced by DOI verses mCPP suggests separate mediation by 5-HT_{2A} and 5-HT_{2C} receptors respectively. Neuropsychopharmaocology 13:1–8
Aulukh CS, Mazzola-Pomietto P, Wozniak, KM, Hill JL, Murphy DL (1994) 1-(2,5-dimethoxy-4-methylphenyl)-2-aminopropane-induced hypophagia and hyperthermia in rat is mediated by serotonin-$_{2A}$ receptors. J Pharm Exp Ther 270:127–152
Barrett AM, McSherry L (1975) Inhibition of drug-induced anorexia in rats by methysergide. J Pharm Pharmacol 27:889–895
Blundell JE (1977) Is there a role for serotonin (5-hydroxytryptamine) in feeding? Int J Obes 1:15–42
Blundell JE (1983) Problems and processes underlying the control of food selection and nutrient intake. In: Wurtman RJ, Wurtman JJ (eds) Nutrition and the Brain, vol 6. Raven Press, NY, pp 163–222
Blundell JE, Hill AJ (1985) Effects of dexfenfluramine in feeding and body weight Relationship with food consumption and palatability. In: Vague J, Guy-Grand B, Bjontroup P (eds) Metabolic complication of human obesities. Elservier, Nordlolland pp 199–206
Blundell JE, Hill AJ (1987) Nutrition, serotonin and appetite: Case study in the evaluation of a scientific idea. Appetite 8:183–194
Blundell JE, Hill AJ (1988) On the mechanism of action of dexfenfluramine: Effect on alliesthesia and appetite motivation in lean and obese subjects. Clin Neuropharmacol 11(S):121–134
Blundell JE, Hill AJ (1989) Do serotoninergic drugs decrease energy intake by reducing fat or carbohydrate intake? Effects of d-fenfluramine with supplemented weight increase diets. Pharmacol Biochem Behav 31:773–778
Blundell JE, Latham CJ (1978) Pharmacological manipulation of feeding behavior: possible influences of serotonin and dopamine on food intake. In: Garattini S, Samanin R (eds) Central mechanisms of anorectic drugs. Raven Press, NY, pp 83–109
Blundell JE, Latham CJ (1979) Serotoninergic influences on food intake: 5-hydroxytryptophan on parameter of feeding behavior in deprived and free-feeding rats. Pharmacol Biochem Behav 11:431–437
Blundell JE, Latham CJ (1980) Characterisation of adjustments to the structure of feeding behaviour following pharmacological treatment: effects of amphetamine and fenfluramine and the antagonism produced by pimozide and methergoline. Pharmacol Biochem Behav 12:717–722
Blundell JE, Lawton CL (1995) Serotonin and dietary fat intake: effects of dexfenfluramine. Met Clin Exper 44:33–37
Blundell JE, Lawton CL, Halford JCG (1995) Serotonin, eating behavior and fat intake. Obes Res 3(4S):471–476
Blundell JE, MacDiarmid JI (1997) Passive overconsumption: fat intake and short-term energy balance. Annals of the New York Acadamy of Sciences 827:392–407
Blundell JE, McArthur RA (1981) Behavioral flux and feeding: continuous monitoring of food intake and food selection, and the video-recording of appetitive and satiety sequences for the analysis of drug action. In: Samanin R, Garattini S (eds) Anorectic agents: mechanisms of action and tolerance. Raven, NY, pp 19–43
Boeles S, Williams C, Campling GM, Goodall EM, Cowen PJ (1997) Sumatriptan decreases food intake and increase plasma growth hormone in healthy women. Psychopharmacology 129:179–182

Bouchard C (1989) Genetic factors in obesity. Medical Clinics of North America 73:67–81
Bouchard C (1991) Heredity and the path to overweight and obesity. Medicine and Science of Sports Exercise 23:285–291
Bradley Nomenclature Committee (1986) Proposals for classification and nomenclature of functional receptors for 5-hydroxytryptamine. Neuropharmacology 25:563–576
Bray GA, York DA (1972) Studies on food intake of genetically obese rats. Am J Physiol 233:176–179
Brewerton TD (1995) Toward a unified theory of serotonin dysregulation in eating related disorders. Psychoneuroendocrinology 20:561–590
Brezinski AA, Wurtman JJ, Wurtman RJ (1990) D-fenfluramine suppresses the increased caloric and carbohydrate intakes and improves the mood of women with pre-menstrual tension. Obestet Gynecol 76:296–301
Caccia S, Bizzi A, Coltro G, Francasso C, Frittoli E, Mennini T, Garattini S (1992) Anorectic activity of fluoxetine and norfluoxetine in rats: Relationship between brain concentrations and in-vitro potencies on monoaminergic mechanisms. J Pharm Pharmacol 44:250–254
Campfield LA, Smith FJ, Guisez Y, Devos R, Burn P (1995) Recombinant mouse on protein: Evidence for a peripheral signal linking adiposity and central neural networks. Science 269:546–549
Cangiano C, Ceci F, Casinco A, Del Ben M, Lavino A, Mauscatioli M, Antonucci F, Rossi-Fanelli F (1992) Eating behavior and adherence to dietary prescription in obese adult subjects treated with 5-hydroxytryptophan. Am J Clin Nutr 56:863–867
Clifton PG (1994) The neuropharmacology of meal patterning. In: Cooper SJ (ed) Ethology and Psychopharmacology. Wiley, Chichester, pp 313–328
Clifton PG, Barnfield AM, Philcox L (1989) A behavioral profile of fluoxetine induced anorexia. Psychopharmacology 107:89–95
Clineschimdt BV, McGuffen JC, Pfleuger AB, Totaro JA (1978) A 5-hydroxytryptamine-like mode of action for 6-chloro-2-[1-piperazinyl]-pyrazine (MK-212). Br J Pharmacol 62:579–589
Coscina DV, Feifel D, Nobrega JN, Currie G (1994) Intra ventricular but not intra paraventricular nucleus metergoline elicits feeding in satiated rats. Am J Physiol 266:1562–1567
Cowen PJ, Clifford EM, Walsh AE, Williams C, Fairburn CG (1996) Moderate dieting causes $5\text{-}HT_{2C}$ supersensitization. Psychological Medicine 26:1156–1159
Currie PJ, Coscina DV (1996) Metergoline potentiates natural feeding and antagonizes the anorectic action of medial hypothalamic 5-HT. Pharmacol Biochem Behav 53:1023–1028
Curzon G (1990) Serotonin and appetite. Ann NY Acad Sci 600:521–530
Curzon G (1991) Effects of tryptophan and 5-hydroxytryptamine receptor subtype agonists on feeding. Adv Exp Med Biol 294:377–388
Curzon G, Gibson El, Oluyomi AO (1997) Appetite suppression by commonly used drugs depends on 5-HT receptors but not on 5-HT availability. Trends Pharmacol Sci 18:21–25
Dourish CT (1995) Multiple serotonin receptors: Opportunities for new treatments for obesity. Obesity Res 3(s4):449–462
Drewnowski A, Kurth C, Holden-Wiltse J, Saari J (1992) Food preference in human obesity: Carbohydrates verses fat. Appetite 18:207–221
Dryden S, Frankish HM, Wang Q, Pickavance L, Williams G (1996a) The serotonergic agent fluoxetine reduces neuropeptide Y levels and neuropeptide Y secretion in the hypothalamus of lean and obese rats. Neuroscience 72:557–566
Dryden S, Frankish HM, Wang Q, Williams G (1996b) Increased feeding and neuropeptide Y (NPY) but not NPY mRNA levels in hypothalamus of the rat following central administration of the serotonin synthesis inhibitor p-chlorophenylalinine. Brain Research 724:232–237

Dryden S, Wang Q, Frankish HM, Williams G (1996c) Differential effects of the 5-$HT_{1B/2C}$ receptor agonist mCPP and the 5-HT_{1A} agonist flesinoxan on hypothalamic neuropeptide Y in the rat: evidence that NPY may mediate serotonin's effects on food intake. Peptides 17:943–949
Dryden S, Williams G (1996) The role hypothalamic peptides in the control of energy balance and body weight. Curr Opin Endocrin Diab 3:51–58
Fernstorm JD, Wurtman RJ (1973) Control of brain 5-HT content by dietary carbohydrates. In: Barchas J, Usdin E (eds) Serotonin and behavior. Academic Press, NY, pp 121–128
Finer N, Finer S, Nauova RP (1992) Drug therapy after very-low-calorie diets. Am J Clin Nutr 1955–85
Fisler JS, Underberger SJ, York DA, Bray GA (1993) D-Fenfluramine in a rat model of dietary fat-induced obesity. Pharmacol Biochem Behav 45:487–493
Fletcher PJ, Coscina DV (1993) Injecting 5-HT into the PVN does not prevent feeding induced by 8-OH-DPAT into the raphe. Pharmacol Biochem Behav 46:487–491
Gaddam JH, Picarelli ZP (1957) Two kind of tryptamine receptor. Br J Pharmaco 12:323–328
Garattini S, Samanin R (1976) Anorectic drugs and brain neurotransmitters. In: Silverstone T (ed) Food intake and appetite. Dahlem Konferenzen, Berlin, pp 82–208
Gibson EL, Kennedy AJ, Curzon G (1993) D-Fenfluramine- and D-norfenfluramine-induced hypophagia: differential mechanisms and involvement of postsynaptic 5-HT receptors. Euro J Pharmacol 242:83–90
Goodall EM, Cowen PJ, Franklin M, Silverstone T (1993) Ritanserin attenuates anorectic endocrine and thermic responses to d-fenfluramine in human volunteers. Psychopharmacology 112:461–466
Goodall EM, Silverstone T (1988) Differential effect of D-fenfluramine and metergoline on food intake in human subjects. Appetite 11:215–228
Goodwin GM, Cowen PJ, Fairburn CJ, Parry-Billings M, Calder PC, Newsholme EA (1990) Plasma concentrations of tryptophan and dieting. Brit Med J 17:839–842
Goodwin GM, Fairburn CG, Cowen PJ (1987) Dieting changes 5-HT function in women but not in men, implications for the aetiology of anorexia nervosa? Psychol Med 17:839–842
Green S, Lawton CL, Wales JK, Blundell JE (1997) Risk factors for overeating: Dex-fenfluramine suppresses the intake of sweet high fat or carbohydrate food in obese women. Int J Obesity 21(3):s64
Greeno CA, Wing R (1996) A double-blind, placebo-controlled trial of the effect of fluoxetine on dietary intake in overweight women with and without binge-eating disorder. Am J Clin Nutr 64:267–273
Grignaschi G, Sironi F, Samanin R (1995) the 5-HT_{1B} receptor mediates the effect of d-fenfluramine one eating caused by intra-hypothalamic injection of neuropeptide Y. Euro J Pharmacol 274:221–224
Grignaschi G, Samanin R (1992) Role of serotonin and catecholamines in brain feeding suppressant effects of fluoxetine. Neuropharmacology 31:445–449
Guy-Grand B (1992) Clinical studies with d-fenfluramine. Am J Clin Nutr 56:173–176
Halaas JL, Gajiwala KD, Maffie M, Cohen SL, Chait BT, Rabinowitz D, Lallone RL, Burley SK, Friedman JM (1995) Weight reducing effects of a plasma protein encoded by the obese gene. Science 269:543–546
Halford JCG, Blundell JE (1993) 5-Hydroxytryptaminergic drugs compared on the behavioral sequence associated with satiety. Br J Pharmacol 100:p95
Halford JCG, Blundell JE (1996a) Metergoline antagonizes fluoxetine induced suppression of food intake but not changes in the behavioral satiety sequence. Pharmacol Biochem Behav 54:745–751
Halford JCG, Blundell JE (1996b) The 5-HT_{1B} receptor agonist CP-94,253 reduces food intake and preserves the behavioral satiety sequence. Physiol Behav 60:933–939
Halford JCG, Blundell JE (1998) Direct and continuous behavioral analysis for the diagnosis of drug action on feeding. Current Protocols in Neuroscience (in press)

Halford JCG, Heal DJ, Blundell JE (1995) effects in the rat of sibutramine on food intake and the behavioral satiety sequence. Br J Pharmacol 144:p387

Halford JCG, Lawton CL, Blundell JE (1997) The 5-HT_2 receptor agonist MK-212 reduces food intake but disrupts the behavioral satiety sequence. Pharmacol Biochem Behav 56:41–46

Halford JCG, Wanninayake SCD, Blundell JE (1998) Behavioral satiety sequence (BSS) for the diagnosis of drug action on food intake. Pharmacol Biochem Behav 61:159–168

Hill AJ, Blundell JE (1986) Model system for investigating the actions of anorectic drugs: Effects of D-fenfluramine on food intake, nutrient selection, food preference, meal patterns, hunger and satiety in human subjects. In: Advances in the Bioscience. Pergamon Press, Oxford, pp 377–389

Hill AJ, Blundell FE (1990) Sensitivity of the appetite control system in obese subjects to nutritional and serotoninergic challenges. Int J Obesity 14:219–233

Hill AJ, Rogers PJ, Blundell JE (1995) Techniques for the experimental measurement of human eating behavior; a practical guide. Int J Obesity 19:361–375

Horton TJ, Drougas H, Brachey A, Reed GW, Hill JO (1995) Fat and carbohydrate overfeeding in humans: different effects on energy storage. Am J Clin Nutr 62:19–29

Hoyer D, Martin G (1997) 5-HT receptor classification and nomenclature: Towards a harmonisation with the human genome. Neuropharmacology 36:419–428

Hutson PH, Donohoe TP, Curzon G (1988) Infusion of the 5-hydroxytryptamine agonists RU-24969 and TFMPP into the paraventricular nucleus of the hypothalamus. Psychopharmacology 97:550–552

Jesperson S, Scheel-Kruger J (1973) Evidence for a difference in mechanism of the action between fenfluramine and amphetamine induced anorexia. J Pharm Pharmacol 22:637–638

Jimerson DC, Wolfe BE, Metzger ED, Finkelstein DM, Cooper TB, Levine JM (1997) Decreased serotonin function in bulimia nervosa. Arch Gen Psychiatry 54:529–534

Kanarek RB (1987) Neuropharmacological approach to studying diet selection. In: Kaufman S (ed) Amino Acids in Health: New Perspectives. Alan R. Liss, NY, pp 383–401

Kennett GA, Dourish CT, Curzon G (1987) 5-HT_{1B} agonists induce anorexia at a post synaptic site. Euro J Pharmacol 141:429–435

Kennett GA, Curzon G (1988a) Evidence that the hypophagia induced by mCPP and TFMPP requires 5-HT_{1C} and 5-HT_{1B} receptors; hypophagia induced by RU-24969 only requires 5-HT_{1B} receptors. Psychopharmacology 96:93–100

Kennett GA, Curzon G (1988b) Evidence that mCPP may have behavioral effects mediated by central 5-HT_{1C} receptors. Br J Pharmacol 94:137–147

Kennett GA, Curzon G (1991) Potencies of antagonists indicate that 5-HT_{1C} receptors mediate 1–3(chlorophenyl)piperazine- induced hypophagia. Br J Pharmacol 103: 2016–2020

Kennett GA, Wood MD, Glen A, Grewal S, Frobes F, Gadre A, Blackburn TP (1994) In vivo properties of SB 200646 A, a selective 5-$HT_{2C/2B}$ receptor antagonist. Br J Pharmacol 111:797–802

Kennett GA, Wood MD, Bright F, Trail B, Riley G, Holland V, Avenel KY, Stean T, Upton N, Bromidge S, Forbes IT, Brown AM, Middledmiss DN, Blackburn TP (1997) SB 242084, a selective and brain potent 5-HT_{2C} receptor. Neuropharmacology 36:609–620

King BH, Brazell C, Dourish CT (1989) MK-212 increase rat plasma ACTH concentration by activation of the 5-HT_{1C} receptor subtype. Neuroscience Letters 105:174–176

Kitchener SJ, Dourish CT (1994) An examination of the behavioral specificity of hypophagia induced by 5-HT1B, 5-HT1 C and 5-HT2 receptor agonists using the post-prandial sequence in rats. Psychopharmacology 113:368–377

Koe BK, Nielson JA, Macor JE, Heym J (1992) Biochemical and behavioral studies of the 5-HT_{1B} receptor agonist, CP-94,253. Drug Dev Res 26:241–250
Koe BK, Weissman A, Welch WM, Browne RG (1983) Sertraline, 1S,4S-N-methyl-4(3,4-dichlorophenyl)-1,2,3,4-tetrahydro-1-naphathylamine, a new uptake inhibitor with selectivity for serotonin. J. Pharmacol Exp Ther 266:686–700
Lafreniere F, Lambert J, Rasio E, Serri (1993) Effects of dexfenfluramine treatment on body weight and post-prandial thermogenesis in obese subjects. A double blind placebo-controlled study. Int J Obesity 17:25–30
Lawton CL, Blundell JE (1992) The effects of D-fenfluramine on intake of carbohydrate supplements is influenced by the hydration of test diets. Physiol Behav 53:517–523
Lawton CL, Blundell JE (1993) 5-HT manipulation and dietary choice: Variable carbohydrate (Polycose) suppression demonstrated only under specific experimental conditions. Psychopharmacology 112:375–382
Lawton CL, Wales JK, Hill AJ, Blundell JE (1995) Serotoninergic manipulation, meal-induced satiety and eating patterns. Obesity Research 3:345–356
Lee MD, Clifton PG (1992) Partial reversal of fluoxetine anorexia by the 5-HT antagonist metergoline. Psychopharmacology 107:359–364
Leibowitz SF (1980) Neurochemical systems of the hypothalamus – control of feeding and drinking behavior and water-electrolyte excretion. In: Morgane PJ, Panksepp J (eds) Handbook of the hypothalamus (vol VI, Part A). Marcel Decker, New York, pp 299–437
Leibowitz SF, Alexander JT, Cheung WK, Weiss GF (1992) Effects of serotonin and serotonin blocker metergoline on meal patterns macronutrient selection. Pharmacol Biochem Behav 45:185–194
Leibowitz SF, Hammer NJ, Chang K (1981) Hypothalamic paraventricular nucleus lesions produce overeating in the rat. Physiol Behav 27:1031–1040
Leibowitz SF, Shor-Posner G (1986) Brain serotonin and eating behavior. Appetite 7S:1–14
Leibowitz SF, Weiss GF, Suh JN (1990) Medial hypothalamic nuclei mediate serotonin's inhibitory effect on feeding behavior. Pharmacol Biochem Behav 37:735–742
Leibowitz SF, Weiss GF, Walsh UA, Viswanath D (1989) Medial hypothalamic serotonin: role in circadian patterns of feeding and macronutrient selection. Brain Res 503:132–140
Lentus K-U, Hinney A, Ziegler K, Rosenkranz K, Wurmser H, Barth N, Jacob K, Comer H, Mayer H, Grzeschik K-H, Schäfer H, Remschmidt H, Pirke KM, Hebebrand J (1997) Evaluation of CYS23SER mutation within the human 5-HT2C receptor gene: No evidence for an association of the mutant allele with obesity or underweight in children, adolescents and young adults. Life Sciences 61(1):9–16
Lightowler S, Kennett GA, Wood MD, Brown AM, Glen A, Blackburn TP, Tulloch IF (1994) Hypophagic effect of fluoxetine in rats is not mediated by inhibition of 5-HT re-uptake or an agonist action at 5-HT_{2C} receptors. Br J Pharmacol 112:359–364
Lightowler S, Wood M, Brown T, Glen A, Blackburn T, Tulloch I, Kennett G (1996) An investigation of the mechanism responsible for fluoxetine-induced hypophagia in rats. Euro J Pharmacol 296:137–143
Lucki I, Kreider MS, Simansky KJ (1988) Reduction of feeding behavior by the serotonin uptake inhibitor sertraline. Psychopharmacology 96:289–295
Luo S, Li ETS (1990) Food intake and selection pattern of rats treated with dexfenfluramine, fluoxetine and RU-24969. Brain Res Bull 24:729–733
MacDiaramid JI, Cade JE, Blundell JE (1997) High and low fat consumers, their macronutrient intake and body mass index: Further analysis of the National Diet Survey of British Adults. Euro J Clin Nutr 50:505–512
MacKenzie RG, Hoebel BG, Ducret RP, Trulson ME (1979) Hyperphagia following intraventricular p-chlorophenylalanine-, leucine- or tryptophan-methyl esters: lack of correlation with whole brain serotonin level. Pharmacol Biochem Behav 10:951–955

McGuirk J, Muscat R, Willner P (1992) Effects of the 5-HT uptake inhibitors femoxetine and paroxetine, and the 5-HT1A agonist eltoprazine, on the behavioral satiety sequence. Pharmacol Biochem Behav 41:801–805

McGuirk J, Silverstone T (1990) The effect of 5-HT re-uptake inhibitor fluoxetine on food intake and body weight in healthy male subjects. Int J Obesity 14:361–372

Macor JE, Burkhart CA, Heym JH, Ives JL, Lebel LA, Newmand ME, Nielsen JA, Ryan K, Schulz DW, Torgersen LK, Koe BK (1990) 3-(1,2,5,6-Tetrahydropyrid-4-yl)pyrrolo[3,2-b]pyrid-5-one; a potent and selective 5-HT1B agonist and rotationally restricted phenolic analogue of 5-methoxy-3-(1,2,5,6-tetrahydropyrid-4-yl)indole. J Med Chem 33:2087–2093

Neill JC, Cooper SJ (1989) Evidence that d-fenfluramine anorexia is mediated by 5-HT_1 receptors. Psychopharmacology 97:213–218

Neill JC, Bendotti C, Samanin R (1990) Studies on the role of 5-HT receptors in satiation and the effect of d-fenfluramine in the runway test. Eur J Pharmacol 190:105–112

Oluyomi AO, Gibson EL, Barnfield AM, Curzon G (1994) D-Fenfluramine and D-norfenfluramine hypophagias do not require increased hypothalamic 5-hydroxytryptamine release. Euro J Pharmacol 264:111–115

Pelleymounter MA, Cullen MJ, Baker MB, Hecht R, Winters D, Boone T, Collins F (1995) Effects of the obese gene product on weight regulation and in ob/ob mice. Science 269:540–543

Peroutka SJ, Snyder SH (1979) Multiple serotonin receptors: different binding of 3H-5-hydroxytryptamine, 3H-lysergic acid diethylamide and 3H-spiroperidol. Mol Pharmacol 16:687–699

Pinder BM, Brogden RN, Sawyer PR, Speight TM, Avery GS (1975) Fenfluramine: A review of the pharmacological properties and therapeutic in obesity. Drugs 10:241–323

Pirke KM, Schweiger U, Laessle RG (1986) Effects on diet composition on affective state in anorexia nervosa and bulimia. Clin Neuropharmacol 9:561–590

Poppitt SD, Prentice AM (1996) Energy density and its role in the control of food intake: Evidence from metabolic and community studies. Appetite 26:153–174

Prats E, Monfar M, Castell J, Iglesias R, Allemany M (1989) Energy intake of rats fed a cafeteria diet. Physiol Behav 45:2263–2272

Raiteri M, Bonanno G, Vallebouona F (1995) In vitro and in vivo effects D-fenfluramine: no apparent relationship between 5-hydroxytryptamine release and hypophagia. J Pharmacol Exp Ther 273:643–649

Ravussin E, Valencia ME, Esparza J, Bennett PH, Schulz LO (1994) Effects of a traditional lifestyle on obesity in Pima Indians. Diabetes Care 17:1067–1074

Rogers PJ, Blundell JE (1984) Meal patterns and food selection during the development of obesity in rats fed a cafeteria diet. Neuroscience Biobeh Rev 8:441–453

Samanin R, Mennini T, Bendotti C, Barone D, Caccia S, Garattini S (1989) Evidence that central 5-HT_2 receptors do not play an important role in anorectic activity of d-fenfluramine in the rat. Neuropharmacology 28:465–469

Sclafani A (1984) Animal models of obesity: Classification and characterisation. Int J Obes 8:491–508

Simansky KJ, Viadya AH (1990) Behavioral mechanisms for the anorectic actions of the serotonin (5-HT) uptake inhibitor sertraline in rats: comparison with directly acting agonists. Brain Res Bull 25:953–960

Sørensen TIA, Price RA, Stunkard AJ, Schulsinger F (1989) Genetics of obesity in adult adoptees and their biological siblings. Brit Med J 298:87–90

Spitzer RL, Devlin M, Walsh BT, Hasin D, Wing RR, Marcus MD, Stunkard AJ, Wadden S, Arga S, Mitchell J, Monas C (1992) Binge eating disorder: A multisite field study of diagnostic criteria. Int J Eating Disorder 11:191–203

Stunkard AJ (1959) Eating patterns in obesity. Psychiatric Quarterly 33:284–292

Stunkard AJ, Berkowitz R, Tratikut C, Reiss E, Young L (1996) D-Fenfluramine treatment of binge eating disorder. Am J Psychiatry 153:1455–1459

Tecott LH, Sun LM, Akana SF, Lowenstien DH, Dallman MF, Julius D (1995) Eating disorder and epilepsy in mice lacking 5-HT_{2C} serotonin receptors. Nature 374:542–546

Teff KL, Young SN, Blundell JE (1989) The effect of protein or carbohydrate breakfasts on subsequent plasma amino acid levels, satiety and nutrient selection in normal males. Pharmacol Biochem Behav 34:410–417

Wadden TA, Bartlett SJ, Foster GD, Greenstein RA, Wingate BJ, Stunkard AJ, Letizia (1995) Sertraline and relapse prevention following treatment by a very low calorie diet: a controlled clinical trial. Obesity Res 3:549–557

Walsh AE, Oldman AD, Franklin M, Fairburn CG, Cowen PJ (1995) Dieting decreases plasma tryptophan and increases the prolactin response to D-fenfluramine in women but not men. J Affective Disorders 33:89–97

Walsh AE, Smith KA, Oldman AD, Williams C, Goodall EM, Cowen PJ (1994) m-Chlorophenylpiperazine decrease food intake in a test meal. Psychopharmacology 116:120–122

Wong DT, Reid LR, Threlkeld PG (1988) Suppression of food intake in rats by fluoxetine: Comparison of anantiomers and effects of serotonin antagonists. Pharmacol Biochem Behav 31:475–479

Wurtman JJ, Wurtman RJ (1977a) Drugs that enhance central serotoninergic transmission diminish elective carbohydrate consumption by rats. Life Science 24:895–904

Wurtman JJ, Wurtman RJ (1977b) Fenfluramine and fluoxetine spare protein consumption while suppressing caloric intake by rats. Science 198:1178–1180

Wurtman JJ, Wurtman RJ, Marks S et al (1985) D-fenfluramine selectively suppresses carbohydrate snaking by obese subjects. Int J Eating Disorders 4:89–99

Wurtman JJ, Wurtman RJ Reynolds S et al (1987) Fenfluramine suppresses snack intake among carbohydrate cravers but not among noncarbohydrate cravers. Int J Eating Disorders 6:687–699

Wurtman RJ, Fernstorm JD (1974) Effects of diet on brain neurotransmitters. Nutr Review 32:193–200

CHAPTER 8

Pharmacological Treatment of Obesity: Outcomes and New Tools

L.A. CAMPFIELD and F.J. SMITH

A. Introduction

Obesity is a complex, increasingly prevalent, and important health problem throughout the world. Human obesity is characterized by increased adipose tissue mass resulting from a complex interaction of genetic predisposition to metabolic efficiency and environmental/lifestyle factors. Recent studies have clearly demonstrated that obesity is a disease with a strong biological basis. Although treatment (e.g., healthy eating (including behavioral modification), physical activity, drugs) is available and most people can achieve medically significant weight loss (5%–10% of initial body weight), *the long-term maintenance of that weight loss* and its associated improvement in health is, unfortunately, *very rare.* However, an explosion of new scientific information concerning the regulation of energy balance and fat mass at several levels has occurred since the discovery of OB protein (also known as leptin). The recent identification of obesity-associated genes and gene products in mice and humans has provided new pathways involved in obesity. This new understanding of the regulation of energy balance has revealed, and will continue to reveal, additional targets for pharmacological intervention that will certainly lead, with a little luck, to innovative medicines to treat obesity. New drugs will be viewed as adjuncts to behavioral and lifestyle change to maintain weight loss and its associated increases in metabolic fitness.

B. Obesity: A Therapeutic Challenge

Obesity is a major health problem throughout the world. It is a complex, multifactorial disease characterized by behavioral, endocrine, and metabolic alterations with an increasing prevalence. Obesity is the most common nutritional disorder in the developed world and is associated with significant chronic metabolic diseases [hypertension, non insulin dependent diabetes mellitus (NIDDM), hypercholesterolemia] as well as stroke, sleep apnea, joint diseases, and certain cancers. Obesity is a cause of significant morbidity and is having an increasing negative impact on the health care systems in both developed and developing world. The overall prevalence of obesity in the United States has increased to 33% and prevalence in children has increased dramatically in the last decade. Thus, obesity remains a poorly managed medical condition that is a major cause of morbidity and mortality (BJÖRNTORP and BRODOFF 1992; ALLISON and PI-SUNYER 1995; THOMAS 1995).

I. Human Obesity: Interaction Between Genetics, Behavior and Environment

Obesity is extremely challenging because it is a multifactorial disease that lies at the interface between the biology of body energy regulation and an environment (physical and sensory) that has been increasingly characterized as "hostile to good health" and even "obesifying" (HILL and PETERS 1998). Many of us in the developed world live with a large and increasing variety of highly palatable food, together with declining incentives and opportunities for physical activity. Today many important issues such as well being, health, and self esteem are closely associated with body weight. This intense focus on body weight has resulted in overt and covert discrimination against obese individuals (BJÖRNTORP and BRODOFF 1992; ALLISON and PI-SUNYER 1995).

1. Fat Mass

Obesity is defined by excessive accumulation of body fat. It is simple to state that obesity is a problem of excessive expansion of fat mass, but that expansion results from a complex and intriguing interaction between genetic predisposition to increased metabolic efficiency and the regulation of energy balance. The most commonly accepted model for the genetics of human obesity is an interaction of genetic predisposition and environmental factors. CLAUDE BOUCHARD and other investigators have estimated the heritability of variations in distribution of body fat to be approximately 35% (BOUCHARD and PERUSSE 1993). To this we could add up to 15% for the presence of "modifier genes" that make some individuals much more susceptible to gaining weight on a palatable high-energy diet than others. Thus, at least 50% of the problem is due to environmental and lifestyle factors. This component of the variance of body fat content is the one that can be modulated by behaviorally-based obesity treatment through healthy eating, physical activity and current drug therapy (CAMPFIELD et al. 1998).

2. Biological Basis

In recent years, additional, and quite compelling evidence has emerged that obesity, basically, is a disease of biological *dysregulation*, particularly in the context of one's biology, one's genetics, and one's adaptation to what has been called an "obesifying" or "obese" environment. The steady-state body weight of an individual is thought to result from an *integration* of *multiple* biological factors, which are, at least, partially *genetically determined*.

Animal and clinical studies have shown that increases or decreases in body weight are *resisted and corrected by very* robust *physiological mechanisms* (LEIBEL et al. 1995). OB protein probably plays a role in this "resetting response" that is responsible for weight regain following weight loss (for review see CAMPFIELD et al. 1997; CAMPFIELD and SMITH 1998). Obese patients who continue to seek treatment after multiple weight regain following multiple periods of weight loss know this by personal experience. Obesity is characterized by the following major pathophysiological alterations:

1. High rates of lipid deposition in adipose tissue
2. Reduced insulin sensitivity of muscle and fat
3. Exaggerated insulin responses to meals
4. Hyperinsulinemia
5. Increased OB protein concentrations
6. Reduced central and peripheral sensitivity to OB protein (CAMPFIELD et al. 1997; CAMPFIELD and SMITH 1998).

We would like to block the progression and, hopefully, reverse these alterations by integrated, multi-modal treatment using a disease management model in which the patient is an active partner. The evidence, particularly from the long-term follow-up of surgical intervention, is quite strong that these goals can be accomplished (PORIES et al. 1995; NASLUND 1996).

Thus, the need for improved medical management of obesity is very clear and compelling and the recent expansion of pharmaceutical and biotechnology research groups dedicated to obesity drug discovery shows that this need is recognized. Improved management will result from the successful implementation of the concept that pharmacological agents for the treatment of obesity should be used as adjuncts to, or partners of, healthy eating and physical activity to promote increased success at weight maintenance (CAMPFIELD et al. 1998). Treatment of obesity can be viewed as a three-legged milking stool or tripod. Healthy eating and physical activity are the two legs that have been in place, but just like a two-legged milking stool, long-term weight control has been unstable. Drug treatment is the missing, third leg of the stool that should provide stable, long-term weight control and improved health for obese individuals.

C. Treatment of Obesity: The Present and the Future

I. Goals

One set of goals of treatment based on body weight is shown in Table 1. The first goal is to prevent further weight *gain*. This is a desirable and achievable goal through healthy eating and physical activity, and its achievement can be helped by drug therapy. Before focusing on the maintenance of weight loss, obese individuals must first be able to avoid gaining more weight. When obese

Table 1. Goals of Obesity Treatment

Prevention of weight gain
Induction of weight loss
Maintenance of weight loss

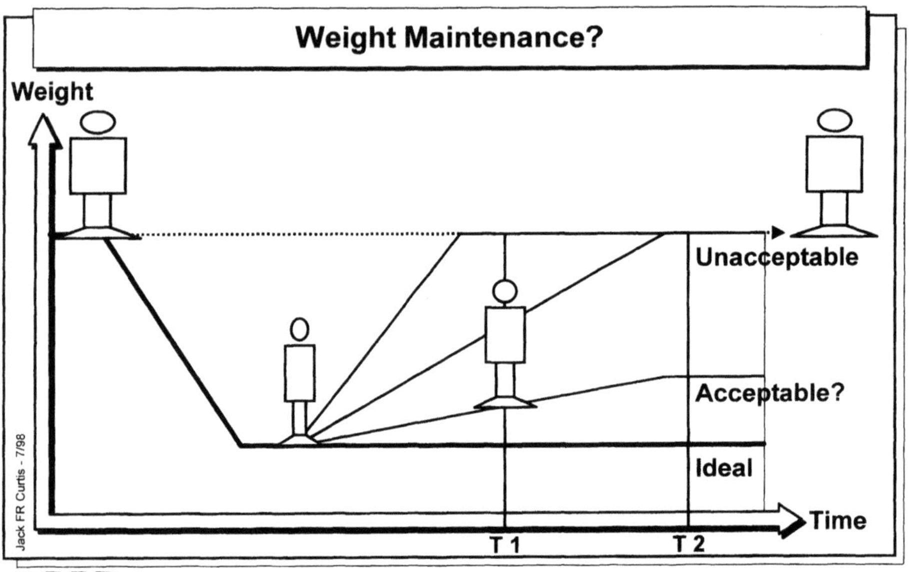

Fig. 1. Idealized body weight curves of obese individuals undergoing weight loss and weight maintenance treatment. The definition of the minimally acceptable weight maintenance is the preservation of a non-zero differential in body weight at the end of active drug treatment. See text for additional details

individuals enter a treatment program, their obvious intention is to lose weight. Thus, they should not be gaining weight during the treatment period, although this has been seen in some studies. The next goal is to induce a 5%–10% weight loss from the initial body weight. Most obese individuals can achieve this goal in a variety of treatment programs (e.g., very low calorie diets). The next, and most important, goal is *to maintain body weight once a 5%–10% weight loss is achieved.* This is the major goal of current and future pharmacological treatments for obesity and it is the most difficult to achieve (CAMPFIELD et al. 1998).

II. Long-Term Weight Maintenance

A schematic body weight profile of an obese individual who presents for treatment at an initial body weight is shown in Fig. 1. Following a period of successful caloric restriction and increased physical activity, body weight decreases to a minimum point at the end of treatment. The goal would be to maintain all or most of the weight loss that was achieved (line labeled "ideal"). Although this would be ideal, it is also considered by most researchers as unrealistic knowing that the biological systems regulating body fat content and body energy balance are actively resisting any change from the pretreatment steady-state. Similarly, regain of all weight lost over a period of one, two, or three years is unacceptable (CAMPFIELD 1996).

Minimally acceptable weight maintenance is a difference in final body weight (and preferably body composition) at the end of active treatment from the initial body weight (and composition) at the beginning of the treatment. This is an appropriate *weight maintenance* goal (line labeled "acceptable" in Fig. 1). However, a range of opinion exists within the research, industrial, and regulatory communities about the magnitude of the minimally acceptable weight differential (CAMPFIELD 1996).

III. Concept of Metabolic Fitness: A Medically Based Alternative Outcome Measure

An alternative, medically based outcome measure for obesity treatment is the "concept of metabolic fitness" (CAMPFIELD 1995, 1996). The concept we are advocating is to focus on, and to track, the "metabolic" health of obese individuals, rather than focus primarily on percentage body fat, body mass index, or body weight. The goal is to be "metabolically fit" and to achieve the absence of any elevated metabolic or biochemical risk factor associated with obesity (Table 2). Obese individuals might have 25% or 35% body fat, but if they do not have any risk factors and their biochemical values are within the normal range, it may be not urgent that they lose weight because their general health is probably fine. However, overweight individuals with only slightly elevated percentage body fat, but with one or more abnormal biochemical or other risk factors, may be in urgent need of losing weight to improve their already impaired health.

The concept of metabolic fitness was based upon, and strongly influenced by, the well established biochemical and behavioral risk factors for cardiovascular disease that have been pioneered and well established by the Framingham longitudinal heart study. Rather than lump all obese individuals into a single group that must lose weight, the metabolic fitness approach puts the primary focus on the actual presence, or increased risk, of one or more obesity-associated disease and views weight loss as a *modality* to improve metabolic health. Many physicians and scientists in the obesity field have embraced this idea and several have applied it to their clinical practice (BLACKBURN et al. 1997).

Within the concept of metabolic fitness there is a spectrum of intervention options ranging from mild to very aggressive (Table 2). The first and least aggressive intervention would be to *reduce significant and abnormal bio-*

Table 2. Concept of "metabolic fitness" as an alternate outcome measure

Definition of metabolic fitness: the absence of any metabolic or biochemical risk factor for diseases associated with obesity
Metabolic fitness would correlate closely with general health status and would improve with risk factor reduction
Metabolic fitness may provide an alternative, medically oriented goal of the treatment of obesity that does not focus on weight
Intervention options to improve metabolic fitness include (from least to most aggressive) Medically significant reduction of risk factors Restoration of abnormal risk factors to normal ranges Reversal of "high normal" or "borderline" parameters Prevention of risk factors in overweight individuals
Improvements in metabolic fitness, and, thus, in health status, may be independent of weight loss (or poorly correlated with weight loss)

chemical and behavioral risk factors that are already present. If an obese person has clinically significant hypertension, hypercholesterolemia, or/and glucose intolerance, then assistance and support are required to help that individual lose weight and maintain that weight loss. This should allow the abnormal biochemical parameters to be reduced back *toward normal.*

A more aggressive approach would be to *reduce the abnormal biochemical risk factors to within normal ranges.* Another level of aggressiveness of treatment would be to attempt to *reduce "high" normal or borderline biochemical values to the middle of the normal range.* The most aggressive intervention would be to *prevent biochemical risk factors in the low to medium normal range from increasing to high normal values.*

Improvements in biochemical risk factors may be dependent upon weight loss or they may be independent of or poorly correlated with weight loss. Many studies of weight loss have shown that during periods of weight loss there is a uniform improvement of risk factors. Some studies also demonstrated an additional complexity in the weight maintenance phase because risk factors increased in some patients. On the other hand, physical activity is known to improve insulin sensitivity, reduce cholesterol levels, and reduce other cardiovascular risk factors without always inducing weight loss.

If we can improve the health of individuals and take the focus away from body weight, percent body fat, or body mass index, sustained change might be easier and our patients might accept better health as a more realistic goal rather than defining success only by dress or belt size or looking like an "ideal stereotype" as represented by multiple images in the media. An additional benefit of using metabolic fitness as the measure of success is that it is objective and can be viewed as such by both the health professional *and the patient.* The physician can display the blood chemistry and blood pressure report graphically in the individual's chart and mark all the abnormal, high normal, and normal values at the beginning of treatment. During each visit the individual's progress can be charted and, as parameters move back toward the normal range, both the health professional and the individual can feel "successful" (CAMPFIELD 1995, 1996).

IV. Drugs for the Treatment of Obesity

Drugs for the treatment of obesity can be classified according to the primary mechanism of action on energy balance. When daily energy intake is exactly balanced by daily energy expenditure, a state of energy balance is achieved and body weight is *constant*. If energy intake exceeds energy expenditure, then a state of *positive* energy balance is achieved and body weight will *increase*. If energy expenditure exceeds energy intake, then a state of *negative* energy balance is achieved and body weight will *decrease*. The goal of all antiobesity drugs is to induce and maintain a state of negative energy balance. This can be achieved by four classes of drugs:

1. Inhibitors of energy (food) intake (appetite suppressants)
2. Inhibitors of fat absorption
3. Enhancers of energy expenditure
4. Stimulators of fat mobilization

Each class has specific advantages. Appetite suppressants have the advantage that they reduce hunger perceptions and energy intake and make compliance to caloric restriction easier. Inhibitors of fat absorption have the advantage of reducing energy intake through a peripheral, gastrointestinal mechanism of action that is specific for fats without any alteration in brain neurochemistry. Enhancers of energy expenditure have the advantage of passively increasing energy expenditure through a peripheral action without behavioral change. Finally, stimulators of fat mobilization have the advantage of passively reducing fat mass and/or decreasing triglyceride synthesis without behavioral change. However, the beneficial actions of all four classes can be easily overcome by increased voluntary food intake (e.g., preferred calorically dense food items) or/and decreased voluntary physical activity (CAMPFIELD et al. 1998).

The major drugs used to treat obesity are shown in Table 3. Until recently, patients and physicians only had a small set of centrally acting appetite suppressants that reduced food intake by modulation of serotonin or/and norepinephrine concentrations within the brain. Modulation of serotonin, norepinephrine, and dopamine neurotransmitters in laboratory animals and humans is followed by changes in food intake (STOCK 1997). It has been clearly demonstrated that inhibition of serotonin re-uptake into serotonin neurons results in appetite suppression in humans (GUY-GRAND et al. 1989). The identification of the specific subtype(s) of serotonin receptors involved in the regulation of energy intake and body energy balance is a major focus of research. The use of these appetite suppressants resulted in an average weight loss of

Table 3. Drugs used to treat obesity (from CAMPFIELD et al. 1998)

Drug	Target	Mechanism	Status
Inhibitors of energy intake (appetite suppressants)			
Fenfluramine serotonin release	Serotonin neurons	Serotonin re-uptake	Withdrawn
Phentermine neurons	Norepinephrine re-uptake	Norepinephrine	FDA approval
Fenfluramine/Phentermine (fen/phen)	Serotonin and norepinephrine neurons	Serotonin re-uptake serotonin release norepinephrine re-uptake	Combination of individually approved drugs (fenfluramine now withdrawn)
Dexfenfluramine (Redux)	Serotonin neurons	Serotonin re-uptake	Withdrawn
Sibutramine (Meridia) neurons	Serotonin neurons norepinephrine re-uptake	Serotonin re-uptake norepinephrine	FDA approval
OB protein (leptin)	OB protein	Activate OB-R receptor in brain and reduce food intake	In clinical trials
Inhibitors of fat absorption			
Orlistat (Xenical)	Pancreatic lipase	Inhibition of fat absorption	Under FDA review
Stimulators of fat mobilization			
OB protein (leptin)	OB protein receptor	Mobilization of fat mass	In clinical trials

approximately 10% of initial body weight and these drugs helped many patients increase their metabolic fitness, reduce risk factors, and improve their health. Many patients report that these drugs helped them control their food intake. Although safety concerns were raised that dexfenfluramine (and fenfluramine) might increase the risk of primary pulmonary hypotension and possible loss of serotonin neurons, it was approved by the FDA in June 1996 based on risk/benefit ratio and extensive clinical experience in Europe. It was the first antiobesity drug approved in the United States in over 20 years.

Unexpectedly, a cluster of reports of heart valve disease (valve leakage seen best on the echo cardiogram) in obese patients treated with fenfluramine and phenteramine or dexfenfluramine rapidly emerged during the summer of 1997. It is not yet known if the heart valve leakage observed in these obese patients resulted from the central or peripheral serotonin system or another unknown mechanism. Following reporting of a large number of cases of heart valve leakage by physicians throughout the United States and analysis of placebo controlled clinical trials using these drugs, fenfluramine and dexfenfluramine were withdrawn from the global market by the manufacturer. The NIH issued a recommendation in October 1997 that all individuals treated with these drugs, alone or in combination with phenteramine, visit their physicians for assessment of their medical condition. The many patients that felt these drugs were helping them were then left without medication to reinforce their attempts at behavioral and lifestyle change.

This situation was reversed by the FDA approval and market entry of sibutramine (Meridia), a combined serotonin and norepinephrine re-uptake inhibitor developed by Knoll Pharmaceuticals (Bray et al. 1996). It is a compound that acts on serotonin and norepinephrine neurons in the brain to suppress appetite. In contrast, orlistat (Xenical), a new treatment developed by Hoffmann-La Roche that is marketed in many countries and is awaiting final approval from the FDA, acts on an entirely different mechanism that does not affect the brain. It specifically targets pancreatic lipase molecules that digests fat into fatty acids and mono-glycerides that can be absorbed into the body. When the lipases are blocked, approximately one-third of the fat passes through the system and is excreted, helping reduce the amount of fat absorbed, thus preventing that fat from being stored by the body (Drent et al. 1995).

These drugs should be followed by future medicines based on OB protein and OB-R receptor pathway that should not only suppress appetite but also reduce the amount of body fat. The discovery other human and mouse obesity genes will probably result in the development of promising new medicines (Campfield et al. 1997; Campfield and Smith 1998).

V. Multiple Drugs with Distinct Mechanisms of Action

A consensus is also emerging that multiple antiobesity drugs, with distinct mechanisms of actions, are not only desirable but also needed, and that combined or multiple drug treatment will play an increasingly important role in future obesity treatment (Campfield et al. 1998). Combined drug treatment of obesity has been popularized by the original combined centrally acting appetite suppressant fenfluramine/phenteramine studies of Weintraub (Weintraub 1992) and became widespread as the popularity of "Fen/Phen" grew. As new drugs with distinct but complimentary mechanisms of action enter the marketplace, it is expected that they will be prescribed sequentially or in combination for obese patients as drugs for other chronic medical conditions are commonly used.

As discussed above, each primary mechanism of action has the disadvantage that changes on the other side of the energy balance equation can overcome or nullify its action. If the combination of a CNS acting (e.g., appetite suppressant) and peripherally acting (e.g., enhancer of energy expenditure) antiobesity drugs demonstrates enhanced efficacy with acceptable safety, it should become widely used. One "ideal" combination would be one drug from each category: an inhibitor of energy intake, an inhibitor of fat absorption, an enhancer of energy expenditure, and a stimulator of fat mobilization. This combination treatment would actively limit energy intake and fat absorption and would actively increase energy expenditure and fat mobilization. Although combined treatment with four separate antiobesity drugs is probably not desirable, perhaps some new drugs may combine two or more of these biological actions. Indeed, if OB protein has the same biological activities in humans as have been established in rats and mice, a single drug that activates the OB protein pathway would have all of these benefits.

Another new trend gaining in appreciation and support is treatment for a very long time, possibly, for life, or at least intermittent treatment during the multiple periods of adult life. The idea would be to reinstitute treatment, or components of the treatment program, when an individual begins to regain weight or becomes more metabolically unfit.

The risk-benefit and quality-of-life analyses of pharmacological treatment of obesity are also important issues. Vigorous dialog and discussion are needed to define, in objective and quantifiable terms, the minimum efficacy required to justify long-term, and possibly chronic, drug treatment for obesity among the various communities involved. Since obesity is a chronic disease, drugs for the treatment of obesity must have acceptable risk/benefit ratios. An additional safety consideration is that, up to the present, women make up the largest group seeking treatment. Thus, undesired effects on reproductive function or hormonal status must be avoided. As new antiobesity drugs are approved, and gain the confidence of health professionals, the intervention studies required to define the minimum efficacy will be performed. The incremental improvements in knowledge about the biology of obesity are expected to lead to incremental improvements in efficacy and safety of new obesity drugs. Hopefully, two generations of drugs from now, efficacy will have increased so that risk/benefit and quality-of-life analyses will be greatly improved.

D. New Tools from Explosion in Obesity Research

The new science of obesity has led to new potential targets for new drugs. If molecules that selectively and specifically interact with these new targets with the appropriate profile to become drugs can be discovered, successfully developed and tested in clinical trials with a minimum of "bad" luck, then new medicines will be available to assist obese individuals to improve their health and quality of life.

I. OB Protein Pathway

The most significant recent discovery was the identification and positional cloning of the *ob* gene (ZHANG et al. 1994). The gene, expressed primarily in fat tissue, produces a protein that circulates in the blood, called OB protein or leptin, a key player in the regulation of body fat and body weight. It was rapidly demonstrated that OB protein had long-lasting biological activity when administered intraperitoneally, subcutaneously, intravenously, and intracerebroventricularly) and that it reversed many of the abnormalities observed in the OB protein deficient, *ob/ob* mice. OB protein also had impressive, long-lasting biological effects when administered to lean mice, rats, and monkeys. Next, it was discovered that obese individuals have increased circulating levels of OB protein in their blood, suggesting that obesity was due to a decreased sensitivity to OB protein, not a deficiency of OB protein. This was followed by the identification and cloning of the receptor that mediates the effects of the OB protein, OB-R, and rapidly three groups confirmed that the *ob-r* gene and the *db* gene, which also causes obesity in mice, are identical. Many laboratories have contributed to determining the wide range of interesting biological activities of OB protein (for reviews see CAMPFIELD et al. 1997; CAMPFIELD and SMITH 1998).

The next step was to determine if obese patients had mutations in the *ob* gene or the OB-R receptor. After screening thousand of obese individuals, mutations in the *ob* and *ob-r* genes have been identified in three different families. Initially, two cousins with severe obesity were identified with mutations in the *ob* gene and, as expected, they have little or no functional OB protein circulating in their blood. They were both normal weight at birth and then rapidly gained weight and became obese, just like the OB protein deficient obese mice (MONTAGUE et al. 1997). Three members of another family, of Turkish descent, have just been identified with a homozygous missense mutation in the *ob* gene that results in low plasma OB protein concentration and severe obesity (STROBEL et al. 1998). Recently, three severely obese sisters in a large consanguineous family of Kabilian origin were found to be homozygous for a splice-site mutation in the *ob-r* gene; this mutation resulted in a truncated form of OB-R (no transmembrane or intracellular domains) with no signaling function. There were no signs of pubertal development in two of the sisters (CLEMENT et al. 1998).These discoveries strongly suggests that OB protein plays an important role in the regulation of body fat and body weight and reproductive function in humans.

II. Mouse and Human Obesity Genes

Although the link between the OB protein pathway and obesity may be the most profound discovery to date in the field of genetics of metabolism, research in this area has implicated several additional genes that may be involved in human obesity.

Mutation in the *agouti* gene was found to be responsible for obesity in a breed of bright yellow mice (BULTMAN et al. 1992). Recent genetic and pharmacological research has made a strong argument that the AGOUTI protein blocks the MC4-R in the brain and causes obesity. In mice lacking the MC4-R, late-onset obesity and altered peripheral metabolism were observed (HUSZAR et al. 1997). This result, combined with experiments with MC4-R agonists (decreased food intake) and antagonists (increased food intake), indicates that the MC4-R lies in a new physiological pathway that normally inhibits food intake and fat storage, and play an important role in late-onset obesity (FAN et al. 1997).

Mutation in the *fat* gene (NAGGERT et al. 1995) makes mice obese by decreasing the amount of an enzyme that is involved in the final stages of processing insulin and POMC and other hormones. This observation took on additional importance when an obese individual was identified with mutations in another protein processing enzyme called human prohormone convertase 1 gene, or PC-1. PC-1 works at the step just before the processing enzyme encoded by the *fat* gene (JACKSON et al. 1997). Together, these findings indicate that one or more fragments of one or more larger protein and/or prohormone must be correctly processed for mice and humans to remain lean.

Although the *tub* gene has been cloned and the TUB protein expressed, the biology of the *tub* gene is presently unknown (KLEYN et al. 1996; NOBEN-TRAUTH et al. 1996). The molecular explanation of the late-onset, mild obesity associated with the *tub* mutation in mice is also unknown.

The longest studied brain neuropeptide involved in energy balance is neuropeptide Y (NPY). NPY is the most widely distributed neuropeptide in the brain with multiple biological effects. It is one of the most potent appetite stimulators in animals and it appears to be one of the mediators of the actions of OB protein in the brain (CLARK et al. 1984). Research is focused on the identification of one or more NPY receptor subtypes (e.g., Y5, Y1) that mediate the effects of NPY on food intake and energy balance.

Two genes which are expressed in peripheral tissue are uncoupling proteins 2 (UCP2) and 3 (UCP3), which are members of the family of protein transporters, that may, when activated, cause increased thermogenesis leading to less calories stored as fat in animals (BOSS et al. 1997; FLEURY et al. 1997;

GIMENO et al. 1997; VIDAL-PUIG et al. 1997). If so, they will be interesting targets for drug discovery.

Each of these genetic findings offers researchers new insight in the biological pathways involved in energy balance. It is reasonable to expect that potential targets for new drugs to treat obesity will emerge from the new knowledge and our increased understanding of this disease (Table 4). These genetic discoveries also emphasize the critical importance of multiple animal models in obesity research and drug discovery. Potential drug for obesity treatment must demonstrate efficacy in obese animal models.

A potential target can be any intervention between the neuropeptide and its associated receptor that would alter the biological responses mediated by the neuronal network, in particular food intake, metabolism, and energy expenditure. They are categorized as inhibitors of energy intake, enhancers of energy expenditure, and stimulators of fat mobilization.

Among the potential inhibitors of energy intake, many pharmaceutical companies have large development programs directed at specific NPY receptor (Y5, Y1) antagonists as an approach to reduce food intake. Among the newer targets, agonists of MC4-R have attracted a lot of attention. If selective MC4-R agonists can be developed, they should be effective inhibitors of food intake. Since AGOUTI protein inhibits the MC4-R, agonists of the agouti-related peptides may also provide an MC4-R based appetite suppressant. Since it is widely believed that the endogenous ligand of the MC4-R is alpha-MSH or another product resulting from POMC processing, increased POMC levels may also result in decreased food intake. A related neuropeptide, melanocyte-concentrating hormone (MCH), has been implicated in the

Table 4. Potential therapeutic targets for drug development in the treatment of obesity (after CAMPFIELD et al. 1998)

Inhibitors of energy intake (appetite suppressants)	GLP-1 agonists
Monoamine neurotransmitter modulators -serotonin dopamine, dopamine, norepinephrine	Bombesin agonists
OB protein	Enhancers of energy expenditure –
OB receptor	Stimulators of UCP2/3 expression/activity
NPY antagonists/ Y5, Y1 receptor antagonists	PKA activator/stimulators
MC-4R agonists	β-3 adrenergic agonists
Agouti-related peptides	Inhibitors of fat absorption
POMC antagonists	Gastrointestinal lipases
MCH antagonist	Stimulators of fat mobilization
CRH antagonists/binding proteins	OB protein
Urocortin antagonists	PKA stimulators
Galanin antagonists	β-3 adrenergic agonists
Orexin antagonists	GH
CCK agonists	

control of food intake. When MCH was injected into the brain of rats, food intake was increased. Once the MCH receptor is cloned, antagonists of this receptor should also reduce food intake. Brain administration of corticotrophin releasing hormone (CRH) and the closely related urocortin result in decreased food intake and administration of OB protein causes increased expression of the *crh* gene. These observations suggest that CRH receptor antagonists and CRH binding proteins provide potential targets for novel appetite suppressants. Brain administration of galanin caused increased food intake in rats; thus, galanin antagonists should reduce food intake. Opioid peptides also increase food intake when administered to the brain and opioid receptor subtype specific antagonists suppress food intake in animals. Another stimulator of food intake in rats, orexin, was recently described. If this target is validated in further studies, orexin receptor antagonists may be another potential target for inhibition of food intake.

Three well-known gastrointestinal hormones are also targets for the development of appetite suppressants. Cholecystokinin (CCK) is released from the intestine by food and has been shown to play an important role in termination of meals. CCK-A receptor agonists have been shown to reduce meal size and food intake. Bombesin is a peptide that also reduced food intake when injected into rats and mice. Thus, a bombesin receptor agonist may be useful to inhibit food intake. Glucagon-like peptide-1 (GLP-1) has been under study as an endogenous stimulator of insulin release from the pancreas in humans. Recently it was shown that brain administration of GLP-1 to rats reduced food intake. If gastrointestinal side effects and non-specific effects can be avoided, GLP-1 receptor agonists may also be useful for reducing energy intake.

Targets for enhancers of energy expenditure consist of two new targets and a target extensively studied by the pharmaceutical industry. One set of targets is UCP-2 and UCP-3 as described above. If these proton transporters lead to increased thermogenesis, then drugs that stimulate UCPs should increase energy expenditure. The other is the well-known enzyme PKA. When regulation of PKA was disrupted in a PKA-subunit RII β knock-out mouse, a genetically lean, obesity resistant mouse resulted. Thus, if PKA could be stimulated by a drug, increased thermogenesis and fat mobilization may result. The remaining target is the very well established β-3 adrenergic receptor. Early non-selective drugs that were β-adrenergic receptor agonists increased thermogenesis in obese humans, but had unwanted side effects. These drugs also increase fat mobilization in animals. Newer, apparently selective, β-3 adrenergic receptor agonists have failed to demonstrate increased thermogenesis in humans. Additional efforts to develop β-3 adrenergic receptor agonists as stimulator of energy expenditure and fat mobilization for the treatment of obesity continue.

Targets for stimulators of fat mobilization include growth hormone (GH) receptor agonists, in addition to OB protein, PKA stimulators, and β-3 adrenergic receptor agonists. In animal studies, administration of GH increased lean muscle mass and reduced fat mass. The role of GH in the treatment of obesity is currently being evaluated in clinical trials (CAMPFIELD et al. 1998).

III. Example: The Concept of Reduced Sensitivity to OB Protein in Obesity

Strong support for the concept of *reduced sensitivity to OB protein in obesity* is provided by the observation of elevated OB protein concentrations in the blood of obese individuals and the experimental result that higher doses are required to effect feeding behavior, metabolism, and body fat in diet-induced obese mice (CAMPFIELD et al. 1995; CONSIDINE et al. 1996). At one extreme of the continuum of OB protein responsiveness is the obese *db/db* mouse, with elevated OB protein levels, which is totally unresponsive to OB protein, while at the other end is the obese *ob/ob* mouse, OB protein-deficient, which is very responsive to OB protein.

Movement along the continuum of OB protein sensitivity as a function of adiposity and back again has been demonstrated in studies of diet-induced obese (DIO) mice. When lean AKR/J mice are fed a high fat, energy dense diet, they become obese with elevated OB protein and insulin concentrations (CAMPFIELD et al. 1995). Our demonstration that these DIO obese mice required higher IP doses of OB protein to reduce food intake contributed to the concept of OB protein resistance. Recently, we have shown that DIO mice have decreased central sensitivity to OB protein (CAMPFIELD et al. 1997). In these studies, we determined the sensitivity to ICV injection of mouse OB protein in DIO mice after changes in diet. These results demonstrate that expansion of the adipose tissue mass, as a result of high fat diet feeding, is associated with decreased brain sensitivity to OB protein injected ICV. Thus these studies with DIO mice indicate that the brain sensitivity to OB protein of the neural network in the brain controlling energy balance is decreased by weight gain and can be reversed by weight loss (CAMPFIELD et al. 1997).

Before the discovery of OB protein, the brain mechanisms controlling body fat mass could be viewed as a "black box" that contained the "decision rules" that determine how much body fat a person stores. We had only a few clues as to what was in the box, which neuronal pathways provided inputs to the box and carried signals from the box before OB protein. Research with DIO mice has shed light on the box and the elucidation of the OB protein pathway has opened part of the box, the light that entered transformed it from black to gray, and we have a much better picture of what is in the box and clearer sense of what we have to do to regulate a person's weight. Since obese patients seem to have decreased brain sensitivity to OB protein, the goal is to identify the molecular sites of the decreased sensitivity.

Once the sites and mechanisms of the decreased brain sensitivity to OB protein are identified, the goal will be to discover a small molecular weight compound, that can be taken orally as a capsule or tablet, that circulates in the blood, and that enters the brain and replaces and overcomes the weak link in the communication or signaling pathway. Discovery of such a drug that overcomes the weak link in the OB protein pathway in obese individuals that prevents the message conveyed by OB protein from getting through to the critical neuronal network would be a great advance. Such a drug would provide an elegant solution to the challenge of obesity treatment.

E. Conclusion

Each advance in our knowledge of obesity and the brain mechanisms involved in OB protein signaling and the other mouse and human obesity-associated genes and gene products will accelerate the development of effective and safe drugs which should assist obese individuals to reduce and then chronically maintain a lower amount of body fat that will result in improved health (CAMPFIELD et al. 1998). However the process of discovering and developing new medicines require a large commitment of resources (human and financial), skill, and considerable luck. The selection of which targets to investigate is greatly improved by additional scientific information. The spark created by a single paper in Science, Nature, or other high impact journal, describing a new target, is not sufficient and has to be followed by a thorough, systematic step-by-step exploration of the mechanism or/and molecules involved. Complementary investigations at several levels, including pharmacological, physiological, and behavioral, are necessary to assess the potential impact of this new target, including its limitations. Deep and systematic new knowledge will certainly lead to more rational target selection that, in turn, will lead to improved drugs for treatment of obesity. The resulting new, innovative medicines will be used as part of sequential or combined treatment programs as adjuncts to lifestyle change to improve metabolic fitness and to improve the health and quality of life for obese individuals. Although the path to innovative medicines for improving the treatment of obesity is a crooked one strewn with many obstacles, the pace of significant discoveries in obesity research is so rapid that the future of obesity treatment seems bright.

References

Allison DB, Pi-Sunyer FX (eds) (1995) Obesity treatment: establishing goals, improving outcomes, and reviewing the research agenda. Plenum Press, New York
Björntorp P, Brodoff BN (eds) (1992) Obesity. J.B. Lippincott, Philadelphia
Blackburn GL, Miller D, Chan S (1997) Pharmaceutical treatment of obesity. Nurs Clin North Am 32:831–848
Boss O, Samec S, Paoloni-Giacobino A, Rossier C, Dulloo A, Seydoux J, Muzzin P, Giabobino JP (1997) Uncoupling protein-3: a new member of the mitochondrial carrier family with tissue-specific expression. FEBS Letters 408:39–42
Bouchard C, Perusse L (1993) Genetics of Obesity. Annu Rev Nutr 13:337–354
Bray GA, Ryan DH, Gordon D, Heidingsfelder S, Cerise F, Wilson K (1996) A double-blind randomized placebo-controlled trial of sibutramine. Obesity Research 4:263–270

Bultman S, Michaud E, Woychik R (1992) Molecular characterization of the mouse agouti locus. Cell 71:1195–1204
Campfield LA (1995) Treatment Options and the Maintenance of Weight Loss. In: Allison DB, Pi-Sunyer FX (eds) Obesity treatment: establishing goals, improving outcomes, and reviewing the research agenda. Plenum Press, New York, pp 93–95
Campfield LA (1996) The role of pharmacological agents in the treatment of obesity. In: Dalton S (ed) Overweight and weight management. Aspen Publishers, Gaithersburg, MD, pp 466–485
Campfield LA, Smith FJ (1998) Overview: neurobiology of OB protein (leptin). Proceedings of the Nutrition Society 57:429–440
Campfield LA, Smith FJ, Burn P (1997) OB Protein: A hormonal controller of central neural networks mediating behavioral, metabolic and neuroendocrine responses. Endocrinol Metab 4:81–102
Campfield LA, Smith FJ, Burn P (1998) Strategies and potential molecular targets for obesity treatment. Science 280:1383–1387
Campfield LA, Smith FJ, Guisez Y, Devos R, Burn P (1995) Recombinant mouse OB protein: evidence for a peripheral signal linking adiposity and central neural networks. Science 269:546–549
Campfield LA, Smith FJ, Yu J, Renzetti M, Simko B, Baralt M, Mackie G, Tenenbaum R, Smith W (1997) Dietary obesity induces decreased central sensitivity to exogenous OB protein (leptin) which is reversed by weight loss. Soc Neurosci Abstr 23:815
Clark JT, Kalra PS, Crowley WR, Kalra SP (1984) Neuropeptide Y and human pancreatic polypeptide stimulate feeding behavior in rats. Endocrinology 115:427–429
Clement K, Valisse C, Lahlou N, Cabrol S, Peeloux V, Cassuto D, Gourmelen M, Dina C, Chambaz J, Lacourte M, Basdevant A, Bougneres P, Lebouc Y, Froguel P, Guy-Grand B (1998) A mutation in the human leptin receptor gene causes obesity and pituitary dysfunction. Nature 392:398–401
Considine RV, Sinha MK, Heiman ML, Kriauciunas A, Stephens TW, Nyce MR, Ohannesian JP, Marco CC, McKee LJ, Baur TL, Caro JF (1996) Serum immunoreactive-leptin concentrations in normal-weight and obese humans. New England Journal Of Medicine 334:292–295
Drent ML, Larsson I, William-Olsson T, Quaade F, Czubayko F, von Bergmann K, Strobel W, Sjostrom L, van der Veen EA (1995) Orlistat (Ro 18-0647), a lipase inhibitor, in the treatment of human obesity: a multiple dose study. Int J Obesity 19:221–226
Fan W, Boston BA, Kesterson RA, Hruby VJ, Cone RD (1997) Melanocortinergic inhibition of feeding behavior and disruption with an agouti-mimetic. Nature
Fleury C, Neverova M, Collins S, Raimbault S, Champigny O, Levi-Meyrueis C, Bouillaud F, Seldin MF, Surwit RS, Ricquier D, Warden CH (1997) Uncoupling protein-2: a novel gene linked to obesity and hyperinsulinemia. Nature Genetics 15:269–272
Gimeno RE, Dembski M, Weng X, Deng N, Shyjan AW, Gimeno CJ, Iris F, Ellis SJ, Woolf EA, Tartaglia LA (1997) Cloning and characterization of an uncoupling protein homolog: a potential molecular mediator of human thermogenesis. Diabetes 46:900–906
Guy-Grand B, Apfelbaum M, Crepaldi G, Gries A, Lefebvre P, Turner P (1989) International trial of long-term dexfenfluramine in obesity. Lancet 2:1142–1144
Hill J, Peters JC (1998) Environmental contributions to the obesity epidemic. Science 280:1371–1374
Huszar D, Lynch CA, Fairchild-Huntress V, Dunmore JH, Fang Q, Berkemeier LR, Gu W, Kesterson RA, Boston BA, Cone RD, Smith FJ, Campfield LA, Burn P, Lee F (1997) Targeted disruption of the melanocortin-4 receptor results in obesity in mice. Cell 88:131–141
Jackson RS, Creemers JW, Ohagi S, Raffin-Sanson ML, Sanders L, Montague CT, Hutton JC, O'Rahilly S (1997) Obesity and impaired prohormone processing asso-

ciated with mutations in the human prohormone convertase 1 gene. Nature Genentics 16:303–306
Kleyn PW, Fan W, Kovats SG, Lee JJ, Pulido JC, Wu Y, Berkemeier LR, Misumi DJ, Holmgren L, Charlat O, Woolf EA, Tayber O, Brody T, Shu P, Hawkins F, Kennedy B, Baldini L, Ebeling C, Alperin GD, Deeds J, Lakey ND, Culpepper J, Chen H, Glucksmann-Kuis MA, Carlson GA, Duyk GM, Moore KJ (1996) Identification and characterization of the mouse obesity gene tubby: a member of a novel gene family. Cell 85:281–290
Leibel RL, Rosenbaum M, Hirsch J (1995) Changes in energy expenditure resulting from altered body weight. New England Journal of Medicine 232:621–628
Montague CT, Farooqi IS, Whitehead JP, Soos MA, Rau H, Wareham NJ, Sewter CP, Digby JE, Mohammed SN, Hurst JA, Cheetham CH, Earley AR, Barnett AH, Prins JB, O'Rahilly S (1997) Congenital leptin deficiency is associated with severe early-onset obesity in humans. Nature 387:903–908
Naggert J, Fricker L, Varlamov O, Nishina P, Rouille Y, Steiner D, Carrol R, Paigen B, Leiter E (1995) Hyperproinsulinemia in obese fat/fat mice associated with carboxypepidase E mutation which reduces enzyme activity. Nature Genetics 10:134–142
Naslund I (1996) Effects of weight reduction on the somatic, psychological and social complications of morbid obesity. In: Angel A, Anderson H, Bochard C (eds) Progress in obesity research: 7. John Libbey, London, pp 679–686
Noben-Trauth K, Naggert J, North M, Nishina P (1996) A candidate for the mouse mutation tubby. Nature 380:534–538
Pories WJ, Swanson MS, MacDonald KG, Long SB, Morris PG, Brown BM, Barakat HA, deRamon RA, Israel G, Dolezal JM, Dohm L (1995) Who would have thought it? An operation proves to be the most effective therapy for adult-onset diabetes mellitus. Ann Surg 222:339–352
Stock MJ (1997) Sibutramine: a review of the pharmacology of a novel anti-obesity agent. Int J Obesity 21:S25–S29
Strobel A, Issad T, Camoin L, Ozata M, Strosberg D (1998) A leptin missense mutation associated with hypogonadism and morbid obesity. Nature Genetics 18:213–215
Thomas PR (ed) (1995) Weighing the options: criteria for the evaluating weight-management programs. Food and Nutrition Board, Institute of Medicine, National Academy Press, Washington, D.C.
Vidal-Puig A, Solanes G, Grujic D, Flier JS, Lowell BB (1997) UCP3: an uncoupling protein homologue expressed preferentially and abundantly in skeletal muscle and brown adipose tissue. Biochem Biphys Res Commun 235:79–82
Weintraub M (1992) Long-term weight control study: conclusions. Clin Pharmacol Ther 51:642–646
Weintraub M (1992) Long-term weight control study: parts I-VII. Clin Pharmacol Ther 51:581–641
Zhang Y, Proenca R, Maffei M, Barone M, Leopold L, Friedman JM (1994) Positional cloning of the mouse *obese* gene and its human homologue. Nature 372:425–431

CHAPTER 9
Popular Diets and Other Treatments of Obesity

R. KONIKOFF and J. DWYER

A. Introduction

In today's global society, obesity is a problem that affects both industrialized and developing countries. Approximately 35% of American women and 31% of American men over 20 are overweight or obese (SHAPE UP AMERICA 1996). The evidence is overwhelming that obesity, measured by a body mass index, or BMI greater than 27.8 for males and 27.2 for females has adverse affects on health and longevity and diet (NATIONAL INSTITUTES OF HEALTH CONSENSUS DEVELOPMENT CONFERENCE STATEMENT HEALTH IMPLICATIONS OF OBESITY).

Health professionals need to be aware of the current treatments for obesity. The essential components are decreased energy intake, increased physical activity, and behavioral modification of lifestyles that affect these factors. They must be prepared to help their patients implement these essential components of all weight reduction efforts. Pharmacology and surgery are additional options that may be needed when obesity is severe.

This chapter focuses on dietary measures. We review the psychological and physiological means by which reducing diets work. We evaluate some popular reducing plans and other regimens and stress the importance of physical activity/exercise/behavior modification. After reading the chapter, health professionals should be better able to distinguish safe and effective from unsafe and ineffective therapies.

B. Causes of Obesity

Obesity is often misunderstood as a moral failing that involves gluttony and sloth. In fact, its roots are in genetic defects involving the misregulation of food intake. The phenotype expression of this predisposition to obesity involves environmental and psychological factors. Imbalances between energy intake and energy output are particularly likely to occur in individuals with certain genetic backgrounds and in certain environments that favor physical inactivity and inexpensive, tasty high calorie food.

The empirical observation that it takes 3500 calories to gain or lose a pound of fat is not as straightforward as it seems. If adipose tissue were solid lipid, it would take only 388g of fat at 9 calories/g to produce 3500 calories. Assuming the fat were hydrated at approximately 20% water, this would mean that about 465g fat tissue would be needed to produce the 3500 calories. Another theoretical calculation takes better account of the fact that adipose tissue is not made of simply lipid and water. It begins with observation that the composition of the tissue of weight loss is approximately 25% lean and 75% fat tissue. The energy lost by catabolism of fat free mass is approximately 1.2 calories/g (since fat free mass is 73% water providing no calories and the rest is protein at 4 calories/g). Adipose tissue has an energy value of 7.2 calories/g since it is approximately 20% water.

Depending on assumptions for the tissue of loss and the degree to which compensatory changes in resting metabolism, energy cost of work, and discretionary physical activity occur, approximately 3307–3613 calories per pound would be necessary.

Excessive body fatness can be reduced in one of three ways: energy intake can be decreased; energy output can be increased; or both can be altered simultaneously to produce a net energy deficit. The size of the energy deficit between energy needs/energy expended determines the slope of the decline in weight over the time. If adherence to dietary measures is perfect, and the energy deficit is constant, the association between energy deficit and loss is linear and both body fat loss and weight loss are directly proportional to it. The amount of weight lost over time is also dependent on the individual's adherence to a weight reducing plan (which is never perfect) and to shifts in water balance (which usually occur over the short run).

The average energy intake for adult American men is about 2800 calories per day (DWYER 1980). The energy deficit to lose a pound of fat is 3500 calories. Yet for every 1000kcal decreased there will usually be a 330 calorie compensatory decrease in energy expenditure which equates to only a net 670kcal per 1000 calorie decrease that will contribute to actual weight loss. Among adults of similar height the excess weight in obese individuals compared to lean individuals consists of 75% fat tissue and 25% fat free mass. That is, only 25% of an obese person's excess body mass is metabolically active. When normal to slightly overweight individuals lose weight their lean tissue declines and their energy needs may fall dramatically because actively metabolizing tissue declines a great deal. When weight is lost among the obese, most of the loss is of fat tissue, and declines in resting metabolism are not so pronounced. The energy stored in excess is about 3500 calories per pound gained (SALTZMAN and ROBERTS 1996). Thus, during periods of energy imbalance, weight should roughly change by 1 lb for every 3500 calories of energy excess or deficit, assuming that there are no shifts in water balance. To lose 1 lb per week an individual would therefore need to consume 500 calories less per day then he need to maintain his current weight, so that a 3500 calorie deficit would occur after 7 days. In the first few weeks on a severely hypocaloric (e.g., deficits of 1000 or more calories/day) reducing diet, a dramatic weight loss often occurs even with such a deficit as the body adjusts to the hypocaloric state and shifts in water balance occur (VAN ITALLIE 1980). Water shifts are more apparent on some hypocaloric regimes, such as very low carbohydrate modified fasts, than on others. Fluid shifts cause a deviation from the linear rate of loss. While these fluid shifts usually are most dramatic at the beginning of a weight reducing plan, they may also occur at other points during weight loss. For example, when large amounts of carbohydrates are eaten by individuals on a very low calorie diet or a protein supplemented modified fast who engage in binge eating, they often gain several pounds of weight in a day.

After weight loss, the energy intake at which the individual will be in energy balance will be lower than that prior to weight reduction because of two factors. First there is a decrease in lean body mass owing to loss of lean tissue as well as fat during weight loss and thus there is less metabolically active tissue. Second is the decreased work of moving the lighter body. Therefore, the success of a reducing diet depends on the maintenance of decreased energy intake and increase in energy output slightly *over* that required prior to weight loss.

C. Criteria for Evaluating the Dietary Components of Weight Loss

One method for evaluating the soundness of hypocaloric diets focuses on seven essential criteria that must be met for healthful weight loss (DWYER and LU 1993). The "seven Cs" of weight loss include: the amount of *Calories* allotted in the eating plan, the *Composition* of the diet, the *Costs* associated with following the diet, the *Consumer* friendliness, how the plan *Copes with Co-existing health problems*, whether the *Components* of sound weight management are included, and if *Continuation* provisions for long term maintenance are included. Weight management schemes that meet the "seven Cs" can be considered safe and effective. Those that do not should not be recommended or only recommended to be undertaken with extensive medical support.

I. Calories

Daily energy needs are expressed in calories (kcal). The calculation of energy needs is helpful for determining what individuals are likely to need to maintain their current weight. Existing energy needs can be estimated using a formula based on the individual's resting metabolic rate (RMR) and a factor to adjust for physical activity (DWYER 1992). The RMR is the amount of energy expended when an individual is lying down or sitting quietly at rest. The RMR for males and females differs owing to differences in lean body mass, and hence gender specific equations are used. Below are the equations for estimating RMR (OWEN et al. 1986, 1987):

Men: 900 + 10 (body weight in kg) = RMR

Women: 800 + 7 (body weight in kg) = RMR

An activity factor multiplier is then applied to the RMR. Three activity factors are used: low, 1.2, moderate, 1.3, and high, 1.6. Table 1 describes the common activities in the three categories. For example, a waiter who walks a mile to and from work every morning, climbs several flights of stairs, is active at his job, and gardens on the weekends would have a moderate activity level. A woman who has a sedentary job as a telephone operator and climbs one flight of stairs at night would have a low activity level. A woman who engages in aerobic exercise and weight training 5 times a week for 2h a day would have a high activity factor (1.6) applied to her RMR.

The individual's usual energy need is a key consideration because the difference between it and the calorie level of the reducing diet will determine the rapidity of weight loss. An energy deficit can result from either eating less and/or increasing the amount of physical activity. In order to lose 1 lb per week a deficit of 500 calories per day from either of these sources is necessary. Slow weight loss is advisable since it is less metabolically stressful; no more than 2lbs per week should be lost (NATIONAL TASK FORCE ON THE PREVENTION AND TREATMENT OF OBESITY 1993). It may also improve the chances of long term maintenance since the dieter has the opportunity to develop lower calorie eating habits.

Table 1. Suggested activity factors for estimating energy expenditure in various activities [Reprinted from: DWYER JT (1992) Assessment of energy intake and expenditure. In: MICOZZI M, MOON T (eds) Macronutrients: investigating their role in cancer. Marcel Dekker, New York]

Activity	Activity Factor Per Unit Time
Resting (sleep, reclining)	1.0 × REE
"Sedentary" activity	1.2 × REE
"Very light" activity: seated and standing activities, painting trades, driving, laboratory work, typing, sewing, ironing, walking, playing cards, or a musical instrument	1.5 × REE
"Light" activity: walking on level surface at 2.5–3 mph, garage work, electrical trades, carpentry, restaurant trades, housecleaning, child care, golf, sailing, ping-pong	2.5 × REE
"Moderate" activity: walking 3.5–4 mph, weeding, hoeing, carrying a load, cycling, skiing, tennis, dancing	5.0 × REE
"Heavy" activity: walking with a load uphill, tree felling, heavy manual digging, basketball, climbing, football, soccer	7.0 × REE

REE, resting energy expenditure.

The next step is to obtain a rough estimate of energy intakes. This can be obtained from asking the patient to keep a diet record, and then calculating usual energy intakes from it. Usually, people underestimate their energy intakes by 10%–20%; this is particularly likely among heavy persons and among females. The caloric deficit can be calculated easily once current energy needs and energy intakes are known. For example, consider a moderately active, 5′ 9″, 176 lb, 55 year-old man who has energy needs of 2200 calories a day, has a usual intake of about 2700 calories, and is gaining weight. His energy deficit would be 500 calories on a diet of about 2200 calories, and he would be expected to lose a little over 1 lb per week if he adhered perfectly to the diet. On a 1700 calorie diet, his energy deficit would be 1000 calories. Therefore he would lose 2 lbs per week (e.g., 2700 calories −1700 = 1000 calories ×7 days = 7000 calories per week/3500 (1 lb) = 2 lbs lost per week) The same 1700 calorie diet for his 154 lb, moderately active wife with energy needs of only 1657 calories per day would have an excess of 43 calories; therefore she would not lose any weight on the same diet. These thermodynamic realities explain the differences often observed in practice between two individuals following the same reducing diet. Reducing diets are categorized by caloric level.

Calorie levels are categorized in three levels (Shape Up America 1996):

1. Moderate deficit diet: 1200 + calories per day for women; 1400 calories per day for men
2. Low calorie diet: 800–1200 calories per day for women, 800–1400 calories per day for men
3. Very low calorie diet: <800 calories per day for both women and men

Treatment options are based on the *least intensive therapy* that will produce the *best possible outcome* for the individual. The possible risks and side effects of each calorie level therefore need to be weighed against the outcome and potential benefits of each treatment. In order to select an optimal treatment the first step is to obtain the patient's weight history, history of attempts to increase physical activity, his/her other lifestyle changes, and a history of diets or other weight control efforts. Some dietary treatments may not be appropriate because they have not worked in the past or because of the risks they impose. Diets that restrict the individual to less than 800 calories per day are reserved for high (BMI > 27 with co-morbidities), and very high risk obese individuals (BMI > 30 without or with co-morbidities). A low calorie diet should be considered for moderately overweight individuals (BMI 25–27) or for high or very high risk persons where other treatments are not appropriate (Mahan and Escott-Stump 1996). A moderate deficit diet, which allows the most flexibility, is appropriate for most healthy people who need to decrease their weight. The decision on which regimen is most appropriate should be made by the individual with advice from knowledgeable healthcare professionals.

1. Fasting

Total fasting, or starvation, is contraindicated for all patients, especially those who are trying to lose weight without medical supervision. Fasting causes excessive loss of lean body mass and can have profound metabolic effects including diuresis, kaliuresis, saliuresis, and nutrient deficiency. Fasting is self-defeating because the loss of lean body mass and caloric deficit decrease the RMR, debilitate the individual, and make weight loss more difficult (Garrow 1995).

2. Very Low Calorie Diets

Very low calorie diets (VLCD) provide between 400 and 800 calories per day. The definition of a very low calorie diet depends on the patient in the diet and what his/her usual diet is. While this VLCD definition is widely accepted, it does not take into account the fact that obese individuals differ markedly in their daily energy requirements. The use of the same diet among different individuals can result in different caloric deficits. Larger individuals should receive more protein to preserve their greater lean body mass. For this reason, some adjustment should be made to account for this. Usually protein intakes are adjusted using desirable body weights as the criterion; otherwise very heavy persons would be prescribed very large amounts of protein. Further upward adjustment in protein intakes is also necessary because the efficacy of protein utilization for maintenance of lean body mass tends to be lower in hypocaloric diets, since some protein is catabolized for energy.

Energy deficits created in VLCD must also be considered. If a short (5-foot), but obese (150-lb) woman's energy needs (adjusted for obesity by accounting for metabolically active excess fat mass) are 2000 calories and she were on a 700 calorie per day program she would have a deficit of 1300 calories. However if a 6-foot, 290-lb man who required 3200 calories were on the same regimen he would have a 2500 calorie deficit. The former patient would lose about 2.5 lbs per week and the latter would lose 5 lbs per week on the same reducing diet. Atkinson has recognized this and suggests that a VLCD is one that provides an individual with less than 10 calories per day per kilogram or 4.3 calories per pound of ideal body weight (Atkinson 1990). "Ideal" or desirable weight is the appropriate healthy weight for height of an individual. It can be calculated roughly for women as 100 lbs for the first 5 feet and 5 lbs for every inch thereafter. Thus for the example, a VLCD would consist of a reducing diet below about 43 calories for the women. For men's ideal body weight the equation is 106 lbs for the first 5 feet and 6 lbs for every inch thereafter. For the man mentioned above a VLCD would consist of a reducing diet below 765 calories per day using Atkinson's rule of thumb.

Patient screening should include evaluations by several members of the health care team: physician, dietitian, and behavioral therapist. Factors to evaluate include medical necessity, degree of overweight, overall health status, psychological status, initial nutritional status, projected length of time on the VLCD, previous weight loss attempts, and the ability to understand and comply with the diet.

a) Patient Selection

A major consideration in deciding who should or should not go on a VLCD is the degree of obesity. Experience in the 1970s demonstrated that there was a greater increase of cardiac problems in lighter individuals, possibly because of problems with both lack of myocardial tissue protein repletion, electrolyte imbalances, or higher rates of body protein loss in leaner individuals (CENTERS FOR DISEASE CONTROL 1979). The American Dietetic Association recommends that VLCD should be used for individuals who are 30%–40% overweight (POSITION OF THE AMERICAN DIETETIC ASSOCIATION 1990).

Patients should be discouraged from self prescribing VLCDs to themselves because of potential health risks that may go unrecognized. A team of physicians, dietitians, psychologists, and nurses should be involved with the patient's care during a VLCD. With appropriate therapy, risks can be minimized and weight loss can be satisfactory. Some of the problems that arise in using VLCD are due to the small amount of food provided, which means that certain nutrients may be low.

VLCDs are effective in bringing about significant weight loss in obese patients; mean weight losses in women are approximately 20 kg in 12–16 weeks and in men 30 kg (WADDEN et al. 1990). These reductions are 2–3 times greater than those produced on a 1000–1500 kcal/day diet during the same time period, as might be predicted from the caloric deficit. The clinical impression is that comparable weight losses are achieved with "home made" protein supplemented modified fasts with appropriate vitamin and mineral supplements and commercially produced protein formula liquid diets.

With VLCD, weight loss is rapid, especially in the initial stages due to sodium loss. Such losses and the diuresis that accompanies them have a positive psychological effect on many patients. VLCD regimens are also beneficial in treating certain co-morbid conditions such as diabetes mellitus, hypertension, and hyperlipidemia (SEGAL 1990). Significant reductions in serum total cholesterol and triglycerides, markedly improved glucose and lowered plasma glucose levels in patients with non-insulin diabetes mellitus, lowered blood pressure in persons with hypertension, improved breathing in those with pulmonary problems, and diminished surgical risk for those needing surgery are often reported. However, improvements in medical conditions are only sustained to the extent that the weight loss is maintained.

b) Contraindications

Patients selected for VLCD should be people who are a minimum of 30% or more overweight and who have failed to lose weight with more conservative approaches (WADDEN and BARTLETT 1992). The risks and side effects of VLCD result from the metabolic effects of the semi-starvation they bring about (CENTERS FOR DISEASE CONTROL 1979) are not trivial and can be life-threatening in some patients. VLCD are not recommended for obese children because they may stunt linear growth. They are also contraindicated for adolescents under 18, pregnant women, lactating women, those over 65 years of age, persons who have had a recent heart attack or who suffer from type I insulin dependent diabetes, or multiple co-morbidities. Table 2 outlines some of the contraindications to using VLCD. Although VLCD diets are safe when used by appropriately selected patients and with careful medical supervision, side effects may still occur.

The Food and Drug Administration requires that every VLCD program contain a minimum of 70g of protein daily as well as the RDA or Adequate Intake for essential vitamins, potassium, magnesium, and sodium. A safe VLCD provides protein, nutrients, (especially calcium, iron, and electrolytes) and provides for the transition back to an eating pattern consisting of ordinary foods (SMOLLER et al. 1988).

c) *Multidisciplinary Approach*

The VLCDs are best provided in the context of a multidisciplinary team that includes a physician, dietitian, and behavioral psychologist. A physician with specialized training in the use of VLCD and medical consultation should be available to the patient throughout the treatment program. Medical personnel should be alert to the signs and symptoms associated with rapid weight loss such as orthostasis, abdominal pain, or joint pain that may be symptoms

Table 2. Contraindications secondary to VLCD use [Adapted from ATKINSON R (1989) Medical management of VLCD in NIDDM. On the cutting edge. Dietitians in Diabetes Care and Education Newsletter 9:3]

1. Absolute contraindications
   a. Malignant arrhythmias
   b. Unstable angina
   c. Protein wasting diseases (i.e., lupus, Cushing's syndrome)
   d. Major system failure
   e. Drug therapy causing protein wasting (i.e., steroids, anti neoplastic agents)
   f. Body weight less than 20% over "desirable" (e.g. BMI < 25)
   g. Pregnancy or lactation
2. Relative Contraindications
   a. Congestive heart failure
   b. Drug therapy with potassium wasting diuretics, adrenergic stimulating agents
   c. History of failed compliance with medical regimes
   d. Body weight less than 30% over "desirable" (BMI < 30)
   e. Substance abuse
3. Cautions
   a. Angina or history of heart disease
   b. Presence of systemic disease
   c. History of psychiatric or emotional disorder
   d. Chronic drug therapy (oral hypoglycemic agents, insulin)

of some of the negative side effects of VLCD (Shape Up America 1996). At a minimum, patients should have their weight, blood pressure, and pulse checked every 2 weeks while on a VLCD program and a physician should be available for monitoring the patient's health.

The dietitian should be available to assist patients while they are gradually transitioning back from the liquid diet to conventional foods and onto a maintenance diet after weight loss has occurred. Patients are usually treated weekly in groups of 8–15 persons that are led by a behavioral psychologist who instructs the participants in the modification of eating and exercise habits.

d) Preparation and Duration of Use

Guidelines for eating patterns should include re-feeding and maintenance as well as diet during the VLCD period. The VLCD should preceded by 2–4 weeks of a well- balanced 1200 calorie diet that allows time for the body to adjust to the caloric deficit and promotes a gradual diuresis. This avoids the rapid sodium and water loss seen with abrupt induction of the VLCD. The pre-VLCD diet also allows the health care team to evaluate the patient's ability to adhere to program guidelines. In the beginning phases of programs where a low carbohydrate program is prescribed, a mild ketosis can occur (Anderson et al. 1992).

The typical course of treatment on a VLCD lasts 20–30 weeks:

1. Pre-VLCD phase
2. The VLCD phase (12–16 weeks)
3. The refeeding period (4–6 weeks)
4. The weight maintenance(4–8 weeks)

Individuals vary in the amount of lean body mass loss they experience while on VLCD. Researchers and physicians experienced in the use of VLCDs recommended that the time on a VLCD be limited to 12–16 weeks, to minimize potentially dangerous effects from losses of lean tissue (Shape Up America 1996).

e) Nutritional Adequacy

The first well-documented use of VLCD diets similar to those seen today was described in the early 1930s (Howard 1989). These early VLCD diets provided between 400–600 calories per day and about 50 g per day of protein; similar regimes are still in use. In the mid 1970s the *Last Chance Diet* was introduced and it was one of the first popular VLCD books to be self prescribed by Americans (Quincy 1991). Unfortunately this diet book was sold with a product that consisted of hydrolyzed collagen, an incomplete protein consisting solely of glycine units that is very low in biological value. The product also did not provide adequate amounts of vitamins, minerals, and electrolytes. During use of various types of VLCD, including this liquid protein product,

60 deaths were reported to the US government, which prompted legal action by the Food and Drug Administration (FDA) to regulate commercial low calorie diet formulas that were designed to be used as a sole source of nutrition (Sours et al. 1981). VLCD use fell, and less restrictive and more adequate regimes such as the *Cambridge Diet* gained popularity. Its formulation was better from the nutritional standpoint. By 1983, five million people had tried the VLCD diet and for about 24% of its users it was the sole source of nutrition for 4 weeks or more (Carmen 1985).

In 1979, a joint FDA, NIH (National Institute of Health), and WHO (World Health Organization) scientific review panel recommended that the term "very low calorie diet" be reserved for a reducing diet providing fewer than 800 calories per day (Life Sciences Research Office 1979). Most commercially prepared VLCDs today provide 70–90g of protein per day. The protein in these commercial products is usually obtained from high biological value sources such as egg or dairy sources and is provided as a powdered formula which is mixed with water. These liquid diets usually provide 30–45 g of carbohydrates, 2–5 g of fat, and 100% of the US RDA for vitamins and minerals. In addition, drinking 2 l of a non-caloric beverage is usually prescribed (Atkinson 1989).

Alternatively, the food source of protein may be from lean meat, fish, or poultry served in a food form known as the "protein sparing modified fast" (PSMF). On most PSMF programs, carbohydrate is prohibited although glucogenic amino acids still provides some glucose. Fat is usually restricted to that present in the dietary protein source. This diet must be supplemented with both 2–5 g of potassium daily and a multivitamin/mineral preparation daily (Segal 1990).

VLCDs are most effective when they are administered as a part of a more general weight control program that includes physical activity, moderate amounts of exercise, behavior modification, and nutrition education to help promote maintenance of weight loss and long term weight control.

f) Adverse Effects of VLCD

Patients using nutritionally adequate VLCDs under careful supervision for 4–16 weeks sometimes report adverse effects, but these are generally minor, well tolerated, and diminish with time if the diet is administered appropriately. Common side effects include extreme fatigue or weakness, dizziness, constipation, hair loss, dry skin, brittle nails, and nausea. More significant side effects include gout, gallstones, and cardiac complications, but these are rare (Kirchner et al. 1987). These adverse reactions are most common during the period of re-feeding (when the patient is gradually reintroduced to solid foods) or during lapses on the regimen. Table 3 illustrates some of the serious potential side effects that have been due to misuse of VLCDs.

Table 3. Serious potential adverse effects due to misuse of very-low calorie diets (VLCD) [Modified from DWYER JT, LU D (1993) Popular diets for weight loss. In: STUNKARD A, WADDEN TA (eds) Obesity; theory and therapy, 2nd edn. Raven Press, New York, chap. 14]

Effect	Starvation	VLCD	Periodic Fasting
Lean body mass decrease	xxx	xx, if unsupervised x, if supervised	x
Linear growth decrease	xxx	xx, if unsupervised x, if supervised	x
Cardiac changes	xx	xx, if unsupervised	x
Disorder in water balance: disease, dehydration	xx	xx	x
Ketosis	xx	xx	x
Electrolyte imbalances	x	x	x
Nutrient deficiencies if not supplemented	xxx	xx	x

xxx, most pronounced; x, least pronounced.

g) Benefits vs Costs

Are the benefits of VLCDs as compared with traditional reducing diets worth it? A 26-week medically supervised VLCD program often costs $2500 and yet provides results 1 year later that are no better than those produced by less expensive approaches. Moreover, in many cases, these costs are not covered by insurance, leaving the individual responsible for the payment. Therefore, this treatment option may be too expensive and not be available to all those that may potentially benefit from it to receive it.

There are many hospital-based VLCD diet programs. Several products and commercial programs are available. The recommended or required foods, daily calorie levels, level of medical supervision, and degree to which physical activity and exercise are included vary from one commercial program to another. The VLCD range in cost and their availability depends on the location of the hospital and on adjustments made for patient income constraints; these adjustments are unfortunately rare. Some programs are entirely hospital-based and include a hospital stay, most are clinic based and patients might visit the center once a week. Table 4 lists some of the available programs.

Several popular diet books that suggest VLCD and do not recommend medical supervision are still in circulation. These include such titles such as *Lean for Life* that provides a diet of approximately 750 calories (GRAFF 1997). Most of the older books such as *The Last Chance Diet*, *Dr. Atkins Diet Revolutions*, (ATKINS 1970), and *Fasting as a Way of Life* are out of print, but unfor-

Table 4. VLCD programs and their adequacy using the seven Cs

Diet	Entry criteria	Cost	Medical supervision	Nutrition education behavior modification, exercise included?	Description/Comments
Optifast	Dieters must be at least 30% above their ideal body weight or at least 50 lbs overweight.	\$250–\$500 per month with the entire program costing \$1500–\$300. Approximately \$40 per pound.	Yes	Yes	Optifast is a medically supervised VLCD weight management program that average 25 weeks in duration. Caloric levels are different for men (800 calories) and women (420 calories) broken down into 50% carbohydrate, 35% protein, and 15% fat. 4 phases of the program: *Initial Phase*: Modified Fast *Refeeding* *Stabilization* *Maintenance*
Medifast	Must be at least 20% above ideal body weight as determined by the Medifast staff.	\$280–\$350 per month during maintenance. Total cost for the 34 week program is \$1880, excluding lab tests.	Yes	Yes, but not seen by a team of professionals	Medifast is primarily doctor based rather than hospital based and participants receive their supplements directly from the physician. One day medical evaluation done to screen patients. 3 phases of the program: *Phase 1*: Weight loss. 480–650 calories. *Phase 2*: Refeeding. Lasts 4–6 weeks. Reintroduces solid foods. Approximately 600–800 calories. *Phase 3*: Medifast – 12 weeks in duration.

Health Management Resources	Must be at least 20% overweight or 40 lbs overweight.	$420 per month during weight loss and $90 per month during maintenance.	Yes. Nationally hospital supervised.	Yes		3 phases: *Phase 1*: Weight loss. Lasts for a minimum of 12 weeks. 520–800 calories per day *Phase 2*: Transition. Lasts 4–6 weeks and focuses on transition to real foods and participants decrease use of supplements *Phase 3*: Maintenance. Lasts a minimum of 12 months.
New Directions	Must have BMI form 27–0 with comorbidities of 30–35 without comorbidities. must have at least 40 lbs to lose.	$400–$480 per month. Depends on location. Program approved for insurance reimbursement.	Yes	Yes		Medically supervised with intensive behavioral component. Sixteen week program with 6 phases: Initial phase: Patient Assessment. Seen by MD, RD, and Behavioral Counselor Reducing Phase: 600–840 kcals per day. Patient Reassessment: reseen by medical professional team Adapting Phase: practicing new eating and exercise habits patients gradually return to regular food. Maintenance 1: First 180 days of maintenance ideas heavily reinforced On-going Maintenance: Ongoing support to help participants practice relapse prevention.

Table 4. (*Continued*)

Diet	Entry criteria	Cost	Medical supervision	Nutrition education behavior modification, exercise included?	Description/Comments
Optitrim	Designed for people who have moderate amounts of weight to lose (less than 40 lbs) and who are not at risk for medical problems.	\$250–\$500 per month totaling approximately \$1500-\$3000 for entire treatment depending on location.	Yes	Yes	Program is clinic based, medically supervised and features low-calorie diet consists of powdered supplements and self-prepared or packaged meals. Patients follow a 950–1200 calorie program. After initial weight loss, patients are gradually transition to predominately self-prepared foods.
Weigh to Live	Must be 30% over ideal body weight and are at medical risk.	\$400–\$450 for medical supervision, powdered formula and 4–6 weeks of maintenance, for additional \$20 per month. This is the least expensive of the VLCD programs.	Yes	Yes	Clinic and hospital based program. Dieters can choose from among three variations during the weight loss phase *Phase 1*: Weight Control – lasts up to 16 weeks. Participants may choose from formula, prepackaged foods, or both, each giving ~650 calories *Phase 2*: Maintenance – goal to establish a precise calorie level. Lasts 4 weeks *Phase 3*: Access – helps dieters maintain weight loss by providing on-going and lifelong support.

tunately they are not off the library shelves, and some patients attempt to use them on their own. The side effects that are seen likely to be greater than in those medically supervised program. While they are not life-threatening, they can produce some undesirable side effects. Therefore, medical supervision is mandatory and self-prescribed VLCD are not recommended.

Some over the counter products (OTC) such as meal replacements are used by individuals in an attempt to construct "their own" VLCD. Common meal replacement products, such as Slim Fast or Ultra Slim Fast, which are readily available at most convenience stores, are amenable to such uses. These products are intended to be used as meal replacements along with fruit, snacks, and an additional meal for dinner. They should not be used as the only source of nutrients. Because of the potential for abuse of Slim Fast, Ultra Slim Fast, and other OTC meal replacement products, they are not recommended for self-prescribed VLCD. Oral nutrition supplements such as Ensure, Sustacal, or Boost provide between 190–239 calories per can and are not designed for such purposes.

h) Conclusions on VLCD

Long-term maintenance of weight loss with VLCD is disappointing, just as it is for all obesity treatments currently available. Even when VLCD that provide all recommended components are employed, most patients regain the weight within 5 years, just as they do on other regimens (Forster et al. 1996). There is no evidence that the use of VLCDs will lead to any greater long-term weight loss than any other dietary treatment. Patients should be encouraged to participate in long-term supportive treatment over the long term after the initial weight loss phase is completed and to focus on changes in lifestyle and behavior that will be most conducive to maintaining a healthy weight when eating a diet of conventional foods.

3. Low Calorie Diets (800–1200 kcal/day)

There is no sharp numerical distinction between what is considered a VLCD and a low calorie diet (LCD). Usually, LCD are considered to be in the range of 800–1200 kcal/day for women and 800–1400 kcal/day for men. The more calories the diet provides, the fewer metabolic complications are likely to be present. LCDs are still below the resting metabolic rate (RMR) of most adults. The average weight loss tends to be 1 lb per week if adherence is good. Most people who adhere to a LCD can expect approximately a 10%–15% loss in initial body weight over 10–20 weeks (Blackburn 1993). Because these diets are still quite hypocaloric, individuals should still obtain physician approval before beginning a LCD. The "Shape-Up America!" program guidelines state that this type of reduced calorie eating plan is appropriate for those individuals who have a BMI > 27 with no co-morbid conditions or those who have a BMI > 25 with co-morbid conditions such as history of gallbladder disease or cardiac abnormalities. Those best suited for a LCD include those who have medical conditions that will be improved with weight loss such as non insulin dependent diabetes, hypertension, and hypercholesteremia (ST JEOR et al. 1993).

Like other types of therapies, weight reduction, when prescribed without a plan for weight maintenance does not work. Weight is gradually regained. If a moderately hypocaloric diet is combined with physical activity and lifestyle change strategies, improved maintenance of the healthier weight can be expected. Studies of outcomes after 10–12 weeks in persons with average weight losses of 10%–15% of body weight show that the percentage of initial loss maintained varies from a low of approximately 40% to a high of 80%–100% (SHAPE UP AMERICA 1996). Higher percentages are invariably associated with intensive post-treatment support.

Low calorie diets may not meet the Recommended Daily Allowances for certain vitamins and minerals such as iron, calcium, zinc, and vitamin B6. Therefore, a multivitamin mineral supplement at RDA levels is recommended. For women, iron and calcium need special attention; the current Adequate Intakes (which satisfy the RDA for this nutrient) range from 1000–1200 mg per day. Supervision and follow-up are very important in a low calorie diet.

Table 4 lists many of the more recent low calorie diet programs that are available in the United States. Current commercial weight loss programs employing LCD include Weight Watchers, Diet Center, Jenny Craig, Nutri/System, and the OUTlook program. Initially, regimes are usually at levels of from 900 kcal/day to about 1100 kcal/day. The beginning phases of most of the programs are also relatively low in carbohydrate (about 150 g) to stimulate diuresis and provide the dieter with a large initial decrease in weight. Some of the programs provide packaged foods at various stages throughout the program. Others have specially designed programs for different types of clients that are based on client preferences during the weight loss phase. Usually, all of the programs increase the calorie levels to about 1200 calories per day and remain there after a very low calorie phase.

Many popular books recommend self administered LCD programs (Table 5). Recent favorites are Dr. Dean Ornish's *Eat More, Weigh Less* program that emphasizes a high carbohydrate and low fat regimen (ORNISH 1997). Ornish's plan recommends that total fat intake be reduced to 10%. From a practical standpoint this means that the caloric density of the diet is decreased by including liberal amounts of fruits, vegetables, and whole grains. Cardiovascular benefits from such a plan are that saturated fat, cholesterol, and sodium are reduced and dietary fiber intakes rise. *The Five Day Miracle Diet* (PUHN 1996) is questionable because it does not include some of the food groups that are major sources of nutrients and makes inappropriate claims, such as that blood glucose will decrease on the diet.

4. Balanced Calorie Deficit Diets

Diets referred to as balanced calorie deficit diets (BCDs) tend to be above 1200 calories/day and are appropriate for all risk groups. On 1200 calories per

Table 5. Recent self-administered diet therapy books and their adequacy using the seven Cs

Diet/Assessment	Calories: low calorie or moderate deficit diet	Composition: satisfies four basic food groups	Consumer friendliness/ cost	Coping with existing health problems	Components of sound weight management: nutrition education, behavior modification, exercise included	Description/Comments: continuation provisions included?
3 Day Energy Fast by Pamela Serure, New York: Harper Collins, 1997 Not recommended	1200+ Moderate deficit diet	No. No meat, bread, or milk. No solid foods	Requires juicer or blender, too	No	Encourages "gentle" exercises and breathing techniques	Three day fast designed to "detoxify" the body. Dieters cleans the impurities and stress from their life. No solid foods consumed. Diet consists of six "blenderized" drinks throughout the day 1600–1800 kcal options. Fasting is not recommended as a means for weight loss and can have serious medical implications
Eat More, Weigh Less by Dean Ornish. Recommended for those who do not object to vegetarian-like diets	1200+ No specific caloric restriction. Moderate deficit diet	Limited: low in meat	Low cost	No	Yes	What was once a diet for recovering heart attack patients is also a way to lose weight by eating a very low fat diet. Ornish recommends a diet with less than 10% total fat. Generally no calorie restriction but some low-fat/fat-free foods can only be eaten in moderation. Weight loss attributed to removal of fat from the diet therefore decreasing the caloric density of foods

Table 5. (*Continued*)

Diet/Assessment	Calories: low calorie or moderate deficit diet	Composition: satisfies four basic food groups	Consumer friendliness/cost	Coping with existing health problems	Components of sound weight management: nutrition education, behavior modification, exercise included	Description/Comments: continuation provisions included?
The Zone by Barry Sears, New York: Harper Collins, 1995 Questionable	1200+ (depends on protein needs). Moderate deficit diet	Yes. Emphasis on equal bread and meat group servings. Low in fruit	Low cost	No	Limited. Exercise tips are provided	Sears recommends a balance of protein and carbohydrates to create a "zone" for weight loss through maximal fat release and burning carbohydrates. Meals are set into equal blocks of carbohydrate, fat, and protein choices. Each block is one serving as specified in the book. Individuals can only eat enough block throughout the day to give them their individual protein amount, which is figure using the lean body mass of the individual. Although this diet's principle is based on protein intake, it actually provides no more than ~1700 calories which will cause weight loss in most individuals
4 Blood Types, 4 Diets by Peter D'Amato Questionable	1500 Moderate deficit diet	Yes. Emphasis of food groups differ with the diet designed for the blood type. Generally encourages low fat options	Low Cost	Yes	Limited. Exercise specific for each blood type	This is a naturopathic diet that is based on the chemistry of individuals blood types. Each type has different food choices that work best with their chemistry. Diet for each blood type provides ~1500 kcal. Diet not based on substantiated research and principles questionable

Five Day Miracle Diet by Adele Puhn Questionable	1200+, but portions are not specified, making it hard to calculate. Moderate deficit diet	Limited. Low in bread and milk groups	Low Cost	No	Yes	Diet to "stabilize blood sugar." Based on unsupported claims that eating specified foods at specific times throughout the day will decrease cravings for carbohydrates and therefore promote weight loss. Diet high in protein. Eating must occur at specific times throughout day. Snacks consist of hard chew foods (carrots, celery) and soft chew foods (oranges, cantaloupe) which are designed to decrease cravings and stabilize blood sugar. This diet does not lower blood sugar and has no scientific support for blood sugar control
Eating By Design by Carrie Wiatt Questionable.	1200+ Depends on the EQ plan. Moderate deficit diet	Yes. Emphasis of food groups vary among personality types and specific eating plan	Low cost	No	Limited	Program individualizes weight loss based on twelve personality types that express overeating in different ways. Gives meal plan based on personality type and energy needs for that individual. Energy needs are individualized calculated using an EQ (energy quotient) figured by genetic tendencies, weight loss attempts, and BMI

Table 5. (*Continued*)

Diet/Assessment	Calories: low calorie or moderate deficit diet	Composition: satisfies four basic food groups	Consumer friendliness/ cost	Coping with existing health problems	Components of sound weight management: nutrition education, behavior modification, exercise included	Description/Comments: continuation provisions included?
Lean for Life by Cynthia Stampfer, California: Griffen Publishing Co. 1997 Not recommended	800–1200 Low calorie	Limited. Low in bread, fruit, and vegetables. High in protein	Need to purchase ketone sticks	No. Following diets may be hazardous to one's health	Limited. Moderate exercise recommended.	Diet designed to induce ketosis by adhering to strict protein diet and consuming under 50 g of carbohydrates per day. Diet follows the Lindora Weight Loss Clinic's (Calf.) VLCD program. 3 phases: Phase 1: Weight Loss 28 day cycle, begins with a 3-day protein fast. ~600–700 kcals/d. Protein fast repeated at beginning of each week ~1200 kcal with carbohydrate restriction Phase 2: Metabolic Adjustment – 2 weeks. Add 2 more proteins and 1 serving carbohydrate ~1400 kcals Phase 3: Maintenance 1600–1700 kcals. Ketosis is not a medically acceptable way to lose weight. It can cause nutrient deficiencies and anyone on a VLCD should be monitored by a physician

Carbohydrate Addicts Diet by Richard and Rachel Heller Questionable	1200+ Moderate deficit diet	Limited. Low in bread group	Low Cost	Yes	Yes	This diet is based on the principle that certain people are genetically predisposed carbohydrate addicts and can not have too many carbohydrates in their diet or they will binge on carbohydrates. Diet provides 5 plans (A–E) that consist of various combination of diet meals and reward meals
Thin for Life by Anne Fletcher Vermont: Chapters, 1994 Recommended	1200+ Moderate deficit diet	Yes	Low cost	Yes	Yes	This diet is one that is based on the testimony of successful diet "masters" who have lost weight and kept it off. The actual diet is a 6-week "non-dieting weight control plan," that takes dieting one food group at a time. There is no set amount of food choices, however, each week there are tips to incorporate that food group as part of a healthy diet
Outsmarting the Female Fat Cell by Debra Waterhouse Questionable	1200+ Moderate Deficit Diet	Yes	Low Cost	Yes	Limited. Suggests exercise 45 min, 3× a week	Suggests that dieting is the reason that women are overweight. Recommends a low fat (20% of total calories) diet and the eating of smaller, more frequent meals

day the distribution of nutrients and the amount of energy provided are greater and more likely to be sufficient than on a LCD or VLCD. BCDs tend to produce fewer complications or contraindications than diets with fewer calories or those that are very low in carbohydrate owing to less physiological side effects. The amount of carbohydrate is higher than it is in the VLCD, alleviating side effects associated with rapid diuresis. Additionally, BCDs are usually easier to adhere to because the regimen is higher in calories. Most BCD programs allow individuals to eat foods that are readily available in the supermarket. In recent years commercially available meal replacement entrees that are used as components of BCD plans have been promoted. Several lines offer "lite" or low fat versions of their regular entrees such as Stouffer's Lean Cuisine, Budget Gourmet Light, and Ultra Slim Fast that can be used on such programs. These entrees range from 190 to 410 calories each. Weight Watchers also offers a single low calorie line that is low in fat and ranges from 200 to 400 calories per entree. The Weight Watcher products are keyed to the food exchange system that the program uses. All of these products provide consumers with a convenient and quick way to prepare healthy meals. However, these frozen entrees can be quite costly.

Ultra Slimfast, one of the frozen entrees, is designed to be part of a reducing program that uses two 220 calorie meal replacements for breakfast and lunch with the addition of fruits for snacks, and a 300–400-kcal frozen entree for dinner, with vegetables or salads, one slice of bread, and a low-fat dessert. If used as recommended, the Ultra Slimfast meal would provide approximately 1200kcal per day and meet all of the RDAs for vitamins and minerals. However, the danger is that not all users will follow the regimen. If an individual only drank the beverages for each meal he would consume less than 840 calories per day and levels of some vitamins and minerals would fall short of the Recommended Daily Allowances.

Overall, balanced calorie diets give the consumer more choice and greater variety than LCD or VLCD. While a 1200-calorie level does provide more than LCD or VLCD, this amount is not enough for every patient For extremely obese individuals and large males, more than 1200 calories per day may be necessary. For example, a 200-lb male who is moderately active and has energy needs of 2500kcal per day would experience a weight loss of greater than 2lbs a week on a 1200kcal per day program. A weight loss that is too rapid (>2lbs per week) may have greater adverse health consequences than a more moderate weight loss, especially for those with co-morbidities.

5. Exercise and Behavior Modification

At the same time a balanced calorie diet is initiated, physical activity and exercise and behavior modification should also begin. Exercise will help increase energy output, and help counter the diet induced decrease in RMR. The physical activity combined with the dietary modifications produces greater weight loss than dieting alone (BALIR 1993).

Behavior modification is also an essential part of a sound weight loss program. It involves identifying eating or related lifestyle behaviors to be modified, setting specific behavioral goals, modifying the determinants of the behavior to be changed, and reinforcing the desired behavior. Patients treated with behavior therapy and a 1200-calorie per day diet with good adherence, maintain approximately 65%–70% of their weight loss one year following treatments (American College of Sports Medicine 1995).

A good way for individuals to receive such personalized dietary and lifestyle care specific to their needs is to make an appointment with a registered dietitian (RD). An RD usually works with a team of health care providers such as psychologist, physician, and exercise physiologist who provides additional expertise to help the individual lose weight.

Many voluntary, non-profit and commercial weight loss programs offer sound nutrition advice combined with group support, a physical activity program, and behavior modification. Some commercial programs that offer sound nutritional therapy are listed in Table 4. The programs that use their own pre-packaged food as a component of the program include Nutri/System and Jenny Craig. These programs sell most of the individual foods that dieters will need each week. Other programs such as Weight Watchers, Diet Center, and Physician's Weight Loss Center also recommend that the dieter use portion controlled foods that are available in the supermarket. There are both advantages and disadvantages to these approaches. Having the entire week's worth of food already prepared is more convenient and may remove some of the choices and temptations to eat large portions that some dieters find difficult to overcome. However, use of these foods does tend to be more expensive than shopping at the supermarket. Patients need to review the various strengths and weaknesses carefully before choosing a program. All weight loss programs have a high drop out rate. For those who persist it is important to determine if the program offers maintenance programs. Some programs even include "lifetime membership" that enables participants to return to the program cost free when they feel it necessary.

Many of the commercial programs mentioned above are quite costly, ranging from $20 to $300 per month (not including food purchased). Less expensive programs include Overeaters Anonymous and TOPS (Take Off Pounds Sensibly). Both of these programs provide psychological support with group counseling and self-help techniques and modest entrance fees.

Table 5 lists recent books recommending a low calorie diets and reviews their strengths and weaknesses using the seven Cs criteria mentioned earlier in this chapter. Table 6 does the same for commercial diet programs. Some sound books with reasonable approaches include *Eat More, Weigh Less*, *Eating By Design* (Wiatt 1995), and *Outsmarting the Female Fat Cell* (Waterhouse 1993). Some diet books are questionable or not recommended. *The Zone* (Sears 1995) by Barry Sears uses a high protein, high fat low carbohydrate approach, and, like *Four Blood Types, 4 Diets*(D'adomo 1996), makes questionable scientific claims. *The Three Day Energy Fast* (Seure 1997) actually

Table 6. Commercial diet programs and their adequacy using the seven Cs

Diet	Entry criteria	Cost	Medical supervision?	Nutrition education, behavior modification, exercise included?	Description
Weight Watchers	Must be at least 5 lbs above minimum ideal body weight	\$14–17 entry fee and \$10–14 per week	Not medically supervised but encourages physician referral	Encouraged but not provided by a multidisciplinary team of professionals	Program designed to promote gradual weight loss of up to 2 lbs per week. Offers three food plan options: *Classic selections, Fat and Fiber*, and *Superstart* based on needs and preferences of dieter. Group meetings occur weekly and discuss different topics each week
Diet Center	18 and over. If 50 lbs or over must get a medical authorization	\$35–50 per week for weight loss. Maintenance \$50–\$200	Some centers have established a link with a local physician but medical supervision is not normally offered	Encouraged but not seen by a multidisciplinary team of professionals	Diet Center follows the food guide pyramid. Not medically supervised Discourages rapid weight loss, recommending no more than 2 lbs lost per week. 4 phases: *Conditioning* *Reducing* *Stabilization* *Maintenance* Also has one-on-one counseling. Recently added a vegetarian program, lactose intolerant program, and "WomanStyle" program

Jenny Craig	14 and over. Those who do not satisfy medical criteria must get a note from a physician	Program fees \$99–\$299, depending on options chosen and time of initiation/ discounts. Pre packaged cuisine averages \$70/week	No	Prescribed but not seen by a multidisciplinary team of professionals	Program uses packaged foods to get dieters started and when clients have reached their desired weight loss, regular food is gradually reintroduced. Phases include: *Program based food* *1/2 program food, 1/2 participants food* *Maintenance with participants food* Not medically supervised. Diets available for diabetic, kosher, hypoglycemic, vegetarian, and breast feeding clients
Nutri/System	14 and over. Physician approval for those with preexisting health problems	Program costs differ from location to location. Average program cost \$99; \$49 a week for food	No	Prescribed but not seen by a multidiscplinary team of professionals	Uses pre-packaged food items throughout diet period. Once clients reach their goal weight they decrease the amount of packaged foods and reintroduce regular foods Phases: Nutri/System offers the option of using herbal Phen/fen, a herbal weight loss agent. Caution is advised given problems with drugs
OUTlook Program. 1000–1500 kcals	BMI of 25–27 with comorbidities. BMI of 27 to <30 with no comorbidities	\$50–\$60 per week	No	Yes. Groups led by behavioral counselors, registered dietitians, and exercise physiologists	Intensive moderate calorie program(1000–1500 kcals) Emphasis on groups education groups led by multidisciplinary team of health professionals. Often includes in-class exercise

Table 6. (*Continued*)

Diet	Entry criteria	Cost	Medical supervision?	Nutrition education, behavior modification, exercise included?	Description
SlimFast/Ultra Slim Fast. Not Recommended	None	$1.79 can	No	No	Ultra Slim fast shake is used as a meal replacement for one or two meals per day. Not medically supervised. The diet is unreasonable and no other lifestyle changes are made. Long-term compliance with this program is not available and diet has little individual options.
Slimmons by Richard Simmons	None	Individual products vary in cost. The Deal-a-Meal food plan is approximately $10.00. Video tapes may be purchased for $20–40. Food products are comparable to same item regular food costs	No	Exercise encouraged in program but, no formal behavioral change component	Program consists of "deal a meal" cards. Not medically supervised. Provides between 1000 and 2000 calories per day. Expected weight loss 1–2 lbs per week. While program gives a specific plan, dieter may follow it or not. Does not give dieter structure

provides about 1500 calories per day in the form of liquids it and is also not recommended.

II. Composition

A good weight loss program incorporates all of the major nutrients including carbohydrate, protein, essential fatty acids, electrolytes, water, vitamins, and minerals. Programs that do not meet the Recommended Dietary Allowances (RDA) or adequate intakes (AI) currently referred to as the dietary reference intakes (DRI) for any nutrient other than energy should be avoided. Amounts in excess of the RDA or AI are *not* better for health. When intakes of a micronutrient exceed 10–100 times the RDA, the upper safer level for nutrient intakes may be exceeded.

1. Carbohydrates

Carbohydrates comprise only 1% of the body but account for over 50% of intakes, suggesting their important role as metabolic fuels even though there is no specific requirement for carbohydrates (Ensiniger et al. 1995). In the fed state, glucose is used as a bodily fuel by most tissues, including the brain. Carbohydrate is especially important in reducing diets. Meeting energy needs takes priority over other functions and thus energy deficits in intake are met by catabolism of glycogen stores, lean and adipose tissue. When dietary energy levels are low or dietary carbohydrate levels are below 50–100 g per day, glycogen stores are drawn upon to maintain blood glucose levels. However glycogen stores are only adequate to sustain blood glucose levels for about 12 h. If deficits persist for a longer time, protein in skeletal muscles and other cells is catabolized, and the glucocogenic amino acids released are converted into glucose and help to sustain blood glucose levels. Since usual intakes of carbohydrate among non-dieters who are in energy balance are well in excess of 50–100 g per day, this rarely occurs. However when energy deficits are present, or dietary carbohydrate is less than these levels, protein catabolism to produce glucose may occur. Dietary intakes that are less than about 800 kcal or 100 g of carbohydrate produce a mild ketosis because the lipid in adipose tissue is the main metabolic fuel available and it is incompletely oxidized (Shils 1994). Glucose continues to be used by the brain although it can gradually adapt to use of ketone bodies. On VLCD, which are less than 50–100 g of carbohydrate, at least 50 g carbohydrate is needed to avoid excessive protein metabolism and to keep ketosis mild. If intakes are 600 calories, this means carbohydrates must be at least 33% of calories. Ketogenic diets are popular because they generate fluid losses while ketosis ensues. Many popular diet books and most VLCD (especially VLCD that are protein supplemented modified fasts and extremely low in carbohydrate) are ketogenic. Table 2 lists the side effects of a mild ketosis. Most of them are manageable if the patient is under medical supervision. However, when patients embark on such a regimen on their own, they may become ill because drug doses may change or side effects are not recognized and dealt with. For this reason books or reducing regimes that advocate ketogenic diets [such as Lean for Life, Dr. Atkins Diet Revolution (ATKINS 1992), the Grapefruit diet, and the Cabbage Soup Diet] are not recommended.

Low calorie and balanced calorie deficit diets should be composed of about 50%–55% (or more) carbohydrate, mostly in the form of complex carbohydrates such as pasta, baked potatoes, and whole grain breads. It is also important to consider dietary fiber, which consists largely of non-bio-available carbohydrates such as cellulose, gums, hemicelluloses, pectins, and pentosans. The National Cancer Institute currently recommends that individuals incorporate 25–35 g of dietary fiber per day in their diets to decrease the risks of certain types of cancers (DIET, NUTRITION, AND CANCER PREVENTION). This can easily be achieved by consuming five or more servings of fruit and vegetables per day and incorporating whole grain foods, beans, and legumes into the diet.

2. Protein

In the hypocaloric state, catabolism of adipose tissue is preferentially favored, and skeletal muscle and other lean body mass tends to be spared. Inevitably some lean body mass is lost in the hypocaloric state however. When protein levels on reducing diets are very low or lacking altogether, catabolism of lean tissues become necessary to meet metabolic needs other than energy production that only the amino acids and nitrogen in protein can provide. If carbohydrate is also low, protein needs rise even further because of the need for gluconeogenesis from the glucogenic amino acids to maintain blood sugar levels (ROUBENOFF and KEHAYIAS 1991). For these reason, fasting, which combines a complete lack of dietary energy, protein, and carbohydrate, causes monumental losses of lean body mass and sharp increases in protein needs. Hence it should be avoided, VLCD also increase protein needs.

Protein needs are calculated based on an individual's ideal weight and started as "adjusted" body weight to avoid overestimates protein needs of obese or understating those of the emaciated patient. For example, a 190-lb (86-kg) woman would require about 70 g of protein per day and so would a more moderate weight woman of 165 lbs (75 kg). On a VLCD, since protein becomes a major source of energy, needs increase to 1.5 g/kg per day (National Research Council 1989). Recommendations may be even higher if the patient has co-morbidities that increase protein needs such as chronic infections, HIV, or certain cancers. However, reducing diets among such chronically ill persons are contraindicated anyway.

Some recent books have suggested much higher protein intakes than the RDA. For example, *The Zone* popularized by Barry Sears recommends that dieters have between 1.0 g and 1.9 g, depending on activity level. This diet promotes food "combinations" instead of focusing on total calories from the diet. Sears makes the claim that the "zone," a favorable metabolic state, is induced by lowering the amount of carbohydrate in the diet and balancing it with more protein and fat and that such a metabolic mix will stimulate insulin production and control food cravings. This focus on stimulating insulin production by large amounts of protein and fat to control appetite is misplaced. Also, very high levels of protein are usually accompanied by high amounts of saturated fat as well as sodium. Insulin homeostasis is achieved not only by diet but by physical activity, and there is little evidence that such activity directly affects appetite. The Zone diet works *not* because the theory is correct, but because it provides 1200–1700 calories at most, enough for the majority of very obese females and most males to lose weight. Although large amounts of protein do not appear to have deleterious effects in persons with normal kidney function, there is no inherent advantage to such high levels in weight loss. *Lean for Life*, a diet book recommending that dieters consume an average of 100 g or more of protein per day, is another book that also advocated such a regimen.

3. Vitamins and Minerals

Generally, the fewer the calories, the more likely it is that intakes of essential nutrients such as potassium, magnesium, calcium, iron, vitamin B6, Folate, and other vitamins and minerals fall short. Usually, diets that are less than 1200 calories will require some vitamin and mineral supplementation in order to meet the RDA or AI.

4. Electrolytes

Sodium and potassium are the two main electrolytes that are of particular concern during weight loss. Hypokalemia, can be a fatal side effect of unsupervised and unsupplemented VLCD (Amatruda 1982). Patients who are at high risk because of illness or medications, or who have electrolyte imbalance problems should be carefully monitored. Electrolytes are important on reducing diets, especially on VLCD. For this reason, unless a commercial formula is used for the VLCD, a potassium supplement is often recommended. Sodium tends to be less of a problems since it is ubiquitous. However, intakes should not be lower than 1000 mg and blood levels should not be allowed to fall below 128 mg/dl.

5. Water

Much of the initial weight loss in the first few days that occurs on hypocaloric diets is fluid loss, and water balance needs to be restored. It is important to maintain adequate hydration during weight loss. VLCD or other low carbohydrate diets that can cause marked diuresis need to be supplemented with additional fluids to provide 1 ml/kg of ideal body weight, since ketosis can also cause dehydration (Amatruda et al. 1982).

III. Costs

Weight loss is a $30–50 billion annual business. Fees for weight loss programs vary enormously. Commercial programs must also be profitable to serve and not-for-profit programs must also break even. The good news for dieters is that there is little association between cost and effectiveness of diet programs. Competition among the various commercial plans is aggressive both in pricing and in attempts to continually upgrade programs in order to achieve the best results. Price also varies by season and by promotions. Table 4 describes some of the programs and the prices participants paid at the time of the survey. The costs of food are usually consistent throughout the year. However, some of the initiation fees may be less in the summer and fall than at the beginning of the new year or in the spring when dieting is at its height.

The most expensive programs tend to be the medically supervised programs that provide all of the food to be eaten, medical services, and group support. Other programs that provide pre-packaged food or which have special interventions such as physical activity and exercise with or without personal trainers are also more expensive. Some programs such as Weight Watchers allow members to rejoin at no cost if they have previously achieved their goal weight with the program.

IV. Consumer Friendliness

Fortunately, there are many ethical and responsibly well run programs. However currently there is no weight control industry standard for disclosure of success rates or that insures that consumers are protected and that regimens are safe. There is no legal course of action that consumers who have complaints can easily turn to. In 1995, Guidelines on Safe and Effective Dieting were developed by the National Academy of Sciences' (NAS), Institute of Medicine (INSTITUTE OF MEDICINE 1995). They recommend choosing a program that incorporates individual needs, conferring with a health professional before beginning any program, and avoiding programs promising weight loss without dieting and exercise since they will not work. Other problems that need to be addressed in the industry are access to the programs by the poor and others who lack adequate health insurance and who need to lose weight.

Currently, there is no industry standard for marketing, either. Some programs use misleading claims or adulterous language to lure clients to their programs. They make unrealistically high claims for losses that are possible, claim that all persons lose the same amount of weight, or conceal the fact that the dissatisfied consumer cannot break long-term financial obligations to the program. Most people cannot lose the weight promised in the amount of time as the person in weight loss commercials. Consumers need to be aware of the problems in the industry and to exercise care in choosing a sound program. The Better Business Bureau, the Food and Drug Administration, and the Federal Trade Commission should be notified if problems arise.

V. Coping with Co-Existing Health Problems

Every program should have adequate provisions for treating pre- or coexisting physical and psychological health problems. Programs should assess motivation and individual readiness for making changes in other areas of health as well as in losing weight. The program should also monitor co-existing health problems, physical activity, and mental health. Individuals who are at high risk of health complications from dieting, especially if self-imposed regimes are used, include the elderly, adolescents, children, pregnant or nursing women, those who are already underweight, those at risk for eating disorders, and the morbidly obese (BMI > 40) who have multiple co-morbidities. In addition, those with significant medical problems such as diabetes or renal disease may have further complications if medication and dosing changes are not made when weight is being lost. For example, persons with diabetes may need to decrease their insulin dosage on reducing diets. When preexisting co-morbidities are present that require treatment, medical supervision during weight reduction is mandatory.

VI. Contains All Essential Components for Sound Weight Management

Sound weight management programs must also cover the transition back to usual diets. The reintroduction of usual foods and reorientation toward weight maintenance can be stressful, especially for those who have been on formula diets or VLCD of other sorts that offer little or no choice. Going back to eating regular foods involves re-education on dietary choice and moderation. The problems encountered in re-adjustments to a regular eating plan after weight loss vary, depending on the caloric level of the diet. Those transitioning from a VLCD will have a much longer transition period, often referred to as "refeeding," than those who were on an LCD. VLCD programs usually incorporate a transition phase to deal with these issues. The patients must implement new eating patterns and exercise habits under supervision, gradually returning to regular food meals. Patients must also be educated in relapse prevention skills to prevent or deal with lapses that occur during refeeding. If these are not addressed they may lead to rapid regain. Even after patients have successfully adjusted to an eating pattern with regular food, most need to have ongoing medical, social, and moral support to continue to avoid crises and to incorporate their new eating habits into a healthier lifestyle. Some need metal health or other counseling as well.

The transition to usual eating patterns is less of a change for those on an LCD. Good programs gradually increase calories to a level where weight losses are maintained and weight stabilizes. Once the individual has successfully lost weight, the most difficult challenge is to maintain the lower, healthier weight. A good program includes nutrition education, low-calorie eating tips, and ideas that are applicable to a variety of social situations. General understanding of nutrition helps in maintaining weight losses on the maintenance regimen.

Maintenance readiness programs should also incorporate exercise as a part of daily life. It is important for the patient to understand the overall health benefits of exercise and its role in maintaining weight loss. During maintenance, the need for dietary moderation, exercise, and increased physical activity becomes even more important to keep weight in control since lean body mass and RMR are lower and the work of moving the lighter body is less.

Behavior modification is also an essential part of a good maintenance program. In addition, the tools for continual motivation and support must be included both in the weight loss and weight maintenance phases. Books that help dieters maintain their weight loss and understand the benefits of adhering to a well-balanced diet include *Thin for Life* (see Table 5) (FLETCHER 1991), *The LEARN program (*LEARN EDUCATION CENTER 1991*)*, and *The Weight Management Survival Guide (* BROWNELL and RODIN 1991*)*.

VII. Continuation Provisions for Long-Term Maintenance

"Maintenance" means much more than maintaining a lower, healthier weight. It also involves the maintenance of the other life style goals such as physical activity, behavior modification, and adherence to medications for treating co-morbidities.

Weight loss and maintenance have the same components but the two stages do differ in their goals. During maintenance, energy intakes should be lower and physical activity levels higher than before weight loss or fat will be regained. Some books that offer practical guidance in healthful lifestyles to control weight over the long term are *Thin for Life*, *Tufts University Guide to Total Nutrition* (GERSHOFF and WHITNEY 1991) and the *Dr. Jean Mayer's Diet and Nutrition Guide* (MAYER and GOLDBERG 1990).

Unfortunately, the "yo-yo" dieting syndrome (weight cycling), a gradual increase upward in weight after weight loss has ceased, is common in many people. Individuals who have success in maintaining their weight appear to share common factors. Keys to success, observed in 160 individuals who successfully maintained their weight loss, include believing in yourself, incorporating physical activity as a part of a healthy lifestyle, breaking the relapse cycle early, and having a good support system. Also, nipping regains in the bud rather than waiting until 10–20 lbs have accumulated is helpful.

1. Physical Activity

A sound weight loss program incorporates exercise and physical activity as a basic components of the program. Many commercial programs offer incentives or rewards for incorporating them.

Physical activity offers ways to increase energy output by moving the body. Exercise helps contribute to the caloric deficit of the dieting individual by increasing caloric output in two ways, First, there is the caloric cost of the exercise. Second the effects of vigorous physical activity on metabolism can last for a several hours after the exercise is completed. In one study of college athletes, resting metabolic rate (RMR) was elevated for as much as 25% 15 min after a strenuous game or workout; the effect was dependent on the duration and intensity of the workout. In another study obese women who walked 45 min before and after eating had an increase in RMR that lasted for up to5 h, especially in the less obese individuals. Exercise also increases lipid utilization rates and many lower serum lipoprotein concentrations (STERN 1993).

When physical activity or exercise are used alone in the treatment of obesity, weight losses are modest and average about 5–7 lbs (ASHLEY and KANNEL). They can produce much larger reductions when combined with a weight reduction diet. The combination of diet and exercise generally produces better results than dieting alone (BALIR 1993).

Most controlled studies show 4–7 lbs of added weight loss from an exercise program when it is it of sufficient frequency, intensity, and duration and added to a low calorie reducing diet (AMERICAN COLLEGE OF SPORTS MEDICINE 1995). There is a direct relationship between level of physical activity, exercise, and amount of weight lost. The problem is that adherence to very vigorous levels of physical activity and exercise programs may be just as difficult as adhering to a weight control diet.

For example, one study in which dieters randomized to exercise or no exercise conditions for 12 weeks found that the exercise group lost an average 6 lbs more than the diet only group (GRUBBS 1993). Most studies show that such benefits as improved maintenance of lost weight, preservation of lean body mass, improvement of cardiovascular, respiratory, and musculoskeletal fitness result from combining exercise with a low calorie diet (STERN 1993). The benefits of exercise are not limited to physiological improvements. Also, improvements in mental status, mood, and self-esteem may result.

D. Surgery

Surgical treatment is a method of last resort, since it is not reversible. It is an option for achieving weight control for the severely obese. Surgical treatment should not be undertaken for cosmetic reasons because risks are too high. Bariatric surgery involves decreasing the size of the gastric reservoir and generates various degrees of associated malabsorption. Eating behavior changes dramatically thereafter. Surgery decreases the caloric intake over the short term because of the discomfort involved in eating only small amounts of food and possibly also from malabsorption with the end result being decreased body weight.

I. Who Should be Considered

The option of surgical treatment is reserved for individuals who have a BMI of over 40 (SUGERMAN et al. 1992). Under certain circumstances less severely obese patients (BMIs between 35 and 40) may also be considered for surgery. Included in this category are patients with high co-morbid conditions such as life threatening cardiopulmonary problems. Other possible indications for patients with BMIs between 35 and 40 include obesity-induced physical problems that interfere with lifestyle. Patients who have a low probability of success with non-surgical measures, for example, by failure in many previous weight control programs, may be considered for surgical treatment. Patients should be well informed, motivated, accept operative risks, and should be able to participate in their treatment and long term follow-up. A decision to elect surgical treatment requires an assessment of the risks and benefits in each individual case.

II. Surgical Options

Different types of surgical treatments are available (NIH CONSENSUS DEVELOPMENT CONFERENCE 1991). Surgeons now use techniques that produce weight loss primarily by limiting how much the stomach can hold. These restrictive procedures are often combined with modified gastric bypass procedures that somewhat limit caloric and nutrient absorption and may lead to altered food choices (BENOTTI and FORSE 1995).

Restriction or decreasing food intake is one form of surgery. Gastric banding, gastric bypass, and vertical-banded gastroplasty are surgeries that limit the amount of food the stomach can hold by closing off or removing parts of the stomach. These procedures also delay emptying of the stomach (gastric pouch).

Restriction operations for obesity include both gastric banding and vertical-banded gastroplasty. Both operations serve only to restrict food intake. They do not interfere with the normal digestive process. Food intake is restricted by creating a small pouch at the top of the stomach where the food enters from the esophagus. After the operation, the person usually can eat only half to a whole cup of food without discomfort or nausea. Also, food has to be well chewed. For most people, the ability to eat a large amount of food at one time is lost, but some patients do return to eating modest amounts of food without feeling hungry. Malabsorption surgeries causes food to be poorly digested and absorbed. In gastric bypass procedures, a surgeon makes a direct connection from the stomach to a lower segment of the small intestine, bypassing the duodenum and some of the jejunum.

Gastric bypass operations combine the creation of small stomach pouches to restrict food intake and construction of bypasses of the duodenum and other segments of the small intestine to cause malabsorption. Patients who have underlying eating disorders often rapidly regain the weight lost unless they are treated for these problems. Gastric bypass operations include Roux-en-Y procedure and biliopancreatic diversion. They tend to produce more weight loss than restriction procedures that only decrease food intake. Patients who have bypass operations generally lose two-thirds of their excess weight with in two years.

III. Risks

The risks associated with surgery need to be recognized before considering it as an option. There is a 10%–20% reoperative surgery rate to correct complications (Stunkard 1996). Abdominal hernias are the most common complications requiring follow up surgery. Less common complications include breakdown of the gastric staple line and stretched stomach outlets. Additionally, there is the risk of developing gallstones. Finally, many dietary deficiencies may develop after surgery, including anemia, osteoporosis, and metabolic bone disease. These can be prevented if vitamin and mineral stores are maintained post-surgery by appropriate diet, supplements, and medications.

E. Pharmacotherapy

Pharmocologic treatments for obesity have recently received much public attention. The new interest in pharmacology has risen partly from the growing awareness of the limitations of behavioral therapy and partly from the revisions of the views of medication (Stunkkard 1996). The basic point is that they are best used only *with* a reducing diet and physical activity/exercise.

Pharmacologic agents currently available fall into two broad categories: those that act via brain catecholamine pathways and those that act via seratonin pathways. The first are commonly referred to as nonadrenergic agents acting directly on the central nervous system and lower food intake by potentiating central adrenergic, dopaminergic, or serotonergenic mechanisms (Zhang 1994). They work by releasing catecholamines from storage sites in central nerve terminals that modulate feeding through the hypothalamus (Rohner-Jeanrenaud and Jeanrenaud 1996). The drugs in this category include phenylpropanolamine, diethylpropion, and mazindol. All of these drugs reduce appetite and decrease food intake (Weiser et al. 1997). They all have stimulant properties but little abuse potential.

Anorectic drugs that promote serotonin neurotransmission have no stimulant or sypathomemetic properties because serotonin, unlike catecholamine, lacks central stimulant activity. Serotonin anorexia differs from that caused by phenylethylamines in that it is characterized by a decreased rate of eating and early termination of meals rather than inhibition of eating to begin with. The drugs in this class include fluoxetine and sibutramine. Two drugs in this category, fenfluramine (Pondamin) and dexfenfluramine (Redux) were recalled by the Food and Drug Administration because they were suspected to cause valvular heart disease in some patients (CLEMENT et al. 1995).

Dexfenfluramine, a serotoninogenic agent commonly known as *Redux*, was also recalled in September 1997. In addition at that time, fenfluramine, commercially known as Pondamin, was also withdrawn because of safety concerns. Fenfluramine was initially approved in 1973 for short term weight loss for morbidly obese patients. It was felt at that time that any negative side effects were minimal compared to the side effects of morbid obesity. However, in 1992 fenfluramine began to be prescribed in combination with phentermine to help dieters lose weight faster but with apparently few side effects. The combination known as phen/fen was never approved by the FDA for combined use; however physicians could legally use them as "off label" combining drugs that had been approved by the FDA for different usage purposes. In July 1997, the Mayo Clinic, reported 33 cases of valvular heart disease that were linked to patients on fenfluramines and dexfenfluramine. Currently, the other half of the phen/fen combination, phentermine, still remains on the market. Experts are mixed in their views on its safety. The FDA advises anyone who has taken either Redux or phen/fen to report to their physician to rule out any health problems associated with the two drugs.

I. New Pharmocologic Agents

There are many new pharmocologic agents that have recently been approved or are still in the research process. They have varying mechanisms that work to increase energy expenditure, increase energy storage, decrease energy intake, and act peripherally acting pharmocologic agents (BRODOFF 1995). Recently, there has been much research exploring the genetic abnormalities that contribute to obesity. In 1994, researchers identified the *ob/ob* gene in mice and identified a protein product of the gene called leptin (Silverton 1992). Leptin is produced in the adipocyte tissue and is thought to act as an afferent satiety signal in a feedback loop that putatively affects the appetite and satiety centers of the brain (Wellman 1992). The overall effect of the loop is to regulate body fat mass. In *ob/ob* mice, the *ob* gene is mutated and no leptin is produced; however, when the *ob* mice were given leptin they stopped eating as much and lost weight (US DEPARTMENT HEALTH AND HUMAN SERVICES). One theory on obesity is that obese individuals have decreased sensitivity to leptin. Drug research is now focusing on these possibilities.

Another theory is that there is a genetic variation in the β3-andrenergic receptor. The receptor is involved in lipolysis and thermogenesis. A recent study by Clements has shown that people with a mutation of the receptor have a greater propensity to be overweight (FINER 1997).

1. Increasing Energy Expenditure

$\beta 3$-adrenergic receptor agents are thought to enhance the uncoupling process in order to dissipate energy as heat. Adrenergic stimulation is known to increase thermogenesis; greater than 40% of this response has been linked to $\beta 3$-adrenergic receptor (HAUPTMAN et al. 1992). Receptors for $\beta 3$-adrenergic antagonists are found primarily on brown adipose tissue (BAT). Initially it was thought that the compounds acted to stimulate BAT heat production in humans, but since humans had little brown fat it was assumed that the effects were slight. However, research has shown that humans skeletal muscle is the main location for thermogenesis induced by this category of drugs. In clinical trials two $\beta 3$-adrenergic receptor compounds did stimulate metabolic rate and therefore caused weight loss. However, the thermogenic effect could not be separated from heat produced by troublesome skeletal muscle tremors. The potential of such a drug may therefore be limited (DRENT et al. 1995).

2. Interfering with Energy Storage

Orlistat is a drug that works by inhibiting gastric and pancreatic lipase, thus interfering with the absorption of dietary fat (CONNOLEY and HEAL 1994). The FDA advisory committee was close to issuing an approval for Orlistat (Xenical) in May 1997. However, its maker, Hoffman-LaRoche, withdrew its application for approval in August of the same year because of two concerns. One major concern with Orlistat is that by interfering with fat absorption, the absorption of fat soluble vitamins (A, D, E, and K) may also be affected. Also in clinical trials Orlistat did reduce the levels of vitamins D, E, and beta-carotene by a small but statistically significant amount. Another concern was that nine of 3000 women enrolled in clinical trials of the drug developed breast cancer compared to only one in the control group (DRENT et al. 1995). This prompted the FDA to take a closer look at the issue and review the drug in March of 1998. The petition was denied. The drug produced a modest weight loss. Patients who followed a well balanced low calorie diet lost an average of 10% of their body weight in a year. Side effects included steatorrhea, and abdominal cramping. The drug underwent mere trials and Hoffman-LaRoche submitted additional evidence. It is now approved for use in the United Kingdom, and now also in the United States.

3. Drugs to Decrease Energy Intake

Sibutramine, marketed as Meridia, is the newest FDA approved drug in 1998. It is recommended for obese patients with a BMI > $30\,kg/m^2$, or BMI > $27\,kg/m^2$ with the presence of other risk factors (CONNOLEY and HEAL 1994). Meridia works as a serotonin and epinepherine reuptake inhibitor. Instead of increasing the release of neurotransmitters, such as serotonin, it works outside the cells to stop neurotransmitters from being reabsorbed. It may also act to stimulate thermogenesis indirectly by activating the $\beta 3$ system in BAT. Meridia was studied among men and women aged 18–65 with BMIs ranging from 27 kg/m^2 to 40 kg/m^2. In one trial, 56% achieved a 5% weight reduction and 30% achieved a 10% weight reduction (NATIONAL TASK FORCE ON OBESITY 1996). Some side effects of sibutramine include dry mouth, headaches, constipation, and insomnia. In some patients it has been shown to increase blood pressure and heart beat. The drugs should not be used in patients with a history of coronary artery disease, stroke, congestive heart failure, or uncontrolled hypertension, or with patients who are on serotonin reuptake inhibitors.

4. Peptides and Neuropeptides

Pharmocologic therapy options also include the use of peptides, which can cause satiety or act as signals which can then regulate energy balance. The mechanisms by which peptides induce satiety are not clearly understood. Cholecystokinin (CCK) is a peptide thought to function by suppressing food intake by producing early satiety. The cardiovascular effects of CCK have not been studied extensively. CCK must be administered parenterally and has a short half life. Another gut peptide is bombesin, which reduces food intake in rodents. In human trials it has shown a satiety effect in lean but not obese subjects (WEISER et al. 1997).

Leptin, the gene product whose defect produces obesity in the *ob/ob* mouse, reduces food intake in the *ob/ob* mouse and in animals who became obese after eating a high fat diet. Leptin decreases neuropeptide Y gene expression which has been shown to stimulate strongly carbohydrate intake. Neuropeptide Y potentially stimulates feeding, an effect most likely due to activity within the hypothalamus in the brain (ROHER-JEANRENAUD 1995). This potential pharmocologic intervention would then occur at specific receptor subtypes that would enable drug therapy to be administered with specificity at the feeding centers of the hypothalamus, thereby decreasing food intake.

5. Conclusions on Pharmocologic Therapy

Pharmacologic treatment requires long-term medication and for the centrally acting agents approvals are currently for months not years. It is only an adjunctive treatment to conventional therapies and long-term lifestyle alterations. Together these therapies may help some patients.

F. Conclusion

There are many available dietary treatments. Many of the existing programs and therapies available are questionable. Safe, effective, scientifically and psychologically sound therapies being distinguished from unsafe and unproven therapies is essential before recommending or participating in any weight loss regimes.

However, comprehensive, long-term treatment is essential. Weight reeducation plans that do not incorporate physical activity and behavior modification are not likely to be helpful.

Acknowledgments. Dr. Dwyer's participation in preparing this article was funded in part by Federal funds from the United States Department of Agriculture's Agricultural Research Service, under contract number 53–3K06–01. The contents of this publication do not necessarily reflect the views or policies of the United States Department of Agriculture nor does mention of trade names, commercial products, or organization imply endorsement by the United States Government.

References

Amatruda JM, Biddle TL, Patton ML, Lockwood DH (1983) Vigorous supplementation of a hypocaloric diet prevents cardiac arrhythmias and mineral depletion. Am J Med 74:1016–1022

Amercian College of Sports Medicine (1995) ACSM's Guide for Exercise Testing and Prescription. Fifth Edition. Philadelphia: Williams and Wilkins

American College of Sports Medicine (1995) ACSM's Guidelines for Exercise Testing and Prescription. Fifth Edition Philadelphia: Williams and Wilkins

Anderson JW, Hamilton CC, Brinkman-Kaplan, V (1992) Benefits and risks of an intensive very low calorie diet program for severe obesity. Am J Gastroenterology 87:6–15

Ashley FW, Kannel WB (1974) Relation of weight to change in atherogenic traits: The Framingham Heart Study. Journal of Chronic Disease 27:103–114

Atkins R (1970) Dr. Atkins Diet Revolution. Avon Press: New York, New York

Atkins R (1992) The New Dr. Atkins Diet Revolution. Avon Press: New York, New York

Atkinson RL (1990) Usefulness and limits of VLCD in the treatment of obesity. In: Oomura Y et al (eds) Progress in Obesity Research 1990. New York: John Libbey and Company, pp 473–480

Atkinson RL (1989) Low and very low calorie diets. Med Clin North Am 73:203–215

Balir SN (1993) Evidence for success of exercise in weight loss and control. Ann Intern Med 119:702–706

Benotti P, Forse RA (1995) The role of gastric surgery in the multidisciplinary management of severe obesity. Am J Surgery 169:361–366

Blackburn Gl (1993) Comparison of medically supervised and unsupervised approaches to weight loss and control. Ann Intern Med 119:714–718

Brodoff BN (1995) Pharmacologic treatment of obesity. American Health Consultants

Brownell KD, Rodin J (1991) The Weight Management Survival Guide. Dallas: Brownell and Hager

Carmen JM (1985) An analysis of user experience with a very low calorie diet. In: Blackburn GL, Bray GA (eds) Management of Obesity by Severe Caloric Restriction. Massachusetts: PSG Publishing, pp 33–49

Centers for Disease Control. Liquid protein diets. Atlanta, Georgia: Centers for Disease Control; 1979. Public Health Service Report EPI-78-11-2

Clement K, Vaisse C, Manning BS, Basdevant A, Guy-Grand B, Ruiz J, Silver KD, Shuldiner AR, Froguel P, Strosberg AD (1995) Genetic variation in the β3-adrenergic receptor: and an increased capacity to gain weight in patients with morbid obesity. N Engl J Med 333:352–354

Connoley JP, Heal DJ (1994) The thermogenic effect of sibutramine (Abstract 52) Proceedings of Pharmacologic Treatment of Obesity; Satellite Symposium of the Seventh International Congress on Obesity; Saint Adele, Quebec

D'Amato P (1996) Four Blood Types, 4 Diets. GP Putnam and Sons: New York
Drent ML, Larsson I, William-Olsson T, Quaade F, Czubayko F, von Bergmann K, Strobel W, Sjostrom L, van der Veen EA (1995) Orlistat (RO 18-0647) a lipase inhibitor, in the treatment of obesity: a multiple dose study. Int J Obes 19:221–226
Dwyer JT, Lu D (1993) Popular diets for weight loss; from nutritionally hazardous to healthful. In: Stunkard AJ, Wadden TA (eds) Obesity; Theory and Therapy, 2nd ed. New York, NY: Raven Press; chapter 14
Dwyer JT (1992) Assessment of Energy Intake and Expenditure. In: Micozzi MS, Moon T (eds) Macronutrients Investigating Their Role in Cancer. New York: Marcel Dekker Inc., p 141
Dwyer JT (1980) Sixteen Popular Diets: brief nutritional analysis In: Stunkard AJ (ed) Obesity. Philadelphia: WB Saunders, pp 276–291
Ensiniger A, Ensiniger ME, Kolande J, Robson J (1995) (eds) Concise Encyclopedia of Foods and Nutrition. CRC Press: Boca Raton, Florida, p 162
Finer N (1997) Present and future pharmacologic approaches. British Medical Bulletin (53)
Fletcher A (1994) Thin for Life. Chapters Publishing, Shelburne, Vermont
Forster GD, Wadden TA, Kendall PA, Stunkard AJ, Vogt RA. Psychological effects of weight loss and regain: a prospective evaluation. Journal of Consulting and Clinical Psychology 64:752–757
Garrow JS (1995) Effect of Energy Imbalance on Energy Stores and Body Weight" In: Brown KD, Fairburn CG (eds) Eating Disorders and Obesity. New York: Guildford Press, Chapter 7
Gershoff S, Whitney C (1991) Editorial Advisory Board of the Tufts University Diet and Nutrition Letter. Tufts University Guide to Total Nutrition. New York: Harper and Row
Graff, CG (1997) Lean for Life. Griffin Publishing: Glendale, California
Grubbs L (1993) The critical role of exercise in weight control. Nurse Practitioner 18(4):20–29
Hauptman JB, Jenuet FS, Hartmann D (1992) Initial studies in humans with novel gastrointestinal lipase inhibitor RO18-0647. Am J Clin Nutr. (55):309S–313S
Howard AN (1989) The development of a very low calorie diets. Int J Obes 13 [Suppl 2]:1–9
http://www.niddk.nih.gov/Gastric/Gastricsurg.html
Institute of Medicine (1995) Weighing the Options: Criteria for Evaluation of Weight Management and weight programs
Kirchner, MA. Schnieder, G, Ertel, NH, Gorman, L (1987) An eight year experience with a very low calorie formula diet for control of major obesity. Int J of Obesity 12:69
Learn Education Center (1991) Weight Control Digest. Dallas; Learn Education Center
Life Sciences Research Office (1979) Research needs in management of obesity by severe caloric restriction. Federation of American Sciences for Experimental Biology Contract Number FDA 223-75-2090, Washington DC
Mahan LK, Escott-Stump S (1996) Energy. In: Krause's Food Nutrition and Diet Therapy, 9th ed. Philadelphia; WB Saunders, 17–30
Mayer J, Goldberg, JP (1990) Dr Jean Meyers Diet and Nutrition guide. New York: Pharos Books
National Institutes of Health (1985) Consensus Development Conference Statement: Health Implications of Obesity. Annals of Internal Medicine 103(6 of pt2): 1073–1077
National Research Council (1989) Recommended Dietary Allowances 10th ed. National Academy Press: Washington DC
National Task Force on Obesity (1996) Pharmocotherapy for Obesity. JAMA 276(23): 1907–1915
National Task Force on the Prevention and Treatment of Obesity (1993) Very Low Calorie Diets. JAMA 270(8) August

National Institutes of Health (1991) Consensus Development Conference Statement: Gastrointestinal Surgery for Severe Obesity. Annals of Internal Medicine 115:956–961
Ornish Dean (1997) Eat More, Weigh Less. Harper Collins; San Francisco, California
Owen OE, Holup JL, D'Allessio DA, Craig ES, Polansky M, Smalley KJ, Kavle EC, Bushman MC, Owen LR, Mozzoli MA (1987) A reappraisal of the caloric requirements of men Am J Clin Nutr 46:875–885
Owen OE, Kavle E, Owen RS et al (1986) A reappraisal of caloric requirements in healthy women . Am J Clin Nutr 44:1–19
Position of the American Dietetic Association (1990) Very-low calorie weight loss diets. J Am Diet Assc 90(5):722–726
Puhn, A (1996) Five Day Miracle Diet. Ballantine Books, New York
Quincy M (1991) Diet Right! The consumer's guide to diet and weight loss programs. Berkley CA: Canari Press
Roher-Jeanrenaud T (1995) A neuroendocrine reappraisal of the dual-center hypothesis: its implications for obesity and insulin resistance. Int J Obesity 517–524
Rohner-Jeanrenaud F, Jeanrenaud B (1996) Obesity, leptin, and the brain. N Engl J Med 334:324–325
Roubenoff R, Rehayias JJ (1991) The meaning and measurement of lean body mass. Nutr. Reviews 49:163–175
Saltzman E, Roberts SB (1996) Effects of energy imbalance on energy expenditure and respiratory quotient in young and old men: a summary of data from two metabolic studies. Aging 8(6):770–778
Sears, Barry (1995) The Zone. Harper Collins: New York New York
Segal M (1990) Modified fast: a sometime solution to a weighty problem, FDA Consumer 24:11–16
Seure, Pamela (1997) The Three Day Energy Fast. Harper Collins: San Francisco, California
Shape up America! (1996) Guidance for the treatment of obesity American Obesity Association. November
Shils M, Olson J, Shike M (eds) (1994) Modern Nutrition in Health and Disease 8th ed. Lea and Febiger: Matvern, PA; chapter 2
Silverton T (1992) Appetite suppressants: a review. Drugs. 34:820
Smoller JW, Wadden TA, Brownell KD (1988) Popular and very-low-calorie diet in the treatment of obesity. In Frankle RT, and Yang M. eds. Obesity and Weight Control. Rockville, Maryland.: Aspen Publications
Sours HE, Fratelli, VP, Brand, CD, Felman, RA, Frobes, Al, Swanson, RC, Paris, AL (1981) Sudden death associated with very low calorie weight regimens. Am J Clin Nutr 34:453–461
Stampfer C (1997) Lean for Life. Griffen Publishing Co: California
St Jeor ST, Brownell KD, Atkinson RL, Bouchard C, Dwyer J, Foreyt JP, Heber D, Kris-Etherton P, Stern JS, Willet W, Wilson GT, Wood PD (1993) Obesity. Workshop III. Circulation 88:1391–1396
Stern J (1993) "Diet and Exercise" In: Stunkard AJ, Wadden TA (eds) Obesity; Theory and Therapy, 2nd ed. New York, NY: Raven Press
Stunkard AJ (1996) Current Views on Obesity. Am J Med 100:230–236
Sugerman H, Kellum J, Engle K, Wolfe L, Starkey JV, Birkenhauer R, Fletcher P, Sawyer MJ (1992) Gastric Bypass for Treating Severe Obesity. Am J Clin Nutr 55:560S–566S
US Department of Health and Human Services (1987) Diet, Nutrition, and Cancer Prevention: A Guide to Food Choices
US Department Health and Human Services (1997) Cardiovalvyoplasty Associated with Exposure to Fenfluramine or Dexfenfluramine: Interim Public Health Recommendations. JAMA 278:197, 1729–1731
Van Itallie TB Dietary approaches to the treatment of obesity (1980) In: Stunkard AJ (ed) Obesity. Philadelphia: WB Saunders, pp 249–261

Wadden TA (1992) Bartlett Very Low Calorie Diets: An Overview and Appraisal In: Wadden TA, Van Itallie TB (eds) Treatment of the Seriously Obese Patient. New York: Guildford Press, pp 44–81
Wadden, TA, Van Itallie, TB, Blackburn, GL (1990) Responsible and irresponsible use of very low calorie diets in the treatment of obesity. JAMA 263:83–85
Waterhouse, Deborah (1993) Outsmarting the Female Fat Cell. Hyperion, New York
Weiser M, Frishman W, Michaelson M, Abden A (1997) The pharmacologic approach to the treatment of Obesity. J Clin Pharm (37):453–457
Wellman PJ (1992) Overview of adrenergic anorectic agents. Am J Clin Nutr 55:193S–198S
Wiatt, Carrie. Eating By Design (1995) Pocket Book; New York, New York
Zhang Y, Proenca R, Maffei M, Barrone M, Freidman JM (1994) Positional cloning of the mouse obese gene and its human homologue Nature. 372:428–432

CHAPTER 10

Diet and Body Weight Reduction

V. DRAPEAU and A. TREMBLAY

A. Introduction

The energy balance equation, which implies that changes in body energy are equal to the difference between mean daily energy intake and expenditure, has been the main guideline to plan dietary regimens in the context of weight-reducing programs for decades. In dietetic practice, low-calorie and very low-calorie diets represent two popular dietary approaches that are based on this guideline and that are still promoted by many agencies and health professionals. These diets permit a substantial energy deficit but clinical experience has also revealed that they do not favor the adherence to healthy food habits and that they are frequently followed by a considerable weight regain (HENSRUD et al. 1994; WEINSIER et al. 1995).

Experimental data collected over the last decades have emphasized the importance of not just considering the energy balance equation in the planning of a dietary prescription for the obese individual. Indeed, it is also important to recommend a macronutrient composition which promotes a satisfactory level of satiety with a minimal energy intake. In addition, evidence tends to show that the body is not passive in this context and that its regulatory processes can reduce over time the impact of a dietary restriction. Our main aim in preparing this chapter is to summarize this literature and to integrate it in a way that permits one to propose dietary strategies to prevent and treat obesity.

B. Macronutrients: Impact on Satiety, Thermogenesis, and Energy Balance

Proteins, carbohydrates, and lipids are not equivalent in their potential to favor thermogenesis. There are also differences in the potential of each macronutrient to inhibit subsequent food intake when they are supplemented independently to the diet. In the context of a weight-reducing program, the optimal diet composition should thus be viewed as a macronutrient mix promoting thermogenesis as well as an inhibition of subsequent energy intake.

Literature pertaining to thc thermic effect of macronutrients clearly shows that proteins are much more thermogenic than carbohydrates which in turn are more thermogenic than lipids. The thermic effect of proteins varies between 20% and 30% of their energy content whereas corresponding values for carbohydrates and lipids are 5%–10% and 2%–3%, respectively (Flatt 1978).

Experimental evidence also suggests that proteins are characterized by a good potential to promote satiety. For instance, it has been shown that a high-protein load can favor satiety to a greater extent than an equicaloric carbohydrate meal or low-protein content preload (Barkeling et al. 1990; Hill and Blundell 1986; Rolls et al. 1988; Stubbs et al. 1996; Teff et al. 1989). Other studies have demonstrated that a high-protein load can also decrease energy intake (Barkeling et al. 1990; Booth et al. 1970; Hill and Blundell 1990). On the other hand, some studies have revealed that the satiating power and the subsequent suppression of energy intake of proteins and carbohydrates are equivalent (Rolls et al. 1988; Stubbs et al. 1995).

From a clinical standpoint, the good potential of proteins to favor both thermogenesis and satiety should represent a conceptual basis justifying an increase in the relative content of proteins in weight-reducing dietary regimens. Accordingly, a recent study demonstrated that mean body weight loss induced by a low-fat, high-protein diet over 6 months exceeded by 3.8 kg that induced by a low-fat, high-carbohydrate diet over the same period of time (Skov et al. 1999). Despite the above observations, public health agencies proposing guidelines for optimal diet therapy of obesity have not yet considered this argument and protein intake recommendations have remained mainly based on the need for amino acids to promote optimal growth in children and to preserve lean body mass in adults. Even if this preoccupation is quite important, further research will be necessary to determine whether the protein fraction of daily energy intake should be increased to facilitate the control of body weight. In the meantime, the above referenced literature provides some support to the idea that in a weight-reducing regimen, absolute protein intake should be maintained to its pre-treatment level to facilitate the tolerance of an energy deficit by the patient.

The effects of carbohydrates on satiety and energy intake have been mainly compared to those of lipids over the last two decades. Many studies have demonstrated that the ability of carbohydrates to suppress subsequent energy intake is greater than that of lipids (Blundell 1993; Rolls et al. 1988, 1994; Stubbs et al. 1995) although others did not provide support to this idea (Foltin et al. 1990, 1992; Rolls et al. 1991). In other studies, the weak potential of high-fat preloads to promote satiety has been emphasized (Blundell 1993; Lawton et al. 1993). This agrees with the numerous studies showing that the free access to high-fat foods is associated with excess energy intake compared to when the diet has a high-carbohydrate content (Lissner et al. 1987; Stubbs et al. 1993, 1995; Tremblay et al. 1989, 1991). In the long term, a high-fat intake predicts an increase in body fatness, as demonstrated by epidemiological (Dreon et al. 1988; Romieu et al. 1988; Tremblay et al. 1989; Tucker and KANO 1992) or prospective (KLESGES et al. 1992) studies. There is also evidence suggesting the existence of individual variations in the response of adiposity to high-fat feeding. Indeed, HEITMAN et al. (1993) showed that individuals with a familial history of obesity are more predisposed to gain body fat when exposed to a high-fat diet. In addition, LISSNER et al. (1987) observed a significant positive correlation between body fatness and variations in ad libitum daily energy intake in a metabolic ward.

The main clinical implication of these results is that a decrease in relative fat intake should permit satiety while being in negative energy balance in obese individuals. Figure 1 summarizes the outcome of some studies in which this issue was investigated. In general, a reduction in dietary fat intake is associated with a decrease in body weight. However, this figure also reveals that, beyond a decrease of 10% units of fat intake, there is no clear additional benefit of a further reduction of fat intake on body weight loss.

When considering the effects of a decrease in fat intake on body weight and fat, it is important to emphasize that this does not necessarily refer to a reduction in all dietary lipids. Experimental data show that polyunsaturated fat is a better promoter of its oxidation than saturated fat (JONES et al. 1992), maybe because of a stimulating effect on sympathetic nervous system activity (TAKEUCHI et al. 1994). In the Québec Family Study (DOUCET et al. 1998), we

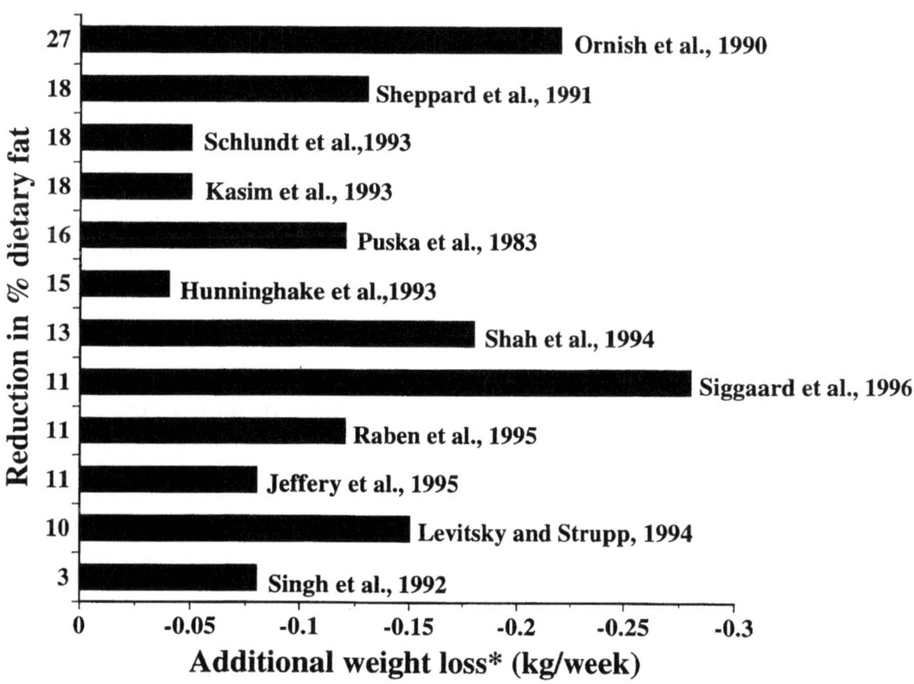

Fig. 1. Impact of reduction in % fat of the diet on body weight. *The additional weight loss is the difference between weight loss observed in response to a low-fat diet and that resulting from either a control diet or a conventional calorie counting

obtained concordant results showing that high polyunsaturated fat intake is not related to the increase in body fat and central fat deposition that is generally observed with high intakes of saturated, monounsaturated and total fat. Others studies have also showed similar results (DREON et al. 1988; ROMIEU et al. 1988).

Recent experimental data tend to show that alcohol intake is also associated with an increase in energy intake. We have recently reported that the overfeeding associated with high-fat intake is additive to the excess energy intake resulting from alcohol consumption (TREMBLAY et al. 1995c) and that this effect is not explained by the high energy density of both fat and alcohol (TREMBLAY and ST-PIERRE 1996). This is in agreement with data showing that alcohol intake produces no inhibitory effect (POPPITT et al. 1996) or at best only a partial compensation (FOLTIN et al. 1993) on subsequent energy intake. Alcohol also transitorily decreases fat oxidation, which favors a positive fat balance (MURGATROYD et al. 1996; SHELMET et al. 1988; SUTER et al. 1992). Accordingly, epidemiological studies have shown that there is a positive correlation between body fat and alcohol intake (CIGOLINI et al. 1996; KROMHOUT 1983; TREMBLAY et al. 1995a). However, this association has not been found in other studies (COLDITZ et al. 1991; FISHER and GORDON 1985; LIU et al. 1994; SONKO et al. 1994).

High-fat intake can also modify the acute effect of exercise on energy balance. In that respect, our research team was the first one to demonstrate that the high fat intake-related overfeeding can totally compensate for the increase in energy expenditure occurring both during and after exercise (TREMBLAY et al. 1994). KING et al. (1995) also showed that high-fat feeding can attenuate the impact of exercise on energy balance.

In summary, these observations demonstrate that high-fat intake should be prevented to facilitate body weight control and to favor fat loss in the obese individual. To achieve this goal, it is also relevant to maintain alcohol intake as low as possible since its impact on fat balance seems to be comparable to that of a high-fat diet. The main question to be further documented in the context of a weight-reducing program is whether a decrease in the fraction of dietary energy as fat should be compensated by an increase in relative carbohydrate intake, protein intake or both. Finally, as discussed in the next section, it is important to go beyond the study of the effects of diet composition on energy balance to take full advantage of the potential of foods to influence energy balance.

C. Diet and Weight Reduction: Effects of Glycemic Index of Foods, Dietary Fibers, and Food-Related Sympathomimetic Agents

The optimal food combination to allow satiety with a negative energy balance is not solely based on macronutrient content of the diet as the unique factor that can facilitate body weight control in the obese. Indeed, other factors may exert a significant impact either by increasing perceived fullness and related signaling, by improving the efficiency of a regulatory process, or by stimulating a component of the regulatory system such as sympathetic nervous system activity. These effects are discussed in the next sections by making reference to the potential role of glycemic index of foods, dietary fibers, and sympathomimetic agents.

I. Glycemic Index of Foods

The classification of foods on the basis of their glycemic response (JENKINS et al. 1981a) was first based on the observation that for a standardized carbohydrate weight of different carbohydrate-containing foods, there was substantial variations in postprandial plasma glucose (OTTO and NIKLAS 1980; SCHAUBERGER et al. 1977). The glycemic index of foods (GI) reflects the plasma glucose response following the ingestion of a food relative to the plasma glucose response after the ingestion of a reference food such as white bread. For instance, a food with a slow rate of starch digestion produces a low but sustained increase in plasma glucose and has a low GI. In contrast, a food with a high GI generally provokes a sharp rise in plasma glucose. A classification based on GI is useful to predict the glycemic response of foods ingested one at a time as well as the response to a mixed meal (WOLEVER et al. 1985).

Removal of fibers from bread, rice, or pasta has little effect on GI (JENKINS et al. 1981a,b). However, in whole foods, cellulose content and the amount of uronic acid in insoluble fibers seem to be more closely related to their GI (WOLEVER 1990).

The impact of variations in GI on satiety and energy balance has been examined in recent studies. As shown in Table 1, a low GI has been found to favor an increase in satiety (HOLT et al. 1992; HOLT and MILLER 1995; KRISHNAMACHAR and MICKELSEN 1987; LAVIN and READ 1995; LEATHWOOD and POLLET 1988). However, this table also reveals that the results of some of these studies do not permit one to associate this change in satiety with a decrease in energy intake (HOLT and MILLER 1995; LAVIN and READ 1995; LEATHWOOD and POLLET 1988). Accordingly, other studies demonstrated that there was no difference in body weight loss between low and high GI diets (JENKINS et al. 1988; WOLEVER et al. 1992).

These observations tend to suggest that variations in GI do not play a major role in the prevention and treatment of obesity. This does not mean however that GI can be ignored in the diet prescription to the obese. Indeed, low GI diets promote a decrease in glycemia which is likely accompanied by a decreased demand on pancreas to produce insulin. Since obese individuals are at greater risk of developing diabetes, the ingestion of low GI foods is justified even if this dietary manipulation has not been shown up to now to have a strong potential to alter energy balance.

Table 1. Impact of different glycemic index foods on satiety, food intake and body weight

Reference	Treatment	Impact of low glycemic index foods		
		Increase satiety	Decrease energy intake	Impact on weight loss
HOLT et al. 1992	Six carbohydrate containing breakfast meals: GI ranging from 43 to 116	Yes	–	–
HOLT and MILLER 1995	High GI: quick-cooking rice and low-amylose puffed rice cakes Low GI: ordinary boiled rice and high-amylose puffed rice	Yes	No (decrease in weight only, 1.4 fold)	–
KRISHNAMACHAR and MICKELSEN 1987	50 g carbohydrate/ portion of carbohydrate-rich foods of different "botanical" origin	Yes	–	–
LEATHWOOD and POLLET 1988	High GI: hachis made with potatoes Low GI: hachis made with bean flakes	Yes	No	–
LAVIN and READ 1995	250 ml glucose drink with (low GI) or without (high GI) the addition of guar gum (2%)	Yes	No	–
JENKINS et al. 1988	Two isocaloric and similar macronutrient content diets with low GI diet or high GI diet	–	–	No difference between two groups
WOLEVER et al. 1992	Two isocaloric (1400 kcal) and similar macronutrient content diets with low GI diet (58) or high GI diet (86)	–	–	No difference between two groups

GI, glycemic index; –, no measurement.

II. Dietary Fibers

Many experimental data support the idea that energy density plays a role in the control of energy intake (DUNCAN et al. 1983; POPPITT and PRENTICE 1996;

PORIKOS et al. 1977; STUBBS et al. 1995; TREMBLAY et al. 1991). To apply this concept in clinical practice, dietary fibers have been frequently used to increase the bulk of meals and to dilute calories of macronutrients. Fibers may also promote chewing, reduce gastric emptying, decrease hunger, and prolong fullness, and decrease digestive and absorptive efficiency. Theoretically, high-fiber foods are thus expected to be more satiating than their energy equivalent low-fiber foods. As shown in Table 2, this effect has been examined in many studies which generally confirmed that high-fiber foods are more potent to induce satiety than low-fiber foods. This table also reveals that there is less consensus in the available literature concerning the reducing effect of high-fiber foods on energy intake. Indeed, many studies tend to confirm that high-fiber foods suppress ad libitum energy intake whereas others did not permit one to detect a difference between low- and high-fiber foods.

The impact of high-fiber dietary regimens on body weight has also been investigated under various conditions. Evidence suggests that obesity is less prevalent in populations consuming high-fiber diets and animals normally eating high-fiber diets become obese when exposed to low-fiber diets (KIMM 1995; LEEDS 1987; VAN ITALLIE 1978). Many studies have also been performed to determine whether dietary fiber supplementation accentuates body weight loss in obese individuals. Figure 2 illustrates the additional weight loss associated with fiber supplementation above that induced by a control low-fiber dietary regimen. If one does not consider the study of WALSH et al. (1983), who used glucomannan fibers (which are highly soluble fibers), this figure reveals that the additional body weight loss seems to be proportional to the increase

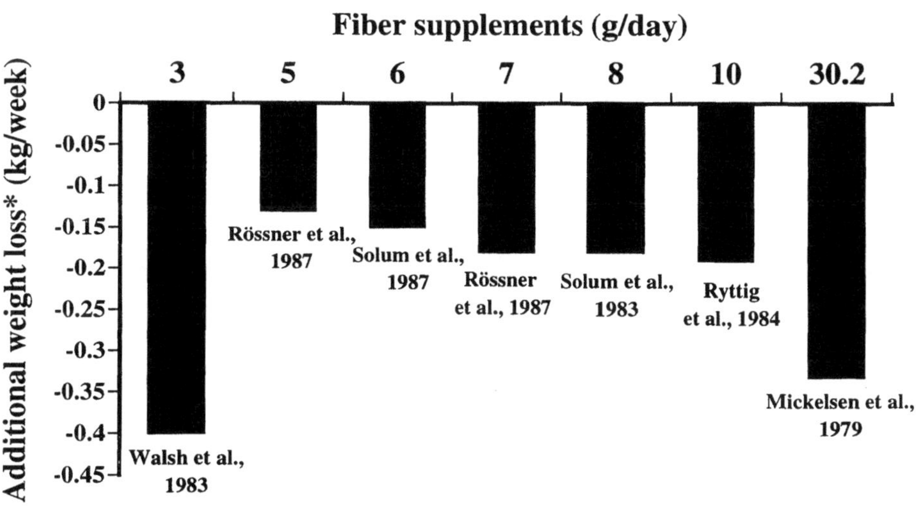

Fig. 2. Impact of different amount of fiber on weight loss. *Values represent the additional weight loss induced by fiber supplementation above that induced by a control treatment

Table 2. Impact of fiber on satiety feeling and food intake

Reference	Subject descriptions	Fiber treatment	Treatment period	Impact on satiety feeling	Impact on food intake
Short-term studies					
HABER 1977	10 subjects, normal weight	Fiber-rich foods: 14 g fibers	1 meal	↑	–
GRIMES 1978	12 M, normal weight	Fiber-rich foods: whole wheat bread	1 meal	↑	↓
BRYSON 1980	5 M, 5 W, normal weight	Fiber-rich foods: whole wheat bread	2 days	n.s.	n.s.
WILMSHURST 1980	12 subjects	Fiber supplement: 2 g guar gum	1 meal	↑	–
BOLTON 1981	10 subjects, normal weight	Fiber-rich foods: fruits vs fruit juice	1 meal	↑	–
COCCHI 1984	10 subjects, normal weight	Fiber-rich foods: bran pasta	1 meal	↑	–
ELLIS 1985	12 NIDDM, 9 normal subjects, normal weight	Fiber supplement: 0, 3, 6, 8 g fibers (guar gum in flour)	1 meal	↑ (with 6 and 8 g fibers)	–
PORIKOS 1986	31 M, normal weight. 19 M, overweight	Fiber-rich foods: 5.2 g fibers (bread)	1 meal	↑	↓ (only in overweight)
BURLEY 1987	16 W, normal weight	Fiber-rich foods: 30 g fibers in foods	1 meal	n.s.	↓ (–590 kcal)
BURLEY 1987	20 W , normal weight	Fiber-rich foods: 12 g fibers	1 meal	↑	n.s.
LEVINE 1989<?1>	a) 14 subjects, normal weight	Fiber-rich foods: 7, 11, 14 g fibers	1 meal	n.s.	↓ (lunch, dinner)
	b) 19 subjects, normal weight	Fiber-rich foods: 11 g fibers (cereals)	1 meal	n.s.	↓ (lunch, dinner)
CYBULSKI 1991	15 W, normal weight	Fiber supplement: 8.7, 5.2, 1.95 g fibers (psyllium)	2 meals	↑ (with 11.7 and 8.2 g fibers)	↓ (with 11.7 and 8.2 g fibers)

BURLEY 1992	18 subjects, normal weight	Fiber supplement: 27.7 g fibers (sugar beet fibers)	1 meal	↑	↓ (−14%)
BURLEY 1993	9 M, 9 W, normal weight	Fiber supplement: 11 g fibers (mycoproteins)	1 meal	↑	↓ (−18%)
RABEN 1994	10 M, normal weight	Fiber supplement: 25.5 g fibers, (peas fibers)	1 meal	↑	–
TOMLIN 1995	10 M, 7 W, BMI = 24–34	Fiber supplement: 7.65 g fibers (liquid fibers)	1 day	↑	↓ (−582 kcal)
DELARGY 1995	12 M, normal weight	Fiber-rich foods: 20 g fibers	1 meal	n.s.	↓ (−512 kcal)
DELARGY 1997	16 M, normal weight	Fiber supplement: 22 g fibers (psyllium or wheat bran)	1 day	↑	↓ (snack time)
Long-term studies					
KAHANER 1976	4 M, 2 W	Fiber-rich foods: 5.4 g fibers/d (All bran)	3 weeks	–	n.s.
SHEARER 1976	73 subjects, obese	Fiber supplement: ≥1.8 g fibers/d (methyl-cellulose or guar gum)	3 weeks	↑	–
SHEARER 1976	60 subjects, obese	Fiber supplement: ≥1.8 g fibers/d (methyl-cellulose or guar gum)	4 weeks	↑	–
DURRANT 1978	13 subjects, obese	Fiber supplement: 1 g fibers/d (methylcellulose)	4 weeks	n.s.	–
RIGAUD 1987	10 M, 10 W normal weight	Fiber supplement: 7.3 g fibers/d (soluble : insoluble = 1 : 4)	4 weeks	↑	–
TURCONI 1995	12 M, 12 W, normal weight	Fiber-rich foods: 10 g fibers/d (All Bran) 6 g fibers/d (Bran Flakes)	2 weeks	↑	n.s.
DE ROOS 1995	24 M, normal weight	Fiber supplement: 32 g fibers/d (resistant starch)	4 weeks	n.s.	n.s.
PASMAN 1997	a) 17 W, overweight	Fiber supplement: 40 g fibers/d (guar gum)	1 week	n.s	↓ (−20%)
	b) 14 W, overweight	6 MJ/d or 4 MJ/d with or without fiber supplement: 20 g fibers/d (guar gum)	↑ 1 week	(only with 4 MJ/d and fiber supplement)	–

M, men; W, women; n.s., nonsignificant; –, no measurement.

in fiber intake. Since, in most of these studies, fiber supplementation was accompanied by dietary restriction or recommendations, it has been frequently proposed that a fiber supplement facilitates compliance to a weight-reducing program.

Since dietary fiber supplementation has no effect on energy expenditure and may even decrease thermic effect of a meal (SCALFI et al. 1987) and fat oxidation (RABEN et al. 1994a), a decrease in energy intake represents the main outcome justifying the incorporation of fiber supplements in reduced-calorie regimens. As discussed above, this may facilitate body weight control in the obese and accentuate the success of a diet therapy of obesity. In addition, the incorporation of fibers in weight-reducing regimens can improve the intestinal regularity which is often disturbed by these regimens (STEVENS 1988).

III. Food-Related Sympathomimetic Agents

A reduced activity of the sympathetic nervous system (SNS), which is more easily detectable in the reduced-obese state (ARONNE et al. 1995), has been proposed as a possible factor predisposing to obesity. Since foods contain substances which act as sympathomimetic agents, it is relevant to address the question as to whether their consumption should be raised in obesity-prone individuals. These compounds include caffeine and capsaicin whose effects on physiological variables and feeding behavior have been largely studied.

As a constituent of coffee, caffeine is part of the diet of most adults. Caffeine increases metabolic rate (ACHESON et al. 1980; ASTRUP et al. 1990; BRACCO et al. 1995; DULLOO and GEISSLER 1989; HOLLANDS et al. 1981; YOSHIDA et al. 1994) and can also increase fat oxidation (ASTRUP et al. 1990; BRACCO et al. 1995) and in vivo lipolysis (VAN SOEREN et al. 1996). However, evidence suggests that the physiological response to caffeine is lower in the obese than in their lean counterparts (ASTRUP et al. 1990; BRACCO et al. 1995).

Experimental data show that women may also be less responsive to caffeine than men. It has been recently reported that the acute increase in energy expenditure induced by caffeine is lesser in women than in men (PERKINS et al. 1994). Moreover, we demonstrated that preprandial ingestion of caffeinated coffee had essentially no effect on spontaneous energy intake at breakfast time in women, whereas it induced a decrease of about 200 kcal in men (TREMBLAY et al. 1988).

Capsaicin is a pungent principle of hot red pepper that has long been used to increase food palatability. Early studies in animals showed that capsaicin increases energy expenditure (KAWADA et al. 1986; WATANABE et al. 1987) and stimulates SNS activity (WATANABE et al. 1988). In humans, HENRY and EMERY (1986) found that a meal containing capsaicin has a greater thermic effect than a control meal not containing capsaicin. Our own experience with capsaicin also reveals that it may produce a significant increase in thermic effect of a meal (YOSHIOKA et al. 1998).

We also recently examined the impact of capsaicin on energy and macronutrient intake in humans (YOSHIOKA et al. 1999). In a first study, we observed that red pepper intake in a meal reduces hunger and decreases ad libitum fat intake at the subsequent meal. Moreover, we investigated the impact of incorporating red pepper in an appetizer on ad libitum energy intake in the remaining part of the meal and in a facultative subsequent snack. The incorporation of red pepper in the appetizer induced an immediate accentuation of the increase in the sympathetic/parasympathetic balance and this was accompanied by a significant decrease in spontaneous energy intake. (YOSHIOKA et al. 1999).

These observations emphasize the potential of food-related sympathomimetic agents to alter energy balance. However, despite this apparent benefit, it is likely premature to promote the inclusion of caffeine and capsaicin in the dietary regimen of obese persons. Indeed, these compounds not only exert beneficial physiological effects; they can disturb the quality of sleep and capsaicin may cause gastrointestinal irritation. Thus, a search for compromise between beneficial effects on energy balance and the above-mentioned side effects is necessary if caffeine, capsaicin, or their combination are proposed to facilitate body weight control in obese individuals.

D. Clinical Implications

An integration of nutritional guidelines which could emerge from the above review of literature is presented in Table 3. As discussed above, these suggestions are proposed to facilitate the occurrence of satiety with a minimal spontaneous energy intake. To reach this goal, we propose an accentuation of some nutritional guidelines. This is the case for proteins which are molecules with a good thermogenic and satiating potential. The fraction of 20%–25% of total energy intake that we propose should not be perceived as the ideal level of intake. It rather reflects a reasonable level of relative intake in a context where menu preparation remains feasible and where absolute protein intake is maintained when spontaneous energy intake is reduced.

The proposed fat intake between 25% and 30% of total energy intake also represents a slight but significant accentuation of the habitual target of public health agencies who recommend 30% energy intake as fat (SANTÉ ET BIEN-ETRE SOCIAL CANADA 1990). Such a suggestion remains applicable under free living conditions and reflects the intent to reduce as much as possible the detrimental impact of high-fat foods on energy balance.

The other guidelines presented in Table 3 are concordant with nutritional suggestions of public health agencies. For instance, an intake of dietary fibers ranging between 25 g/day and 30 g/day represents the intake level that is

Table 3. Proposed nutritional guidelines for weight reducing-weight maintenance dietary regimens

Nutrients	Nutritional guidelines	Practical recommendations
Protein	20%–25% of total energy as protein	↑ Lean meats and low-fat milk products ↑ Legumes
Carbohydrate	45%–50% of total energy as carbohydrate Dietary fibers (25–30 g/day) High intake of low glycemic index foods	↑ Whole grain products ↑ Vegetables and fruits ↑ Unrefined foods ↑ Legumes
Fat	25%–30% of total energy as fat High intake of polyunsaturated fat	↑ Lean meats and low-fat milk products ↑ Fish ↑ Grains and seeds ↑ Oils: safflower, linseed, soy, sunflower, wheat germ, corn ↑ Margarine: sunflower, corn, soy (non-hydrogenated)
Alcohol	Low in alcohol	

general considered as suitable (GÉLINAS et al. 1991; WHITNEY et al. 1991). Furthermore, it also corresponds to the sum of reported intake in population studies (SANTÉ ET BIEN-ETRE SOCIAL CANADA 1985; WHITNEY et al. 1991) plus the amount of supplemented fibers in some studies (Fig. 2) to induce weight loss in obese individuals.

For the clinical dietitian, the main challenge related to guidelines presented in Table 3 is their integration in palatable menus to which obese individuals may adhere with facility. Even if the proposed guidelines are applicable under free-living conditions, long-term adherence remains uncertain. It is thus important to perform more research aimed at developing successful interventions allowing compliance to diets promoting satiety with less energy intake.

E. Importance of Realistic Goals

The engagement in a weight-reducing program represents for many obese individuals an attempt to reach a body weight level comparable to that of lean people. However, many studies as well as clinical experience generally reveal that even a successful weight loss of 10–20 kg is generally not sufficient to reduce the adiposity of the obese within the range of their lean counterparts. It is thus important to fix realistic goals which take into account the impact of body weight loss on the regulatory processes affecting energy balance.

It is now well established that, even when the dietary program is optimal, body fat loss is associated to adaptations which make more difficult over time the compliance to a dietary restriction. The effects which are currently documented can be summarized as follows:

1. A decrease in plasma free-fatty acids concentration since an increase in FFA can increase fat oxidation (GROOP et al. 1991, 1992) and progressively attenuate a positive fat balance (SCHUTZ et al. 1992)
2. A decrease in sympathetic nervous system activity (ARONNE et al. 1995) which is expected to result in a decrease in thermogenesis and an increased hunger level (HIMMS-HAGEN 1984; ROTHWELL and STOCK 1981)
3. A decrease in plasma insulin for a given plasma glucose level. Since insulin at euglycemia increases SNS activity (BERNE et al. 1992; ROWE et al. 1981), postprandial energy expenditure (TREMBLAY et al. 1995b), and long-term body weight stability (SCHWARTZ et al. 1995), a decrease in insulinemia likely complicates adherence to a dietary restriction in the reduced-obese state
4. A decrease in plasma leptin whose variations are correlated with changes in body fat (CONSIDINE et al. 1995; HAMILTON et al. 1995; LONNQUIST et al. 1995; MAFFEI et al. 1995), SNS activity (GIRARD 1997; PELLEYMOUNTER et al. 1995), energy expenditure (GIRARD 1997; NICKLAS et al. 1997), spontaneous physical activity (SALBE et al. 1997), and energy intake (CAMPFIELD et al. 1995; GIRARD 1997)

These observations suggest that following a successful dietary restriction, the reduced-obese individual becomes over time more vulnerable to a fattening environment because of a progressive decrease in the impact of mechanisms involved in the regulation of energy balance. The corollary of this argument is that the dietary regimen which will have allowed a substantial fat loss should be maintained on a permanent basis. If this is not the case, fat regain will likely be the price to pay to restore a new fat balance under uncontrolled dietary conditions. Under such circumstances, it is also relevant to point out that fat regain may exceed the amount of fat lost in response to the weight-reducing program. This makes reference to the "yo-yo phenomenon" which has not been definitively established in humans and the demonstration of which would require a protocol of testing which might be non-ethical for obese individuals. It is thus useful to refer to other experimental models providing results potentially applicable to humans. This is potentially the case of caribous, which were tested under two different winter settlements. As illustrated in Fig. 3, animals passing winter under normal settlement conditions (Southampton Island) maintained a constant adiposity after both winter and summer. On the other hand, those which were exposed to a rigorous settlement (Coats Island) displayed exceptionally low fatness levels at the end of winter and a very high compensation in fatness when they recovered free access to food over summer time (OUELLET et al. 1997). In the latter case, body fat mass at the end of summer was about twice that observed in animals which had not experienced food deprivation during winter.

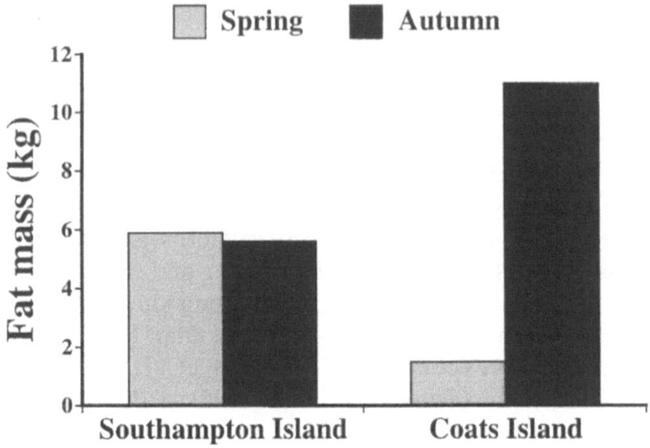

Fig. 3. Comparison of dissectible fat mass in adult female caribous between Southampton Island and Coasts Island populations (see text for more details). Adapted from OUELLET et al. (1997)

The results obtained in caribous are concordant with those obtained in an explorer who experienced marked weight loss and regain during and after a cross country skiing expedition in Greenland (ST-PIERRE et al. 1996). Indeed, following a weight cycle involving substantial changes in body weight, his daily energy expenditure measured in a respiratory chamber was reduced by 1.4 MJ compared the value obtained before the expedition. Even if these observations do not reflect the impact of a well controlled weight cycle in obese individuals, they nevertheless encourage one to be cautious with the long term stability of body weight. For the obese person this means that realistic weight loss goals should be fixed, i.e., that the weight loss to be achieved could be maintained without discomfort in the reduced-obese state.

F. Conclusions

Current literature related to the impact of diet on energy balance reveals that many nutritional factors influence feeding behavior and energy intake and expenditure. The challenge which is imposed on health professionals and scientists interested by this issue is to characterize and prepare diets which favor satiety while decreasing energy intake relative to expenditure. This is obviously not by prescribing low-calorie diets which impose a non-macronutrient specific energy restriction that this objective will be reached. Available evidence is perceived as sufficiently solid to justify reconsideration of some nutritional guidelines to populations, particularly regarding the increase in protein intake and a reemphasis of the importance to reduce fat intake in obese individuals subjected to dietary restriction. It is also relevant to recommend the planning of body weight and fat losses which are commensurate with the potential of regulatory systems of energy balance to sustain body weight stability in a reduced-obese state.

References

Acheson KJ, Zahorska-Markiewicz B, Pittet P, Anantharaman K, Jéquier É (1980) Caffeine and coffee: their influence on metabolic rate and substrate utilization in normal weight and obese individuals. Am J Clin Nutr 33:989–997

Aronne LJ, Mackintosh R, Rosenbaum M, Leibel RL, Hirsch J (1995) Autonomic nervous system activity in weight gain and weight loss. Am J Physiol 269:R222–R225

Astrup A, Tourbo S, Cannon S, Hein P, Breum L, Madsen J (1990) Caffeine: a double-blind placebo-controlled study of its thermogenic, metabolic, and cardiovascular effects in healthy volunteers. Am J Clin Nutr 51:759–767

Barkeling B, Rössner S, Bjorvell H (1990) Efficiency of a high-protein meal (meat) and a high-carbohydrate meal (vegetarian) on satiety measured and by automated computerised monitoring of subsequent food intake, motivation to eat and food preferences. Int J Obes 14:743–751

Berne C, Fagius J, Pollare T, Hemjdahl P (1992) The sympathetic response to euglycemic hyperinsulinemia. Diabetologia 35:873–879

Blundell J (1993) Dietary fat and control of energy intake: evaluating the effects of fat on meal size and postmeal satiety. Am J Clin Nutr 57:772S–778S

Bolton RP, Heaton KW, Burroughs LF (1981) The role of dietary fiber in satiety, glucose, and insulin with fruit and fruit juice. Am J Clin Nutr 34:211–217

Booth DA, Chase A, Campbell AT (1970) Relative effectiveness of protein in the late stages of appetite suppression in man. Physiol Behav 5:1299–1302

Bracco D, Ferrarra JM, Arnaud MJ, Jéquier É, Schutz Y (1995) Effects of caffeine on energy metabolism, heart rate, and methylxanthine metabolism in lean and obese women. Am J Physiol 32:E671–E678

Bryson E, Dore C, Garrow JS (1980) Whole meal bread and satiety. J Hum Nutr 34:113–116

Burley VJ, Blundell JE, Leeds AR (1987a) The effect of high and low-fibre lunches on blood glucose, plasma insulin levels and hunger sensations. Int J Obes 11 [Suppl 2]:12

Burley VJ, Leeds AR, Blundell JE (1987b) The effect of high and low-fibre breakfasts on hunger, satiety and food intake in a subsequent meal. Int J Obes 11 [Suppl 1]:87–93

Burley VJ, Paul AW, Blundell JE (1992) Investigation of the effects of sugar beet fibre-supplemented breakfast on energy intake in young non-obese subjects. Int J Obes 16:53

Burley VJ, Paul AW, Blundell JE (1993) Influence of a high-fibre food (myco-protein) on appetite: effects on satiation (within meals) and satiety (following meals). Eur J Clin Nutr 47:409–418

Campfield LA, Smith FJ, Guisez Y, Devos R, Burn P (1995) Recombinant mouse OB protein: evidence for a peripheral signal linking adiposity and central neural networks. Science 269:546–549

Cigolini M, Targher G, Bergamo Andreis IA, Tonoli M, Filippi F, Muggeo M, De Sandre G (1996) Moderate alcohol consumption and its relation to visceral fat and plasma androgens in healthy women. Int J Obes 20:206–212

Cocchi M, Siniscalchi C, Billi CG, Sciarretta G, De Mutti R, Ruffilli E (1984) Alimentary fibre effect on the feeling of satiety: subjective evaluation and biochemical data modifications. Giorn Clin Med 65:99–108

Colditz GA, Giovannucci E, Rimm ER, Stampfer MJ, Rosner B, Speizer YE, Gordis E, Willett WC (1991) Alcohol intake in relation to diet and obesity in women and men. Am J Clin Nutr 54:49–55

Considine RV, Shina MK, Heiman ML, Kriauciunas A, Stephens TW, Nyce MR, Ohanessian JP, Marco CC, McKee LJ, Bauer TL et al (1995) Serum immunoreactive-leptin concentrations in normal-weight and obese subjects. New Engl J Med 334:292–295

Cybulski KA, Lachaussée J, Kissileff HR (1991) The threshold for satiating effectiveness of psyllium in a nutrient base. Physiol Behav 51:89–93

de Roos N, Heijnen M-L, de Graaf C, Woestenenk G, Hobbel E (1995) Resistant starch has little effect on appetite, food intake and insulin secretion of healthy young men. Eur J Clin Nutr 49:532–541

Delargy HJ, Burley VJ, O'Sullivan KR, Fletcher RJ, Blundell JE (1995) Effects of different soluble: insoluble fibre ratios at breakfast on 24-h pattern of dietary intake and satiety. Eur J Clin Nutr 49:754–766

Delargy HJ, O'Sullivan KR, Fletcher RJ, Blundell JE (1997) Effects of amount and type of dietary fiber (soluble and insoluble) on short-term control of appetite. Int J Food Sci 48:995–1000

Doucet E, AlmJras N, White MD, Després J-P, Bouchard C, Tremblay A (1998) Dietary fat composition and human adiposity. Eur J Clin Nutr 52:2–6

Dreon DM, Frey-Hewitt B, Ellsworth N, Williams PT, Terry RB, Wood PD (1988) Dietary fat: carbohydrate ratio and obesity in middle-aged men. Am J Clin Nutr 47:995–1000

Dulloo AG, Geissler CA (1989) Normal caffeine consumption: influence on thermogenesis and daily energy expenditure in lean and postobese human volunteers. Am J Clin Nutr 49:44–50

Duncan KH, Bacon JA, Weinsier RL (1983) The effects of high and low energy density diets on satiety, energy intake, and eating time of obese and nonobese subjects. Am J Clin Nutr 37:763–767

Ellis PR, Apling EC, Leeds AR, Peterson DB, Jepson EW (1985) Guar bread and satiety: effects of an acceptable new product in overweight diabetic patients and normal subjects. J Plant Foods 6:253–262

Fisher M, Gordon T (1985) The relation of drinking and smoking habits to diet: the Lipid Research Clinics Prevalence Study. Am J Clin Nutr 41:623–630

Flatt JP (1978) The Biochemistry of energy expenditure. In: Bray G (ed) Recent advances in obesity research. Food and Nutrition Press, Wesport, pp 100–116

Foltin RW, Fischman MW, Moran TH, Rolls BJ, Kelly TH (1990) Caloric compensation for lunches varying in carbohydrate content by humans in a residential laboratory. Am J Clin Nutr 52:969–980

Foltin RW, Kelly TH, Fischman MW (1993) Ethanol as an energy source in humans: comparison with dextrose containing beverages. Appetite 20:95–110

Foltin RW, Rolls BJ, Moran TH, Kelly TH, McNelis AL, Fischman MW (1992) Caloric, but not macronutrient, compensation by humans for required-eating occasions with meals and snack varying in fat and carbohydrate. Am J Clin Nutr 55:331–342

Gélinas MD, Dubost-Bélair M, Bernier P, et coll (1991) Régime riche en fibres. In: Connolly G (ed) Manuel de nutrition clinique. Corporation professionnelle des diététistes du Québec, Montréal, pp 16.7.1–16.7.7

Girard J (1997) Is leptin the link between obesity and insulin resistance? Diabetes and Metabolism 23:16–24

Grimes DS, Gordon C (1978) Satiety value of whole meal and white bread. Lancet 2:106

Groop LC, Bonadonna RC, Shank M, Petrides AS, DeFronzo RA (1991) Role of free fatty acids and insulin in determining free fatty acid and lipid oxidation in man. J Clin Invest 87:83–89

Groop LC, Bonadonna RC, Simonson DC, Petrides AS, Shank M, DeFronzo RA (1992) Effect of insulin on oxidative and nonoxidative pathways of free fatty acid metabolism in human obesity. Am J Physiol 263:E79–E84

Haber GB, Heaton KW, Murphy D, Burroughs L (1977) Depletion and disruption of dietary fibre. Effects on satiety, plasma-glucose and serum-insulin. Lancet 2:679–682

Hamilton BS, Paglia D, Kwan AYM, Deitel M (1995) Increased obese mRNA expression of on mental fat cells from massively obese humans. Nat Med 1:953–956

Heitmann BL, Lissner L, Sorensen TIA, C. B (1993) Dietary fat intake promotes weight gain in predisposed individuals. A prospective population study of Swedish women. Int J Obes 17 [Suppl 2]:108

Henry CJK, Emery B (1986) Effect of spiced food on metabolic rate. Hum Nutr: Clin Nutr 40:165–168

Hensrud DD, Weinsier RL, Darnell BE, Hunter GR (1994) A prospective study of weight maintenance in obese subjects reduced to normal body weight without weight-loss training. Am J Clin Nutr 60:688–694

Hill AJ, Blundell JE (1986) Macronutrients and satiety: the effects of high-protein or high-carbohydrate meal on subjective motivation to eat and food preferences. Nutr Behav 3:133–144

Hill AJ, Blundell JE (1990) Comparison of the action of macronutrients on the expression of appetite in lean and obese humans. Ann NY Acad Sci 597:529–531

Himms-Hagen J (1984) Thermogenesis in brown adipose tissue as an energy buffer. Implications for obesity. New Engl J Med 311:1549–1558

Hollands MA, Arch JRS, Phil D, Cawthorne MA (1981) A simple apparatus for comparative measurements of energy expenditure in human subjects: the thermic effect of caffeine. Am J Clin Nutr 34:2291–2294

Holt SHA, Bran J, Soveny C, Hansky J (1992) The relationship of satiety to postprandial glycaemic, insulin and cholecystokinin responses. Appetite 18:129–141

Holt SHA, Miller JB (1995) Increased insulin responses to ingested foods are associated with lessened satiety. Appetite 24:43–54

Hunninghake DB, Stein EA, Dujovne CA, Harris WS, Feldman EB, Miller VT, Tobert JA, Laskarzewski PM, Quiter E, Held J et al (1993) The efficacy of intensive dietary therapy alone or combined with lovastatin in outpatients with hypercholesterolemia. New Engl J Med 328:1213–1219

Jeffery RW, Hellerstedt WL, French SA, Baxter JE (1995) A randomized trial of counseling for fat restriction versus calorie restriction in the treatment of obesity. Int J Obes 19:132–137

Jenkins DJA, Wolever TMS, Buckley G, Lam KY, Giudici S, Kalmusky J, Jenkins AL, Patten PL, Bird J, Wong GS et al (1988) Low-glycemic-index starchy foods in the diabetic diet. Am J Clin Nutr 48:248–254

Jenkins DJA, Wolever TMS, Taylor RH, Barker H, Fielder H, Baldwin JM, Bowling AC, Newman HC, Jenkins AL, Goff DV (1981a) Glycemic index of foods: a physiological basis for carbohydrate exchange. Am J Clin Nutr 34:362–366

Jenkins DJA, Wolever TMS, Taylor RH, Barker H, Fielder H, Gassull MA (1981b) Lack of effect of refining on the glycemic response to cereals. Diabetes Care 4:509–513

Jones PJH, Ridgen JE, Phang T, Birmingham CL (1992) Influence of dietary fat polyunsaturated to saturated ratio on energy substrate utilization in obesity. Metabolism 41:396–401

Kahaner N, Fuchs H-M, Floch MH (1976) The effect of dietary fiber supplementation in man: I. Modification of eating habits. Am J Hum Nutr 29:1437–1442

Kasim SE, Martino S, Kim PN, Khilnani S, Boomer A, Depper J, Reading BA, Heilbrun LK (1993) Dietary and anthropometric determinants of plasma lipoproteins during a long-term low-fat diet in healthy women. Am J Clin Nutr 57:146–53

Kawada T, Hagihara K-I, Iwai K (1986) Effects of capsaicin on lipid metabolism in rats fed high-fat diet. J Nutr 116:1272–1278

Kimm SYS (1995) The role of dietary fiber in the development and treatment of childhood obesity. Pediatrics 96:1010–1014

King NA, Blundell JE (1995) High-fat foods overcome the energy expenditure due to exercise after cycling and running. Eur J Clin Nutr 49:114–123

Klesges RC, Klesges LM, Haddock CK, Eck LH (1992) A longitudinal analysis of the impact of dietary intake and physical activity on weight change in adults. Am J Clin Nutr 55:818–22

Krishnamachar S, Mickelsen O (1987) The influence of different carbohydrate sources on blood glucose levels and satiety: effect of physical activity in blood glucose response. Hum Nutr: Food Sci Nutr 41F:29–39

Kromhout D (1983) Energy and macronutrient intake in lean and obese middle-aged men (the Zutphen study). Am J Clin Nutr 37:295–299

Lavin JH, Read NW (1995) The effect on hunger and satiety of slowing the absorption of glucose: relationship with gastric emptying and postprandial blood glucose and insulin responses. Appetite 25:89–96

Lawton CL, Burley VJ, Wales JK, Blundell JE (1993) Dietary fat and appetite control in obese subjects; weak effects on satiety. Int J Obes 17:409–416

Leathwood P, Pollet P (1988) Effect of slow release carbohydrates in the form of bean flakes on the evolution of hunger and satiety in man. Appetite 10:1–11

Leeds AR (1987) Dietary fibre: mechanisms of action. Int J Obes 11 [Suppl 1]:3–7

Levine AS, Tallman JR, Grace MK, Parker SA, Billington CJ, Levitt MD (1989) Effect of breakfast cereals on short-term food intake. Am J Clin Nutr 50:1303–1307

Levitsky DA, Strupp BJ (1994) Imprecise control of food intake on low-fat diets. In: Fernstrom JD, Miller GD (eds) Appetite and body weight regulation: sugar, fat and macronutrients. CRC Press, Boca Raton, 179–190

Lissner L, Levitsky DA, Strupp BJ, Kalkwarf HJ, Roe DA (1987) Dietary fat and regulation of energy intake in human subjects. Am J Clin Nutr 46:886–892

Liu S, Serdula MK, Williamson DF, Mokdad AH, Byers T (1994) A prospective study of alcohol intake and change in body weight among US adults. Am J Epidemiol 140:912–20

Lonnqvist F, Arner P, Nordford L, Schalling L (1995) Overexpression of the obese (ob) gene in adipose tissue of human subjects. Nat Med 1:950–953

Maffei M, Halaas J, Ravussin E, Pratley RE, Lee GH, Zhang Y, Fei H, Kim S, Lallone R, Ranganathan S et al (1995) Leptin levels in human and rodent: measurement of plasma leptin and OB RNA in obese and weight reduced subjects. Nat Med 1:1155–1161

Mickelsen O, Makdani DD, Cotton RH, Titcomb ST, Colmey JC, Gatty R (1979) Effects of a high-fiber bread diet on weight loss in college-age males. Am J Clin Nutr 32:1703–1709

Murgatroyd PR, Van de Ven MLHM, Goldberg GR, Prentice AM (1996) Alcohol and the regulation of energy balance: overnight effects on diet-induced thermogenesis and fuel storage. Br J Nutr 75:33–45

Nicklas BJ, Toth MJ, Poehlman ET (1997) Daily energy expenditure is related to plasma leptin concentrations in older african-american women but not in men. Diabetes 46:1389–1392

Ornish D, Brown SE, Scherwitz LW, Billings JH, Armstrong WT, Ports TA, McLanahan SM, Kirkeeide RL, Brand RJ, Gould KL (1990) Can lifestyle changes reverse coronary heart disease? Lancet 336:129–133

Otto H, Niklas L (1980) Différence d'action sur la glycémie d'aliments contenant des hydrates de carbone. Conséquence pour le traitement diététique du diabète sucré. Méd Hyg 38:3424–3429

Ouellet J-P, Heard CD, Boutin S, Mulders R (1997) A comparison of body condition and reproduction of caribou on two predator-free arctic islands. Can J Zool 75:11–17

Pasman WJ, Wauters MAJ, Westerterp-Plantenga MS, Saris WHM (1997) Effect of one week of fibre supplementation on hunger and satiety or energy intake. Appetite 29:77–87

Pelleymounter MA, Cullen MJ, Baker MB, Hecht R, Winters D, Boone T, Collins F (1995) Effects of the obese gene product on body weight regulation in ob/ob mice. Science 269:540–543

Perkins KA, Sexton JE, Epstein LH, DiMarco A, Fonte C, Stiller RL, Scierka A, Jacob RG (1994) Acute thermogenic effects of nicotine combined with caffeine during light physical activity in male and female smokers. Am J Clin Nutr 60:312–319

Poppitt SD, Eckhardt JW, McGonagle J, Murgatroyd PR, Prentice AM (1996) Short-term effects of alcohol consumption on appetite and energy intake. Physiol Behav 60:1063–1070

Poppitt SD, Prentice AM (1996) Energy density and its role in the control of food intake: Evidence from metabolic and community studies. Appetite 26:153–174

Porikos K, Hagamen S (1986) Is fiber satiating? Effects of a high fiber preload on subsequent food intake of normal-weight and obese young men. Appetite 7:153

Porikos KP, Booth G, Van Itallie TB (1977) Effect of covert nutritive dilution on the spontaneous food intake of obese individuals: a pilot study. Am J Clin Nutr 30:1638–1644

Puska P, Nissinen A, Vartianinen E, Dougherty R, Mutanen M, Iacono JM, Korhonen HJ, Pietinen P, Leino U, Moisio S et al (1983) Controlled, randomized trial of the effect of dietary fat on blood pressure. Lancet 1:1–5

Raben A, Christensen NJ, Madsen J, Holst JJ, Astrup A (1994a) Decreased postprandial thermogenesis and fat oxidation but increased fullness after a high-fiber meal compared with a low-fiber meal. Am J Clin Nutr 59:1386–94

Raben A, Jensen ND, Marckmann P, Sandström B, Astrup A (1995) Spontaneous weight loss during 11 weeks' ad libitum intake of a low fat/high fiber diet in young, normal weight subjects. Int J Obes 19:916–923

Raben A, Tagliabue A, Christensen NJ, Madsen J, Holst JJ, Astrup A (1994b) Resistant starch: the effect on postprandial glycemia, hormonal response, and satiety. Am J Clin Nutr 60:544–551

Rigaud D, Ryttig K, Leeds AR, Bard A, Apfelbaum M (1987) Effects of moderate dietary fibre supplement on hunger rating, energy input and faecal energy output in young, healthy volunteers. Int J Obes 11:73–78

Rolls BJ, Hetherington M, Burling VJ (1988) The specificity of satiety: the influence of foods of different macronutrient content on the development of satiety. Physiol Behav 43:145–153

Rolls BJ, Kim S, McNelis AL, Fischman MW, Foltin RW, Moran TH (1991) Time course of effects preloads high in fat or carbohydrate on food intake and hunger ratings in humans. Am J Physiol 260:756–763

Rolls BJ, Kim-Harris S, Fischman MW, Foltin RW, Moran TH, Stoner SA (1994) Satiety after preloads with different amounts of fat and carbohydrate: implications for obesity. Am J Clin Nutr 60:476–87

Romieu I, Willett WC, Stampfer MJ, Colditz GA, Sampson L, Rosner B, Hennekens CH, Speizer FE (1988) Energy intake and other determinants of relative weight. Am J Clin Nutr 47:406–12

Rössner S, Zweigbergk DV, Öhlin A, Ryttig K (1987) Weight reduction with dietary fibre supplements: results of two double-blind randomized studies. Acta Med Scan 222:83–88

Rothwell NJ, Stock MJ (1981) Regulation of energy balance. Ann Rev Nutr 1:235–56

Rowe JW, Young JB, Minaker KL, Steven AL, Pallotta J, Lansberg L (1981) Effect of insulin and glucose infusions on sympathetic nervous system activity in normal man. Diabetes 30:219–225

Ryttig KR, Larsen S, Haegh L (1984) Behandling av lett til moderat overvektige personer. Tidsskr Nor Laegeforen 104:989

Salbe AD, Nicholson M, Ravussin E (1997) Total energy expenditure and the level of physical activity correlate with plasma leptin concentrations in five-year-old children. J Clin Invest 99:592–595

Santé et Bien-Etre social Canada (1985) Rapport du comité consultatif d'experts sur les fibres alimentaires. pp 2, 14–22, 28–32

Santé et Bien-Etre social Canada (1990) Recommandations sur la nutrition. Centre d'édition du gouvernement du Canada, Ottawa

Scalfi L, Coltorti A, D'Arrigo E, Carandente V, Mazzacano C, Dipalo M, Contaldo F (1987) Effect of dietary fiber on postprandial thermogenesis. Int J Obes 11 [Suppl 1]:95–99
Schauberger G, Brinck UC, Guldner G, Spaethe R, Niklas L, Otto H (1977) Exchange of carbohydrates according to their effect on blood glucose. Diabetes 26:415
Schlundt DG, Hill JO, Pope-Cordle J, Arnold D, Virts KL, Katahn M (1993) Randomized evaluation of a low fat ad libitum carbohydrate diet for weight reduction. Int J Obes 17:623–629
Schutz Y, Tremblay A, Weinsier RL, Nelson KM (1992) Role of fat oxidation in the long-term stabilization of body weight in obese women. Am J Clin Nutr 55:670–674
Schwartz MW, Boyko EJ, Kahn SE, Ravussin E, Bogardus C (1995) Reduced insulin secretion: an independent predictor of body weight gain. J Clin Endocrinol Metab 80:1571–1576
Shah M, McGovern P, French S, Baxter J (1994) Comparison of a low-fat, ad libitum complex-carbohydrate diet with a low-energy diet in moderately obese women. Am J Clin Nutr 59:980–984
Shearer R (1976) Effects of bulk-producing tablets on hunger intensity in dieting patients. Curr Ther Res 19:433–441
Shelmet JJ, Reichard GA, Skutches CL, Hoedtke RD, Owen OE, Boden G (1988) Ethanol causes acute inhibition of carbohydrate, fat, and protein oxidation and insulin resistance. J Clin Invest 81:1137–1145
Sheppard L, Kristal AR, Kushi LH (1991) Weight loss in women participating in a randomized trial of low-fat diets. Am J Clin Nutr 54:821–828
Siggaard R, Raben A, Astrup A (1996) Weight loss during 12 week's ad libitum carbohydrate-rich diet in overweight and normal-weight subjects at a Danish work site. Obes Res 4:347–356
Singh RB, Rastogi SS, Verma R, Laxmi B, Singh R, Ghosh S, Aniaz M (1992) Randomized controlled trial of cardioprotective diet in patients with recent acute myocardial infarction: results of one year follow-up. Br J Med 304:1015–1019
Skov AR, Toubro S, Ronn B, Holm L, Astrup A (1999) Randomiced trial on protein vs carbohydrate in adlibitum fat reduced diet for the treatment of obesity. Int J Obes 23:528–536
Solum TT (1983) Fibertabletter, Dumo Vital-hjelpemiddel til a oppn > vektreduksjon. Tidsskr Nor Laegeforen 103:1707
Solum TT, Ryttig KR, Solum E, Larsen S (1987) The influence of a high-fiber diet on body weight, serum lipids and blood pressure in slightly overweight persons. Int J Obes 11 [Suppl 1]:67–71
Sonko BJ, Prentice AM, Murgatroyd PR, Goldberg GR, van de Ven MLHM, Coward WA (1994) Effect of alcohol on postmeal fat storage. Am J Clin Nutr 59:619–625
St-Pierre S, Roy B, Tremblay A (1996) A case study on energy balance during an expedition through Greenland. Int J Obes 20:493–495
Stevens J (1988) Does dietary fiber affect food intake and body weight? J Am Diet Assoc 88:939–945
Stubbs JR, Harbon GH, Murgatroyd PR, Prentice AM (1995) Covert manipulation of dietary fat and energy density: effect on substrate flux and food intake in men eating ad libitum. Am J Clin Nutr 62:316–329
Stubbs RJ, Murgatroyd PR, Goldberg GR, Prentice AM (1993) Carbohydrate balance and the regulation of day-to-day food in humans. Am J Clin Nutr 57:897–903
Stubbs RJ, van Wyk MCW, Johnstone AM, Harbron CG (1996) Breakfast high in protein, fat or carbohydrate: effect on within-day appetite and energy balance. Eur J Clin Nutr 50:409–417
Suter PM, Schutz Y, Jéquier É (1992) The effect of ethanol on fat storage in healthy subjects. New Engl J Med 326:983–987

Takeuchi H, Matsuo T, Tokuyama K, Shimomura Y, Suzuki M (1994) Diet-induced thermogenesis is lower in rats fed a lard diet than in those fed a high oleic acid safflower diet, a safflower oil diet or a linseed oil diet. J Nutr 125:920–925
Teff KL, Young SN, Blundell JE (1989) The effect of protein or carbohydrate breakfasts on subsequent plasma amino acid levels, satiety and nutrient selection in normal males. Pharmacol Biochem Behav 34:829–837
Tomlin J (1995) The effect of liquid fibre on feeding behaviour. Eur J Clin Nutr 49 [Suppl 3]:246–249
Tremblay A, Alméras N, Boer J, Kranenbarg EK, Després JP (1994) Diet composition and postexercise energy balance. Am J Clin Nutr 59:975–979
Tremblay A, Buemann B, Thériault G, Bouchard C (1995a) Body fatness in active individuals reporting low lipid and alcohol intake. Eur J Clin Nutr 49:824–831
Tremblay A, Lavallée N, Alméras N, Allard L, Després JP, Bouchard C (1991) Nutritional determinants of the increase in energy intake associated with a high-fat diet. Am J Clin Nutr 53:1134–1137
Tremblay A, Masson E, Leduc S, Houde A, Després J-P (1988) Caffeine reduces spontaneous energy intake in men but not in women. Nutr Res 8:553–558
Tremblay A, Nadeau A, Després J-P, Bouchard C (1995b) Hyperinsulinemia and regulation of energy balance. Am J Clin Nutr 61:827–830
Tremblay A, Plourde G, Després JP, Bouchard C (1989) Impact of dietary fat content and fat oxidation on energy intake in humans. Am J Clin Nutr 49:799–805
Tremblay A, St-Pierre S (1996) The hyperphagic effect of high-fat and alcohol persists after control for energy density. Am J Clin Nutr 63:479–82
Tremblay A, Wouters E, Wenker M, St-Pierre S, Bouchard C, Després J-P (1995c) Alcohol and high-fat diet: a combination favoring overfeeding. Am J Clin Nutr 62:639–644
Tucker LA, Kano M (1992) Dietary fat and body fat: a multivariate study of 205 adult females. Am J Clin Nutr 56:616–622
Turconi G, Bazzano R, Caramella R, Lanzola E (1995) The effects of high-fibre cereal-based breakfast meals on daily dietary intakes and bowel function. Eur J Clin Nutr 49 [Suppl 3]:312–316
Van Itallie TB (1978) Dietary fiber and obesity. Am J Clin Nutr 31:S43–S52
Van Soeren MV, Mohr T, Kjaer M, Graham TE (1996) Acute effects of caffeine ingestion at rest on humans with impaired epinephrine response. Am J Physiol 80:999–1005
Walsh DE, Yaghoubian V, Behforooz A (1983) Effect of glucomannan on obese patients: a clinical study. Int J Obes 8: 289–293
Watanabe T, Kawada T, Iwai K (1987) Enhancement by capsaicin of energy metabolism in rats through secretion of catecholamine from adrenal medulla. Agri Biol Chem 51:75–79
Watanabe T, Kawada T, Kurosawa M, Sato A, Iwai K (1988) Adrenal sympathetic efferent nerve and catecholamine secretion excitation caused by capsaicin in rats. Am J Physiol 255:E23–E27
Weinsier RL, Neslon KM, Hensrud DD, Darnell BE, Hunter GR, Schutz Y (1995) Metabolic predictors of obesity. J Clin Invest 95:980–985
Whitney EN, Cataldo CB, Rolfes SR (1991) The carbohydrates: sugar, starch, and fiber. In: Jarpey M (ed) Understanding normal and clinical nutrition. West Publishing, United States, pp 71–99
Wilmshurst P, Crawley CW (1980) The measurement of gastric transit time in obese subjects using Na and the effect of energy content and guar gum on gastric emptying and satiety. Br J Nutr 44:1–6
Wolever TM (1990) Relationship between dietary fiber content and composition in foods and the glycemic index. Am J Clin Nutr 51:72–75
Wolever TMS, Jenkins DJA, Vuksan V, Jenkins AL, Wong GS, Josse RG (1992) Beneficial effect of low-glycemic index diet in overweight NIDDM subjects. Diabetes Care 15:562–564

Wolever TMS, Nuttall FQ, Lee R et al (1985) Prediction of the relative blood glucose response of mixed meals using the white bread glycemic index. Diabetes Care 8:418–428

Yoshida T, Sakane N, Umekawa T, Kondo M (1994) Relationship between basal metabolic rate, thermogenic response to caffeine, and body weight loss following combined low calorie and exercise treatment in obese women. Int J Obes 18:345–350

Yoshioka M, St-Pierre S, Drapeau V, Dionne I, Suzuki M, Tremblay A (1999) Effects of red pepper on feeding behavior and energy intake. Brit J Nutr 82:115–123

Yoshioka M, St-Pierre S, Suzuki M, Tremblay A (1998) Effects of red pepper added to high-fat and high-carbohydrate meals on energy metabolism and substrate utilization in Japanese women. Brit J Nutr 80:503–510

CHAPTER 11

The Surgical Treatment of Morbid Obesity

D. M. Smith, W. J. Pories, and K. G. MacDonald Jr

A. Introduction

Morbid obesity is a serious health problem that is associated with significant morbidity and mortality (Sjostrom 1992a,b; Drenick et al. 1980). It is an increasingly serious health problem in the United States, where about 50 million Americans, over one-third of the adult population, are obese. Of these, at least 12 million are morbidly obese, exceeding their ideal body weight by more than 100 pounds (Kuczmarski et al. 1994; National Task Force on Prevention and Treatment Of Obesity 1994). In minority populations, obesity rates are out of proportion; nearly 50% of African American, Mexican American, and Native American women are overweight.

Much of the morbidity and mortality associated with morbid obesity, called "comorbidities," include cardiovascular disease, non-insulin-dependent diabetes mellitus (NIDDM), hypertension, stroke, certain types of cancer, osteoarthritis, obstructive sleep apnea, biliary disease, and hyperlipidemia (Pories 1997; MacDonald et al. 1997). Many of these comorbidities have additional associated effects. For example, diabetes mellitus is a major cause of atherosclerotic cardiovascular disease, renal disease, retinopathy, and neuropathy. Kral summarized it well when he concluded that the morbidly obese are severely handicapped in every way: physically, emotionally, economically, and socially (Kral 1992). Morbid obesity is indeed a serious disease.

B. Complications of Morbid Obesity

In addition to the physical and social handicaps from their massive bulk, many of the morbidly obese also suffer from serious health problems. Hypertension is the most common complication associated with morbid obesity, occurring in 59% of patients. A principal cause of death in the morbidly obese is directly related to hypertension-induced cardiovascular disease, usually in the form of stroke, acute coronary thrombosis, or arrhythmias. The role of obesity in hypertension is emphasized by the observation that weight reduction alone lowers blood pressure in over one-half of hypertensive obese patients.

One-third of the morbidly obese have adult-onset or Type II diabetes. Another one-third demonstrate impaired glucose tolerance, as defined by the World Health Organization. The diabetes seen in the morbidly obese seems to be identical to that seen in the general population with equivalent abnormal metabolic disorders, morbidities, and mortality rates. In these patients, though, Type II diabetes is not caused by a lack of insulin; in fact, they generally exhibit significant hyperinsulinemia. The insulin resistance in these obese individuals is caused not by alterations in the structure of the insulin receptors, but by defects in the metabolism of the cell membrane, the cytoplasm, and the nucleus. Recent studies in human liver, muscle, and several types of fat continue to suggest a complex disorder with decreased availability of insulin receptors on the cell membrane, decreased insulin receptor kinase activity, inadequate generation of second messengers, and diminished activity of specific glucose transport proteins (ELTON 1993). When diabetes complicates morbid obesity, mortality increases to 40% (DUBLIN and MARKS 1952).

The heart is especially burdened (BENOTTI et al. 1992). Obese patients have an increased cardiac output, stroke volume, central blood volume, plasma volume, and total blood volume; they also have decreased peripheral resistance (BENOTTI et al. 1992). If they are also hypertensive, the peripheral resistance is elevated as well. As the left ventricular volume is augmented in response to the high stroke volume, afterload is correspondingly increased. Not surprisingly, left ventricular function becomes impaired easily in morbidly overweight patients, regardless of arterial pressure.

Pulmonary insufficiency develops in almost all the morbidly obese to some degree as the expiratory reserve volume falls with the continuing gain in weight. In those who develop the Pickwickian hypoventilation syndrome, the mortality exceeds 30%. The lungs expand poorly because the weight of the chest wall, breasts, and abdominal organs limits the function of the intercostal and diaphragmatic muscles. Accordingly, the morbidly obese show significant decreases in their vital capacities, expired respiratory volume, and ventilatory capacity, along with sharp increase in their carbon monoxide diffusing capacity. Sleep apnea is also frequently seen in the obese, resulting in fatigue, morning headaches, and inappropriate daytime somnolence. If untreated, pulmonary artery hypertension and right heart failure can result.

The incidence of cholelithiasis is increased in the obese. The Nurses Health Study and the Framingham Heart Study both showed increased symptomatic gallstones in the obese or those with a history of dieting (MACLURE et al. 1989; FRIEDMAN et al. 1966). With more severe obesity, the risk of developing gallstones increases to 3.7–7.4 times that seen among women with a BMI of less than 24 kg/m^2.

The morbidly obese often have difficulty conceiving; their infertility and frequent amenorrhea generally reverse after surgically induced weight loss. Pregnancy in obese women is associated with an increased risk of preeclampsia, hypertension, poor fetal weight gain, diabetes mellitus, and wound infections.

Morbid obesity also exerts profound hormonal effects as a result of the excessive production of estrogen by the large volume of fat cells. In morbidly obese women, this hormonal change results in dysfunctional uterine bleeding, amenorrhea, and an inadequate luteal phase. In men, the high estrogen blood concentrations are associated with low serum testosterone and low testosterone-estradiol-binding globulin levels. In both sexes, the hypothalamic-pituitary-gonadal axis remains normal. The increased incidence of breast cancer in obese women may well be due to these high levels of estrogen.

The debilitating symptoms of degenerative arthritis are compounded by severe stress placed on articulating joint surfaces due to the patient's excessive weight. The skeleton, stressed by the massive and poorly distributed weight, undergoes early joint and bone deterioration.

Other complications of obesity include gout, skin diseases, proteinuria, pseudotumor cerebri, increased hemoglobin concentration, and immunologic impairment with increased risk of certain cancers. In fact, morbid obesity probably affects every organ system.

Not only is the presence of excess fat related to the aforementioned complications, but the distribution of fat also plays a role. One may ask the question, "Who lives longer, apples or pears?" Fat is not symmetrically distributed throughout the body but is typically more concentrated in the abdomen of the male and in the hips of the female. Five prospective studies examined this relation of fat distribution to morbidity and mortality (Bray 1979). Despite which method was used as an indicator of abdominal fat (waist to hip circumference ratio (WHR), the subscapular skinfold, or a combination of skin folds), all five studies found a clear-cut and highly significant increase in the risk of death and/or in the risk of diabetes, hypertension, or stroke in those individuals with large bellies and narrow hips (apples) vs those with a more gynecoid distribution (pears) (Lapidus et al. 1994). Fat distribution was a more important risk factor for morbidity and mortality than degree of overweight, with a relative risk ratio of two or more. The data for WHR, divided into fifths for a cohort of residents of Gothenburg, Sweden (Lapidus et al. 1984), showed that the quintile with the lowest WHR had a much greater chance of remaining free of myocardial infarction and a long-term survival compared with the quintile with the highest WHR. This effect was independent of total fatness. Before fully accepting the concept that "pears outlive apples," however, we need to heed Juczmarski's warning (Juczmarski 1989) that the definitions of waist and hip measures are inexact and vary throughout the literature. Future studies with computed tomography (CT) scans, ultrasonography, magnetic resonance imaging (MRI), and more accurately defined circumference measurements are badly needed to clarify this important question.

Of greater immediate concern to patients than the physical consequences, however, are the psychologic and socioeconomic consequences of morbid obesity, the major reason many patients seek surgical treatment. Fat people are frequently objects of public scorn and malicious ridicule. They are viewed as lacking self-esteem and being slovenly by nature, with insufficient willpower to curtail excessive eating. Their obese physiques are the antithesis of the lean, trim, and muscular body habitus so highly prized in today's exercise-conscious, Western society.

The pressures felt by the obese may differ depending on their race. GREENBERG and LAPORTE reported that Euro-American men prefer thinner figures more than do African-American men, resulting in a decreased pressure on African-American women to diet and remain thin. This is reflected in the attitude African-American women have about their bodies. They tend to be less concerned about their weight and to see themselves as thinner than Euro-American women do. Greenberg and LaPorte point to the lower prevalence of eating disorders among African-American women as another indicator of the pressures on Euro-American women to remain thin. FURNHAM and BAGUMA found no difference in opinion between British and Ugandan students of what they felt was an attractive weight. Thus, the phenomenon of thin equals attractive may be unique to Western society.

Obese patients are often unable to fit into armchairs, find suitable clothing, obtain access to public toilets, and enter public conveyances. If they can enter an automobile, they may be unable to get out. Employers usually consider the morbidly obese poor candidates because of their unfavorable appearance, their inability to fit into office furniture or factory environments, and their high absenteeism due to illness. In relationships with their peers, the severely obese make few friends and seldom find satisfactory marital or sexual partners. Obesity also limits the availability of educational opportunities. Frequently the object of jokes, the morbidly obese play the role of the jolly fat person, hiding their misery in public and soothing it in private by eating even more. Finally, our patients frequently mention their inability to meet their parental and other social roles. Their children and other family members are ashamed to be seen with them at school, at athletic events, and in social situations. The environment of the morbidly obese is neither happy nor filled with opportunities.

Obesity is a costly disease. Much of the money is spent by the obese, but society too pays a price for obesity. According to Marketdata Enterprises, in 1993, nearly $33 billion was spent in the United States in the pursuit of diets, slimming programs, and the management of obesity. Additional billions are spent by pharmaceutical companies pursuing research to formulate various dietary supplements, to find the compounds that could safely alter metabolic rate, to introduce dietary fats that cannot be digested, or to change the appetite control center. The cost of such efforts is eventually passed on to everyone who deals with these companies. Even the smallest towns have stores that specialize in clothing for the "stout or larger woman." In terms of the nation's health, control of obesity, if achieved, would probably have an effect at least equal to that of controlling smoking.

Furthermore, the cost of diabetes cannot be overestimated. Health care for diabetes in 1992 was estimated to be $105 billion, with each diabetic spending an average of $9493, compared with $2604 spent by a nondiabetic individual (American Medical News, May 9, 1994). Most of these expenditures are for the 95% of diabetics who have the adult-onset form of the disease, an illness almost always associated with obesity.

The largest price is paid by the morbidly obese. Not only do they have higher health care costs, but they often cannot meet the costs of daily living; they are often unemployed and indigent. It is time to recognize, as the 1991 National Institutes of Health (NIH) Consensus Conference on Obesity concluded, that morbid obesity is a serious, disabling, and common disease. It like other diseases, deserves treatment and insurance coverage for therapy.

C. Etiology of Obesity

Obesity is more complex than just "fat=intake>energy requirements." It remains a poorly misunderstood syndrome. Although the observation that obesity runs in families is widely accepted, the cause of that obesity, whether genetic or environmental, continues to be disputed.

The genetic argument rests on several findings. First, most of the offspring of thin parents are thin, and fat offspring mostly come from fat parents. Also, concordance rates for obesity in monozygotic twins were approximately twice those at lesser degrees of overweight and even height at greater degrees of overweight. There is additional evidence that the body mass index (BMI) of adoptees is strongly related to the BMI of the biological parents and not of the adoptive parents. Finally, the observation that the body weight of animals is remarkably constant is described in the set-point theory.

The set point theory states that body weight is set at a certain point, perhaps controlled at the hypothalamic level, and is defended at that level when circumstances change in the environment. In studies such as the Minnesota starvation study (Keys et al. 1950), where volunteers lost 25% of their body weight, and the Vermont prison overfeeding study (Sims and Horton 1986), where men gained 50 or more pounds, body weight rapidly returned to normal when subjects were allowed to control their own food intake, demonstrating the body's insistence on the stability of weight. DuBois (1983) suggested that the set-point is based on a gene that allows the storage of energy in the form of fat during periods of plenty, favoring the survival of fat individuals during times of famine. For such a gene to be fully expressed, one has to be in a situation where unlimited amounts of food are continuously available. The best example is that of the desert rat, which gorges after each rain and starves during dry spells. If placed in a cage with food available at all times, the rat becomes so fat it is eventually unable to move. The Pima Indians, with their high prevalence of obesity, may have such a thrifty genotype developed through centuries of desert existence. Studies of lipectomy also support the theory of regulation of body weight, because the removal of fat from one adipose depot is followed by a compensatory increase in fat in other depots.

Other investigators have proposed the idea of an appetite control center in the brain, perhaps located in the ventromedial hypothalamus. Studies of parabiotic rats (surgically produced Siamese twins) showed that hypothalamic lesions caused obesity in one twin while causing a decrease in food intake and even the risk of death by starvation in the other twin. The body weight of the artificial twins seems to be regulated as a unit.

Genetics may play a role in morbid obesity by contributing to an unusually inefficient utilization of food, a decreased energy expenditure from lessened activity, and/or an altered metabolism, such as a reduced thermogenic response to food or attenuated loss of heat through the thickened subcutaneous fat. The examinations by Ravussin and colleagues of the energy expenditure in Pima Indians suggest that some persons with similar physical characteristics are more energy efficient and, thus, more capable of weight gain than others (RAVUSSIN et al 1988).

Similarly, there is evidence that environmental factors are also strong. For example, adoptees who were raised in a rural environment tended to be more overweight than those who were raised in an urban setting. Studies of Danish draftees showed no change in the average weight of these men from 1943 to 1960, but there was an eightfold increase in the number with severe obesity in the following 12 years, certainly too short a time for a change in the gene pool. Stunkard, in his classic Salmon lecture, concluded that both etiologies play a role (STUNKARD 1988). Currently, most authorities concur that both genetic and environmental factors play a role but that genetic influences appear to be stronger.

D. Definition and Quantification of Obesity

I. Definition of Obesity

Many indices have been developed for describing obesity due to an inability to quantitate it unequivocally. Every approach has advantages, making it useful for quantifying the state of obesity as well as its pitfalls, often hindering its universal application.

Height and weight standards continue in popularity among physicians because of their simplicity and well-established use. The 1983 Metropolitan Life Insurance Company tables can be used to determine ideal body weight (IBW) for various heights (METROPOLITAN LIFE INSURANCE COMPANY 1983). IBWs can be derived from the midpoints for the average frame as listed in the tables. These standards, however, have been criticized for their skewed sampling, with under-representation of lower socioeconomic levels, minorities, and the elderly, and for their arbitrary definition of body frame sizes. For example, African American girls have been reported to be heavier than Caucasian girls as early as age 9 (NATIONAL HEART,

Lung, and Blood Institute Growth and Health Study Research Group 1992).

The 1983 Metropolitan Life Insurance Company tables can also be used to show the number of pounds required to achieve body mass indices (BMI) of 24, 35, and 40 (Metropolitan Life Insurance Company 1983). The BMI, defined as weight in kilograms divided by height in meters squared (kg/m^2), provides a better means of comparison of obese individuals than weight alone. The formula emphasizes the relative fatness of individuals while factoring in the effects of height. The optimal value for the BMI lies between 2 and 27, according to the Ministry of National Health and Welfare of Canada. The National Academy of Sciences, however, emphasizes that the BMI rises with age; it subdivides optimal BMIs according to age, recommending values of 19–24 for young adults and 24–29 for those 65 and older.

In epidemiologic studies, the BMI correlates remarkably well with mortality, nullifies the gender differentiation required by the insurance tables, and serves as a reasonable surrogate for assessing body fat. Obesity is considered morbid obesity when the BMI exceeds 35, a value usually equivalent to the individual's exceeding the IBW by 100 lbs or more. Some use the term superobese to refer to the morbidly obese who exceed their IBW by more than 200 lbs. The IBW values of 1983 are significantly lower than the weights currently considered optimal in the United States.

Though the BMI is a useful tool clinically, it can be misleading in muscular individuals. For example, an All-American running back on the East Carolina University football team who was capable of running the 100-yard dash in less than 10 s had a BMI of 47 (height of 5′8″; weight of 308 lbs). Rather than being obese, he was very muscular, with only 7% body fat by hydroimmersion studies. Basing qualifications of individuals for the military or police force solely on the height-weight tables or BMI would be inappropriate for muscular individuals. Hydrodensitometry, or underwater weighing, is strongly recommended before proceeding with evaluation of these individuals' physiques.

Measures of waist-to-hip ratios (WHRs) are another means of classifying obesity. They are an easy and affordable means of measuring the distribution of fat. WHRs become increasingly inexact with increasing degrees of morbid obesity, though, making them less useful for those in that category.

Other indices of relative adiposity include skinfold thickness, total body potassium, total body water, uptake of fat-soluble inert gases, energy balance, nitrogen balance, and various combinations of height vs weight calculations. Except for special research protocols, no clinical advantage is evident for any of these approaches.

While the most accurate measure of obesity is likely hydrodensitometry, this is clinically difficult to measure. Evans and colleagues (Krolewski and Warram 1994) developed a head-above-water approach, which measures body composition in a clinically acceptable fashion. They showed an inverse correlation ($r = -0.21$) between percent fat and relative fat-free weight (rel FFW =

actual FFW/idealFFW). As body fat increases, FFW is added to support the increased load during ambulation. This relationship begins to fail at high body fat levels because the fat-free weight actually decreases. Mobility is impaired in the morbidly obese, which leads to atrophy in the fat-free muscle compartment. The relative FFW will be decreased for those individuals since the actual FFW will begin to decrease.

E. Non-Surgical Treatment of Obesity

I. Dieting

Although dieting remains the most useful method of weight control for individuals who are mildly or moderately overweight, it is generally ineffective for the morbidly obese. In most cases, even when massively obese patients are aided by groups such as Weight Watchers, by psychotherapy, by diuretics, by thyroid preparations, or by anorectic agents such as amphetamines, the lost pounds are usually regained along with a few extra ones as soon as the intense weight-reducing regimen has ceased. In fact, weight loss is usually disappointingly low. In a classic study of 100 patients subjected to an intensive weight-reducing regimen by four dietitians, 77% lost less than 10 lbs, and only one patient lost more than 20 lbs (SCHUMACHER et al. 1979). STUNKARD and McLAREN (1979) reported similarly disappointing results. In their review of the literature on the failure rate of nonsurgical management of the grossly obese, they found that only 25% of the patients lost as much as 20 lbs, and only 5% lost 40 lbs or more. In their own series of 100 consecutive patients, only 12% lost 20 lbs, 1% (one patient) lost 40 lbs, and 28% failed to return after the first visit. After pooling the data from several studies, BRAY (1979) found that the overall success rate over 1 year with diet alone is about 25%, with diet and medication about 40%, and with behavior modification nearly 50%. For individuals losing 40 lbs but still weighing 250 or 300 lbs, such losses are inadequate. Bray concluded that the long-term success rate by any of the nonoperative methods is not much more than 10%–15%. The 1992 NIH Technology Assessment Conference on Methods for Voluntary Weight Control came to the same conclusion when it noted that in the morbidly obese, non-surgical methods were not effective long-term; within 5 years of commencing any medical regime, almost all participants gained back at least the weight lost (COWAN 1992).

II. Medication

Although the use of drugs for the treatment of obesity has been disappointing in the past, advances are being made in the development of new agents and in the application of multiagent protocols. Appetite-suppressant drugs (amphetamines) have been in clinical use for over 50 years, but they have a long history of abuse. Other drugs include fenfluramine, phenylpropanolamine, and mazindol (Bunker 1994). Walsh et al., in a preliminary report, saw no improvement in weight reduction for patients treated with a 3-month course of dexfenfluramine, but there was no rebound weight gain during their 30-month follow-up. Also, the beneficial effects they saw on type II diabetes was lost after discontinuing the drug. The use of fenfluramine alone as well as the combination of fenfluramine and phentermine were reviewed by Goldstein and Potvin (1994). The most weight lost was achieved by Stunkard et al. and Craighead et al. with fenfluramine; 15.3 kg on average was lost while patients were also under extensive nutritional counseling. The combination therapy showed the most weight lost to be 14.3 kg. Not only were these medications of questionable benefit in achieving adequate weight loss in the severely obese, but increasing reports of serious cardiac valvular side effects resulted in the abrupt withdrawal of dexfenfluramine and fenfluramine from the market.

III. Wiring the Teeth

Wiring the teeth deserves mention as another unsuccessful approach to morbid obesity. Patients with fractured and immobilized jaws almost always lose weight, even though such loss could be prevented by the use of appropriate liquid nutrient formulas. Similarly, the morbidly obese lose weight if their jaws are wired shut. The approach has fallen into disuse for three reasons: (1) many morbidly obese patients are edentulous and, therefore, not suitable candidates; (2) they tolerate wiring poorly because of their personality patterns and generally demand removal of the wires within a few days; and (3) they resume previous eating habits and quickly return to their original weight as soon as the wires are removed, even if a major change in appearance is apparent (Kark 1980). There may still be limited indications for wiring the teeth in poor-risk obese patients who are in cardiac or pulmonary failure and are being prepared for surgery.

F. Surgical Treatment of Obesity

As diets are largely ineffective in the management of morbid obesity, most authorities now agree with the recommendation of the 1991 National Consensus Conference on Surgery for Obesity that operation is the treatment of choice (NIH Consensus Development Conference Program and Abstracts 1991). The evidence showing surgery superior to other treatment methods has accumulated slowly. The history of bariatric operations is a story of various procedures that were developed beginning in the early 1950s, enthusiastically adopted, and rapidly dropped without further follow-up as soon as the next operation or modification was created.

I. Historical Perspective

The progress of obesity operations is interesting not only because of its historical value but because it represents a remarkable and useful record of physiologic and metabolic studies in the gut of man.

1. Intestinal Bypass

KREMEN et al. (1954) first reported the idea of reducing the length of functional intestine for therapeutic reasons. In 1954, after careful study in dogs, they reported a patient in whom an end-to-end jejunoileostomy was performed for weight reduction. In the discussion of that paper, Sandblom mentioned that Henrikson of Gothenburg, Sweden, had resected "an appropriate amount of small intestine" because of obesity and had induced weight loss, but he encountered difficulty in achieving nutritional balance. In 1956, Payne and De Wind (PAYNE and DE WIND 1969) initiated the first clinical program of operations for obesity. Their initial procedure was designed to bypass most of the small intestine and half of the colon by an end-to-side anastomosis of the proximal 36 cm of jejunum to the mid-transverse colon. When Payne and associates reported their series of patients in 1963, it was evident that the procedure produced dramatic weight loss but that liver failure was prohibitive. One patient died, and each of the remaining nine required reoperation. In six of these individuals, the original intestinal continuity was restored. In three patients, however, in recognition of the importance of the terminal ileum, end-to-side jejunoileostomies were performed. This last modification was associated with late regaining of weight due to reflux into the bypassed bowel, increasing the absorptive surface area unpredictably. As a result, other groups, most notably Scott's (SCOTT et al. 1977) and Buchwald's (BUCHWALD et al. 1975), developed the end-to-side jejunoileostomy which drained the bypassed small intestine into the colon. Scott joined 30 cm of proximal jejunum to 20 cm of terminal ileum, whereas Buchwald utilized corresponding 40-cm and 4-cm segments. In the massively obese who weigh over 158.9 kg or 350 lbs, Scott employed 30-cm and 15-cm segments. Scott's version, the most widely applied form of the intestinal bypass, is shown in Fig. 1.

Good results in terms of weight loss were reported in approximately 80% of the patients who underwent the intestinal bypass with average weight loss of 45 kg (100 lbs). Generally, the more obese and the younger individuals lost more weight. By the second or third year, however, almost all patients reached a plateau and many regained their weight as adaptations for better absorption occurred in their intestines. Operations with more than 64 cm (25 inches) of intestine in continuity produced unsatisfactory weight loss.

Jejunoileal bypass was associated with a high complication rate. The early in-hospital mortality and complication rates in the six series of intestinal bypass operations demonstrated that operative mortality varied from 1% to 8%. The major complications included wound problems (10%–20%), pulmonary emboli (1%–4%), and hepatic failure (1%–2%). BROLIN et al.

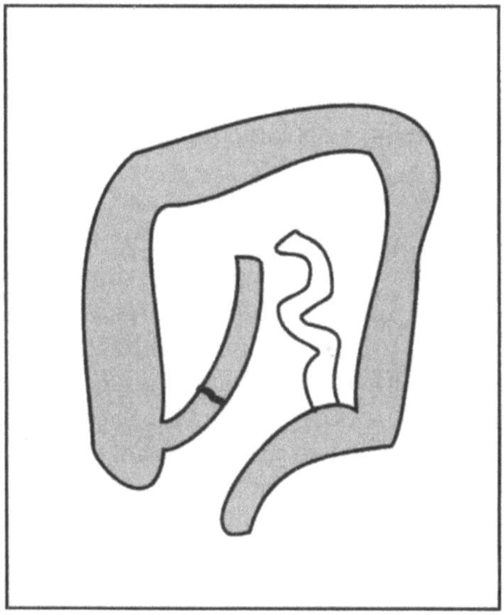

Fig. 1. Scott's version of the intestinal bypass in which 30 cm of proximal jejunum are joined to 20 cm of distal ileum. The isolated segment of small bowel drains into the colon

Table 1. Deaths reported in the literature from early intestinal bypass operations as reported by BROLIN et al. (1992)

Complication	Deaths
Liver failure	5
Pulmonary embolism	8
Cardiac failure	2
Surgical technique[a]	7
Other	7

[a] Includes anastomotic leaks and sepsis.

(1992) reviewed 989 patients collected in the literature and reported the following causes for the 29 deaths (3%) during the first postoperative month (Table 1).

They also reported a long-term follow-up on 1500 patients and found serious metabolic problems. Some resulted from the severe loss of minerals associated with diarrhea, including hypocalcemia, hypokalemia, and hypomagnesemia, as well as iron and zinc deficiencies. Hypoproteinemia, anemia, and vitamin B_{12} and folic acid deficiencies were common.

The diarrhea usually began about the fifth postoperative day and soon reached 12–20 liquid movements per day, gradually decreasing to 6–10 semiformed movements per day. Serum electrolyte, mineral, and vitamin levels had to be carefully monitored and corrected in order to avoid serious and occasionally uncontrollable deficiencies.

Bypass enteritis was a syndrome likely caused by pathologic bacterial colonization of the bypassed bowel from the colon, characterized by diarrhea, abdominal pain, fever of up to 39°C, and occasionally even pneumatosis intestinalis.

Urinary calculi developed in 8% of patients following intestinal bypass, possibly as a result of hyperoxaluria and dehydration. Some patients developed renal failure severe enough to require surgical restoration of bowel continuity. The enhanced oxalate absorption was thought secondary to increased fecal fat excretion. In the normal gut, fecal oxalate combines with calcium to form complexes that are poorly absorbed. Following intestinal bypass, however, fatty acids are present in large amounts and combine with calcium more readily than they combine with oxalate, effectively increasing the amount of soluble oxalate for absorption.

Liver disease was the most feared complication of intestinal bypass, resulting in a number of fatalities. Abnormalities in liver function occurred in about 40% of patients with an intestinal bypass. These changes were unpredictable and dangerous, appearing as early as 3 weeks or as late as several years postoperatively. Nausea, vomiting, jaundice, and enlargement of the liver were reported frequently, but ascites and anasarca were rare. Hepatic coma and death occurred in 1% of all bypass patients. Although many patients showed improvement 6–12 months after the operation, in 3%–5% the changes were progressive and associated with marked fibrosis. In severe cases, the terminal changes were histologically indistinguishable from alcoholic cirrhosis.

Hepatic deterioration progressed in some patients in spite of restoration of good nutrition by hyperalimentation. O'LEARY et al. (1974) demonstrated that liver failure following intestinal bypass can be caused by an overgrowth of bacteroides, an organism that produces a hepatotoxic endotoxin in the excluded intestinal limb. Antibiotics specific for bacteroides, such as metronidazole, prevented liver damage in animal models.

In addition to hepatic dysfunction and urinary and biliary calculi, a large number of case reports cited a variety of fatigue, lethargy, muscle cramps, uncontrollable nausea and bloating, tuberculosis, and nontuberculous granulomas. Besides the various metabolic problems, patients developed mechanical complications, including obstruction of the bypassed small intestine and intussusception of the blind loop into the colon. These two problems could be particularly puzzling and dangerous due to difficulties in diagnosis. Roentgenographic findings were often not helpful due to the absence of characteristic air-fluid levels in the obstructed bowel.

Perhaps the most disappointing aspect of the intestinal bypass was that the long-term weight control was poor. Twenty percent of the patients failed to lose weight satisfactorily, and some lost no weight at all. A significant number regained some of the lost weight during the second and third years after the operation, and some of these patients required another procedure, in which an additional segment of the now adapted small bowel was removed.

In summary, following intestinal bypass, most patients lost one-third of their total body weight with improvement of insulin resistance, hypertension, cardiac failure, pulmonary function, and hyperlipidemia. Unfortunately, the long-term complications were serious: persistent diarrhea, hypokalemia, profound hypomagnesemia and hypocalcemia, arthralgias, neurologic signs, enteropathies, intussusceptions, avitaminoses, trace element deficits, cholelithiasis, renal disease, and liver failure. Over one-half of the patients required rehospitalization, and many required reversal of the bypasses to prevent death from hepatic failure. Equally sad was that most patients adapted their intestinal function and began to regain their weight in the second year. GRIFFEN et al. (1983) finally laid the procedure to rest after a review of the results: "Jejunoileal bypass is not an appropriate operation for morbidly obese patients and should be abandoned."

2. Gastric Operations

The most important advance in bariatric surgery occurred in 1966, when MASON and ITO (1969) devised the gastric bypass, designed to interfere with food intake rather than with absorption. The procedure and various modifications were rapidly adopted when others confirmed that the gastric bypass was not only safer, but also as effective as the intestinal bypass in producing weight loss in the morbidly obese.

In the intervening decades, there have been a host of new gastric procedures, which can be divided into four types: gastroplasties, gastric bypasses, external constricting prostheses, and artificial bezoars or gastric balloons. All four procedures reduce the gastric reservoir and, therefore, limit intake. Comparisons between the procedures are shown in Table 2.

The gastroplasties, gastric bypasses, and banding procedures also delay gastric emptying with a small gastric outlet. The gastric bypasses also add variable degrees of malabsorption by excluding the chyme from the antrum, duodenum, and variable lengths of proximal jejunum. These four types of gastric bariatric procedures are shown in Fig. 2.

Table 2. Comparison of gastric procedures

	Banding procedures	Gastric bypasses	Gastroplasties	Gastric balloons
Reduction of gastric reservoir and limitation of food intake	✓	✓	✓	✓
Delays gastric emptying due to small gastric outlet	✓	✓	✓	
Produces varying degrees of malabsorption		✓		

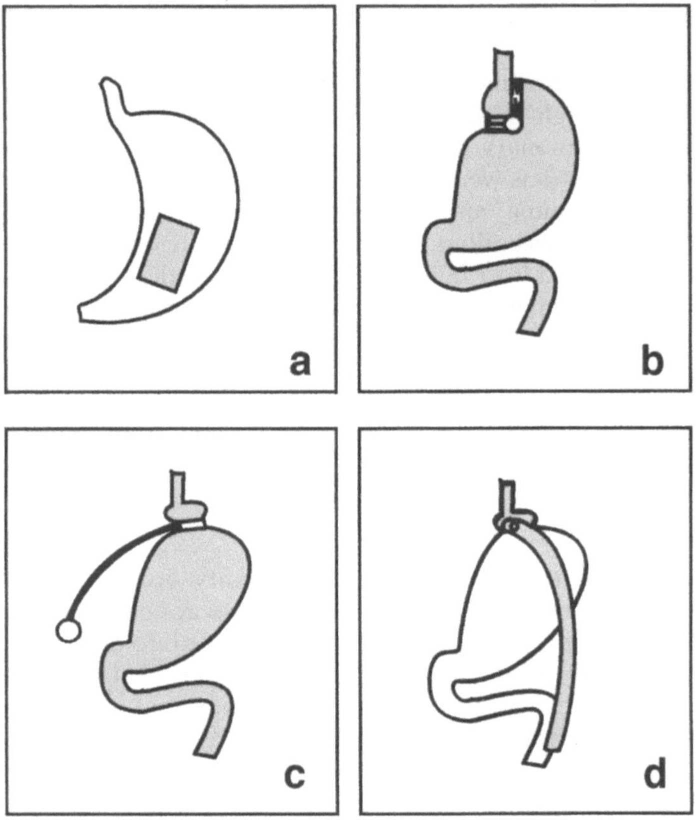

Fig. 2. Four gastric operations for weight control. All limit intake by decreasing the reservoir. The gastric balloon (**a**) proved to be a dismal failure, not only because it did not reduce weight reliably but because of its complications, which included ulceration and obstruction. The vertical banded gastroplasty (**b**) and gastric banding (**c**) limit intake and delay emptying with stapling devices or by plastic bands. The gastric bypass (**d**) not only limits intakes and delays emptying, it also excludes a part of the foregut from the digestive process

Although bariatric surgical procedures are still developing, they are no longer experimental approaches. Two operations are now well-established and recognized bariatric procedures – the vertical banded gastroplasty (VBG) and the gastric bypass (GB). Two additional operations – Silastic gastric banding (SGB) and the biliopancreatic diversion (BPD) – are increasingly used.

There is a strong advantage in having several operations available. Morbidly obese patients differ, and "one size does not fit all." For example, patients who are addicted to sweets will lose more weight following gastric bypass than gastroplasty, because the bypass procedures cause dumping and the gastric restrictive procedures do not. Furthermore, there are some early indications that results of the gastric bypass in the superobese can be improved by lengthening the biliopancreatic afferent limb of the Roux-en-Y and shortening the common channel (SCRUGGS et al. 1993). Finally, bariatric operations may require a staged approach (as recommended by KRAL et al. (1989)) in which simpler and less-demanding procedures are done first, with the understanding that if these prove to be inadequate, additional operations will be needed.

3. Restrictive Procedures

Mason's first gastric bypass was a major technical challenge which caused other alternatives to be developed that could also provide a small gastric pouch and a limited outlet, without requiring gastric division or an intestinal anastomosis. These gastroplasties evolved around the same central principle: the partition of the stomach with a stapler leaving a small passage about 1 cm in diameter in the center of the staple line or at the greater or lesser curvature. Others solved the problem with a gastro-gastrogastrostomy, forming a small anastomosis between the two separated gastric pouches. All these eventually failed due to dilation of the opening or breakdown of the staple line.

4. Vertical Banded Gastroplasty (VBG)

Mason, the real pioneer in bariatric operations, persevered through a number of model procedures and developed the VBG in 1980, which until recently was the most commonly performed bariatric operation in the United States. Ease of performance and other benefits, such as antireflux properties (DEITEL and ILVES 1992), contributed to the popularity of this procedure.

Although the reader is advised to refer to Mason's detailed description (MASON 1992), the procedure is summarized below.

a) Description of the Operation

The abdomen is entered through a midline incision and carefully explored. If no contraindications are found, a 32-French Ewald tube is passed into the stomach and positioned against the lesser curvature. A Penrose drain is then passed around the esophagus, just above the cardia, and the gastrohepatic ligament is entered. The fingers of the left hand are inserted into the lesser sac. A passage is made between the gastric neurovascular bundle and the gastric wall for the future collar. The Penrose drain is brought through, closely encompassing only esophageal and gastric tissue, excluding the vessels and branches of the vagus nerves. A 25-mm-diameter end-to-end (EEA) stapler is used to create a window adjacent to the Ewald tube, which is positioned against the lesser curvature. The window may require reinforcement with absorbable sutures, especially if there is any doubt about the doughnut specimens excised by the stapler. The linear four-row stapler is then applied through the gastric window parallel to the Ewald tube to produce the pouch. The final pouch volume is adjusted to values of 9–25 ml, with an average of 15 ml, as measured through the tube. The pouch outlet is reinforced with a 1.5 × 7 cm strip of Marlex mesh marked and sewn to create a collar 5 cm in circumference. At the end of the procedure, the integrity of the pouch is tested with insufflation of air with the abdomen filled with saline. The mesh is then covered with omentum. The abdomen is closed with absorbable suture, and the skin is reapproximated with staples.

b) Outcomes

Mason reported a combined series of 1000 patients from the University of Iowa Hospitals and Clinics in Iowa City and the St. Francis Memorial Hospital in San Francisco. The operative mortality was 0.33%. Three-year results showed sustained weight control, with a loss of 35 kg (54% excess weight loss) in 45 patients whose initial weight was greater than 225% of ideal. The report suffers somewhat in that the size of the collar – the gastric pouch outlet – was changed during the series, and no data were provided regarding the intensity of the follow-up. DEITEL et al. (1986) reported a series of 233 patients and confirmed Mason's excellent results with a follow-up of 94%, but for only 3 years. More recent studies (MACLEAN et al. 1993), with division of the stomach rather than staple partition and a follow-up of at least 4 years, revealed a substantial failure rate, with 36% of patients requiring reoperation for stenosis or failure to lose weight.

c) Concerns

Surgeons are no longer as enthusiastic about VBG as they were in the previous years for several reasons. First, weight loss is about 10%–15% less than that attained with the GB operation. Second, on longer follow-up, a significant number of patients require reoperation with revision to a GB because of erosion of the Marlex band into the gastric lumen, stenosis of the outlet due to fibrosis or rotation of the band, or staple-line dehiscense. Finally, the VBG, in contrast to the GB, allows the intake of sweets (BROLIN et al. 1994). Even so, the VBG continues to be an acceptable and safe bariatric operation.

With the silicone ring vertical gastroplasty (SRVG) of WILLBANKS (1991), Laws and Eckhout offer a simpler alternative for the reinforcement of the stoma. The approach is technically easier than the VBG, since it avoids the need for the through-and-through EEA hole in the stomach and uses an easily placed 8-French radiopaque silicone ring. Although his follow-up was limited, Willbanks was able to report an operative mortality of 0.04% (1 of 2147) and only a 15%–25% failure rate to maintain weight loss at 46 to 55 months.

5. Gastric Banding

The application of external prostheses to the stomach offers another way to limit the size of the reservoir. The concept of using gastric banding for the treatment of obesity originated in 1976, when Wilkinson (WILKINSON and PELOSO 1981) performed the first operation, using a strip of Marlex mesh to pull the stomach into an hourglass configuration. When this approach failed to produce adequate weight loss, Wilkinson went on to wrap the entire stomach, but this operation failed as well. MOLINA (1984) continued to explore banding with nylon and later with Dacron bands, placing these primarily by palpation through a small skin incision, in a attempt to reduce the trauma of the operation. The most successful approach appears to be Kuzmak's (KUZMAK 1991) adjustable silicone gastric banding (ASGB), in which a 1-cm-wide soft, radiopaque, partly inflatable silicone band with a subcutaneous reservoir is used to develop a pouch, which is measured with an inflatable balloon. The stoma is precisely sized to a diameter of 12–13 mm with a banding instrument equipped with an electronic sensor.

The procedure has become increasingly popular throughout the world ever since the introduction of the adjustable band and the demonstration that the operation could be performed with laparoscopy (CATONA et al. 1993).

a) Description of the Operation

An upper midline incision is made into the subcutaneous tissue. At about 10 cm below the rib cage, subcutaneous fat is dissected from the right rectus sheath. The sheath is cut transversely, and part of the rectus muscle is transected to prepare the space for the reservoir. The abdomen is then entered through the midline. The avascular part of the hepatogastric ligament is opened, and blunt dissection is used to form a tunnel under the stomach and through the avascular suspensory ligament of the fundus. A Penrose drain is placed around the stomach for traction, and the fundus is further mobilized to obtain the correct size of the pouch. A small opening is then made next to the lesser curvature, 3 cm below the gastroesophageal junction and medial to the gastric vessels and branches to the vagus nerve. A 1-cm-wide soft band of silicone, reinforced with two layers of Dacron mesh and fitted with a 5-cm inflatable segment connected to the reservoir, is threaded through the opening and around the stomach. When the band is placed around the stomach, the anesthesiologist inserts the calibrating tube, and the pouch is calibrated by inflating the balloon with 20 ml of saline solution. The silicone band is then tightened with a buckle; an electronic sensor is used to determine the desired diameter of the stoma. When the correct tightness is reached, the band is sutured, and the buckle and the redundant part of the band are excised. The previously mobilized greater curvature is then sutured with three or four 3–0 silk sutures over the band to the pouch. As in all bariatric operations, attention to detail is essential, and reference to the author's original description and use of his materials is strongly advised.

b) Outcomes

Kuzmak's 7-year experience in 311 patients combines two groups: an initial cohort of 173 treated with a nonadjustable band and the remainder treated with the adjustable modification. Although percentage of follow-up is not defined, the group of 138 adjustable patients lost a mean of 64.3% of excess body weight at 36 months, compared to only 49.4% excess weight lost in the nonadjustable group. Slightly more than 28% of the patients reached a weight that was less than 30% above their estimated ideal weight. Operative mortality was 0.6% for the first group and 0.7% for the second group. Stoma complications occurred in 8%; eight stomas required revision, and two bands eroded and required replacement. Kuzmak reported no intra-abdominal infections.

c) Concerns

Similar results were reported by LISE et al. (1994). Others, however, report less satisfactory results. Granstrom and Backman (NATIONAL INSTITUTES OF HEALTH 1986), in 72 patients followed for 12–54 months after gastric banding, had 2 operative deaths, and 22 of their patients required 33 reoperations. The difference may well be due to technique or to the difference in the band itself.

The operation also has the limitations of the other gastroplasties in that it can be defeated by those addicted to sweets. Also, weight loss is about 10%–15% less than that following operations that include a malabsorptive component. Nevertheless, gastric banding offers a simple, safe, and reversible procedure that is adjustable to the patient's needs. Wider application of this procedure is likely.

II. Malabsorption Procedures

1. Gastric Bypass

The GB has undergone continuous modification since the initial description by Mason in 1963. The critical areas of the procedure have proved to be the volume of the proximal gastric pouch, partitioning of the stomach, the diameter of the gastrojejunostomy, and the length of the jejunal limbs.

The proximal gastric pouch has been reduced over the years to the currently accepted 20–30 ml. Some dilation can be expected with time so that, after about a year, most patients can eat about half a hamburger.

Partitioning of the stomach continues to be a troublesome challenge, because failure of the staple lines, with consequent transit of food into the distal gastric pouch, is responsible for the majority of bariatric operative failures. A number of approaches have been tried to partition the stomach securely: single staple lines, close and separated double staple lines, triple superimposed staple lines, and division of the stomach. Single staple lines, a double row of staples placed with one application of a TA90 instrument, were abandoned early on because of their almost universal failure rate. Double staple lines, four rows of staples placed with a wider TA90, have a failure rate of almost 20% in our long-term follow-up, with some breakdowns occurring as late as 10 years after the original surgery. The 7 mm strip of gastric tissue between the rows of staples likely allows formation of mucoceles, which would then initiate staple-line dehiscences by draining into the pouch and distal stomach. Two methods of partitioning seem to produce reliable results: the triple superimposed application of the single TA90 stapling device developed by Sugerman (personal communication, 1993), with a failure rate of 1%–3%, and the division of the stomach, a far more challenging technique with a significant danger of a leak or fistula formation (Cucchi 1999).

The delay in gastric emptying by the gastrojejunostomy is most effective when the diameter of that anastomosis measures 8–10 mm. Smaller stomata frequently become occluded and require repeated dilatations; larger stomata are followed by inadequate weight loss. Dilatation of the anastomosis can be minimized with a double-layered continuous suturing technique with a 3–0 polypropylene suture.

The length of the alimentary and the biliopancreatic jejunal limbs can also play a significant role in the success of the operation. Most bariatric surgeons now use a length of 40–75 cm for the alimentary limb; limbs as long as 150 cm may be used for the superobese (Brolin et al. 1992). Scruggs et al. (1993), for example, recently reported satisfactory results with two modifications – a regular gastric bypass (R-GBP), with an alimentary limb of 90 cm and an afferent limb of 60 cm, and an extended gastric bypass (X-GBP), with a similar alimentary proximal limb of 90 cm but a much longer afferent limb that leaves only 180 to 240 cm of common channel. The longer loops produce greater weight loss but do so with the increased danger of serious malnutrition.

a) Description of the Operation

The abdomen is entered through a vertical midline incision. Exposure is provided by a mechanical retractor. The incision is carried down through the skin for about 1 cm. Firm traction on both sides of the incision will then split the fat precisely in the midline, preventing considerable blood loss. The linea alba and peritoneum are next opened in the exposed midline.

If the exploration demonstrates no contraindications, the upper stomach is isolated by inserting the index finger gently into the angle at the cardia to the left of the esophagus. At this point there is a weak, thin area of the posterior peritoneum, which is easily entered by the dissecting finger. The dissection is gently continued behind the esophagus and cardia, and the finger is brought out not at the right side of the esophagus but between the ascending branches of the left gastric artery, 2.5–3.0 cm below the esophagogastric junction. Before the finger is pushed through the tissue at the lesser curvature, the area should be carefully examined to make certain that a large enough avascular space is available to accommodate the stapling instrument. It is occasionally necessary to divide one or two vessels with clamps and cautery to achieve an adequate passage. The opening should be large enough to admit two fingers. A large Malecot catheter from which the bulbous end has been cut is passed through the tunnel. The lower jaw of the stapler is then pushed into the end of the catheter and pulled through the tunnel with manual assistance. The small pouch of approximately 20–30 ml capacity is demarcated by the stapler. The anterior wall of the stomach can be pulled up more easily than the posterior wall, and this mobility facilitates creation of the anastomosis. The stapler is then fired, isolating the gastric pouch with two parallel staple-lines. Two additional applications of the stapler are then fired directly over one another to crush the tissue, hopefully forming a strong scar with limited possibility of future breakdown.

Because the staples may not close the ends of the staple lines, it is wise to place a figure-of-eight stitch of 3–0 non-absorbable suture material at each end of the staple-line. These serve also as guy sutures to stabilize the stomach during the anastomosis. The jejunum is tagged with a silk suture to mark the proximal end. The bowel is then extended to its maximum length and divided at its apex (about 30–40 cm beyond the ligament of Treitz) with the GIA stapler. The mesentery is not divided; only the small hole needed to transmit the lower blade of the GIA instrument is required. As an additional precaution, we oversew each end with a continuous suture of 3–0 absorbable suture for hemostasis and extra protection against a leak. A small opening 3×3 cm is made through an avascular window of the gastrocolic ligament into the lesser sac. This allows the finger to make an opening through the mesocolon in the thin avascular area to the left of the ligament of Treitz.

The distal segment of the divided jejunum is then gently guided through the opening in the mesocolon, through the lesser sac, and around the greater curvature up to the proximal gastric pouch. A single layer of 3–0 monofilament polypropylene suture is placed to approximate the jejunal loop to the proximal gastric pouch and to form the posterior outer layer of the two-layer anastomosis. Opposing holes 4–5 mm in length are made with the cautery into the pouch and jejunum. The holes are spread to about 1–1.2 cm in size with a curved clamp. Blood or intestinal contents may leak out with this maneuver, and it is advisable to hold a sucker nearby to catch these fluids. A full-thickness mucosa-to-mucosa anastomosis is created with a second double-armed suture and carried around two-thirds of the circumference. The No. 18 nasogastric double-lumen sump tube is advanced through the partially completed anastomosis. The anterior half of the inner suture line is then completed with full-thickness by closely spaced stitches similar to those placed in the back of the anastomosis.

The outer layer is completed in a similar manner. The final anastomosis should fit snugly around the tube but allow easy advancement and withdrawal of the tube through the suture line. When it is clear that no suture has transfixed the tube (all too easy to do), the tube is taped to the nose so that half the holes are on either side of the anastomosis.

The opening in the mesocolon is closed around the Roux-en-Y loop with several non-absorbable sutures to prevent an internal hernia. Approximately 60–70 cm distal to the gastrojejunostomy, the jejuno-jejunostomy is then performed with linear and transverse staplers. Care should be taken to oversew all staple-lines to prevent postoperative bleeding, particularly from the intraluminal staple-line. After irrigation and careful exploration for retained sponges and adequate hemostasis, the wound is closed.

b) Outcomes

Published reports of the GB demonstrate similar statistics in most series. The results described here in detail are included because the series offers the longest and most rigorous follow-up in the literature. In the 14-year study, personal contact was maintained with all but 9 of 608 patients, for a follow-up rate of 98%. The 608 morbidly obese patients were all treated with the Greenville modification of Mason's GB operation: a proximal gastric pouch of approximately 20–30 ml, an 8–10-mm gastroenterostomy, and Roux-en-Y limbs varying from 40 cm to 60 cm (these measurements are critical but are presented as honest ranges, given the limitations of working with tissues). The first 519 gastric partitions were done with a four-row TA90 stapling device. Subsequently, to avoid further staple line failures, either the stomach is divided or triple superimposable layers of double-row TA90 staples are used. Except for the different approaches for partitioning, the operations were identical and can therefore be evaluated as one cohort. The operations were performed on 404 white and 102 black women and 87 white and 15 black men. The average age at the time of the operation was 37.3 (range 14–64) years.

The operation produced significant and durable weight loss. An average maximum weight loss of 70% of excess body weight occurred approximately 2 years after operation. Not only has the mean weight of these patients at 14 years dropped 100 lbs below their operative weights, but also the maximum weights have diminished even more significantly from 615 lbs to 270 lbs. Gastric bypass effectively reduced the proportion of body fat. Measurements of body composition with hydrodensitometry (underwater immersion weighing) in 220 randomly selected patients demonstrated that the percentage of fat in females fell from a preoperative mean of 50.92% to 38.46% and in males from 46.70% to 31.93%. At the end of 5 years, mean weight loss was 58% of excess body weight; after 10 years, it was 55%; and after 14 years, 49%.

The gastric bypass produces remarkable weight control in diabetics. Our series of 232 morbidly obese patients with Type II diabetes (MacDonald and Long 1997) attained a maximum 62.4% loss of excess body weight 1 year after gastric bypass. Further, the weight loss was durable, in fact, more durable than achieved by any other weight loss therapy. Even after 14 years, there was only a modest gain with the mean loss remaining at approximately 50% for up to 14 years of follow-up. In terms of actual weight, the group ranged from a preoperative mean weight of 314 lbs to a minimum of 206 lbs at 1 year after surgery. The mean weight then remained within a stable range of 199–224 lbs for up to 14 years.

As significant as the weight loss is the control of Type II diabetes mellitus, or non-insulin dependent diabetes mellitus (NIDDM). Before operation, 164 of the 608 (27%) had NIDDM, and another 166 (27.3%) proved to have impaired glucose tolerance (IGT). The normalization of glucose metabolism occurs with surprising speed, even before there is significant weight loss. After 10 days, there is no further evidence of diabetes in most patients, even in those who required over 100 units of insulin preoperatively. Following operation, only 26 of those with NIDDM and 2 with IGT still had diabetic blood values. Seven died during follow-up, yielding a denominator of 323 patients. Thus 295 of 323 (91%) continue to be restored to euglycemia by the GB. No other therapy for diabetes, medical or surgical, has ever reported such a high rate of success in controlling the hyperglycemia and hyperinsulinemia associated with diabetes.

One method of comparing control of diabetes in the surgical vs the control group with the data that were available to us was to compare the percentage of patients being treated with either oral hypoglycemics or insulin at the initiation of the study with the percentage of those requiring medical management at the time of our last contact with them. The percentage of patients in the control group who were taking either oral hypoglycemics or insulin increased from 56.4% at the initiation of the study to 87.5% at the time of last contact, a significant increase (p=0.0003). Conversely, the percentage of surgical patients requiring medical management of diabetes fell from 31.8% preoperatively to only 8.6% at time of last contact, a significant decrease (p = 0.0001).

Before operation, 353 (58.1%) of the patients had hypertension. After operation, this rate was reduced to 14%. The GB effectively reduced the proportion of body fat. Measurements of body composition with hydrodensitometry (underwater immersion weighing) in 220 of the patients demonstrated that the females fell from a preoperative mean of 50.92% fat to 38.46% fat, and males fell from 46.70% to 31.93% fat. In addition to these improvements, patients generally demonstrated improvement in cardiopulmonary function, in their disabilities from arthritis, and in fertility.

The GB produced long-term improvement in the health and physical functioning of the morbidly obese, but the positive emotional and social changes proved to be temporary. As reflected by the RAND scale measures, the emotional improvement seen in a group of patients returned to preoperative levels by 3 years after operation. Only general health and vitality, both measures of physical performance, remained improved. The patients felt better and looked better, but those with inadequate preoperative personalities seemed to return to that state in spite of physical improvements. Whether counseling or other psychiatric intervention can alter this regression bears investigation. Relevant to these findings is the report by RYDEN et al. (1989) who found depression to be more common in those patients who had excellent results after bariatric operations and concluded that "the marked weight loss leads to problems of adaptation, which in turn may trigger depressive reactions." KOPEC et al. (1994) reported that in a series of 45 patients who had undergone the VBG, 44% reported an improved social life after operation, 46% stated that it remained the same, and 9% thought that their social lives had deteriorated.

Improvement in daily function following bariatric operations was also documented by DEITEL et al. (1991), who studied 44 morbidly obese patients who met the 17 Canadian criteria for severe disability before their operations. In these individuals, the disabilities disappeared in 22, became mild in 17, and became moderate in 3. Two were lost in the follow-up.

Complications diminished with increasing experience. For the whole series, however, perioperative mortality was 9 of 608 (1.5%), with 6 dying of sepsis and 3 from pulmonary embolism. Perioperative morbidity during the first 30 days included the complications shown in Table 3.

The total mortality over the 14 years was 35 of 608, with the 26 late deaths divided into two groups – 12 from emotionally related causes and 14 from more natural etiologies as shown in Table 4. The emotionally related deaths included three suicides, three cases of cirrhosis due to a return to drinking, one death from bulimia, one from pernicious anemia due to a refusal to take vitamin B12, and four, perhaps more questionable, from auto accidents. The other late deaths included four from cardiac causes, three from cancer, and one each from atherosclerosis, pneumonia, AIDS, peritonitis, pulmonary embolus, and two from unknown causes, presumably arrhythmias. The most frequent late complications are included in Table 5.

Dumping developed in 70.6% of the patients. Although the syndrome is sometimes listed as a complication, it is actually a desired side effect which prevents these patients from gorging on sweets after operation. In spite of the complications and the limitations in food intake, only two of these patients

Table 3. Perioperative complications ($n = 608$)

Complication	*n*	%
Mortality (first 30 days)	9	1.5
Operative complications		
Splenic tear	15	2.5
Subphrenic abscess, leak	15	2.5
Anastomotic stenosis	18	3.0
Re-operated	17	2.8
Wound Complications		
Severe	18	3.0
Mild	54	8.9
Seromas	35	5.8

Table 4. Complications: late deaths (n = 608; 14-year period)

Cause of death	n
Cancer	2
Atherosclerosis	1
Pneumonia	1
MI or arrhythmia	4
AIDS	1
Sepsis	1
Peritonitis	1
Suicide	3
Cirrhosis	3
Bulimia	1
Auto accidents	4
Pernicious anemia	1
Alcoholic hepatitis	1
Pulmonary embolus	1

Total late deaths = 25; emotionally related deaths = 13.

Table 5. Late complications (n = 608; 14-year period)

Complication	n	%
Vitamin B12 deficiency	239	39.9
Incisional hernia	143	23.9
Depression	142	23.7
Staple-line failure	90	15.0
Gastritis	79	13.2
Cholecystitis	68	11.4
Anastomotic problems	59	9.8
Dehydration, malnutrition	35	5.8
Dilated pouch	19	3.2
Readmission for various reasons	229	38.2

have requested and undergone reversal of the GB. In those patients in whom the operation failed due to pouch or anastomotic dilation or to staple-line failure, all but one requested that the defect be corrected as quickly as possible before they regained their previous weight. Few operations are performed that have such a high level of patient satisfaction.

Gastric bypass surgery significantly improves the long-term mortality rate in morbidly obese diabetic patients, largely by reducing the number of deaths from cardiovascular causes. This improvement in cardiovascular disease could be due to other benefits of the surgery such as improvement in pulmonary artery hypertension due to obstructive sleep apnea and improvement in hypertension. Inasmuch as the hyperinsulinemia associated with Type II diabetes has been associated with the progression of coronary artery disease

(KELLUM et al. 1990), the greater reduction in insulin levels achieved with gastric bypass compared to operations that do not bypass the foregut, such as vertical banded gastroplasty (NASLUND et al. 1986), provides additional benefit.

The percentage of control patients requiring medical management of their diabetes increases significantly over the follow-up period, from 56.4% to 87.5%, whereas among surgical patients on medication for diabetes, the percentage dropped from 31.8% preoperatively to 8.5% at the time of last contact. These results demonstrate the progressive nature of the disease when obesity is not treated and reconfirms the effective control of diabetes with gastric bypass in the majority of patients.

c) Concerns

The GB is becoming the procedure of choice among bariatric surgeons because it produces durable weight loss, can be performed with a remarkably low mortality and complication rate in a very difficult group of patients, and provides excellent control of comorbidities such as diabetes and hypertension. However, two concerns remain: staple-line dehiscence and the stabilization of weight before reaching ideal levels, especially in the superobese.

The attainment of normal weights by the superobese may not be a feasible goal, even with staged procedures. Recent attempts at improving weight loss by increasing the length of the limbs comes close to the hazards of the intestinal bypass. The follow-up of these trials is still short; time will tell if the degree of malnutrition in these patients becomes a serious problem. A return to the ideal body weight may not, in fact, be appropriate, because correction of diabetes, hypertension, and other comorbidities occurs even with losses of 10%–20% of body weight (BUCHWALS and CAMPOS 1991).

2. Biliopancreatic Bypass

The biliopancreatic bypass represents a combination of the gstric bypass and the intestinal bypass, devised to produce an even greater weight loss than is achievable with gastric bypass alone (ADAMI et al. 1987). The operation, shown in Fig. 3, includes a gastric resection, a cholecystectomy, and a diversion of the biliopancreatic stream into a short common channel to inhibit digestion and absorption. Because of the gastrectomy, the operation is not reversible. The procedure has fewer complications than the intestinal bypass, perhaps because it avoids the bacterial overgrowth seen with totally excluded loops of small bowel. Weight loss is excellent, but recent reports of hepatic failure after the biliopancreatic bypass (GRIMM et al. 1993; LANGDON et al. 1992) suggest that the operation should be used sparingly, perhaps being saved for those patients who have failed less demanding procedures. Close lifelong follow-up is essential to avoid serious nutritional problems.

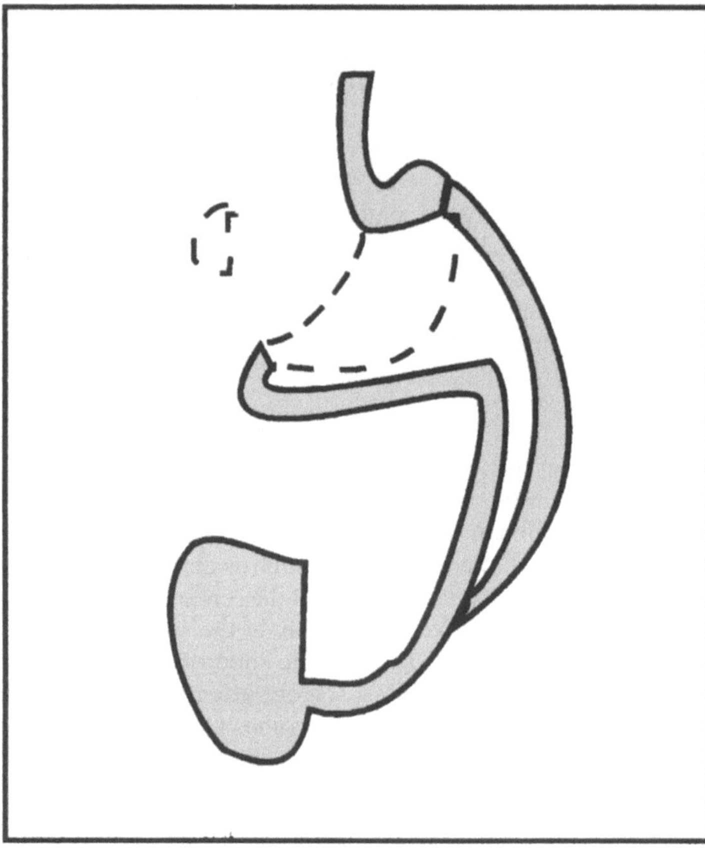

Fig. 3. Scopinaro's biliopancreatic bypass. The operation, the most radical of the bariatric procedures, involves a gastric resection, a cholecystectomy, and a diversion of the biliopancreatic stream into a short, common channel

3. Gastric Balloon

The gastric balloon is included in this chapter for completeness, so that others will not have to re-explore this idea. The balloon was introduced by GARREN (1984) on the premise that this artificial bezoar would produce early satiety, and thus, diminish intake. The device enjoyed rapid acceptance and widespread application due to massive publicity and the scheduling of multiple well-attended national courses. Within a few years, thousands had been implanted. The Food and Drug Administration took the device off the market because of its lack of long-term effectiveness, the unjustifiable cost ($7000 a year), and the high complication rates (including obstruction, ulceration, and perforation).

III. Comparison of Gastric Bypass, Gastroplasties, and Gastric Banding

Several prospective studies compared the old gastroplasties – those performed prior to the VBG – with gastric bypasses; in each, the GB proved to be the better procedure. Whether the stoma was on the lesser or the greater curvature or in the middle of the staple line, the gastric pouches enlarged and the stomas stretched so that, within 1 or 2 years, the 20% mean weight loss was soon regained. NASLUND et al. (1986), in a series of 57 patients followed for 2 years, found that weight loss at 1 year was significantly greater and failures significantly fewer in the 29 who underwent the GB. In a prospective and randomized series of 87 patients (PORIES et al. 1982), where GB was compared with gastric partition (gastric pouch stapling with gastrogastrostomy), GB patients lost 15% more of their original weight at 12 months and 21% at 18 months, with fewer failures and a similar complication rate, than those patients who had the gastric partition.

ZIMMERMAN et al. (1992) compared the results after GB and the Silastic-ring vertical gastroplasty in a small series of 35 patients in a 1-year retrospective study. They concluded that the weight loss was roughly comparable between the two procedures but that the Silastic-ring patients had a much greater problem with vomiting and food intolerance (76% vs 7%).

Only one well-controlled series comparing the GB with the true Mason's VBG has been published. SUGERMAN et al. (1989) stopped the randomization at 9 months after 20 patients had undergone each procedure, because greater weight loss ($p < 0.05$) was noted after the GB. The difference increased ($p < 0.01$) with each 3-month interval through 3 years. In a later study, the same authors obtained their best results by assigning sweets eaters to the GB and recommending that VBG not be performed in patients who are addicted to sweets.

MACLEAN et al. (1993) in a prospective randomized trial between the VBG and vertical gastric bypass in 106 patients, reported a success rate of 39% with the gastroplasty and of 58% for the bypass, due to failure of the staple lines. Their success rate, defined as a return of the patient to 150% of ideal body weight, rose to 83% when the gastric partition was performed with division rather than staples.

In summary, all three procedures, GB, Mason's VBG, and gastric banding, have been shown to be effective therapies for morbid obesity. Each has been reported to produce effective weight loss, to reverse hyperglycemia and hypertension, and to provide significant rehabilitation. All can be done with surprisingly low operative mortalities (<1% in good risk patients and <2% in those with severe comorbidities) and acceptable morbidity rates. Willbanks (personal communication, 1994) summarized the comparison nicely: "When statistically valid data become available, these will show the vertical gastroplasties to have a slightly lower complication rate and the gastric bypass to be slightly more effective and consistent." Perhaps the most important conclusion is that bariatric operations, like other operative therapies, need to be individualized to fit different patients; and in some, several staged operations will be necessary, as first recommended by KRAL (1989).

G. Management of the Morbidly Obese Patient

I. Patient Selection

Most groups operating for morbid obesity use similar criteria for the selection of their patients and, in general, manage their patients in the same way. Patients are considered to be candidates if their BMI is 40, and they have no comorbidities such as arthritis, diabetes, hypertension, the Pickwickian syndrome, sleep apnea, or other obesity-related disabling conditions; if their BMI is 35, and their obesity is complicated by these diseases; if they have realistic expectations of the outcomes after the procedure; and if they can tolerate the operation. The requirement by many insurance companies that patients provide evidence of physician-supervised diets is silly; virtually all morbidly obese have dieted all their lives and repeatedly failed. Bariatric operations are generally denied or delayed if the patient is physically unable to tolerate the procedure, owing to causes such as irreversible cardiac disease; there is a history of alcohol or substance abuse within the preceding 5 years; the patient has inadequate intelligence to understand the procedure and its consequences; there is a previous pattern of severe depression or suicide attempts; the patient has unrealistic expectations and does not agree to long-term follow-up; and, occasionally, if there is an irreconcilably hostile and nonsupportive family.

II. Preoperative Evaluation

The preoperative evaluation proceeds slowly in order to ensure that the patient is well-educated about the procedure and to emphasize that the operation is a serious undertaking. The protocol begins with initial interviews, which include the family, if possible. Usually two visits suffice to determine whether the patient is an appropriate candidate, and if so, the preoperative work-up is begun. The evaluation includes a complete blood count, urinalysis, SMA-12, electrolytes, a glucose tolerance test if the patient is not a known diabetic, electrocardiogram, posteroanterior and lateral chest roentgenograms, and pulmonary function tests with arterial blood gases. All patients are screened for psychopathology by the project psychologist or psychiatrist. Members of the family, if available, are counseled regarding the procedure. For the ongoing studies, a series of psychometric tests, special tests for glucose metabolism, and immersion hydrostatic weighing are performed. If indicated by the results of the work-up, patients may undergo more sophisticated studies, such as upper gastrointestinal series, fiberoptic endoscopy, echocardiography, stress testing, and cardiac angiography. Bariatric operations are a major undertaking so it is important patients are well prepared.

III. Preoperative Preparation

Patients are generally admitted on the morning of the procedure. More complicated cases, such as patients with cardiorespiratory failure, may require several days of inpatient preparation to achieve optimal preoperative status. A cephalosporin is given intravenously for prophylaxis the morning of the procedure and for 2 days thereafter. Serious health problems need to be stabilized before surgery. For example, skin lesions need to clear as much as possible and chronic problems such as asthma, chronic pulmonary infections, and diabetes and hypertension, need to be stabilized by adjusting or prescribing medications.

IV. Postoperative Care

In general, the postoperative care of bariatric patients resembles that of following standard abdominal procedures and is, therefore, not described in detail, except to emphasize the need for prophylactic antibiotics, careful monitoring of vital signs, fluid balance, and glucose metabolism. Appropriate furniture and equipment to manage these massive individuals is also required.

The first 24 h are particularly critical because of the great seriousness of a leak or intra-abdominal infection in these individuals. If the pulse remains over 120, if there is a rise in temperature over 102°F, or if the patient looks ill in spite of normal vital signs, emergency exploration and addition of other antibiotics may be needed. Barium swallows may be helpful but are not always reliable. Several patients with anastomotic leaks have demonstrated normal passage of barium without extravasation. Neglect of a perforation or intra-abdominal infection is associated with a high mortality rate. If there is doubt, it is best to proceed with the operation because an unnecessary exploration is a lot safer than a missed perforation.

Patients usually spend the first night on an intermediate unit with nurses who are familiar with bariatric care. Patients eat nothing by mouth until they pass flatus, usually on the third day. They are then begun on half-strength Ensure Plus (60 ml t.i.d.) with water (30 ml qh.) on the fourth day. Full-strength Ensure Plus with water in the same doses are given on the fifth and last day. After discharge, the patients are maintained on full fluid diets for 2 weeks and then cautiously progressed to a full diet by the end of 6 weeks.

Most patients gradually return to their previous diet in terms of variety, but with a marked reduction in quantity, because they fill up quickly and the gastric pouch empties slowly. Most patients do not tolerate carbohydrates well because of the induced dumping. Meats may present difficulties, so patients need to start slowly with fish, progress to chicken, and finally eat red meat. By the end of 3 months, most patients eat a small, but well-balanced diet.

The most common early complications seen in the clinic are wound abscesses and, as might be expected, these occur most commonly in diabetic patients. The wound infections generally present as red bulges that drain spontaneously or that can be drained through a small 1- to 2-cm opening of the incision. It is not necessary to open the whole wound or significant lengths of the incision because this may lead to long-term wound care due to delays in healing.

V. Long-Term Follow-up

Patients generally do remarkably well and are a delight to follow with their new body images, their freedom from diabetes, and their new lives. Daily long-term intake of liquid or chewable total mineral and vitamin products is essential, including 600 mg per day of B_{12}, the recommended minimum dose. Omission of these supplements may lead to severe anemias, Wernicke-Korsakoff syndrome, or other neuropathies. Weight gain exceeding 12% above the lowest postoperative weight is generally the result of staple line breakdown, pouch or anastomotic dilatation, or compulsive snacking. Abdominal pain is most commonly due to cholecystitis, although some patients may develop marginal ulcers, which fortunately clear quickly with H2 blockers. Recurrent vomiting usually signals overeating but may be due to stenosis of the gastrojejunostomy. Such strictures can almost always be relieved with one or two dilatations of the anastomosis. Finally, these patients require considerable emotional support from their referring physician, the surgical staff, and their own families. Monthly support groups are quite useful.

References

Abenhaim L, Moride Y et al (1996) Appetite Suppressant Drugs and the Risk of Primary Pulmonary Hypertension. NEJM 335:9, 609–616

Adami G, Gianetta E, Barreca A et al (1987) body composition after "very-little-stomach" bilio-pancreatic bypass. Eur Surg Res 19:91

Benotti PN, Bistrian B, Benotti JR et al (1992) Heart disease and hypertension in severe obesity: The benefits of weight reduction. Am J C Nutr 55:586S

Bray GA (1979) Surgical treatment of morbid obesity. Trans Assoc Live Insur Med Directors 62:107

Bray GA (1979) Surgical treatment of morbid obesity. Trans. Assoc. Life Insur Med Directors 62:107

Brolin RE, Kenler HA, Gorman JH et al (1992) Long-limb gastric bypass in the super-obese: A prospective randomized study. Ann Surg 215:387

Brolin RE, Kenler HA, Gorman JH et al (1992) Long-limb gastric bypass in the super-obese: A prospective randomized study. Ann Surg 215:387

Brolin RE, Robertson LB, Kenler HA et al (1994) Weight loss and dietary intake after vertical banded gastroplaty and Roux-en-Y gastric bypass. Ann Surg 220:782

Buchwald H, Campos CT (1991) Remedial operations following surgery for morbid obesity. Obes Surg 1:337

Buchwald H, Varco RL, Moore RB et al (1975) Intestinal bypass procedures. Curr Prob Surg Apr:1

Bunker A the use of drug therapy in the treatment of obesity: A status report. Parts I and II. The Bariatrician, Fall 1993 and Winter 1994

Catona A, Gossenberg M, La Manna A et al (1993) Laparoscopic gastric banding: Preliminary series. Obes Surg 3:207

Cowan GS (1992) The non-surgical treatment of serious obesity. Obes Surg 2:216
Cucchi BS, Pories WJ, MacDonald KG (1999) Gastro-gastric fistulas, a complication of divided gastric bypass surgery. Ann Surg 1995
Deitel M, Ilves R (1992) Mechanism of antireflux in vertical banded gastroplasty. Problems Gen Surg 9:390
Deitel m, Jones BA, Petrov I et al (1986) Vertical banded gaastroplasty: Results in 233 patients. Can J Surg 29:322
Deitel M, Shahi B, Deitel FH (1991) Effect of weight loss in the morbidly obese patient with severe disability. Obes Surg 1:419
Drenick FJ, Gurunanjappa SB, Seltzer FSA et al Excessive mortality and causes of death in morbidly obese men. JAMA 1980;243:443–445
Dublin LI, Marks HH (1952) Mortality among Insured Overweights in Recent Years, New York, Recording and Statistical Recording Press
Dubois A (1983) Obesity and gastric emptying. Gastroenterology 84:875
Elton CW, Tapscott EB, Pories WJ, Dohm GL (1993) Effect of moderate obesity on glucose transport in human muscle. Horm Metab Res 26:181
Friedman GD, Kannel WB, Dawber TR (1966) The epidemiology of gallbladder disease: Observations in the Framingham study. J Chrom Dis 19:273–292
Furnham Adrian, Baguma Peter. Cross-cultural differences in the evaluation of male and female body shapes. Internat J of Eating Disorders 15:1, 81–89
Garren LW (1984) Intragastric balloon in the treatment of morbid obesity. Presented at Symposium on Surgical Treatment of Obesity. Los Angeles
Goldstein DJ. Potvin JH (1994) Long-term weight loss: the effect of pharmacologic agents. Am J Clin Nutr 60:647–657
Greenberg DR, LaPorte DJ. Racial Differences in Body Type Preferences of Men for Women, Internat. J of Eating Disorders 19:3, 275–278
Griffen WO Fr., Blivins FA, Bell RM (1983) The decline and fall of the jejunoileal bypass. Surg Gynecol Obstet 147:301
Grimm IS, Schindler W, Haluszka O (1993) Steatohepatitis and fatal hepatic failure after biliopancreatic diversion. Am J Gastroenterol 88:321
Juczmarski RJ (1989) The assessment of body fat distribution in population based surveys. In Worship on Basic and Clinical Aspects of regional Fat Distribution. Washington DC, National Institutes of Health p 47
Kark AE (1980) Jaw wiring Am J Clin Nutr 33:420
Kellum JM, Kuemmerly JF, Oderisio PM et al (1990) Gastrointestinal hormone responses to meals before and after gastric bypass and vertical banded gastroplasty. Ann Surg 211:763–770
Keys A, Borzek J, Henschel A et al (1950) The Biology of human starvation. Minneapolis, University of Minnesota Press
Kopek E, Gertler R, Ramsey-Stewart G, Beaumont PJV (1994) Psychosocial outcome and long-term weight loss after gastric restrictive for morbid obesity. Obes Surg 4:336
Kral JG, Strauss RJ, Wise L (1989) Perioperative risk management in obese patients. In: Deitel M (ed) Surgery for the Morbidly Obese Patient. Philadelphia Lea & Febiger
Kral JG (1989) Surgical treatment of obesity. Med Clin North Am 73:251
Kral JG (1989) Surgical treatment of obesity. Med Clin North Am 73:251
Kral JG (1992) Overview of surgical techniques for treating obesity. Am J Clin Nutr 55:552S
Kremen AJ, Linnerk JH, Nelson CH (1954) An experimental evaluation of the nutritional importance of the proximal and distal small intestine. Ann Surg 140:439
Krolewski AS, Warram JH (1994) Epidemiology of late complications of diabetes. In: Kahn CR, Weir GC (eds) Joslyn's Diabetes Mellitus, 13th ed. Philadelphia: Lea & Febiger, pp 605–619
Kuczmarski RJ, Flegal KM, Campbell SM (1994) Increasing prevalence of overweight among adults; The national health and nutrition examination surveys, 1960 to 1991. JAMA 272:205

Kushner RF (1993) Body weight and mortality. Nutr Rev 51:127
Kuzmak LI (1991) A review of seven years' experience with silicone gastric banding. Obes Surg 1:403
Langdon DE, Leffingell T, Rank D (1992) Hepatic failure after biliopancreatic diversion. Am J Gastroenterol 87:775
Lapidus L, Bengtsson C, Bjorntorp P (1994) The quantitative relationship between "the metabolic syndrome" and abdominal obesity in women. Obes Res 2:372
Lapidus L, Bengtsson C, Larsson B et al (1984) Distribution of adipose tissue and risk of cardiovascular disease and death: Twelve-year follow-up of participants in the population study of women in Gothenburg, Sweden Br Med J 289:1257
Lise M, Favrett F, Belluco C et al (1994) Stoma adjustable silicone gastric banding: Results in 111 consecutive patients. Obes Surg 4:274
MacDonald KG, Long SD et al (1997) The gastric bypass operation reduces the progression and mortality of non-insulin-dependent diabetes mellitus; J of Gastrointestinal Surgery, vol 1, no. 3, pp 213–220
MacLean LD, Rhode BM, Forse RA (1993) A gastroplasty that avoids stapling in continuity. Surgery 113:380
MacLean LD, Rhode BM, Sampalis J et al (1993) Results of the surgical treatment of obesity. Am J Surg 165:155
Maclure KM, Hayes KC, Dolditz GA, Stampfer MJ, Speizer FE, Willet WC (1989) Weight, diet, and the risk of symptomatic gallstones in middle-aged women. N Engl J Med 321:563–569
Mason EE (1969) Ito C Gastric bypass. Ann Surg 170:329
Mason EE (1992) Gastric surgery for morbid obesity. Surg Clin North Am 72:501
Metropolitan Life Insurance Company (1983) New weight standards for men and women. Statistical Bulletin Metropolitan Life Insurance Company 64:2–9
Molina M (1984) Gastric banding, an experience with more than 500 cases. Presented at Symposium on Surgical Treatment of Obesity, Los Angeles
Naslund I, Wickbom G, Christorffersson E et al (1986) A prospective randomized comparison of gastric bypass and gastroplasty. Acta Chir Scand 152:681
National Heart, Lung, Blood Institute Growth and Health Study Research Group (1992) Obesity and cardiovascular disease risk factors in black and white girls: The NHLBI growth and health study. Am J Public Health 82:1613
National Institutes of Health (1986) Consensus development statement. Health implications of obesity. Washington DC
National Task Force on Prevention and Treatment of Obesity (1994) Towards prevention of obesity: research directions. Obes Res 2:571
NIH consensus development conference program and abstracts. Gastrointestinal surgery for severe obesity (1991) Bethesda Md March 25–27
O'Leary JP, Maher JW, Hollenbeck JI et al (1974) Pathogenesis of hepatic failure after obesity bypass. Surg Forum 25:356
Payne JH, De Wind LT (1969) Surgical treatment of obesity. Am J Surg 118:141
Pories WJ, Flickinger EG, Meelheim HD et al (1982) The effectiveness of gastric bypass over gastric partition in morbid obesity. Ann Surg 196:389
Pories WJ, in Sabiston DC, Lyerly HK (1997) Textbook of Surgery. WB Sanders Co., Philadelphia, pp 933–946
Ravussin E, Lillioja S, Knowler WC et al (1988) Reduced rate of energy expenditure as a risk factor for body weight gain. N Engl J Med 318:467
Ryden O, Olsson WA, Danielsson BA (1989) Weight loss after gastroplasty: Psychologic sequelae in relation to clinical and metabolic observations. J Am Coll Nutr 8:15
Schumacher N, Groth B, Kleinseck J et al (1979) Successful weight control for employees. J Am Diet Assoc 74:466
Scott HW, Jr, Dean RH, Shull HJ et al (1977) Results of jejunoileal bypass in two hundred patients with morbid obesity. Surg Gynecol Obstet 145:661

Scruggs DM, Cowan SM, Fr, Klesges L (1993) Weight loss and caloric intake after regular and extended gastric bypass. Obes Surg 3:233
Scruggs DM, Cowan SM, Jr, Klesges L (1993) Weight loss and caloric intake after regular and extended gastric bypass. Obes Surg 3:233
Sims EAH, Horton EW (1986) Endocrine and metabolic adaptation to obesity and starvation. Am J Clin Nutr 21:1455
Sjostrom L (1992a) Morbidity of severely obese subjects. Am J Clin Nutr 55:508S–515S
Sjostrom L (1992b) Mortality of severely obese subjects. Am J Clin Nutr 55:516S–523S
Stunkard A, McLaren H (1979) The results of treatment for obesity. Arch Intern Med 103:79
Stunkard AJ (1988) the Salmon lecture: Some perspectives on human obesity; Its causes. Some perspectives on human obesity: Treatment. Bull NT Acad Med 64:902
Sugerman HJ, Londrey GL, Kellum JM (1989) Weight loss with vertical banded gastroplasty and Roux-en-Y gastric bypass for morbid obesity with selective versus random assignment. Am J Surg 157:93
Sztuke-Fournier, Primary Pulmonary Hypertension and Long-term use of Appetite Suppressants, Canadian Med Assn J 156;1:89–90
Walsh JP, Davis TM, Stewart GO, Stein GR, Findlater P, Dexfenfluramine in Type II diabetes: effect on weight and diabetes control. Med J of Australia 160;18, p528
Wilkinson LH, Peloso OA (1981) Gastric (reservoir) reduction for morbid obesity. Arch Surg 116:602
Willbanks O (1991) Silicone elastomere ring vertical gastroplasty: Extended follow-up. Obes Surg 1:369
Zimmerman V, Campos CT, Buchwald H (1992) Weight loss comparison of gastric bypass and Silastic ring vertical gastroplasty. Obes Surg 2:47

Section III
Pharmacological Targets

CHAPTER 12

Leptin: The Adipocyte Signal in the Control of Body Weight

J.F. CARO and M.E. TRAUTMANN

A. Introduction

The poorly understood loop that regulates body weight has at least three systems:

1. The messenger system, which is in the periphery, informs the brain of the amount of fat stores
2. The translation system, which is in the brain, receives the information from the periphery and compares it with an internal standard of fat stores (the "set point" or "lipostat")
3. The effector system (see Fig. 1).

When fat stores deviate from the internal standard, compensatory efforts come into play to return body weight to its status quo (CARO et al. 1996).

These three systems offer a myriad of targets for the development of intelligent anti-obesity drugs. It is likely, however, that intervention in more than one of the systems will be necessary for the effective treatment of obesity. Here, we will only discuss leptin (ZHANG et al. 1994), the recently discovered key player of the messenger system. Furthermore, we will only focus on the knowledge that led to the discovery of leptin, the physiology of leptin in relationship to body weight regulation, and some of the potential roadblocks for leptin to become a treatment for obesity.

B. Discovery of Leptin

It has been known for many years that the hypothalamus plays an important role in the maintenance of body weight through balancing food intake with energy expenditure. Lesions in the ventromedial hypothalamus cause hyperphagia and decreased energy expenditure leading to obesity. KENNEDY (1953) had proposed the existence of a lipostat which senses the amount of energy in body stores (i.e., the amount of fat which maintains body weight at a certain setpoint) by balancing energy expenditure with food intake. Parabiosis experiments in which two animals were surgically connected to produce a common circulatory system have suggested the involvement of factors in the blood-

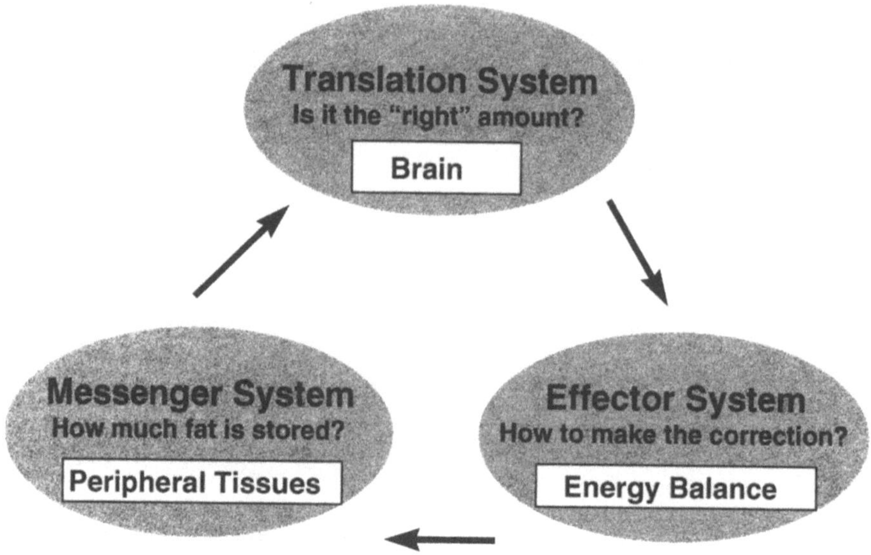

Fig. 1. The defense of the "set point" of body weight in lean and obese individuals. The messenger, translation, and effector systems interact with each other. The treatment of obesity may need to modify one or more of the three systems

stream. Lesions in the ventromedial hypothalamus in one of the animals led to hyperphagia and obesity in the treated animal but to death by starvation in the adjoined untreated animal (HERVEY 1959). This suggested the overproduction of a satiety factor by the treated animal.

Parabiotic experiments with genetically obese mice from different strains further supported this theory. Connection of an obese *ob/ob* mouse to a lean (*ob/+*) mouse suppressed the obesity in the *ob/ob* mouse (HAUSBERGER 1958). In contrast, in another genetically obese mouse, the *db/db* mouse, obesity could not be corrected by parabiosis with a lean mouse. Instead, the lean mouse would stop eating completely and starve to death. It was concluded that the *ob/ob* mice lack a hormone which suppresses food intake so that transfer of blood from a normal mouse is able to normalize the phenotype whereas *db/db* mice (and similarly obese *fa/fa* Zucker rats) are unresponsive to this hormone and rather overproduce this factor which leads to starvation of the parabiotic partner.

The *ob* gene was discovered by positional cloning from *ob/ob* mice (ZHANG et al. 1994). The mouse *ob* gene consists of 3 exons and 2 introns and is transcribed into a 4.5-kb mRNA. The ob hormone is a 167 amino acid protein including a 21 amino acid signal peptide which is required for secretion (Fig. 2). The circulating hormone which was later called leptin (from the Greek word leptos for "thin") is a 16-kDa, 146 amino acid protein. Crystallization of an analogue revealed the three-dimensional structure of a four helix-bundle protein

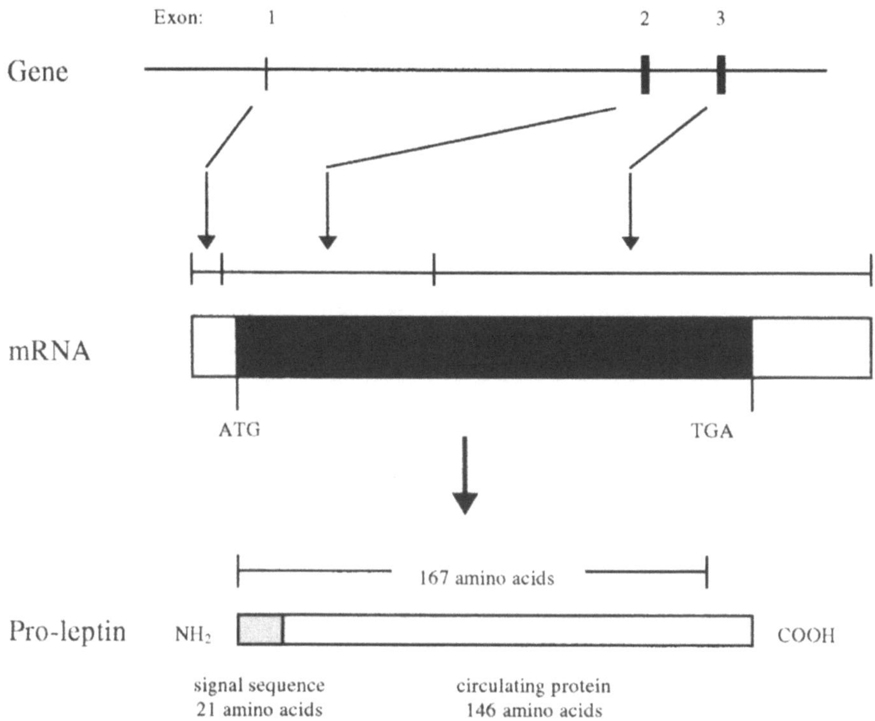

Fig. 2. Leptin synthesis in the adipocyte. The *ob* gene consists of three exons, two of which contain the coding region. The human gene is transcribed into a 3.5 kb message (*coding region in black*) that codes for a pro-hormone of 167 amino acids. The 21 amino acid signal sequence (*gray*) is removed prior to release from the cell. The protein in the circulation is 16 kDa. (Reprinted from International Journal of Biochemistry and Cell Biology, 29, Robert V. Considine, Jose F. Caro, Leptin and the regulation of body weight, 1255–1272, 1997, with permission from Elsevier Science)

(Zhang et al. 1997) with a structural similarity to other cytokines (Fig. 3). In different strains of the *ob/ob* mouse two mutations have been found in the leptin gene: a nonsense mutation in codon 105 leads to a 20-fold increase in expression of an abnormal ob mRNA in the original strain, and in another strain, a mutation in the promoter region prevents synthesis of ob mRNA. The human gene is located on chromosome 7q31.3 (Halaas et al. 1995). The promoter region spans approximately 3 kb and contains multiple CCAAT/enhancer-binding protein (C/EBP) and SP-1 motifs, a glucocorticoid response element and several cAMP response element binding sites (Gong et al. 1996). It is now known that hormones such as insulin (Cusin et al. 1995; Kolaczynski et al. 1996), glucocorticoid (De Vos et al. 1995; Slieker et al. 1996), and adrenergic agents (Collins and Surwit 1996; Trayhurn et al. 1995) regulate leptin secretion (Fig. 4). But additional factors seem to control leptin secretion since in cell culture, even after 3 days, adipocytes from obese patients

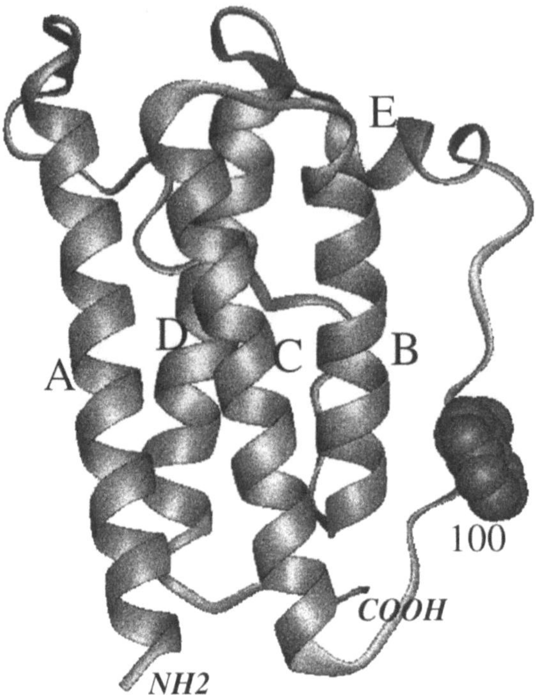

Fig. 3. Ribbon diagram of leptin. The view is perpendicular to the four-helical bundle axis. The four α-helices A, B, C, D; the bend C-terminal 3_{10} helical extension of helix D; the additional short helical segment E; the connecting loops AB, BC, and CD. The N and C termini are indicated. The disulphide bridge between the C-terminal Cys 146 and Cys 96 in the AB loop is represented with the ball-and-stick model. (Repinted with permission from Nature, Zhang F, Basinski MB, Beals JM, Briggs SL, Churgay LM, Clawson DK, DiMarchi RD, Furman TC, Hale JE, Hsiung HM, Schoner BE, Smith DP, Zhang XY, Wery JP, Schevitz RW. Crystal structure of the obese protein leptin-E100, 387:206–209, 1997, Macmillan Magazines Ltd.)

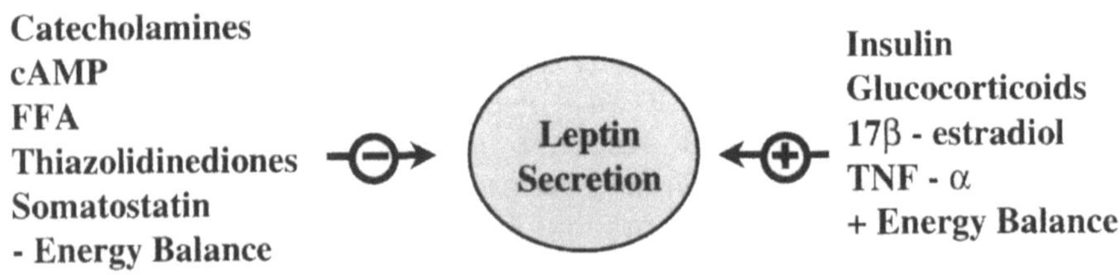

Fig. 4. Regulation of leptin secretion. The hormones or compounds on the *left* inhibit leptin secretions, whereas those of the *right* stimulate leptin secretion

continue to release more leptin than those from lean subjects (Considine et al.1996; Nolan et al. 1996). Cell size seems to correlate with leptin secretion and adipocyte metabolism may also be involved in the regulation of secretion.

C. The Leptin Receptor

The receptor for leptin was cloned from a choriod plexus cDNA library (Tartaglia et al. 1996) (Fig. 5). Several forms of the leptin receptor exist through alternative splicing of a common mRNA precursor (Lee et al. 1996). The longest form consists of a single membrane spanning protein which shares considerable homology with the glycoprotein 130 (GP130) subunit of the interleukin-6 (IL-6) receptor and belongs to the cytokine class I receptor family. This form is most abundantly expressed in the hypothalamus which is proposed to be a major site of leptin action (Mercer et al. 1996). The structure of this receptor isoform contains a large extracellular domain, a short transmembrane domain, and a fairly short intracellular portion which contains two sequences for the binding of Janus kinase (JAK) (Lee et al. 1996). JAK belongs to the tyrosine kinase family and is able to phosphorylate tyrosine residues in the distal leptin receptor which allows the binding of signal transducer and activator of transcription protein 3 (STAT3) (Vaisse et al. 1996). Bound STAT3 is then phosphorylated by JAK again enabling the signaling and activation of specific gene transcription. If the long form of the leptin receptor is absent as in the obese db/db mouse it appears that no effective signaling occurs (Vaisse et al. 1996). This form of the receptor has been identified in humans and can have sequence polymorphisms of unknown significance (Considine et al. 1996a).

D. Role of Leptin in Monogenic Forms of Obesity

It is clear that obesity in the *ob/ob* mouse is due to leptin deficiency (Zhang et al. 1994). Leptin replacements in *ob/ob* mice results in normalization of the metabolic disturbances of hyperglycemia, hyperinsulinemia, hypertriglyceridemia, low metabolic rate, and body weight. In contrast, in the *db/db* mouse and the *fa/fa* rat, which possess mutations in the leptin receptor, leptin levels are elevated and administration of leptin does not improve obesity. In a number of animal species the weight lowering effect of exogenous leptin was demonstrated (Campfield et al. 1995; Pelleymounter et al. 1995; Halaas et al. 1995; Weigle et al. 1995; Stephens et al. 1995; Schwartz et al. 1996).

Recent findings of leptin-deficient and leptin receptor-deficient syndromes in humans have confirmed the importance of leptin for the control of body weight as well as the requirement of leptin for induction of puberty. Two extremely obese children in a highly consanguineous family were the first described cases of leptin deficiency (Montague et al. 1997). This was caused by a homozygous frame shift mutation involving the deletion of a guanine

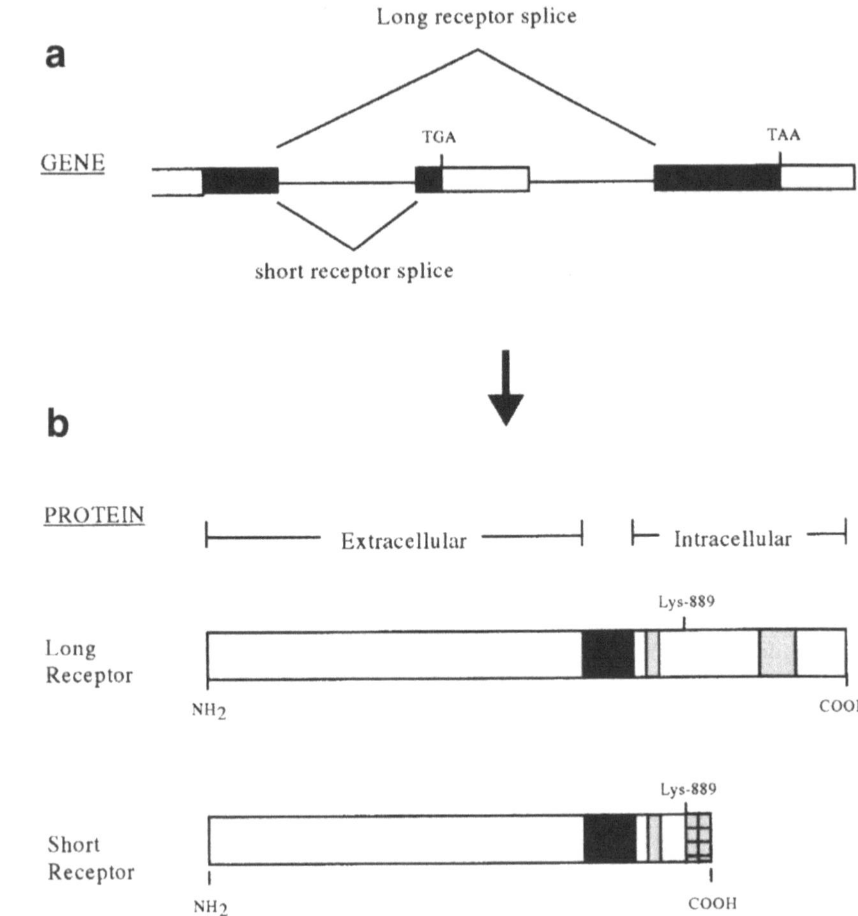

Fig. 5a,b. The two major forms of the leptin receptor. A long (OB-Rb) and a short (Ob-Ra) form of the leptin receptor have been detected in mice. **a** The mRNA for the two receptors is generated by alternative splicing of the same gene (*coding regions in black*; note that mapping of the 5' end of the gene is not complete). **b** The amino acid sequence of both the extracellular and transmembrane domain (*black*) of the two receptors is identical to lysine 899, after which they are completely dissimilar. The long form of the receptor contains interaction sites (*gray boxes*) for JAK and STAT (Reprinted from International Journal of Biochemistry and Cell Biology, 29, Robert V. Considine, Jose F. Caro, Leptin and the regulation of body weight, 1255–1272, 1997, with permission from Elsevier Science)

nucleotide in codon 133 of the leptin gene. Despite an extreme body fat mass their serum leptin levels were extremely low. These children had a normal birth weight but developed obesity shortly after birth. Parents and the heterozygote siblings had normal weight or mild obesity. In another family leptin deficiency was detected in two adults and a child (STROBEL et al. 1998). In these cases a missense mutation due to a C vs T substitution in codon 105 of the leptin gene led to a premature stop codon so that the mutant protein was synthesized, but could not be secreted. This is identical to the appearance of the premature stop codon in the *ob/ob* mouse. The subjects have extreme obesity with a body mass index of greater than 46 kg/m^2 in the adults. All subjects were reported to be hyperphagic. They have elevated circulating insulin concentrations and the 34-year-old female is hyperglycemic. The female patient had primary amenorrhea and the male adult had the clinical features of hypogonadism and had never entered puberty. In these individuals normal body temperature was reported despite decreased sympathetic tone.

A mutation in the leptin receptor resulted in a very similar phenotype with a body mass index of 52–71 kg/m^2 (CLEMENT et al. 1998). In these patients an abnormal leptin receptor mRNA was found which resulted from skipping of exon 16. Exon 16 encodes for the first 830 amino terminal amino acids of the extracellular receptor domain. The mutant receptor protein lacks both the transmembrane and the intracellular domains of the long-form of the leptin receptor. The 13- and 19-year-old teenagers had had no pubertal development and showed reduced growth hormone and thyrotropin secretion.

It is expected that leptin administration in the subjects with extreme leptin deficiency would result in weight loss as leptin normalizes the metabolic syndrome in the *ob/ob* mouse. In contrast, as in the *db/db* mouse, leptin administration will not result in weight loss in patients with mutations in the leptin receptor. Presently, the prevalence of leptin and leptin receptor mutations in the general population is not known. However, these mutations must be extremely rare (CONSIDINE et al. 1995, 1996b).

E. Role of Leptin in Polygenic Forms of Obesity

Except for the *ob/ob* mouse and the rare patients described by MONTAGUE et al. (1997) and STROBEL et al. (1998), who have well-defined mutations in the leptin gene, obesity in humans and animals is associated with hyperleptinemia. Figure 6 shows the relationship between serum leptin and percentage of body fat from 500 individuals with a wide range of body weight (CONSIDINE et al. 1996c). The data demonstrated strong positive correlation between serum leptin concentration and body fat. Thus, the adipocyte appears to send the correct message to the brain on the amount of fat stores. If the action of leptin in humans is similar to that in rodents, appetite should decrease and energy expenditure increase, which should result in weight loss. The finding of increased serum leptin concentrations in obese subjects suggests decreased sensitivity to leptin. Figure 7 illustrates several of the possible sites of abnormal leptin action in humans. We now know that the great majority of obese subjects do not have a defect in the production of leptin. However, there may exist intravascular defects, such as leptin antibodies, leptin antagonists, or increased production of leptin-binding proteins to limit the concentration of

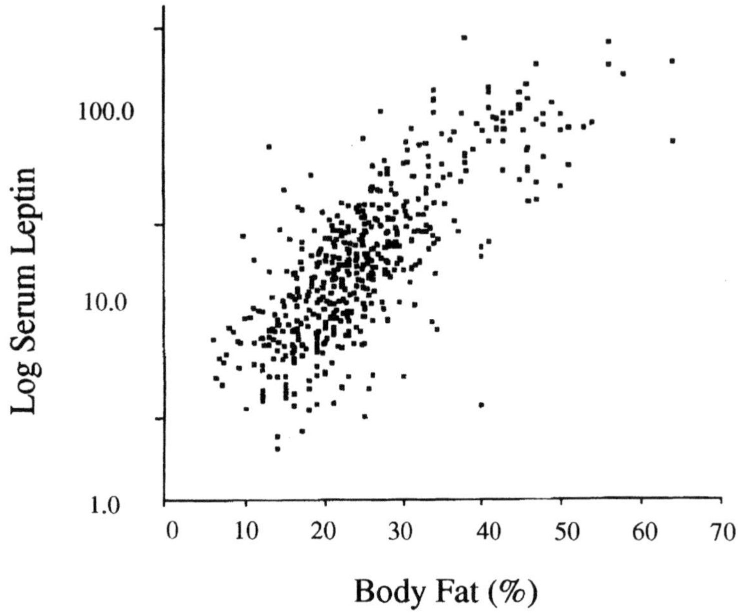

Fig. 6. Relations between percent body fat and log serum leptin concentration. (Reprinted with permission from Diabetes, 1996, Caro JF, Sinha MK, Kolaczynski JW, Zhang PL, Considine RV, Leptin: the tale of an obesity gene, 45:1455–1462)

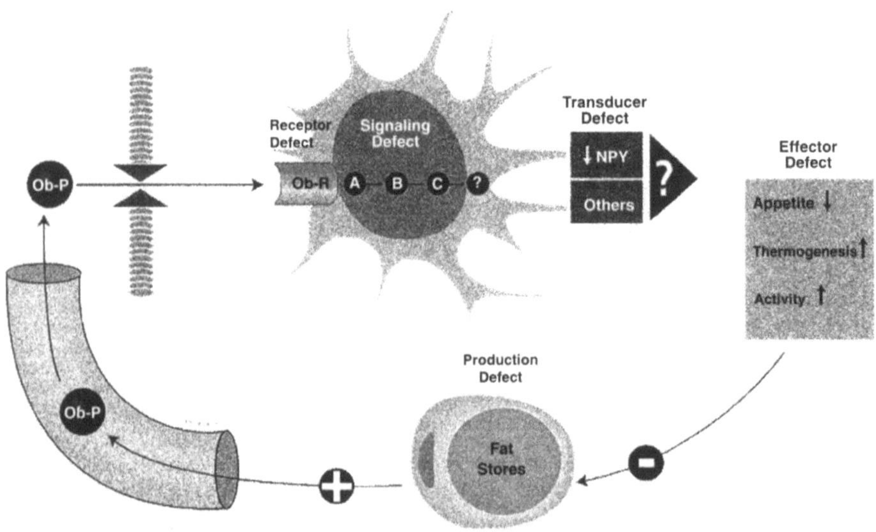

Fig. 7. Possible defects of leptin action in human obesity. They include production, intravascular transport, receptor, signaling transducer, and effector defects, which are discussed in the text. (Reprinted with permission from Diabetes, 1996, Caro JF, Sinha MK, Kolaczynski JW, Zhang PL, Considine RV, Leptin: the tale of an obesity gene, 45:1455–1462)

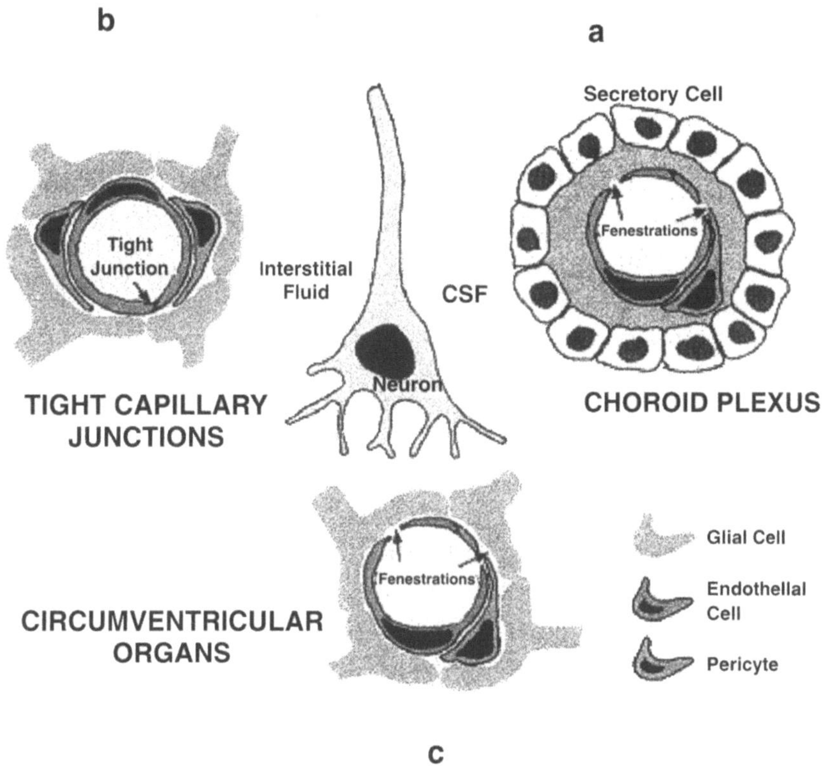

Fig. 8a–c. Potential sites of leptin entry into the brain. **a** Fenestrated capillaries in the choroid plexus allow molecules to diffuse to the secretory cells that are responsible for production of central spinal fluid. **b** Most areas of the brain are supplied with capillaries that have tight junctions in the endothelial lining. Entry via these capillaries must be by facilitated and transcytosis. **c** Capillaries in the circumventricular organs are fenestrated and the lining glial cells do not present a permeability barrier. Leptin would enter these areas of the brain by passive diffusion (Reprinted with permission ??? from Endocrinology and Diabetes, 1998, Stephens TW, Caro JF, To be lean or not to be lean. Is leptin the answer? 106:1–15)

free leptin that reaches the brain. It should be noted that a leptin binding activity has been detected in plasma such that the majority of protein appears bound in lean subjects and free in obese subjects (Sinha et al. 1996). The physiological or analytical significance for this binding activity is presently unknown, since the ability of current immunoassay techniques to detect leptin has not been sufficiently characterized. Hormones like IGF-1 and CRH are inactivated by binding to plasma proteins, whereas, growth hormone may be more active when bound (Stephens et al. 1998; Turyn et al. 1997).

In obesity, an excess of leptin may not easily reach the hypothalamic leptin receptor because of the blood-brain barrier. Figure 8 illustrates the three potential sites for leptin entry into the brain. Most areas of the brain are supplied with capillaries that have tight junctions in the endothelial lining. Entry via these capillaries must be via facilitated transport and transcytosis (GOLDEN et al. 1997). There are discrete areas in the brain, such as the circumventricular organs and the choroid plexus, where capillaries are fenestrated. In the choroid plexus, the choroid epithelium with tight junctions would not allow leptin, a 146 amino acid residue protein, to enter the cerebrospinal fluid (CSF) unless a transport mechanism is also in place. One of the circumventricular organs is the median eminence of the hypothalamus. Here, intravascular leptin would freely diffuse into the interstitial space because the capillary wall is fenestrated. However, evidence suggest that the full-length leptin receptor in the arcuate nucleus of the hypothalamus is excluded from the peripheral circulation by the blood-brain barrier.

In order to test the hypothesis that leptin transport into the brain is a rate-limiting step in leptin action, leptin in CSF and serum was measured in 31 individuals with a wide range of body weight (CARO et al. 1996). Mean serum leptin was 318% higher in obese individuals than in lean individuals. However, the CSF leptin concentration in obese individuals was only 30% higher than in lean subjects. Consequently, the leptin CSF/serum ratio in lean individuals was 43-fold higher than in obese individuals. The relationship between CSF leptin and serum leptin was best described by a logarithmic function (see Fig. 9).

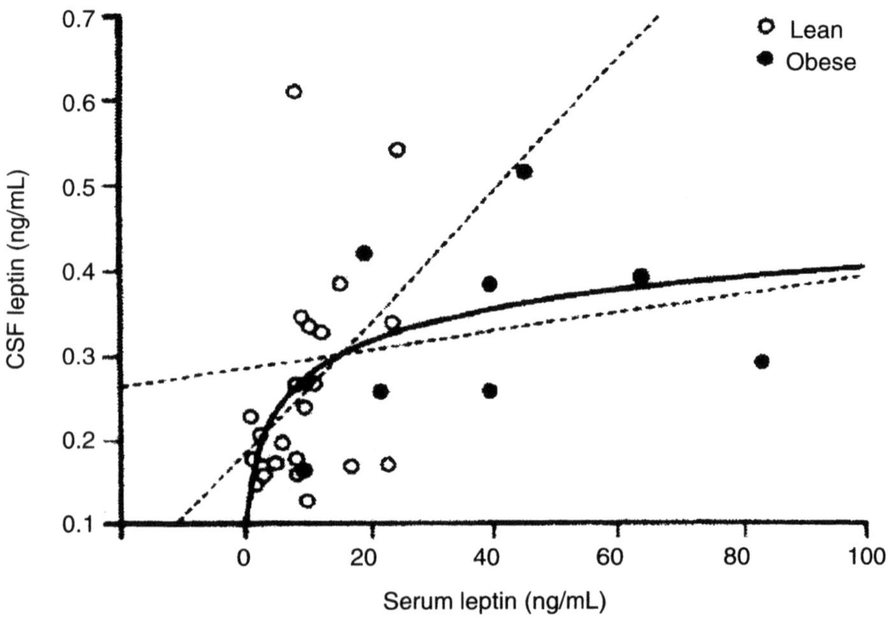

Fig. 9. Relation between CSF and serum leptin concentration. *Solid line*, logarithmic function; *broken line*, linear regression (Reprinted with permission from The Lancet Ltd, 1996, Jose F. Caro, Decreased cerebrospinal-fluid/serum leptin ratio in obesity: a possible mechanism for leptin resistance, 348:159–162)

If the leptin concentration of the CSF can be equated to the hypothalamic interstitial leptin concentration, then it is possible to understand why obese people do not have the expected response to their endogenous hyperleptinemia. Brain transport of leptin might be a rate-limiting step in leptin action. There are two implications: first, beyond a certain leptin level, which in obese patients was about 25 ng/ml, a level which is exceeded in 100% of patients with morbid obesity, an increase in leptin production by the growing fat mass would be futile. The idea of severe leptin resistance in these patients might be more apparent than real. In fact, severe hyperleptinemia might down-regulate the leptin transporters and make the situation worse. An inverse relationship between serum leptin and leptin receptor number in the choroid plexus is suggested by binding studies in *ob/ob* and *db/db* mice (LYNN et al. 1996; MALIK et al. 1996). The hypothalamic leptin receptor is not down-regulated in human obesity, perhaps because the interstitial leptin concentration may be only mildly elevated, at least compared with that of serum leptin. The second implication is that exogenous administration of leptin to treat obesity might be ineffective if endogenous leptin has already saturated its transporters. Overall, the work of several laboratories (BANKS et al. 1996; SCHWARTZ et al. 1996; CARO et. al. 1996) suggest a saturable leptin transport system. The reduced efficacy of brain leptin transport in obese individuals may provide a mechanism for leptin resistance.

The most provocative experimental data consistent with the above hypothesis was recently generated by VAN HEEK et al. (1997). This longitudinal study was designed to determine whether diet-induced obesity in mice produces resistance to peripheral and/or central leptin treatment. Obesity was induced in two strains of mice by exposure to a 45% fat diet. Both groups became resistant to peripherally administered leptin. The peripheral dose of leptin was up to 25 times higher than that found previously to be very effective in *ob/ob* mice. This dose is at lease 5000 times higher than the endogenous level of leptin in these diet-induced obese mice. In sharp contrast, central administration of leptin to peripherally leptin-resistant mice resulted in a robust response to leptin, after a single intracerebroventricular infusion. Therefore, a peripheral dose of leptin that is 4000 times greater than the centrally active dose does not overcome leptin resistance in these mice.

Peripheral leptin resistance in diet-induced obesity in rodents has been confirmed (WIDDOWSON et al. 1997). However, submaximal concentration of intraventricular injection of leptin has also demonstrated an inhibition in rats with diet-induced obesity. Thus diet-induced obesity in the rat is associated with both peripheral leptin resistance and central leptin insensitivity (WIDDOWSON et al. 1997).

Since we know that the great majority of obese patients have no mutations in the leptin receptor (SINHA et al. 1996), human obesity appears to be characterized by a post-leptin receptor defect. How distal to the leptin receptor might this defect be? In vitro and in vivo studies demonstrate that leptin activates cytokine-like signal transduction mechanisms by stimulating the classic JAK-STAT pathway via the long leptin receptor isoform. Flier's laboratory (BJORBAEK et al. 1998) has recently demonstrated that a member of the suppressors of cytokine signaling (SOCS-3) blocked leptin-induced signal transduction. Furthermore, SOCS-3 mRNA in the arcuate nucleus and the dorsomedial hypothalamus is increased in A^y/a mice, a model of murine obesity with both peripheral and central leptin resistance. Whether SOCS-3 overexpression is involved in the mechanism of leptin resistance in man is not know at this time.

As shown in Fig. 7, it is possible that leptin transport, leptin receptor, and its signaling cascade are all normal and the defect resides in the leptin transducer system. This includes, as shown in Fig. 10, the fast growing list of neuropeptides that are influenced by leptin and are believed to be involved in body weight regulation. Excluded in Fig. 10 and in this chapter are other

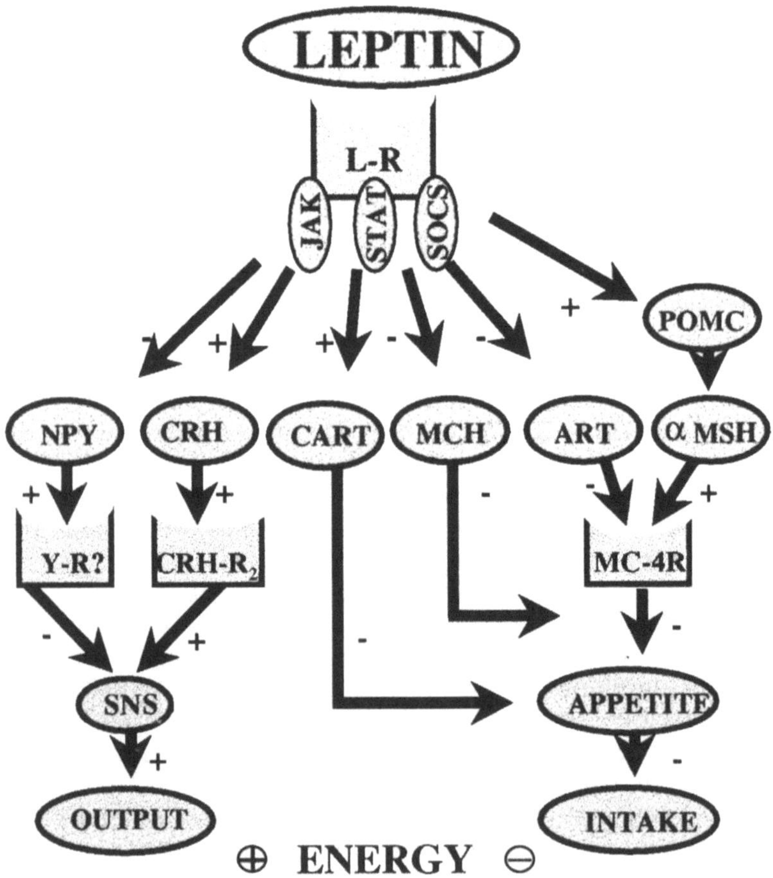

Fig. 10. Interaction of leptin with different hypothalmic neuropeptides. Notice the multiple post-leptin receptor mechanisms which induce weight loss

important pathways now known to be influenced by leptin but which are not directly involved in body regulation, i.e., GnRH and gonadotropin secretion, GH axis, etc.

As shown in Fig. 10, NPY is one candidate for the role of mediator of leptin's effects since it is the most abundant neuropeptide in the brain and is involved in the regulation of food intake and carbohydrate preference and, metabolic and lipogenic rate. Anatomical studies demonstrate NPY neurons are in the appropriate areas of the hypothalamus. NPY in the hypothalamus is largely produced in neurons of the arcuate nucleus. From here, axons project to several areas of the brain including the paraventricular nucleus and lateral hypothalamus. NPY is the only known peptide that can induce obesity through prolonged central administration with or without increased feeding. Indeed, STEPHENS et al. (1995) demonstrated that chronic leptin administration decreased hypothalamic neuropeptide Y mRNA expression in the arcuate nucleus and directly suppressed NPY release from isolated perifused normal rat hypothalamus. On the other hand, CRH plays the inverse role from NPY in that it inhibits feeding and stimulates sympathetic neural activity. Data suggesting an increase in CRH mRNA as a result of leptin treatment indicate the possibility of this peptide also playing a role in regulation of food intake and obesity by leptin (SCHWARTZ et al. 1997).

In Fig. 10 the importance of melanocortin receptor 4 in body regulation can be seen (FLIER et al. 1998). Two findings have placed the MC4 receptor at a pivotal position in the central pathways for energy homeostasis. Targeted deletion of the MC4 receptor produced a syndrome of obesity similar to that of Agouti mice, minus the pigmentary defect that required de-regulated expression of Agouti in skin acting on MC1 receptors. Additionally, the observations of decreased feeding after central administration of agonists (e.g., α-MSH) and the increased feeding after administration of a synthetic receptor antagonist provide pharmacological evidence for the role of the MC4 receptor. What is the endogenous ligand for this receptor? Recent studies have produced a number of surprises. α-MSH, produced from proopiomelanocortin (POMC) precursors, appears to fit the role of agonist to decrease feeding. In the arcuate nucleus (which also expresses NPY), approximately 30% of the POMC neurons express mRNA for the leptin receptor long form and arcuate POMC mRNA expression is regulated positively by leptin. Thus, one can envision a loop in which rising leptin (with obesity) drives increased arcuate POMC expression, which then projects α-MSH containing axons to MC4R-expressing cell bodies elsewhere in the hypothalamus, causing decreased food intake.

But the MC4 receptor pathway is not restricted to a single agonistic ligand, as a newly discovered neuropeptide acts as an antagonist on this same receptor. Agouti-related transcript (ART) is a homolog of agouti whose expression is largely restricted to the arcuate nucleus, where it is markedly increased in leptin-deficient *ob/ob* mice. ART antagonizes α-MSH at the MC4-R, and its forced ectopic expression in transgenic mice produces obesity. The output from the feeding inhibitory MC4 receptor may be determined by the ratio of agonist (α-MSH) and antagonist (ART) at MC4 receptor containing neurons.

Melanin-concentrating hormone (MCH) is also inhibited by leptin as shown in Fig. 10 (reviewed by FLIER and MARATOS-FLIER 1998). Overexpression of MCH mRNA was seen in *ob/ob* mice. Fasting increased expression both in control and *ob/ob* animals. Intracrebroventricular administration of MCH to rats stimulated feeding. It is of interest that MCH and α-MSH have opposing actions in rat brain. However, unlike α-MSH and ART, which antagonize each other by acting on the same receptor, MCH has no affinity for known melanocortin receptors and acts via an unidentified receptor.

The most recently discovered peptide regulated by leptin is CART (cocaine and amphetamine regulated transcript) (KRISTENSEN et. al. 1998). CART is a satiety factor and is closely associated with the actions of two important regulators of food intake, leptin and neuropeptide Y. Food-deprived animals show a pronounced decrease in expression of CART mRNA in the arcuate nucleus. In animal models of obesity with disrupted leptin signaling, CART mRNA is almost absent from the arcuate nucleus. Peripheral administration of leptin to obese mice stimulates CART mRNA expression. When injected intracerebroventricularly into rats, recombinant CART peptide inhibits both normal and starvation-induced feeding and completely blocks the feeding response induced by neuropeptide Y.

It is likely that the list of brain peptides will continue to grow. What is most certain, however, is the lack of knowledge of the role that the known peptides play in the mechanism of leptin resistance in human obesity.

Finally, the defect in leptin action in human obesity might be far from leptin itself or even far from the transduction system. The defect might reside in the effector system (Fig. 7). The distal biochemical and behavioral mechanisms that control appetite, physical activity, and thermogenesis may turn out to be the critical defect(s).

Acknowledgments. The authors would like to thank Dr. Robert Campos for his critical review and Mrs. Rory Seib for excellent preparation of the manuscript.

References

Banks WA, Kastin AJ, Huang W, Jaspan JB, Maness LM (1996) Leptin enters the brain by a saturable system independent of insulin. Peptides 17:305–311

Bjorbaek C, Elmquist J, Frantz D, Shoelson S, Flier J (1998) Identification of SOCS-3 as a Potential Mediator of Central Leptin Resistance. Mol Cell 1(4): 619–625

Campfield LA, Smith FJ, Guisez Y, Devos R, Burn P (1995) Recombinant mouse OB protein: evidence for a peripheral signal linking adiposity and central neural networks. Science 269:546–549

Caro JF, Kolaczynski JW, Nyce MR, Ohannesian JP, Opentanova I, Goldman WH et al (1996) Decreased cerebrospinal-fluid/serum leptin ratio in obesity: a possible mechanism for leptin resistance. The Lancet 348:159–161

Caro JF, Sinha MK, Kalcynski JW et al (1996) Leptin: the tale of an obesity gene Diabetes 45:1455–1462

Clement K, Vaisse C, Lahlou N, Cabrol S, Pelloux V, Cassuto D, Gourmelen M, Dina C, Chambaz J, Lacorte JM, Basdevant A, Bougnères P, Lebouc Y, Froguel P, Guy-Grand B (1998) A mutation in the human leptin receptor gene causes obesity and pituitary dysfunction Nature 392:398
Collins S, Surwitt RS (1996) Pharmacologic manipulation of Ob expression in a dietary model of obesity. J Biol Chem 371:1–4
Considine RV, Considine EL, Williams CJ, Hyde TM, Caro JF (1996a) The hypothalamic leptin receptor in humans: identification of incidental sequence polymorphisms and absence of the db/db mouse and fa/fa rat muations Diabetes 45: 992–994
Considine RV, Considine EL, Williams CJ, Nyce MR, Zhang P, Opentanova I, Ohannesian JP, Kolaczynski JW, Bauer TL, Moore JH, Caro JF (1996b) Mutation screening and identification of a sequence variation in the human ob gene coding region. Biochem Biophys Res Commun 220, pp 735–9
Considine RV, Sinha MK, Heiman, ML, Kriauciunas A, Stephens TW, Nyce MR, Ohannesian JP, Marco CC, McKee LJ, Bauer TL, Caro JF (1996c) Serum immunoreactive-leptin concentrations in normal-weight and obese humans. N Engl J Med 334:292–295
Considine RV, Considine EL, Williams CJ, Nyce MR, Zhang PL, Opentanova I, Ohannesian JP, Kolaczynski JW, Bauer TL, Moore JH, Caro JF (1996) Mutation screening and identification of a sequence variation in the human Ob gene coding. Biochem Biophys Res Comm 220:735
Considine RV, Considine EL, Williams CJ, Nyce MR, Magosin SA, Bauer TL, Rosato EL, Colberg J, Caro JF (1995) Evidence Against Either a Premature Stop Codon or the Absence of Ob Gene mRNA in Human Obesity. J Clin Invest 95:2986
Cusin I, Sainsbury A, Doyle P, Rohner-Jeanrenaud F, Jeanrenaud B (1995) The ob gene and insulin. A relationship leading to clues to the understanding of obesity. Diabetes 44:1467–1470
De Vos P, Saladin R, Auwerx J, Staels B (1995) Induction of ob gene expression by corticosteroids is accompanied by body weight loss and reduced food intake. J Biol Chem 270:15958–15961
Flier J, Maratos-Flier E (1998) Obesity and Hypothalamus: Novel Peptides for New Pathways Cell 92:437–440
Golden PL, Maccagnan TJ, Pardriidge WM (1997) Human blood-brain barrier leptin receptor. J Clin Invest 99:14–18
Gong DW, Bi S, Pratley RE, Weintraub BD (1996) Genomic structure and promoter analysis of the human obese gene. J Biol Chem 271:3971–3974
Halaas JL, Gajiwala KS, Maffei M, Cohen SL, Chait BT, Rabinowitz D, Lallone RL, Burley SK, Friedman JM (1995) Weight reducing effects of the plasma protein encoded by the obese gene. Science 269:543–546
Hausberger (1958) Parabiosis and transplantation experiments in hereditarily obese mice. Anat Rec 130:313
Hervey GR (1959) The effects of lesions in the hypothalamus in parabiotic rats. J Physiol 145:336–352
Kolacyzinski JW, Considine RV, Ohannesian J, Marco C, Opentanova I, Nyce MR, Myint M, Caro JF (1996) Responses of leptin to short-term fasting and refeeding in humans: a link with ketogenesis but not ketones themselves. Diabetes 45:1511–1515
Kennedy GC (1953) The role of depot fat in the hypothalamic control of food intake in the rat. Proc Roy Soc 140:578–592
Kristensen P, Judge M, Thim L, Ribel U, Christjansen K, Wulff B, Clausen J, Jensen P, Madsen O, Vrang N, Larsen P, Hastrup S (1998) Hypothalamic CART is a new anorectic peptide regulated by leptin. NATURE 393:72–75
Lee GH, Proenca R, Montez JM, Carroll KM, Darvishzadeh JG, Lee JI, Friedman JM (1996) Abnormal splicing of the leptin receptor in diabetic mice. Nature 379: 632–635

Lynn RB, Cao GU, Considine RV, Hyde TM, Caro JF (1996) Autoradiographic localization of leptin binding in the choroid plexus of ob/ob and db/db mice. Biochem Biophys Res Commun 219:884
Malik KF, Young WS (1996) Localization of binding sites in the central nervous system for leptin (OB protein) in normal, obese (ob/ob) and diabetic (db/db) C57BL/6J mice. Endocrinology 137:1497–1500
Mercer JG, Hoggard N, Williams LM, Lawrence CB, Hannah LT, Trayhurn P (1996) Localization of leptin receptor mRNA and the long form splice variant (Ob-Rb) in mouse hypothalamus and adjacent brain regions by in situ hybridization. FEBS Lett 387:113–116
Montague CT, Farooqi IS, Whitehead JP, Soos MA, Rau H, Wareham NJ, Sewter CP, Digby JE, Mohammed SN, Hurst JA, Cheetham CH, Earley AR, Barnett AH, Prins JB, O'Rahilly S (1997) Congenital leptin deficiency is associated with severe early-onset obesity in humans. Nature 387:903–907
Pelleymounter MA, Cullen MJ, Baker MB, Hecht R, Winters D, Boone T, Collins F (1995) Effects of the obese gene product on body weight regulation in ob/ob mice. Science 269:540–543
Schwartz MW, Baskin DG, Bukowski TR, Kuijper JL, Foster D, Lasser G, Prunkard DE, Porte D, Woods SC, Seeley RJ, Weigle DS (1996) Specificity of leptin action on elevated blood glucose levels and hypothalamic neuropeptide Y gene expression in ob/ob mice. Diabetes 45:531–535
Schwartz MW, Peskind E, Raskin M, Boyko EJ, Porte D (1996) Cerebrospinal fluid leptin levels: relationship to plasma levels and to adiposity in humans. Nat Med 2:589–593
Schwartz MW, Seeley RJ (1997) Neuroendocrine responses to starvation and weight loss. New England Journal of Medicine 336:1802–1811
Sinha MK, Opentanova I, Ohannesian JP, Heiman ML, Hale J, Becker GW, Bowsher RR, Stephens TW, Caro JF (1996) Evidence of Free and Bound Leptin in Human Circulation: Physiological significance in lean and obese subjects and during short-term fasting. J Clin Invest 98:1277
Slieker LJ, Sloop KW, Surface PL, Kriauciunas A, LaQuier F, Manetta J, Bue-Valleskey J, Stephens TW (1996) Regulation of expression of ob mRNA and protein by glucocorticoids and cAMP. J Biol Chem 271:5301–5304
Stephens TW, Basinski M, Bristow PK, Bue-Valleskey JM, Burgett SG, Craft L, Hale J, Hoffman J, Hsiung HM, Kriaciunas A, MacKellar W, Rosteck PR Jr, Schoner B, Smith D, Tinsley FC, Zhang XY, Heiman M (1995) The role of neuropeptide Y in the antiobesity action of the obese gene product. Nature 377:530–532
Strobel A, Issad T., Camoin L, Ozata M, Strosberg AD (1998) A leptin missense mutation associated with hypogonadism and morbid obesity, Nature Genetics, vol 18, No. 3, pp 213–215
Tartaglia LA, Dembski M, Weng X, Deng N, Culpepper J, Devos R, Richards GJ, Campfield LA, Clark FT, Deeds J et al (1995) Identification and expression cloning of a leptin receptor, OB-R, cell 83:1263–1271
Trayhurn P, Duncan JS, Rayner DV (1995a) Acute cold-induced suppression of (obese) gene expression in white adipose tissue of mice; mediation by the sympathetic system. Biochem J 84:729–733
Turyn D, Dominic F, Sotelo A, Bartke A (1997) Growth hormone-binding protein enhances growth hormone activity in vivo. American Journal of Physiology 273:E549–556
Vaisse C, Halaas JL, Horvath CM et al (1996) Leptin activation of Stat3 in the hypothalamus of wild type and ob/ob mice but not db/db mice. Nat Gen 14:95–97
Van Heek M, Compton DS, France CF, Tedesco RP, Fawzi AB, Granziano MP et al (1997) Diet-induced obese mice develop peripheral, but not central, resistance to leptin. J Clin Invest 99:385–390
Weigle DS, Bukowski TR, Foster DC, Holderman S, Kramer JM, Lasser G, Lofton-Day CE, Prunkard DE, Raymond C, Kujiper JL (1995) Recombinant ob protein

reduces feeding and body weight in the ob/ob mouse. J Clin Invest 96:2065–2070
Widdowson, PS, Upton R, Buckingham R, Arch, and Williams G (1997) Inhibition of Food Response to Intracerebroventricular Injection of Leptin Is Attenuated in Rats with Diet-Induced Obesity. Diabetes 46:1782–1785
Zhang F, Basinski MB, Beals JM, Briggs SL, Churgay LM, Clawson DK, Dimarchi RD, Furman TC, Hale JE, Hsiung HM, Schoner BE, Smith DP, Zhang XY, Wery JP, Schevitz RW (1997) Crystal structure of the obese protein leptin-E100: Nature 387:206–209
Zhang Y, Proenca R., Maffei M et al (1994) Positional cloning of the mouse obese gene and its human homologue. Nature 372:425–432

CHAPTER 13

Central Nervous System Neuropeptides Involved in Obesity

M. ROSSI and S. R. BLOOM

A. Introduction

In recent years obesity has emerged as a major threat to the health of mankind. At the same time, advances into appetite regulation and weight control has led to and helped our understanding of the pathophysiology of this complex system. Opinion has therefore changed from consideration of obesity as a self-inflicted problem to a view that it is a complicated disease with an associated increase in morbidity and mortality. In this chapter we review the central nervous system (CNS) neuropeptides which have been implicated as components of the systems regulating body weight.

The concept of brain circuits regulating food intake was first put forward at the turn of the century. It was noticed that humans who had suffered damage to their hypothalamus became obese. The hypothesis was put to the test in the1940s by HETHERINGTON and RANSON 1940). They produced lesions in discrete hypothalamic nuclei in rodents who subsequently overate and they termed this medial hypothalamic obesity. The notion that physiological control for energy balance may be centred in the hypothalamus was thus established. The ventromedial nucleus (VMN) of the hypothalamus was later identified as the "satiety centre" and the lateral hypothalamus (LH) as the "feeding centre". These nuclei involve multiple neural circuits with specific neuropeptides, neurotransmitters and their respective receptors. Much research has been devoted to understanding the neurotransmitter and neuroendocrine systems that act within the brain to co-ordinate control of energy homeostasis. Neuropeptide Y (NPY), glucagon-like peptide 1 (GLP-1), the melanocortins, galanin and cholecystokinin (CCK), are a few of the hypothalamic neuropeptides that have been implicated in regulating food intake or satiety.

Although we will cover the aspect of CNS neuropeptides in this highly complex control mechanism, it is clear that peripheral inputs are also involved. Although the role of leptin has been covered extensively in chap. 12, it will be referred to throughout in context to its interaction with the neuropeptide systems discussed.

B. Neuropeptide Y

NPY is a 36-amino acid member of the pancreatic polypeptide family (more recently renamed the NPY family) (Michel et al. 1998) which was first isolated and sequenced from porcine brain in 1982 (Tatemoto 1982). Its name derives from the presence of a tyrosine (Y) residue at both the amino and carboxy terminal ends. NPY is highly genetically conserved, showing little change in amino acid sequence across a variety of species. This suggests a critical physiological role. NPY is one of the most highly abundant neuropeptides in the mammalian CNS with particularly high concentrations found in the hypothalamus. In the peripheral nervous system NPY is usually colocalised in synaptic vesicles with norepinephrine (Lundberg et al. 1983). In the hypothalamus NPY originates mostly in the neurons in the arcuate nucleus (ARC) with projections predominantly to the paraventricular nucleus (PVN) and the dorsomedial nucleus (DMN). The PVN is considered a focus for appetite modulation since it has extensive autonomic nervous system connections and is the most effective site within the hypothalamus to elicit adrenergic stimulated feeding (Leibowitz and Rossakis 1979).

One of the earliest observations suggesting a possible physiological role for NPY was that when administered intracerebroventricularly (ICV) in rats, it produced powerful and prolonged increased food intake (Clark et al. 1984). This was later found as being most effective when given in the perifornical area just caudal to the PVN (Stanley and Leibowitz 1984). When administered chronically NPY produces hyperphagia, decreased thermogenesis and obesity (Stanley et al. 1986). Despite discovery of other neuropeptides which have since been implicated in energy balance control, none have been found to elicit such a potent response to food intake after one injection or produce obesity when administered chronically. Furthermore, injection of NPY antibody ICV blunted the feeding response induced by fasting (Lambert et al. 1993). NPY peptide content in the rat PVN showed a diurnal peak at the onset of darkness, the period in which rodents ingest most of their food intake (Jhanwar et al. 1990). All this evidence supports a role for NPY action in daily feeding.

Several studies have examined different rodent models of obesity and/or diabetes for hypothalamic neuropeptide gene expression and peptide content. Many neuropeptides are altered in different rodent models of altered feeding, but remarkably NPY is the only one consistently found to be increased when compared to its control (Wilding et al. 1993; Sanacora et al. 1990; Dryden et al. 1995; Williams et al. 1989). In some genetic models, the NPY content is elevated at onset of the development of the hyperphagic behaviour (Beck et al. 1992). It would seem reasonable to suggest that altered hypothalamic NPY activation may be directly responsible for the phenotype seen in these animals, although distinguishing between cause or effect is problematic. With the identification of leptin, it has been possible to examine the pathophysiology of the *ob/ob* phenotype to a greater extent. Daily administration of leptin to the *ob/ob* mouse, either peripherally or directly into the cerebrospinal fluid, resulted in weight loss not only due to decreased food consumption but also to an increase in thermogenesis (CAMPFIELD et al. 1995; HALAAS et al. 1995). Normalisation of hyperglycaemia and hyperinsulinaemia also occurred. A direct interaction between leptin and NPY was suggested on finding that leptin treatment in *ob/ob* mice resulted in a reduction of ARC NPY mRNA levels which was not seen in pair-fed control or *db/db* mice (with leptin receptor defects) (SCHWARTZ et al. 1996). Furthermore, dual in situ hybridisation studies have shown coexpression of the leptin receptor and NPY gene expression in specific neurons of the ARC (MERCER et al. 1996).

In the rat, hypothalamic NPY has been implicated not only in feeding, but in the regulation of growth (McDONALD et al. 1985; PIERROZ et al. 1996), sexual function (KALRA and KALRA 1996) and the stress response (INOUE et al. 1989; INUI et al. 1990; SUDA et al. 1993). Peripherally it also acts as a vasoconstrictor (ALLEN et al. 1985; PERNOW et al. 1987). To date, six NPY receptor subtypes have been identified (GERALD et al. 1995, 1996; WEINBERG et al. 1996; LUNDELL et al. 1995; LARHAMMAR et al. 1992; GRUNDEMAR et al. 1991). The different receptor subtypes have been characterised by their ability to bind fragments and analogues of the NPY family members, i.e. NPY, peptide YY (PYY) and pancreatic polypeptide (PP) (Table 1). The identification of a specific "feeding receptor" would clearly have important implications for the therapeutic management of obesity.

Table 1. Binding characteristics of six identified NPY receptors showing their differing affinity for NPY analogues and fragments

Receptor	Binding characteristics: high affinity	Binding characteristics: reduced affinity
Y1	NPY (0.2 nmol/l)	C-terminal fragments
	[Pro^{34}]NPY	NPY(3–36) and NPY(13–36)
	PYY (0.7 nmol/l)	Rat and human PP (>100 nmol/l)
Y2	NPY (0.7 nmol/l)	[Pro^{34}]NPY
	C-terminal fragments NPY(3–36) & NPY(13–36)	Rat and human PP (>1000 nmol/l)
	PYY (0.7 nmol/l)	
Y3	NPY	PYY
Y4	Human, bovine and rat PP (0.05 nmol/l)	NPY
	PYY	NPY(13–36)
Y5	Human PP	Rat PP
	PYY	
	NPY (0.6 nmol/l)	
	NPY(3–36)	
	[D-Trp^{32}]NPY weak but selective agonist	
Y6	NPY	Human PP
	PYY	

Previously it was speculated that the Y_1 receptor was responsible for NPY's stimulation of feeding with [Pro^{34}]NPY producing a robust feeding response (Stanley et al. 1992). However, the C-terminal fragment NPY (2–36), with a much reduced Y_1 affinity, is capable of inducing a feeding response (Kalra et al. 1991) and, since only low levels of Y1 receptor had been found at the PVN (Dumont et al. 1993; Mikkelsen and Larsen 1992), this implied that another "Y_1-like" receptor was also involved. Recently the human and rat Y_5 receptor were cloned. The receptor mRNA was detected in most hypothalamic nuclei including the PVN and the LH (Gerald et al. 1996). The pharmacological profile of this receptor, when expressed in a stably transfected cell line system, showed a similar order of potency to that of the analogues and fragments when tested ICV on the feeding response in rodents. Together, this data led Gerald et al. (1996) to conclude that it was "highly suggestive" that the Y_5 receptor was the specific "feeding receptor". Further studies by Schaffhauser et al (1997) supported these findings. Antisense oligonucleotides were targeted against the Y_5 receptor and elicited an inhibition to fast-induced and NPY-induced feeding. In addition they developed a non-peptide Y_5 receptor selective antagonist CGP 71683 A (IC_{50} 1.4 nmol/l, affinity to Y_1, Y_2 and Y_4 receptor subtypes 7500, 1100 and 5800nmol/l respectively) which when administered intraperitoneally reduced fast induced and NPY induced feeding.

Some doubts however exist that the Y_5 receptor is really the "feeding receptor". A comprehensive NPY fragment feeding study from our own laboratory showed ICV administration of high doses of the partial Y_5 agonist, NPY(13–36), to be ineffective at stimulating feeding in rodents (Fig. 1) and the Y_1 receptor antagonist, BIBP-3226, to be an inhibitor of NPY induced feeding (O'Shea et al. 1997). Gerald et al. (1996) had shown activation of the Y_5 receptor with NPY(13–36) at an EC_{50} of 20 nmol/l and were unable to block NPY inhibition of forskolin stimulated adenylyl cyclase by BIBP-3226 at the Y_5 receptor in transfected cells. Small et al. (1997), also from our laboratory, carried out a comprehensive NPY analogue study measuring adrenocorticotrophic hormone (ACTH) release in response to ICV administration of the analogues. In addition to showing that NPY(13–36) significantly increased plasma ACTH, other data was presented suggesting that the Y_5 receptor more closely resembled a NPY receptor subtype controlling ACTH release rather than the "feeding receptor". Further controversies were highlighted from publications in Nature Medicine in 1998. Y_5 receptor gene deleted mice have been developed and up to the age of 18 months were phenotypically normal although with increasing age mild obesity was observed, an effect more pronounced in the males than the females (Marsh et al. 1998). These mice responded normally to low doses of ICV NPY but at higher doses a blunted response was seen. In contrast, Y_1 receptor gene deleted mice were also bred and now the female mice were 30% heavier than their wild type control by 30 weeks, an effect much less pronounced than that seen in the males (Pedrazzini et al. 1998). Interestingly they had a decrease in their metabolic

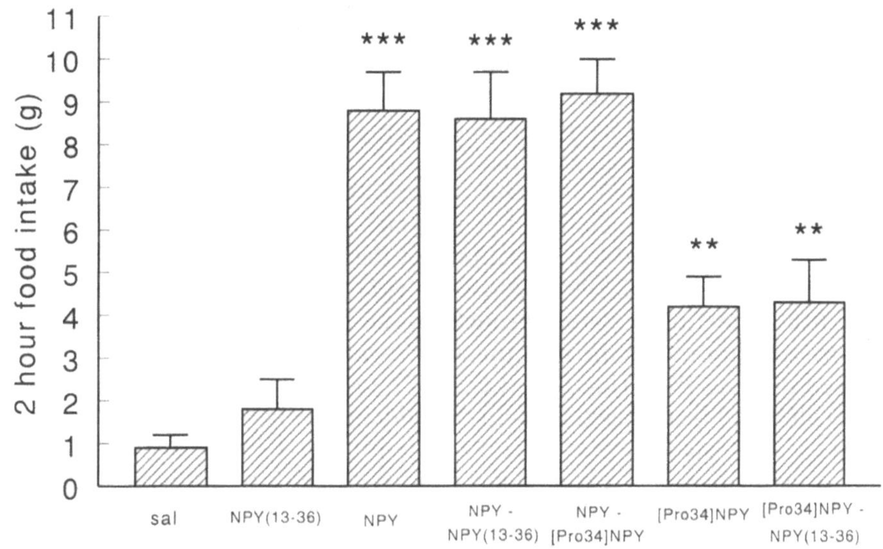

Fig. 1. Two hour food intake ICV following administration of specific NPY Y_1 and Y_2 receptor agonists and the partial Y_5 agonist, NPY(13–36). Peptides were administered either alone or in combination. The dose of NPY and [Pro^{34}]NPY was 7.2 nmol, the dose of NPY(13–36) was 14.4 nmol, $^{**}p < 0.01$, $^{***}p < 0.001$ vs saline

rate in their active dark period with a mild reduction in daily food intake also observed. Again the mice responded to ICV NPY. Together these data suggest that both receptors play some role in NPY's control of food intake but perhaps a further receptor awaits identification. The answer is not yet clear. Undoubtedly this will remain a very active and exciting area of research for the coming years.

Despite all the evidence implicating NPY as a major neurotransmitter in the regulation of body weight it is not the sole agent involved. The complexities of the "feeding circuit" were highlighted in 1996 when the results of the NPY knockout mice were reported as displaying normal feeding behaviour, body weight and a normal hyperphagic response following food deprivation (ERICKSON et al. 1996). The NPY knockout mice were crossed with *ob/ob* mice and the resultant leptin and NPY mice were less obese, had reduced food intake and increased energy expenditure (ERICKSON et al. 1996). It is possible that other systems are capable of overriding the NPY side of the circuit during maturation of the normal rodents, although these mechanisms may be inadequate in the *ob/ob* mice.

C. Glucagon-Like Peptide 1

Glucagon-like peptide 1 (7–36) amide (GLP-1) is a member of the secritin/glucagon family of bioactive peptides including glugagon, glucose-dependent insulinotropic polypeptide, pituitary adenylate cyclase activating polypeptide and vasoactive intestinal peptide, all of which have closely related amino acid sequences and can stimulate insulin secretion. It is synthesised by post-translational processing from the pre-proglucagon gene in the intestinal L-cells and the CNS (Fig. 2), differing from the processing occurring in the pancreas (Varndell et al. 1985; Kauth and Metz 1987). In the brain GLP-1 immunoreactivity has been found in neuronal cell bodies of the nucleus of the solitary tract (NTS) and the dorsal part of the medullary reticular nucleus in the brain stem (Jin et al. 1988; Kreymann et al. 1989). GLP-1 immunoreactive nerve fibres have been found throughout the brain with the highest concentrations seen in the midline hypothalamic structures; in particular, projections are seen into the PVN (Larsen et al. 1997). Its presence in the synaptosomes and its ability to be released calcium-dependently by potassium-induced depolarisation suggested its role as a neurotransmitter (Kreymann et al. 1989). Cloning and functional expression of the rat GLP-1 receptor was reported from pancreatic islets in 1992 (Thorens 1992). Using autoradiography or affinity cross-linking, GLP-1 binding sites were identified in hypothalamic nuclei including the PVN, VMN and DMH, and other brain areas involved in feeding (Calvo et al. 1995; Goke et al. 1995; Turton et al. 1996).

Although peripherally GLP-1 had been established as an insulinotropic hormone (Kreymann et al. 1987; Holst et al. 1987), until recently its central

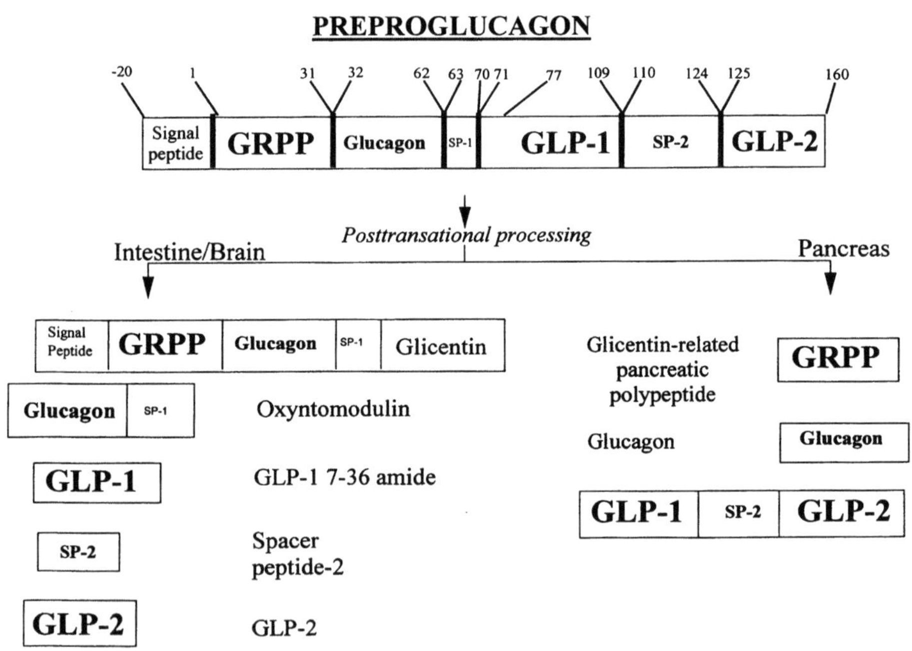

Fig. 2. Processing of preproglucagon gene showing altered post translational processing in the intestine, brain and pancreas

role remained undetermined. In 1996 our laboratory and others published evidence suggesting GLP-1 was a physiological regulator of feeding in the rat (TURTON et al. 1996; TANG et al. 1996). When administered ICV to fasted rats it caused a dose-dependent reduction in feeding; this was not seen when administered intraperitoneally. Exendin-4 is a 39 amino acid peptide isolated from the venom of the reticulate Gila monster (*Heloderma suspectum*) and has sequence homology to GLP-1 (ENG 1992). The truncated fragment, exendin (9–39) was found to be a highly-selective GLP-1 receptor antagonist (GOKE et al. 1993). When administered ICV it was found to block the inhibitory effect of GLP-1 on food intake (TURTON et al. 1996). Exendin (9–39) alone had no influence on fast-induced feeding but more than doubled the food intake in satiated rats (Fig. 3a). When GLP-1 or exendin (9–39) were administered ICV immediately before NPY, the former reduced food intake whereas the antagonist caused an increase compared to treatment with NPY alone (Fig. 3b). Further studies were performed examining expression of the immediate early gene *c-fos* (TURTON et al. 1996), an established marker of neuronal activation (SAGAR et al. 1988). Injection of GLP-1 ICV exhibited dense expression of *c-fos* in the PVN and central nucleus of the amygdala (Fig 4). Administration of exendin(9–39) or saline did not activate *c-fos* in any of the brain areas examined. Together all this data strongly implied that GLP-1 is a physiological mediator of satiety.

It has been suggested by some investigators that GLP-1 produces its anorectic effect solely via conditioned taste aversion (THIELE et al. 1997; PANKSEPP et al. 1996). This remains a point of some controversy since acute blockade of endogenous GLP-1 can immediately lead to increased food intake. Cholecystokinin (CCK) is a neuropeptide that has also been shown to cause learned taste aversion (DEUTSCH and HARDY 1977). As will be discussed later in this chapter, there is much evidence to suggest that CCK also acts as a specific satiety signal, suggesting that the ability to cause conditioned taste aversion should not automatically negate the role of a peptide as a satiety signal.

GLP-1 has been administered chronically by repeated injections ICV over seven days (TURTON et al. 1997). Rats continued to eat less and lose weight over this period compared to controls demonstrating lack of tolerance to its central effects. In addition, ICV exendin (9–39) has also been administered chronically in conjunction with NPY (MEERAN et al. 1996). Over a 7-day period animals receiving exendin (9–39) plus NPY gained 50% more weight than those receiving NPY alone. These studies suggest that GLP-1 is involved in the regulation of body weight as well as feeding in the rat.

Further detailed studies mapping *c-fos* immunohistochemistry within the rat brain following ICV GLP-1 have also identified the NTS, area postrema, lateral parabrachial nucleus and ARC as areas with increased activation (VAN et al. 1996). Within the PVN, the *c-fos* activation occurring in the parvicellular region co-expressed with 80% of the corticotrophin releasing hormone (CRH) neurons (LARSEN et al. 1997). CRF is yet another proposed hypothal-

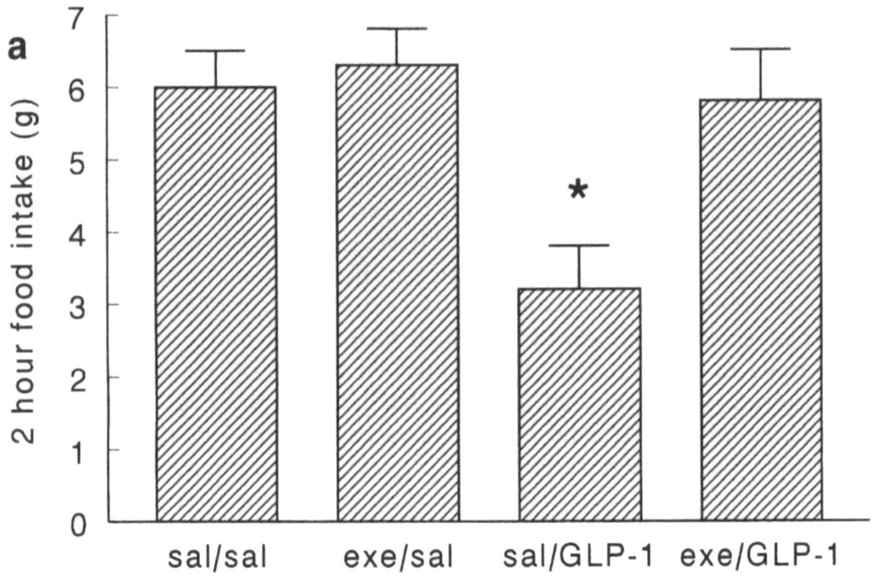

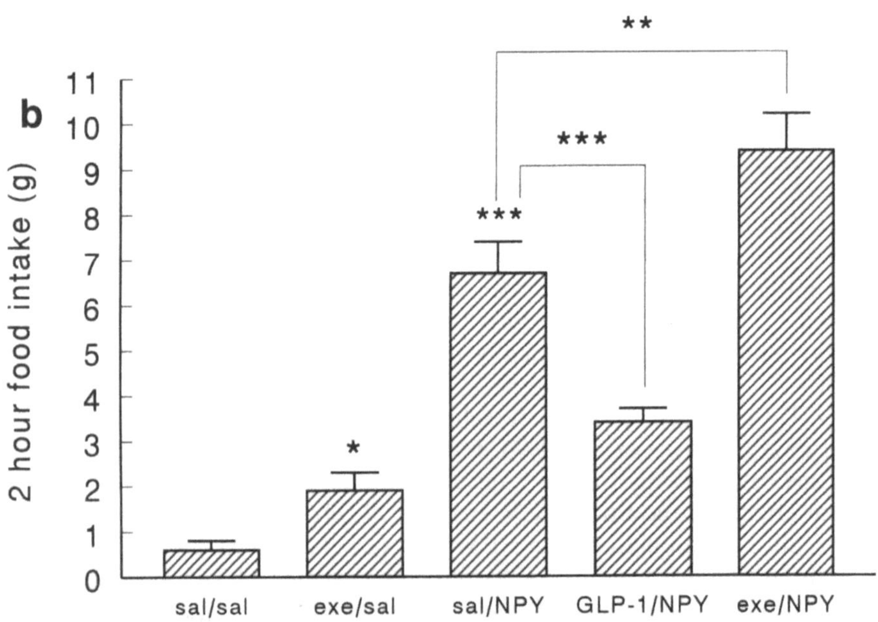

Fig. 3. a Two hour food intake in 24 h fasted rats injected with ICV 0.9% saline (sal) or 100:g exendin(9–39) (exe), immediately followed by 0.9% saline or 3:g GLP-1. $*p < 0.05$. **b** Two hour food intake in rats fed ad libitum injected ICV at the beginning of the light phase with 0.9% saline or 10:g NPY, immediately followed by 0.9% saline, 10:g GLP-1 or 100:g exendin(9–39). $*p < 0.05$, $**p < 0.01$, $***p < 0.001$

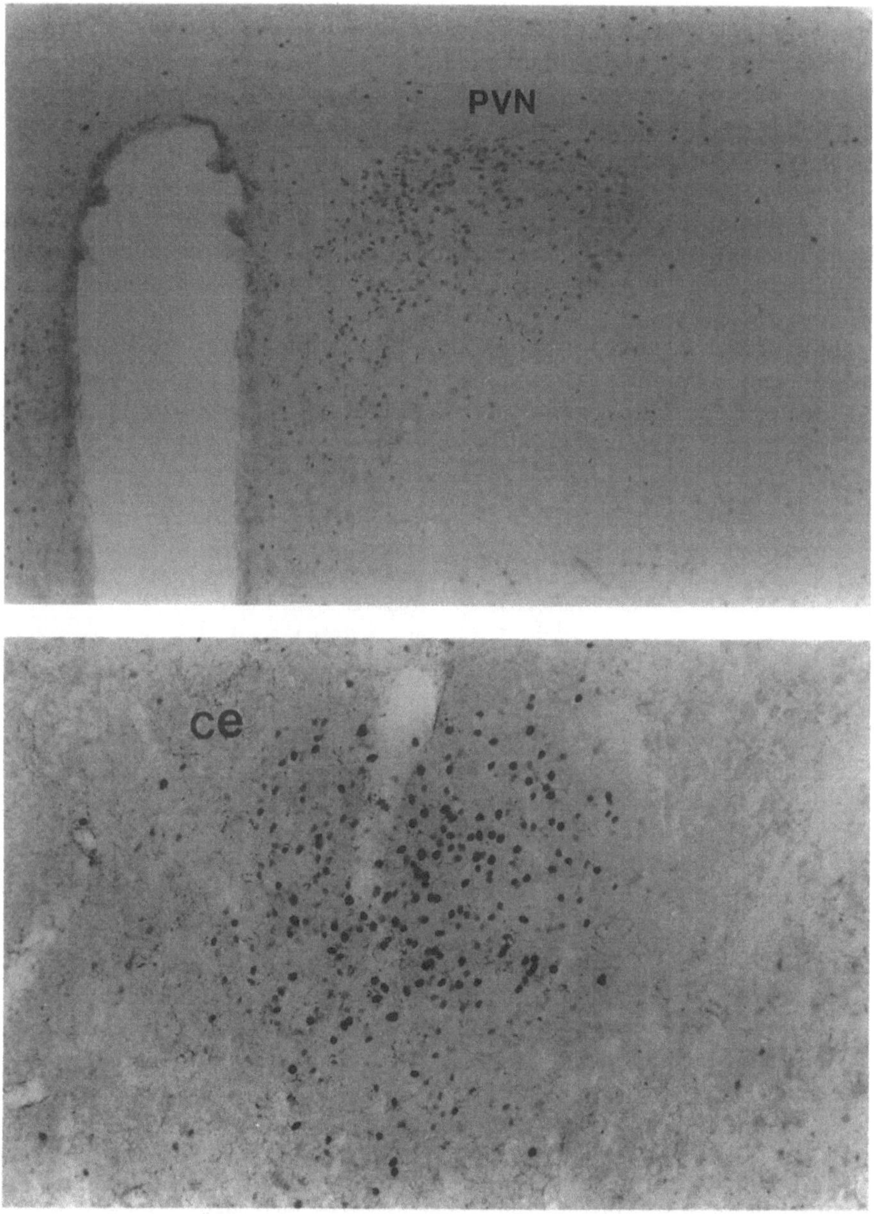

Fig. 4. Activation of *c-fos* following ICV 10:g GLP-1. Significant activation was observed in the paraventricular nucleus (PVN) and central nucleus of the amygdala (ce) when compared to control (saline). Administration of exendin(9–39) did not activate *c-fos* above control in any of the brain areas examined (not shown)

amic satiety factor (see Sect. H) and therefore this suggested that the inhibitory effect on feeding of GLP-1 was in part be due to stimulation of CRF neurons. Recently, we have shown that GLP-1 neurons are a potential target for leptin in its control of feeding (Goldstone et al. 1997). Dual in situ hybridisation found the majority of GLP-1 expressing neurons in the NTS also expressed the long isoform leptin receptor (OB-Rb) (Fig. 5). ICV exendin (9–39), when given to leptin pre-treated rats at the onset of the dark phase, blocked the reduction in food intake and body weight seen in the leptin-saline treated group. No effect of exendin (9–39) on food intake was seen when given to saline pre-treated rats (Goldstone et al. 1997).

In a similar fashion to the NPY knock out mouse, results of the GLP-1 receptor knock out mice were published in 1996 showing normal body weight and feeding behaviour (Scrocchi et al. 1996). This again illustrates the complex nature of CNS neuropeptide involvement in feeding, with the possibility

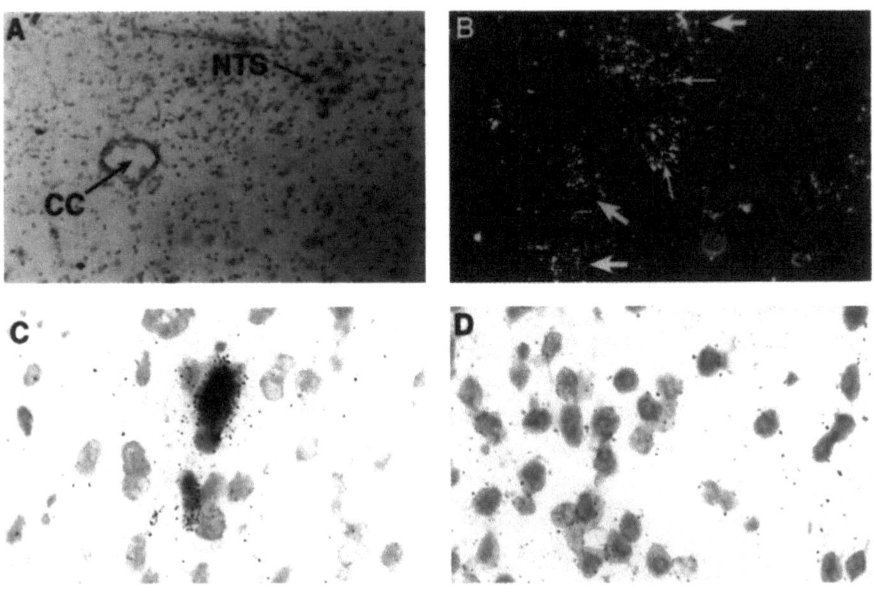

Fig. 5A–D. In situ hybridisation carried out on mouse hindbrain with ^{35}S-labelled leptin receptor (Ob-Rb) and digoxigenin-labelled preproglucagon ribroprobes. Photomicrographs of: **A,C,D** bright; **B** dark field images of the nucleus of the solitary tract (*NTS*), close to the central canal (*CC*), generated following hybrization of adjacent sections with either both antisense probes (**A**, **B**, **C**) or both sense probes (**D**). Photograph **A** reveals a cluster of NTS neurons that express preproglucagon mRNA (*coloured product*). Silver grains (dark spots in bright field and light spots in dark field) representing Ob-Rb mRNA are not visible at this magnification. Photograph **B** and **C** are at higher magnification at the NTS and show overlap in expression of both preproglucagon and Ob-Rb mRNA. Photograph **B** shows cells that express both mRNA species (*large arrows*) as well as though that express Ob-Rb mRNA only (*small arrows*). Reproduced by kind permission from JG Mercer and Elsevier Publications

of gene disruption allowing compensatory mechanisms to develop. The existence of multiple physiological control circuits regulating appetite control was again highlighted.

D. Melanocortins and *Agouti*

The melanocortins are peptides derived from the propiomelanocortin (POMC) polypeptide precursor and possess either melanotropic (melanocyte stimulating) or adrenotrophic properties (Fig. 6). The four major melanocortins are ACTH, α-, β- and γ-melanocyte stimulating hormone (MSH). These peptides are mainly produced in the anterior and intermediate lobes of the pituitary. However POMC expression has also been detected in numerous tissues including the ARC of the hypothalamus, testes, ovary, placenta, liver, thymus and lymphocytes (SMITH and FUNDER 1988).

Interest in these peptides for many years was divided between scientists' studying aspects of the hypothalamic pituitary adrenal axis or those preoccupied with hormonal control of pigmentation. Studies of the MSH receptor, now termed MC1-R, found that it was solely expressed on melanocytes and had much higher affinity for α-MSH than any of the other melanocortins (WONG et al. 1974). It was found that mutations of MC1-R in animals resulted in alterations in the regional distribution of eumelanin (brown-black) and phaeomelanin (yellow-red) pigmentation in their coats with resultant changes

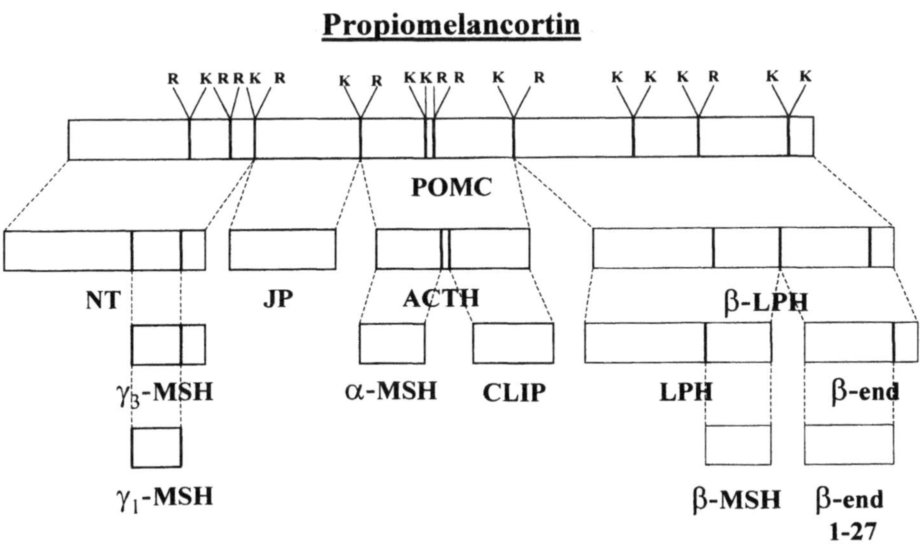

Fig. 6. Processing of propiomelanocortin (*POMC*) gene. N-terminal fragment (*NT*), joining peptide (*JP*), adrenocorticotrophin hormone (*ACTH*), lipotropin (*LPH*), melanocyte stimulating hormone (*MSH*), corticotrophin-like intermediate peptide (*CLIP*), endorphin (*end*)

in colour (Robbins et al. 1993). Dominant mutation resulted in a black pigmentation in the coat. *Agouti* was the name given to a genetic locus on mouse chromosome 2 that was also found to be capable of altering the levels of these pigments (Woychik et al. 1990). Expression of this allele was exclusively within the hair follicle and dominant expression resulted in animals with yellow or red coat colour. The locus was named after a South American rodent, the agouti, which has a distinctive yellow-red banded colour in its coat. Mice with dominant yellow mutations were found to have ectopic *agouti* gene expression in most tissues, not just the hair follicles, and were also surprisingly found to result in adult onset obesity, be hyperinsulinaemic and hyperglycaemic (Silvers 1979). An understanding of why this occurred was the initiation of a whole new area in obesity research.

Since 1992 five melanocortin receptors (MC-R 1 to 5) have been cloned, all having differing affinity for the POMC products (Mountjoy et al. 1992; Roselli et al. 1993; Gantz et al. 1993; Chhajlani et al. 1993; Griffon et al. 1994). MC2-R (previously known as ACTH-R) is expressed in the adrenal cortex and adipocytes and is uniquely activated by ACTH. Both MC3-R and MC4-R are expressed in rat hypothalamic nuclei and limbic systems. MC3-R expression is found in some discrete hypothalamic and brain stem nuclei but is also expressed peripherally in the placenta, heart and gut tissues. MC4-R has only been found to be expressed in the brain with mRNA being localised in multiple sites in virtually every brain region including moderate to high levels in the PVN, VMH, DMH and LH (Mountjoy et al. 1994). MC4-R binds both α- and β-MSH with high affinity. Interestingly, the MC5-R has been found to have extremely wide distribution throughout peripheral tissues. Receptor expression, although relatively low, is seen in sites including muscle, liver, spleen, skin, ovary and testis.

The *agouti* gene was cloned in 1992 (Bultman et al. 1992) and its 131 amino acid gene product, agouti, was soon identified (Miller et al. 1993). It had already been demonstrated that a product of the hair follicle was capable of acting in a paracrine fashion to inhibit the MSH induced synthesis of eumelanin from melanocytes (Silvers 1979). With the identification of agouti it was therefore postulated that it might in fact be an antagonist at MC1-R. Lu et al. (1994) found that the agouti protein was not only a high affinity antagonist of MC1-R but also of MC4-R, without any effect seen at MC3-R or MC5-R. It was therefore conceivable that in these obese mice agouti, being ectopically expressed, was antagonising a receptor at the level of the hypothalamus thereby altering hunger and satiety physiological circuits. This was confirmed when mice lacking MC4-R were generated (Huszar et al. 1997). They were found to be phenotypically identical to the *agouti* mouse, exhibiting hyperphagia, obesity and hyperinsulinaemia. Similarly, a synthetic MC4-R and MC3-R agonist and antagonist were synthesised which, when administered ICV to mice, decreased and increased feeding respectively (Fan et al. 1997). Together this data confirmed another important player in the appetite control circuit.

The Human *agouti* gene was identified in 1994 (KWON et al. 1994). It was found to be 85% identical to the mouse gene and had the potential to encode a protein of 132 amino acids termed agouti signalling protein, ASIP. This novel gene was found to be expressed at high levels in heart, adipose tissue, testis and ovary but no expression at all was seen in the CNS (KWON et al. 1994; WILSON et al. 1995). In vitro studies found ASIP to be a potent antagonist of the hMC1-R, hMC2-R and hMC4-R, and a relatively weak antagonist of the hMC3-R and hMC5-R (YANG et al. 1997). It was postulated that ectopic expression of ASIP could conceivably lead to human obesity; however the differing tissue expression between human and mouse agouti, and their altered affinities at the MC-Rs, suggested that agouti may be regulating a different function in Man.

Further excitement within this area came with the isolation and characterisation of agouti related transcript, Agrt, the first member of a new gene family to be identified with homology to agouti (SHUTTER et al. 1997; OLLMANN et al. 1997). Both human and mouse Agrt mRNA were found to be normally expressed in high levels in the hypothalamus and adrenal gland. In particular, in situ hybridisation studies in mouse brain showed CNS Agrt expression to be restricted to the median eminence and the ARC, in the latter closely resembling the pattern of expression seen with NPY (SHUTTER et al. 1997). Expression was eight- to tenfold increased in the *ob/ob* and *db/db* mice compared to wild type control. Sequence analysis of Agrt predicted a 132 amino acid protein, termed agouti related protein Agrp, with 25% homology to human agouti. Recombinant human Agrp was made and found to be a potent antagonist at MC3-R and MC4-R, being a hundred-fold more potent than agouti (OLLMANN et al. 1997; FONG et al. 1997). Transgenic mice over expressing human Agrt were constructed and phenotypically resembled *agouti* mice, although they retained the wild type coat colour (OLLMANN et al. 1997; GRAHAM et al. 1997). Therefore a new gene and its protein product, related to agouti and having antagonistic effects on the MC-Rs, had been identified which, in initial studies, appeared could be of physiological importance in mammalian appetite and body weight homeostasis. Future studies remain eagerly awaited.

E. Melanin Concentrating Hormone

Rat melanin concentrating hormone (MCH) is a cyclic 19 amino acid peptide identical to human MCH (VAUGHAN et al. 1989; PRESSE et al. 1990). This high conservation of sequence suggests a critical physiological role. MCH was first isolated from salmon pituitaries and its name derived from the finding that it could regulate skin colour in teleost fish by induction of melanin aggregation within the melanocytes (KAWAUCHI et al. 1983). In mammals MCH perikarya are prominent in the lateral hypothalamus and the zona incerta (FELLMANN et al. 1987; SKOFITSCH et al. 1985) with extensive neuronal projections throughout the central nervous system (Bittencourt et al. 1992). The presence of high levels of MCH immunoreactivity in the synaptosomes and release of MCH by potassium and calcium stimulation suggested a role as a neurotransmitter (Sekiya et al. 1988).

In lower vertebrates MCH is of functional importance in causing skin paling, antagonising the melanin dispersing action of α-MSH (Rance and Baker 1979). In mammals there is good evidence to support the role of MCH in the central regulation of feeding behaviour. Qu et al. (1996) found MCH mRNA levels to be increased in the hypothalamus of *ob/ob* mice compared to wild type and *ob/+* controls. Expression was further enhanced in all groups of mice in the fasted state. Hypothalamic MCH mRNA expression was also examined in Wistar rats by Bluet-Pajot et al. (1995) who found a diurnal rhythm pattern, being highest during the beginning of the dark phase which correlates with the rats feeding period. In another study, mice homozygous for the fat mutation, the *fat/fat* mouse, had hypothalamic MCH peptide content measured and this was found to be up to three times higher (Rovere et al. 1996). Our laboratory has also found hypothalamic MCH peptide content to be elevated in the obese leptin receptor deficient Zucker *fa/fa* rat (Rossi et al. 1997).

A dose dependent effect of MCH on food intake was seen when administered ICV in Wistar rats (Rossi et al. 1997). With twice daily administration, MCH repeatedly caused a 2-h increase in food intake for 5 consecutive days, after which time the effect was lost. Daily food intake and body weights were not altered, suggesting a reproducible but short term effect on feeding, with possible down regulation of the receptor involved. Despite identification of specific MCH receptors on mouse melanoma cells (Drozdz et al. 1995), absence of a satisfactory radioligand has hindered research on MCH receptor(s) structure and function. Interestingly an antagonistic effect between MCH and α-MSH has been seen not only on skin coloration but also centrally in response to auditory stimuli or grooming and locomotor activities (Miller et al. 1993; Sanchez et al. 1997). This could suggest that MCH may not be acting through a specific receptor but one shared with α-MSH, such as MC4-R. However, receptor autoradiography on rat brain sections using iodinated [Nle^4, D-Phe^7]α-MSH (NDP-MSH), which binds with high affinity to MC4-R, found that unlabelled MCH was unable to compete for the I^{125} NDP-MSH binding sites (Rossi et al. 1998) (Fig. 7). It is therefore likely that with regards to the hypothalamic effects of MCH, a unique receptor system, separate to the melanoncortins, is functioning.

F. Opioid Peptides

An interaction between the endogenous opioid system and feeding behaviour was first suggested in the 1940s on observing individuals with morphine addiction (Hebb 1949). It was much later, on finding that the non-selective opioid

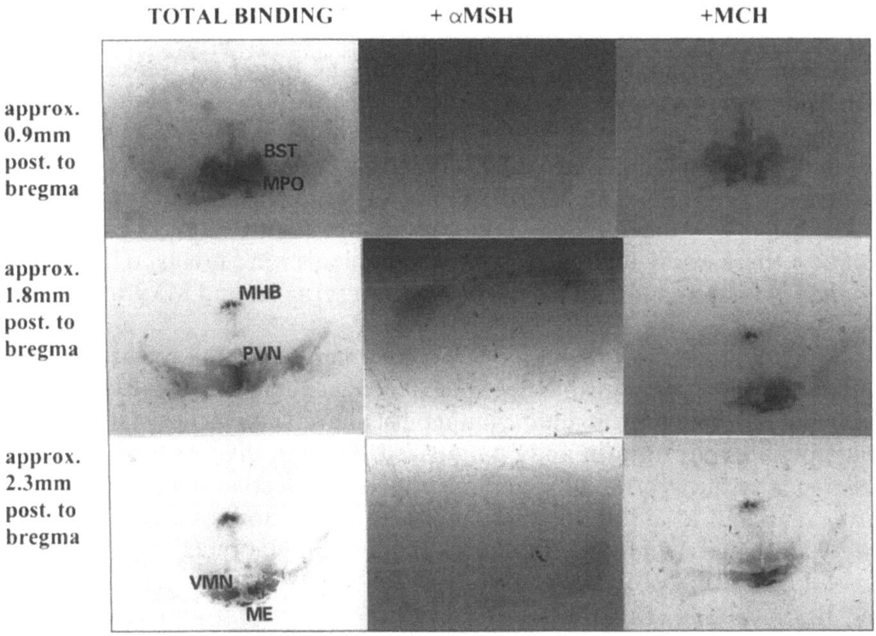

Fig. 7. Receptor autoradiography of MSH binding sites in rat brain. Coronal rat brain sections were incubated with I^{125} NDP-MSH indicating total binding. The binding was displaced with the addition of unlabelled αMSH. However, addition of unlabelled MCH was unable to cause displacement of the I^{125} NDP-MSH binding sites. Binding sites demonstrated in these sections include the bed nucleus of stria terminalis (*BST*), medial preoptic nucleus (*MPO*), medial habuncular nucleus (*MHB*), paraventricular nucleus (*PVN*), ventromedial nucleus (*VMN*) and the median eminence (*ME*)

antagonist naloxone decreased food intake, that this theory was confirmed (Holtzman 1974). In particular, evidence implicated the POMC product β-endorphin (β-End), the prodynorphin product dynorphin (Dyn) and the pro-enkephalin A product met-enkephalin (met-Enk). The cell bodies of the β-End are clustered in the ARC and innervate numerous hypothalamic nuclei including the PVN, VMN and DMN (Cuello 1983; Mezey et al. 1985). The hypothalamus also displays a large number of met-Enk and Dyn cell bodies with distributions including those areas covered by the β-End nerve terminals (Yang et al. 1983; Fallon and Leslie 1986). Dyn can act as an endogenous κ and δ agonist whereas β-End is probably the natural ligand for the μ receptors.

Dyn and β-End were found to increase feeding when administered ICV (Walker et al. 1980; Morley and Levine 1981; McKkay et al. 1981; de Caro et al. 1979). β-End was also active in the PVN (Leibowitz and Hor 1982), and administration of its antibodies into the periventricular nucleus caused a reduction in feeding (Gonzalez et al. 1984). β-End peptide levels, but not Dyn, were found to be increased in the VMN and DMN of *ob/ob* mice, this was not a consistent finding in other models of obesity (Khawaja et al. 1991; Roane et al. 1988; McLaughlin et al. 1985). Gene expression of the peptide precursors of β-End, Dyn and met-Enk were all found to decrease on food restricting rodents, which was opposite to the situation seen with most other orexigenic peptides (Kim et al. 1996). The role of the opioids was further complicated by the finding that when administered peripherally a reduction of feeding was seen (King et al. 1979). However, it did appear that stress-induced hyperphagia was mediated by endogenous opioids (Teskey et al. 1984). In addition, there was evidence to suggest that opioids were involved in the maintenance of fatty and highly palatable diets (Apfelbaum and Mandenoff 1981; Marks and Kanarek 1980).

Much research in this area has concentrated on trying to discern which opioid receptor is responsible for which feeding effect (Morley 1987). Studies involving opioid agonists and/or antagonists have shown that distinct opioid receptor subtypes differentially modulate feeding in diverse feeding models. Feeding is reduced after ICV administration of selective μ, μ_1 and κ antagonists in fed and food deprived rats, although only μ and κ antagonists reduce the hyperphagia seen following electrical brain stimulation (Levine et al. 1990; Arjune and Bodnar 1990; Arjune et al. 1990, 1991; Simone et al. 1985; Islam and Bodnar 1990; Papadouka and Carr 1994). Only μ and μ_1 antagonists reduce stress-induced hyperphagia (Koch and Bodnar 1993; Hawkins et al. 1992). Both μ and κ receptors appear to be involved in ingestion of simple carbohydrates (Koch and Bodnar 1994), stimulation of reward centres after food restriction (Carr and Papadouka 1994) and maintenance of feeding (Papadouka and Carr 1994). Whereas only μ receptors regulate intake of complex carbohydrates and the selective δ antagonists only reduce feeding following saccharin intake (Beczkowska et al. 1993).

An interaction between NPY and the opioids has also been suggested since naloxone has been found to reduce NPY-induced feeding (Levine et al. 1990). This effect on NPY feeding appears to involve the μ and κ receptors (Kotz et al. 1993; Lambert et al. 1993). In particular, studies by Kotz et al. (1995) suggested it was opioidergic pathways from the NTS that acted on NPY from the PVN to reduce not only on feeding but also brown fat thermogenesis.

G. Galanin

Galanin was first isolated from porcine intestine in 1983 and in most species has been found to be a 29 amino acid peptide (Tatemoto et al. 1983). Similarly to NPY it was found to be densely concentrated in the CNS with particularly high concentrations in the hypothalamus in areas including the ARC, PVN and DMN (Melander et al. 1986). On injection of galanin ICV a stimulation of feeding was found to occur (Kyrkouli et al. 1986), an effect which was later found to be specific to the PVN (Kyrkouli et al. 1990), and thus a possible link of this peptide with feeding behaviour was established.

Interestingly, injection of galanin appears to cause a preferential selection of the animal to increase its consumption of fat foods (TEMPEL et al. 1988). This was supported by findings that hypothalamic injections of a specific galanin antagonist and antisense oligonucleotides to galanin mRNA selectively inhibited consumption of fat (LEIBOWITZ and KIM 1992; AKABAYASHI et al. 1994). It has been suggested that galanin may be mediating its feeding effect via a release of norepinephrine (KYRKOULI et al. 1992). In addition, an interaction with NPY has been suggested since, at the level of the ARC and PVN, NPY neurons were found to be in close contact with galanin producing neurons (HORVATH et al. 1996). Furthermore, electron microscope studies have demonstrated intraneural connections between β-End and galanin, thereby implicating a possible association with the opioid system (HORVATH et al. 1995). Specific opioid receptors have also been shown to block the increase in feeding seen following ICV galanin (BARTON et al. 1996). Unlike NPY, but similarly to MCH, chronic administration of galanin ICV was unable to cause a sustained increase in daily food and body weight (SMITH et al. 1994).

H. Other Neuropeptides

A long list of neuropeptides now exists which have now been implicated by some workers in the role of feeding in either a stimulatory or inhibitory fashion (Table 2). CCK, CRH, the orexins and the intriguingly named cocaine and amphetamine regulated transcript (CART) are worthy of further mention in this review.

CCK is among the first of the group of peptides that were initially discovered in the gut and later localised to the CNS (BEINFELD and PALKOVITS 1982). Since CCK fulfils the role of a neurotransmitter (SAITO et al. 1980; PINGET et al. 1979) within the CNS it may therefore be having a direct effect. Its secretion from the small intestinal cells in response to food ingestion has been proposed as a post-prandial satiety signal (LEWIS and WILLIAMS 1990; WELLER et al. 1990). Of the two CCK receptors identified the CCK_A receptor

Table 2. Peptides that have been demonstrated to alter feeding when given intracerebroventricularly

Increase Feeding	Decrease Feeding
NPY	GLP-1
MCH	α-MSH
β-endorphin	CCK
Dynorphin	CRH
Galanin	Neurotensin
Orexin A & B	Calcitonin gene-related peptide
	Bombesin
	Vasoactive intestinal peptide
	Amylin

is thought to mediate the control food intake, since administration of its antagonist increased feeding (Smith and Gibbs 1994) and the naturally occurring rat with lack of CCK_A receptors is obese (Miyasaka et al. 1994). Activation of vagal afferent fibres directly relaying to specific brain sites appears to be the mode of effect (Ritter et al. 1989; Reidelberger 1992). However, a direct action of central CCK has also been suggested since gastric distension is associated with its release in the hypothalamus and when CCK is given into discrete hypothalamic nuclei a suppression of feeding is observed (Reidelberger 1992; Schick et al. 1989, 1990). Interestingly, elegant studies by Barrachina et al. (1997) have suggested that CCK may mediate some of the feeding suppression effects of leptin. Also of note is the development of butabindide, a novel inhibitor of tripeptidyl peptidase II, a CCK inactivating petidase, which when given to mice starved or at the beginning of their feeding cycle was found to reduce feeding (Rose et al. 1996). Human clinical trials are now in progress.

CRH has not only been implicated in stress-induced anorexia but also appears to have a role independent to the hypothalamic-pituitary-adrenal axis. ICV administration of CRH suppresses feeding (Morley and Levine 1982). This is sustained with chronic administration, causing weight loss coupled with increased activation of the sympathetic nervous system (Arase et al. 1988). The feeding effect was found to be specific to the PVN (Krahn et al. 1988) and is still demonstrable following hypophysectomy in rodents (Koob and Bloom 1985). Interestingly, CRH suppression of feeding was more potent when given ICV to animal models of obesity suggesting a possible causal effect for their increase in weight (Arase et al. 1989a,b). In 1995 urocortin was identified as a new member of the CRH family (Vaughan et al. 1995). Its distribution in mammalian brain was found in areas known to express the CRH receptor type II (CRH_2) and it bound with high affinity to this receptor. When given ICV urocortin was later found to have a specific and potent dose dependent effect on feeding suppression (Spina et al. 1996). CRH_2 therefore appeared to be responsible for the CRH feeding effects and urocortin was postulated as the possible endogenous ligand. This system is further complicated by the existence of CRH-binding protein (CRH-BP) which binds with high affinity not only to CRH, but also urocortin, causing biologically inactivation (Potter et al. 1992). Since Heinrichs et al. (1996) demonstrated a specific reduction in weight gain in the obese Zucker rat following chronic administration of a CRH-BP ligand inhibitor, a further possible target for treatment of obesity was identified.

Two novel hypothalamic peptides regulating feeding were identified in 1998. Although further studies remain to be carried out, it is now known that orexin-A and orexin-B derive from the same precursor in the LH, zona incerta and posterior hypothalamic areas and cause an increase in feeding when administered ICV (Sakurai et al. 1998). Interestingly this distribution is very similar to that of MCH. Two G-protein coupled receptors were also identified which displayed exclusive distribution to mammalian brain. Furthermore, like

MCH, expression of the orexin precursor was found to increased on fasting rats. The new gene CART was first identified when it was found to be up-regulated in rodent brains following acute administration of cocaine and amphetamine (DOUGLASS et al. 1995). Kristensen et al. (1998) found ARC CART mRNA to be decreased in obese *ob/ob* mice, Zucker rats and normal fasted rodents. Leptin administration to the *ob/ob* mice restored the CART mRNA expression to control levels. Recombinant CART caused a reduction in feeding when administered ICV to normal, fasted and NPY stimulated rats. Furthermore, injection of CART antiserum caused an increase in night time feeding.

We have covered a brief synopsis of the ever-expanding scientific publications on hypothalamic neuropeptides and their role in the regulation of feeding behaviour. Research is presently moving at a phenomenal pace with major new insights becoming apparent every month. What is emerging is a complex system involving many players with elaborate interactions. Many of the neuropeptides discussed have been shown to have a physiological role in appetite control. Perhaps the most exciting player at present is the melanocortin system. However, our understanding of neuroendocrine systems that act within the brain to control energy homeostasis will undoubtedly continue to increase dramatically over the next decade. The reward will hopefully be that by the year 2010 we may be able to offer patients treatment for their obesity and, with time, reduce the prevalence of the morbid conditions associated with it.

List of Abbreviations

ACTH	adrenocorticotrophic hormone
Agrt	agouti related transcript
Agrp	agouti related protein
ASIP	agouti signalling protein
ARC	arcuate nucleus
β-End	β-endorphin
CART	cocaine and amphetamine regulated transcript
CNS	central nervous system
CCK	cholecystokinin
CRH	corticotrophin releasing hormone
CRH-BP	CRH-binding protein
DMN	dorsomedial nucleus
Dyn	dynorphin
GLP-1	glucagon-like peptide 1
ICV	intracerebroventricularly
LH	lateral hypothalamus
MSH	melanocyte stimulating hormone
MCH	melanin concentrating hormone

met-Enk	met-enkephalin
NPY	neuropeptide Y
NTS	nucleus of the solitary tract
PP	pancreatic polypeptide
PVN	paraventricular nucleus
PYY	peptide YY
POMC	propiomelanocortin
VMN	ventromedial nucleus

References

Akabayashi A, Koenig JI, Watanabe Y, Alexander JT, Leibowitz SF (1994) Galanin-containing neurons in the paraventricular nucleus: a neurochemical marker for fat ingestion and body weight gain. Proc Natl Acad Sci USA 91:10375–10379

Allen JM, Raine AE, Ledingham JG, Bloom SR (1985) Neuropeptide Y: a novel renal peptide with vasoconstrictor and natriuretic activity. Clin Sci 68:373–377

Apfelbaum, M, Mandenoff, A (1981) Naltrexone suppresses hyperphagia induced in the rat by a highly palatable diet. Pharmacol Biochem Behav 15:89–91

Arase K, Shargill NS, Bray GA (1989a) Effects of corticotropin releasing factor on genetically obese (fatty) rats. Physiol Behav 45:565–570

Arase K, Shargill NS, Bray GA (1989b) Effects of intraventricular infusion of corticotropin-releasing factor on VMH-lesioned obese rats. Am J Physiol 256:R751–R756

Arase K, York DA, Shimizu H, Shargill N, Bray GA (1988) Effects of corticotropin-releasing factor on food intake and brown adipose tissue thermogenesis in rats. Am J Physiol 255:E255–E259

Arjune D, Bodnar RJ (1990) Suppression of nocturnal, palatable and glucoprivic intake in rats by the kappa opioid antagonist, nor-binaltorphamine. Brain Res 534:313–316

Arjune D, Bowen WD, Bodnar RJ (1991) Ingestive behavior following central [D-Ala2, Leu5, Cys6]-enkephalin (DALCE), a short-acting agonist and long-acting antagonist at the delta opioid receptor. Pharmacol Biochem Behav 39:429–436.

Arjune D, Standifer KM, Pasternak GW, Bodnar RJ (1990) Reduction by central beta-funaltrexamine of food intake in rats under freely-feeding, deprivation and glucoprivic conditions. Brain Res 535:101–109

Barrachina MD, Martinez V, Wang L, Wei JY, Tache Y (1997) Synergistic interaction between leptin and cholecystokinin to reduce short-term food intake in lean mice. Proc Natl Acad Sci USA 94:10455–10460

Barton C, York DA, Bray GA (1996) Opioid receptor subtype control of galanin-induced feeding. Peptides 17:237–240

Beck B, Burlet A, Bazin R, Nicolas JP, Burlet C (1992) Early modification of neuropeptide Y but not of neurotensin in the suprachiasmatic nucleus of the obese Zucker rat. Neurosci Lett 136:185–188

Beczkowska IW, Koch JE, Bostock ME, Leibowitz SF, Bodnar RJ (1993) Central opioid receptor subtype antagonists differentially reduce intake of saccharin and maltose dextrin solutions in rats. Brain Res 618:261–270

Beinfeld MC, Palkovits M (1982) Distribution of cholecystokinin (CCK) in the rat lower brain stem nuclei. Brain Res 238:260–265

Bittencourt JC, Presse F, Arias C, Peto C, Vaughan J, Nahon JL, Vale W, Sawchenko PE (1992) The melanin-concentrating hormone system of the rat brain: an immuno- and hybridization histochemical characterization. J Comp Neurol 319:218–245

Bluet Pajot MT, Presse F, Voko Z, Hoeger C, Mounier F, Epelbaum J, Nahon JL (1995) Neuropeptide-E-I antagonizes the action of melanin-concentrating hormone on stress-induced release of adrenocorticotropin in the rat. J Neuroendocrinol 7:297–303
Bultman SJ, Michaud EJ, Woychik RP (1992) Molecular characterization of the mouse agouti locus. Cell 71:1195–1204
Calvo JC, Yusta B, Mora F, Blazquez E (1995) Structural characterization by affinity cross-linking of glucagon-like peptide-1(7–36)amide receptor in rat brain. J Neurochem 64:299–306
Campfield LA, Smith FJ, Guisez Y, Devos R, Burn P (1995) Recombinant mouse OB protein: evidence for a peripheral signal linking adiposity and central neural networks [see comments]. Science 269:546–549
Carr KD, Papadouka V (1994) The role of multiple opioid receptors in the potentiation of reward by food restriction. Brain Res 639:253–260
Chhajlani V, Muceniece R, Wikberg JE (1993) Molecular cloning of a novel human melanocortin receptor [published erratum appears in Biochem Biophys Res Commun 1996 Jan 17;218(2):638]. Biochem Biophys Res Commun 195:866–873
Clark JT, Kalra PS, Crowley WR, Kalra SP (1984) Neuropeptide Y and human pancreatic polypeptide stimulate feeding behavior in rats. Endocrinology 115:427–429
Cuello AC (1983) Central distribution of opioid peptides. Br Med Bull 39:11–16
de Caro G, Micossi LG, Venturi F (1979) Drinking behaviour induced by intracerebroventricular administration of enkephalins to rats. Nature 277:51–53
Deutsch JA, Hardy WT (1977) Cholecystokinin produces bait shyness in rats. Nature 266:196
Douglass J, McKinzie AA, Couceyro P (1995) PCR differential display identifies a rat brain mRNA that is transcriptionally regulated by cocaine and amphetamine. J Neurosci 15:2471–2481
Drozdz R, Siegrist W, Baker BI, Chluba de Tapia J, Eberle AN (1995) Melanin-concentrating hormone binding to mouse melanoma cells in vitro. FEBS Lett 359:199–202
Dryden S, Pickavance L, Frankish HM, Williams G (1995) Increased neuropeptide Y secretion in the hypothalamic paraventricular nucleus of obese (fa/fa) Zucker rats. Brain Res 690:185–188
Dumont Y, Fournier A, St PS, Quirion R (1993) Comparative characterization and autoradiographic distribution of neuropeptide Y receptor subtypes in the rat brain. J Neurosci 13:73–86
Eng J (1992) Exendin peptides. Mt Sinai J Med 59:147–149
Erickson JC, Clegg KE, Palmiter RD (1996) Sensitivity to leptin and susceptibility to seizures of mice lacking neuropeptide Y Nature 381:415–421
Erickson JC, Hollopeter G, Palmiter RD (1996) Attenuation of the obesity syndrome of ob/ob mice by the loss of neuropeptide Y Science 274:1704–1707
Fallon JH, Leslie FM (1986) Distribution of dynorphin and enkephalin peptides in the rat brain. J Comp Neurol 249:293–336
Fan W, Boston B, Kesterson RA, Hruby VJ, Cone RD (1997) Role of melanocortinergic neurons in feeding and the agouti obesity syndrome. Nature 385:165–168
Fellmann D, Bugnon C, Risold PY (1987) Unrelated peptide immunoreactivities coexist in neurons of the rat lateral dorsal hypothalamus: human growth hormone-releasing factor1–37-, salmon melanin-concentrating hormone- and alpha-melanotropin-like substances. Neurosci Lett 74:275–80
Fong TM, Mao C, MacNeil T, Kalyani R, Smith T, Weinberg D, Tota MR, Van der Ploeg LH (1997) ART (protein product of agouti-related transcript) as an antagonist of MC-3 and MC-4 receptors. Biochem Biophys Res Commun 237:629–631
Gantz I, Konda Y, Tashiro T, Shimoto Y, Miwa H, Munzert G, Watson SJ, DelValle J, Yamada T (1993) Molecular cloning of a novel melanocortin receptor. J Biol Chem 268:8246–8250

Gantz I, Miwa H, Konda Y, Shimoto Y, Tashiro T, Watson SJ, DelValle J, Yamada T (1993) Molecular cloning, expression, and gene localization of a fourth melanocortin receptor. J Biol Chem 268:15174–15179
Gerald C, Walker MW, Criscione L, Gustafson EL, Batzl Hartmann C, Smith KE, Vaysse P, Durkin MM, Laz TM, Linemeyer DL, Schaffhauser AO, Whitebread S, Hofbauer KG, Taber RI, Branchek TA, Weinshank RL (1996) A receptor subtype involved in neuropeptide-Y-induced food intake. Nature 382:168–171
Gerald C, Walker MW, Vaysse PJ, He C, Branchek TA, Weinshank RL (1995) Expression cloning and pharmacological characterization of a human hippocampal neuropeptide Y/peptide YY Y2 receptor subtype. J Biol Chem 270:26758–26761
Goke R, Fehmann HC, Linn T, Schmidt H, Krause M, Eng J, Goke B (1993) Exendin-4 is a high potency agonist and truncated exendin-(9–39)-amide an antagonist at the glucagon-like peptide 1-(7–36)-amide receptor of insulin-secreting beta-cells. J Biol Chem 268:19650–19655
Goke R, Larsen PJ, Mikkelsen JD, Sheikh SP (1995) Distribution of GLP-1 binding sites in the rat brain: evidence that exendin-4 is a ligand of brain GLP-1 binding sites. Eur J Neurosci 7:2294–2300
Goldstone AP, Mercer JG, Gunn I, Moar KM, Edwards CM, Rossi M, Howard JK, Rasheed S, Turton MD, Small C, Heath MM, O'Shea D, Steere J, Meeran K, Ghatei MA, Hoggard N, Bloom SR (1997) Leptin interacts with glucagon-like peptide-1 neurons to reduce food intake and body weight in rodents. FEBS Lett 415: 134–138
Gonzalez Y, Fernandez TM, Sanchez FF, del Rio J (1984) Antagonism of diazepam-induced feeding in rats by antisera to opioid peptides. Life Sci 35:1423–1429
Graham M, Shutter JR, Sarmiento U, Sarosi I, Stark KL (1997) Overexpression of Agrt leads to obesity in transgenic mice. Nat Genet 17:273–274
Griffon N, Mignon V, Facchinetti P, Diaz J, Schwartz JC, Sokoloff P (1994) Molecular cloning and characterization of the rat fifth melanocortin receptor. Biochem Biophys Res Commun 200:1007–1014
Grundemar L, Wahlestedt C, Reis DJ (1991) Neuropeptide Y acts at an atypical receptor to evoke cardiovascular depression and to inhibit glutamate responsiveness in the brainstem. J Pharmacol Exp Ther 258:633–638
Halaas JL, Gajiwala KS, Maffei M, Cohen SL, Chait BT, Rabinowitz D, Lallone RL, Burley SK, Friedman JM (1995) Weight-reducing effects of the plasma protein encoded by the obese gene. Science 269:543–546
Hawkins MF, Cubic B, Baumeister AA, Barton C (1992) Microinjection of opioid antagonists into the substantia nigra reduces stress-induced eating in rats. Brain Res 584:261–265
Hebb DO (1949) The organization of behavior. John Wiley & Sons 171–206
Heinrichs SC, Lapsansky J, Behan DP, Chan RK, Sawchenko PE, Lorang M, Ling N, Vale WW, De SE (1996) Corticotropin-releasing factor-binding protein ligand inhibitor blunts excessive weight gain in genetically obese Zucker rats and rats during nicotine withdrawal. Proc Natl Acad Sci USA 93:15475–15480
Hetherinton AW, Ranson SW (1940) Hypothalamic lesions and adiposity in the rat. Anat Record 78:149–172
Holst JJ, Orskov C, Nielsen OV, Schwartz TW (1987) Truncated glucagon-like peptide I, an insulin-releasing hormone from the distal gut. FEBS Lett 211:169–174
Holtzman SG (1974) Behavioral effects of separate and combined administration of naloxone and d-amphetamine. J Pharmacol Exp Ther 189:51–60
Horvath TL, Kalra SP, Naftolin F, Leranth C (1995) Morphological evidence for a galanin-opiate interaction in the rat mediobasal hypothalamus. J Neuroendocrinol 7:579–588
Horvath TL, Naftolin F, Leranth C, Sahu A, Kalra SP (1996) Morphological and pharmacological evidence for neuropeptide Y-galanin interaction in the rat hypothalamus. Endocrinology 137:3069–3078

Huszar D, Lynch CA, Fairchild-Huntress V, Dunmore JH, Fang Q, Berkemeier LR, Gu W, Kesterson RA, Boston BA, Cone RD, Smith FJ, Campfield LA, Burn P, Lee F (1997) Targeted disruption of the melanocortin-4 receptor results in obesity in mice. Cell 88:131–141

Inoue T, Inui A, Okita M, Sakatani N, Oya M, Morioka H, Mizuno N, Oimomi M, Baba S (1989) Effect of neuropeptide Y on the hypothalamic-pituitary-adrenal axis in the dog. Life Sci 44:1043–1051

Inui A, Inoue T, Nakajima M, Okita M, Sakatani N, Okimura Y, Chihara K, Baba S (1990) Brain neuropeptide Y in the control of adrenocorticotropic hormone secretion in the dog. Brain Res 510:211–215

Islam AK, Bodnar RJ (1990) Selective opioid receptor antagonist effects upon intake of a high-fat diet in rats. Brain Res 508:293–296

Jhanwar UM, Beck B, Burlet C, Leibowitz SF (1990) Diurnal rhythm of neuropeptide Y-like immunoreactivity in the suprachiasmatic, arcuate and paraventricular nuclei and other hypothalamic sites. Brain Res 536:331–334

Jin SL, Han VK, Simmons JG, Towle AC, Lauder JM, Lund PK (1988) Distribution of glucagonlike peptide I (GLP-I), glucagon, and glicentin in the rat brain: an immunocytochemical study. J Comp Neurol 271:519–532

Kalra SP, Dube MG, Fournier A, Kalra PS (1991) Structure-function analysis of stimulation of food intake by neuropeptide Y: effects of receptor agonists. Physiol Behav 50:5–9

Kalra SP, Kalra PS (1996) Nutritional infertility: the role of the interconnected hypothalamic neuropeptide Y-galanin-opioid network. Front Neuroendocrinol 17:371–401

Kauth T, Metz J (1987) Immunohistochemical localization of glucagon-like peptide 1. Use of poly- and monoclonal antibodies. Histochemistry 86:509–515

Kawauchi H, Kawazoe I, Tsubokawa M, Kishida M, Baker BI (1983) Characterization of melanin-concentrating hormone in chum salmon pituitaries. Nature 305:321–323

Khawaja XZ, Chattopadhyay AK, Green IC (1991) Increased beta-endorphin and dynorphin concentrations in discrete hypothalamic regions of genetically obese (ob/ob) mice. Brain Res 555:164–168

Kim EM, Welch CC, Grace MK, Billington C, Levine AS (1996) Chronic food restriction and acute food deprivation decrease mRNA levels of opioid peptides in arcuate nucleus. Am J Physiol 270:R1019–R1024

King MG, Kastin AJ, Olson RD, Coy DH (1979) Systemic administration of Met-enkephalin, (D-Ala2)-Met-enkephalin, beta-endorphin, and (D-Ala2)-beta-endorphin: effects on eating, drinking and activity measures in rats. Pharmacol Biochem Behav 11:407–411

Koch JE, Bodnar RJ (1993) Involvement of mu1 and mu2 opioid receptor subtypes in tail-pinch feeding in rats. Physiol Behav 53:603–605

Koch JE, Bodnar RJ (1994) Selective alterations in macronutrient intake of food-deprived or glucoprivic rats by centrally-administered opioid receptor subtype antagonists in rats. Brain Res 657:191–201

Koob GF, Bloom FE (1985) Corticotropin-releasing factor and behavior. Fed Proc 44:259–263

Kotz CM, Grace MK, Billington CJ, Levine AS (1993) The effect of norbinaltorphimine, beta-funaltrexamine and naltrindole on NPY-induced feeding. Brain Res 631:325–328

Kotz CM, Grace MK, Briggs J, Levine AS, Billington CJ (1995) Effects of opioid antagonists naloxone and naltrexone on neuropeptide Y-induced feeding and brown fat thermogenesis in the rat. Neural site of action. J Clin Invest 96: 163–170

Krahn DD, Gosnell BA, Levine AS, Morley JE (1988) Behavioral effects of corticotropin-releasing factor: localization and characterization of central effects. Brain Res 443:63–69

Kreymann B, Ghatei MA, Burnet P, Williams G, Kanse S, Diani AR, Bloom SR (1989) Characterization of glucagon-like peptide-1-(7–36)amide in the hypothalamus. Brain Res 502:325–331

Kreymann B, Williams G, Ghatei MA, Bloom SR (1987) Glucagon-like peptide-1 7–36: a physiological incretin in man. Lancet 2:1300–1304

Kristensen P, Judge ME, Thim L, Ribel U, Christjansen KN, Wulff BS, Clausen JT, Jensen PB, Madsen OD, Vrang N, Larsen PJ, Hastrup S (998) Hypothalamic CART is a new anorectic peptide regulated by leptin. Nature 393:72–76

Kwon HY, Bultman SJ, Loffler C, Chen WJ, Furdon PJ, Powell JG, Usala AL, Wilkiso W, Hansmann I, Woychik RP (1994) Molecular structure and chromosomal mapping of the human homolog of the agouti gene. Proc Natl Acad Sci USA 91:9760–9764

Kyrkouli SE, Stanley, BG, Leibowitz SF (1986) Galanin: stimulation of feeding induced by medial hypothalamic injection of this novel peptide. Eur J Pharmacol 122:159–160

Kyrkouli SE, Stanley BG, Leibowitz SF (1992) Differential effects of galanin and neuropeptide Y on extracellular norepinephrine levels in the paraventricular hypothalamic nucleus of the rat: a microdialysis study. Life Sci 51:203–210

Kyrkouli SE, Stanley BG, Seirafi RD, Leibowitz SF (1990) Stimulation of feeding by galanin: anatomical localization and behavioral specificity of this peptide's effects in the brain. Peptides 11:995–1001

Lambert PD, Wilding JP, al DA, Bohuon C, Comoy E, Gilbey SG, Bloom SR (1993) A role for neuropeptide-Y, dynorphin, and noradrenaline in the central control of food intake after food deprivation. Endocrinology 133:29–32

Lambert PD, Wilding JP, al DA, Gilbey SG, Bloom SR (1993) The effect of central blockade of kappa-opioid receptors on neuropeptide Y-induced feeding in the rat. Brain Res 629:146–148

Larhammar D, Blomqvist AG, Yee F, Jazin E, Yoo H, Wahlested C (1992) Cloning and functional expression of a human neuropeptide Y/peptide YY receptor of the Y1 type. J Biol Chem 267:10935–10938

Larsen PJ, Tang CM, Holst JJ, Orskov C (1997) Distribution of glucagon-like peptide-1 and other preproglucagon-derived peptides in the rat hypothalamus and brainstem. Neuroscience 77:257–270

Larsen PJ, Tang CM, Jessop DS (1997) Central administration of glucagon-like peptide-1 activates hypothalamic neuroendocrine neurons in the rat. Endocrinology 138:4445–4455

Leibowitz SF, Hor L (1982) Endorphinergic and alpha-noradrenergic systems in the paraventricular nucleus: effects on eating behavior. Peptides 3:421–428

Leibowitz SF, Kim T (1992) Impact of a galanin antagonist on exogenous galanin and natural patterns of fat ingestion. Brain Res 599:148–152

Leibowitz SF, Rossakis C (1979) Mapping study of brain dopamine- and epinephrine-sensitive sites which cause feeding suppression in the rat. Brain Res 172:101–113

Levine AS, Grace M, Billington CJ (1990) The effect of centrally administered naloxone on deprivation and drug-induced feeding. Pharmacol Biochem Behav 36:409–412

Levine AS, Grace M, Billington CJ, Portoghese PS (1990) Nor-binaltorphimine decreases deprivation and opioid-induced feeding. Brain Res 534:60–64

Lewis LD, Williams JA (1990) Regulation of cholecystokinin secretion by food, hormones, and neural pathways in the rat. Am J Physiol 258:G512–G518

Lu D, Willard D, Patel IR, Kadwell S, Overton L, Kost T, Luther M, Chen W, Woychik RP, Wilkison WO, Cone RD (1994) Agouti protein is an antagonist of the melanocyte-stimulating-hormone receptor. Nature 371:799–802

Lundberg JM, Terenius L, Hokfelt T, Goldstein M (1983) High levels of neuropeptide Y in peripheral noradrenergic neurons in various mammals including man. Neurosci Lett 42:167–172

Lundell I, Blomqvist AG, Berglund MM, Schober DA, Johnson D, Statnick MA, Gadski RA, Gehlert DR, Larhammar D (1995) Cloning of a human receptor of the NPY receptor family with high affinity for pancreatic polypeptide and peptide YY. J Biol Chem 270:29123–29128
Marks KR, Kanarek RB (1980) Morphine selectively influences macronutrient intake in the rat. Pharmacol Biochem Behav 12:427–430
Marsh DJ, Hollopeter G, Kafer KE, Palmiter RD (1998) Role of the Y5 neuropeptide Y receptor in feeding and obesity. Nature Medicine 4:718–721
McDonald JK, Lumpkin MD, Samson WK, McCann SM (1985) Neuropeptide Y affects secretion of luteinizing hormone and growth hormone in ovariectomized rats. Proc Natl Acad Sci USA 82:561–564
McKay LD, Kenney NJ, Edens NK, Williams RH, Woods SC (1981) Intracerebroventricular beta-endorphin increases food intake of rats. Life Sci 29:1429–1434
McLaughlin CL, Baile CA, Della FM (1985) Meal-stimulated increased concentrations of beta-endorphin in the hypothalamus of Zucker obese and lean rats. Physiol Behav 35:891–896
Meeran K, O'Shea D, Turton MD, Heath MM, Gunn I, Choi SJ, Edwards CMB, Small CJ, Ghatei MA, Bloom SR (1996) Chronic blockade of hypothalamic GLP-1 receptors causes obesity. Diabetes 45:170 A
Melander T, Hokfelt T, Rokaeus A (1986) Distribution of galaninlike immunoreactivity in the rat central nervous system. J Comp Neurol 248:475–517
Mercer JG, Hoggard N, Williams LM, Lawrence CB, Hannah LT, Morgan PJ, Trayhurn P (1996) Coexpression of leptin receptor and preproneuropeptide Y mRNA in arcuate nucleus of mouse hypothalamus. J Neuroendocrinol 8:733–735
Mezey E, Kiss JZ, Mueller GP, Eskay R, O'Donohue TL, Palkovits M (1985) Distribution of the pro-opiomelanocortin derived peptides, adrenocorticotrope hormone, alpha-melanocyte-stimulating hormone and beta-endorphin (ACTH, alpha-MSH, beta-END) in the rat hypothalamus. Brain Res 328:341–347
Michel MC, Beck-Sickinger A, Cox H, Doods HN, Herzog H, Larhammar D, Quirion R, Schwartz T, Westfall T (1998) XVI. International union of pharmacology recommendations for the nomenclature of neuropeptide Y, peptide YY, and pancreatic polypeptide receptors. Pharmacol Reviews 50:143–150
Mikkelsen JD, Larsen PJ (1992) A high concentration of NPY (Y1)-receptor mRNA-expressing cells in the rat arcuate nucleus. Neurosci Lett 148:195–198
Miller CL, Hruby VJ, Matsunaga TO, Bickford PC (1993) Alpha-MSH and MCH are functional antagonists in a CNS auditory gating paradigm. Peptides 14:431–440
Miller MW, Duhl DM, Vrieling H, Cordes SP, Ollmann MM, Winkes BM, Barsh GS (1993) Cloning of the mouse agouti gene predicts a secreted protein ubiquitously expressed in mice carrying the lethal yellow mutation. Genes Dev 7:454–467
Miyasaka K, Kanai S, Ohta M, Kawanami T, Kono A, Funakoshi A (1994) Lack of satiety effect of cholecystokinin (CCK) in a new rat model not expressing the CCK-A receptor gene. Neurosci Lett 180:143–146
Morley JE (1987) Neuropeptide regulation of appetite and weight. Endocr Rev 8:256–287
Morley JE, Levine AS (1981) Dynorphin-(1–13) induces spontaneous feeding in rats. Life Sci 29:1901–1903
Morley JE, Levine AS (1982) Corticotrophin releasing factor, grooming and ingestive behavior. Life Sci 31:1459–1464
Mountjoy KG, Mortrud MT, Low MJ, Simerly RB, Cone RD (1994) Localization of the melanocortin-4 receptor (MC4-R) in neuroendocrine and autonomic control circuits in the brain. Mol Endocrinol 8:1298–1308
Mountjoy KG, Robbins LS, Mortrud MT, Cone RD (1992) The cloning of a family of genes that encode the melanocortin receptors. Science 257:1248–1251
O'Shea D, Morgan DG, Meeran K, Edwards CM, Turton MD, Choi SJ, Heath MM, Gunn I, Taylor GM, Howard JK, Bloom CI, Small CJ, Haddo O, Ma JJ, Callinan

W, Smith DM, Ghatei MA, Bloom SR (1997) Neuropeptide Y induced feeding in the rat is mediated by a novel receptor. Endocrinology 138:196–202
Ollmann MM, Wilson BD, Yang YK, Kerns JA, Chen Y, Gantz I, Barsh GS (1997) Antagonism of central melanocortin receptors in vitro and in vivo by agouti-related protein. Science 278:135–138
Panksepp J, Belekedal MYV, Walter M (1996) Potent suppressive effects of the putative satiety agent GLP-1 on social emotive behaviours. Soc Neurosci Abst 22:15.5
Papadouka V, Carr KD (1994) The role of multiple opioid receptors in the maintenance of stimulation-induced feeding. Brain Res 639:42–48
Pedrazzini T, Seydoux J, Konstner P, Aubert J, Grouzmann E, Beermann F, Brunner H (1998) Cardiovascular response, feeding behaviour and locomotor activity in mice lacking the NPY Y1 receptor. Nature Medicine 4L722–726
Pernow J, Lundberg JM, Kaijser L (1987) Vasoconstrictor effects in vivo and plasma disappearance rate of neuropeptide Y in man. Life Sci 40:47–54
Pierroz DD, Catzeflis C, Aebi AC, Rivier JE, Aubert ML (1996) Chronic administration of neuropeptide Y into the lateral ventricle inhibits both the pituitary-testicular axis and growth hormone and insulin-like growth factor I secretion in intact adult male rats. Endocrinology 137:3–12
Pinget M, Straus E, Yalow RS (1979) Release of cholecystokinin peptides from a synaptosome-enriched fraction of rat cerebral cortex. Life Sci 25:339–342
Potter E, Behan DP, Linton EA, Lowry PJ, Sawchenko PE, Vale WW (1992) The central distribution of a corticotropin-releasing factor (CRF)-binding protein predicts multiple sites and modes of interaction with CRF. Proc Natl Acad Sci USA 89:4192–4196
Presse F, Nahon JL, Fischer WH, Vale W (1990) Structure of the human melanin concentrating hormone mRNA. Mol Endocrinol 4:632–637
Qu D, Ludwig DS, Gammeltoft S, Piper M, Pelleymounter MA, Cullen MJ, Mathes MF, Przypek J, Kanarek R, Maratos-Filer E (1996) A role for melanin-concentrating hormone in the regulation of feeding behaviour. Nature 380:243–246
Rance T, Baker BI (1979) The teleost melanin-concentrating hormone – a pituitary hormone of hypothalamic origin. Gen Comp Endocrinol 37:64–73
Reidelberger RD (1992) Abdominal vagal mediation of the satiety effects of exogenous and endogenous cholecystokinin in rats. Am J Physiol 263:R1354–R1358
Ritter RC, Ritter S, Ewart WR, Wingate DL (1989) Capsaicin attenuates hindbrain neuron responses to circulating cholecystokinin. Am J Physiol 257:R1162–R1168
Roane DS, Iadarola MJ, Porter JR (1988) Decreased [3H]-naloxone binding and elevated dynorphin-A(1–8) content in Zucker rat brain. Physiol Behav 43: 371–374
Robbins LS, Nadeau JH, Johnson KR, Kelly MA, Roselli RL, Baack E, Mountjoy KG, Cone RD (1993) Pigmentation phenotypes of variant extension locus alleles result from point mutations that alter MSH receptor function. Cell 72:827–834
Rose C, Vargas F, Facchinetti P, Bourgeat P, Bambal RB, Bishop PB, Chan SM, Moore AN, Ganellin CR, Schwartz JC (1996) Characterization and inhibition of a cholecystokinin-inactivating serine peptidase. Nature 380:403–409
Roselli RL, Mountjoy KG, Robbins LS, Mortrud MT, Low MJ, Tatro JB, Entwistle ML, Simerly RB, Cone RD (1993) Identification of a receptor for gamma melanotropin and other proopiomelanocortin peptides in the hypothalamus and limbic system. Proc Natl Acad Sci USA 90:8856–8860
Rossi M, Small CJ, Goldstone AP, Ghatei MA, Bloom SR (1997a) Melanin concentrating hormone (MCH) concentration is increased in the hypothalamus of obese zucker rats. British Endocrine Society (Abstract)
Rossi M, Choi SJ, O'Shea D, Miyoshi T, Ghatei MA, Bloom SR (1997b) Melanin concentrating hormone acutely stimulates feeding, but chronic administration has no effect on body weight. Endocrinology 138:351–355
Rossi M, Gunn I, Beak SA, Goldstone AP, Morgan DG, Small CJ, Ghatei MA, Smith DM, Bailey CJ, Bloom SR (1998) Melanin concentrating hormone (MCH)

interacts with the leptin system, independently to the melanocortins. European Congress of Endocrinology (Abstract)
Rovere C, Viale A, Nahon J, Kitabgi P (1996) Impaired processing of brain proneurotensin and promelanin-concentrating hormone in obese fat/fat mice. Endocrinology 137:2954–2958
Sagar SM, Sharp FR, Curran T (1988) Expression of c-fos protein in brain: metabolic mapping at the cellular level. Science 240:1328–1331
Saito A, Sankaran H, Goldfine ID, Williams JA (1980) Cholecystokinin receptors in the brain: characterization and distribution. Science 208:1155–1156
Sakurai T, Amemiya A, Ishii M, Matsuzaki I, Chemelli RM, Tanaka H, Williams SC, Richarson JA, Kozlowski GP, Wilson S, Arch JR, Buckingham RE, Haynes AC, Carr SA, Annan RS, McNulty DE, Liu WS, Terrett JA, Elshourbagy NA, Bergsma DJ, Yanagisawa M (1998) Orexins and orexin receptors: a family of hypothalamic neuropeptides and G protein-coupled receptors that regulate feeding behavior. Cell 92:573–585
Sanacora G, Kershaw M, Finkelstein JA, White JD (1990) Increased hypothalamic content of preproneuropeptide Y messenger ribonucleic acid in genetically obese Zucker rats and its regulation by food deprivation. Endocrinology 127:730–737
Sanchez M, Baker BI, Celis M (1997) Melanin-concentrating hormone (MCH) antagonizes the effects of alpha-MSH and neuropeptide E-I on grooming and locomotor activities in the rat. Peptides 18:393–396
Schaffhauser AO, Stricker KA, Brunner L, Cumin F, Gerald C, Whitebread S, Criscione L, Hofbauer KG (1997) Inhibition of food intake by neuropeptide Y Y5 receptor antisense oligodeoxynucleotides
Schick RR, Harty GJ, Yaksh TL, Go VL (1990) Sites in the brain at which cholecystokinin octapeptide (CCK-8) acts to suppress feeding in rats: a mapping study. Neuropharmacology 29:109–118
Schick RR, Yaksh TL, Roddy DR, Go VL (1989) Release of hypothalamic cholecystokinin in cats: effects of nutrient and volume loading. Am J Physiol 256:R248–R254
Schwartz MW, Baskin DG, Bukowski TR, Kuijper JL, Foster D, Lasser G, Prunkard DE, Porte D, Woods SC, Seeley RJ, Weigle DS (1996) Specificity of leptin action on elevated blood glucose levels and hypothalamic neuropeptide Y gene expression in ob/ob mice. Diabetes 45:531–535
Scrocchi LA, Brown TJ, MaClusky N, Brubaker PL, Auerbach AB, Joyner AL, Drucker DJ (1996) Glucose intolerance but normal satiety in mice with a null mutation in the glucagon-like peptide 1 receptor gene. Nat Med 2:1254–1258
Sekiya K, Ghatei MA, Lacoumenta S, Burnet PW, Zamir N, Burrin JM, Polak JM, Bloom SR (1988) The distribution of melanin-concentrating hormone-like immunoreactivity in the central nervous system of rat, guinea-pig, pig and man. Neuroscience 25:925–930
Shutter JR, Graham M, Kinsey AC, Scully S, Luthy R, Stark KL (1997) Hypothalamic expression of ART, a novel gene related to agouti, is up-regulated in obese and diabetic mutant mice. Genes Dev 11:593–602
Silvers WK (1979) The coat colors of mice. Springer-Verlag (ed)
Simone DA, Bodnar RJ, Goldman EJ, Pasternak GW (1985) Involvement of opioid receptor subtypes in rat feeding behavior. Life Sci 36:829–833
Skofitsch G, Jacobowitz DM, Zamir N (1985) Immunohistochemical localization of a melanin concentrating hormone-like peptide in the rat brain. Brain Res Bull 15:635–649
Small CJ, Morgan DG, Meeran K, Heath MM, Gunn I, Edwards CM, Gardiner J, Taylor GM, Hurley JD, Rossi M, Goldstone AP, O'Shea D, Smith DM, Ghatei MA, Bloom SR (1997) Peptide analogue studies of the hypothalamic neuropeptide Y receptor mediating pituitary adrenocorticotrophic hormone release. Proc Natl Acad Sci USA 94:11686–11691

Smith AI, Funder JW (1988) Proopiomelanocortin processing in the pituitary, central nervous system, and peripheral tissues. Endocr Rev 9:159–179
Smith BK, York DA, Bray GA (1994) Chronic cerebroventricular galanin does not induce sustained hyperphagia or obesity. Peptides 15:1267–1272
Smith GP, Gibbs J (1994) Satiating effect of cholecystokinin. Ann N Y Acad Sci 713: 236–241
Spina M Merlo PE, Chan RK, Basso AM, Rivier J, Vale W, Koob GF (1996) Appetite-suppressing effects of urocortin, a CRF-related neuropeptide. Science 273: 1561–1564
Stanley BG, Kyrkouli SE, Lampert S, Leibowitz SF (1986) Neuropeptide Y chronically injected into the hypothalamus: a powerful neurochemical inducer of hyperphagia and obesity. Peptides 7:1189–1192
Stanley BG, Leibowitz SF (1984) Neuropeptide Y: stimulation of feeding and drinking by injection into the paraventricular nucleus. Life Sci 35:2635–2642
Stanley BG, Magdalin W, Seirafi A, Nguyen MM, Leibowitz SF (1992) Evidence for neuropeptide Y mediation of eating produced by food deprivation and for a variant of the Y1 receptor mediating this peptide's effect. Peptides 13:581–587
Suda T, Tozawa F, Iwai I, Sato Y, Sumitomo T, Nakano Y, Yamada M, Demura H (1993) Neuropeptide Y increases the corticotropin-releasing factor messenger ribonucleic acid level in the rat hypothalamus. Brain Res Mol Brain Res 18:311–315
Tang CM, Larsen PJ, Goke R, Fink JA, Jessop DS, Moller M, Sheikh SP (1996) Central administration of GLP-1-(7–36) amide inhibits food and water intake in rats. Am J Physiol 271:R848–R856
Tatemoto K (1982) Neuropeptide Y: complete amino acid sequence of the brain peptide. Proc Natl Acad Sci USA 79:5485–5489
Tatemoto K, Rokaeus A, Jornvall H, McDonald TJ, Mutt V (1983) Galanin – a novel biologically active peptide from porcine intestine. FEBS Lett 164:124–128
Tempel DL, Leibowitz KJ, Leibowitz SF (1988) Effects of PVN galanin on macronutrient selection. Peptides 9:309–314
Teskey GC, Kavaliers M, Hirst M (1984) Social conflict activates opioid analgesic and ingestive behaviors in male mice. Life Sci 35:303–315
Thiele TE, Van DG, Campfield LA, Smith FJ, Burn P, Woods SC, Bernstein IL, Seeley RJ (1997) Central infusion of GLP-1, but not leptin, produces conditioned taste aversions in rats. Am J Physiol 272:R726–R730
Thorens B (1992) Expression cloning of the pancreatic beta cell receptor for the glucoincretin hormone glucagon-like peptide 1. Proc Natl Acad Sci USA 89:8641–8645
Turton MD, Edwards CMB, Meeran K, O'Shea D, Gunn I, Heath MM, Ghatei MA, Bloom SR (1997) Chronic intracerebroventricular administration of GLP-1 reduces body weight in rats. Diabetes 46:185A
Turton MD, O'Shea D, Gunn I, Beak SA, Edwards CM, Meeran K, Choi SJ, Taylor GM, Heath MM, Lambert PD, Wilding JP, Smith DM, Ghatei MA, Herbert J, Bloom SR (1996) A role for glucagon-like peptide-1 in the central regulation of feeding. Nature 379:69–72
Van DG, Thiele TE, Donahey JC, Campfield LA, Smith FJ, Burn P, Bernstein IL, Woods SC, Seeley RJ (1996) Central infusions of leptin and GLP-1-(7–36) amide differentially stimulate c-FLI in the rat brain. Am J Physiol 271:R1096–R1100
Varndell IM, Bishop AE, Sikri KL, Uttenthal LO, Bloom SR, Polak JM (1985) Localization of glucagon-like peptide (GLP) immunoreactants in human gut and pancreas using light and electron microscopic immunocytochemistry. J Histochem Cytochem 33:1080–1086
Vaughan J, Donaldson C, Bittencourt J, Perrin MH, Lewis K, Sutton S, Chan R, Turnbull AV, Lovejoy D, Rivier C et al (1995) Urocortin, a mammalian neuropeptide related to fish urotensin I and to corticotropin-releasing factor. Nature 378: 287–292
Vaughan JM, Fischer WH, Hoeger C, Rivier J, Vale W (1989) Characterization of melanin-concentrating hormone from rat hypothalamus. Endocrinology 125: 1660–1665

Walker JM, Katz RJ, Akil H (1980) Behavioral effects of dynorphin 1–13 in the mouse and rat: initial observations. Peptides 1:341–345
Weinberg DH, Sirinathsinghji DJ, Tan CP, Shiao LL, Morin N, Rigby MR, Heavens RH, Rapoport DR, Bayne ML, Cascieri MA, Strader CD, Linemeyer DL, MacNeil DJ (1996) Cloning and expression of a novel neuropeptide Y receptor. J Biol Chem 271:16435–16438
Weller A, Smith GP, Gibbs J (1990) Endogenous cholecystokinin reduces feeding in young rats. Science 247:1589–1591
Wilding JP, Gilbey SG, Bailey CJ, Batt RA, Williams G, Ghatei MA, Bloom SR (1993) Increased neuropeptide-Y messenger ribonucleic acid (mRNA) and decreased neurotensin mRNA in the hypothalamus of the obese (ob/ob) mouse. Endocrinology 132:1939–1944
Williams G, Gill JS, Lee YC, Cardoso HM, Okpere BE, Bloom SR (1989) Increased neuropeptide Y concentrations in specific hypothalamic regions of streptozocin-induced diabetic rats. Diabetes 38:321–327
Wilson BD, Ollmann MM, Kang L, Stoffel M, Bell GI, Barsh GS (1995) Structure and function of ASP, the human homolog of the mouse agouti gene. Hum Mol Genet 4:223–230
Wong G, Pawelek J, Sansone M, Morowitz J (1974) Response of mouse melanoma cells to melanocyte stimulating hormone. Nature 248:351–354
Woychik RP, Generoso WM, Russell LB, Cain KT, Cacheiro NL, Bultman SJ, Selby PB, Dickinson ME, Hogan BL, Rutledge JC (1990) Molecular and genetic characterization of a radiation-induced structural rearrangement in mouse chromosome 2 causing mutations at the limb deformity and agouti loci. Proc Natl Acad Sci USA 87:2588–2592
Yang HY, Panula P, Tang J, Costa E (1983) Characterization and location of Met5-enkephalin-arg6-phe7 stored in various rat brain regions. J Neurochem 40:969–976
Yang YK, Ollmann MM, Wilson BD, Dickinson C, Yamada T, Barsh GS, Gantz I (1997) Effects of recombinant agouti-signaling protein on melanocortin action. Mol Endocrinol 11:274–280

CHAPTER 14

β3 Adrenergic Receptors as a Therapeutic Target for Obesity

J.G. GRANNEMAN

A. Introduction

The physiological actions of catecholamines are mediated via α- and β-adrenergic receptors (AR). Activation of adipocyte βAR results in lipid mobilization and release from white adipose tissue, and oxidation of fatty acids in brown fat. The net effect of adipocyte βAR activation is the reduction of triglyceride stores in white fat with a concomitant increase in thermogenesis and thermogenic capacity of brown fat. Because of the potential significance of adipocyte βAR in overall energy balance, much attention has been devoted to their characterization. Pioneering work by HARMS et al. (1974) suggested the existence of βAR with atypically low affinity for standard βAR antagonists in adipocytes; however the notion of an "atypical" adipocyte βAR was not widely accepted at the time owing to the lack of selective pharmacological agents. Thus, available radioligands failed to detect these atypical receptors and no agonists were available that would selectively activate them (BAHOUTH and MALBON 1988). This state of affairs changed dramatically nearly 15 years ago when researchers at Beecham Pharmaceuticals identified novel compounds that potently induced thermogenesis in rats via interactions with an atypical βAR (ARCH et al. 1984). These agents placed the concept of an atypical adipocyte βAR on solid pharmacological ground, and ultimately provided the impetus for the molecular cloning of the β3AR. Perhaps more significantly, the agents discovered by Beecham were found to be highly effective in rodent models of obesity and adult-onset diabetes, holding out the promise that the "atypical" adipocyte βAR would be an important therapeutic target for the treatment of human disease.

The cloning of the β3AR genes and cDNAs from several species (especially human and rodents) has provided the important molecular reagents for addressing the significance of β3AR as a therapeutic target for the treatment of obesity and related disorders. This analysis has revealed species differences in the expression, regulation, and pharmacological properties of the β3AR which have important implications regarding the potential utility of β3AR agonists for treatment of obesity in humans. The aim of this chapter is to review the molecular and pharmacological properties of β3AR, and to evaluate their potential as a therapeutic target for treatment of obesity in man.

B. β3AR Genes

β3AR genes and cDNAs have been isolated from various species including humans (EMORINE et al. 1989; GRANNEMAN et al. 1992; LELIAS et al. 1993), rats (MUZZIN et al. 1991; GRANNEMAN et al. 1991), mice (NAHMIAS et al. 1991), rhesus monkey (WALSTON et al. 1997), and cow (PIETRI-ROUXEL et al. 1995). The β3AR is a member of the G-protein-coupled receptor superfamily, and as such, the encoded protein is predicted to span the plasma membrane seven times (STROSBERG 1997). Overall, the β3AR is approximately 65% similar to the β1 and β2-AR within a given species, with the highest degree of similarly among subtypes being found in the predicted transmembrane regions.

The β3AR gene was initially thought to contain a single protein coding-exon (EMORINE et al. 1989; EMORINE et al. 1992); however, subsequent analysis of genes and cDNAs of several species demonstrated that the β3AR genes, unlike β1 and β2-AR, contain multiple exons (GRANNEMAN et al. 1992, 1993; BENSAID et al. 1993; VAN SPRONSEN et al. 1993; LELIAS et al. 1993). In general, β3AR are encoded by two exons, with the first exon encoding all of the transmembrane domains and much of the carboxyl tail. The final amino acids (e.g., 6 in the human and 12 in the rat) are encoded by a second exon. An interesting feature of the β3AR genes is the large degree of divergence in the region encoding the carboxyl tail. The amino acids sequences and patterns of RNA splicing vary distinctly, even among closely related species (VAN SPRONSEN et al. 1993; GRANNEMAN and LAHNERS 1994; WALSTON et al. 1997). For example, the protein-coding second exon of the human β3AR gene is homologous to the 3' *non*coding region of the rodent genes (Fig. 1). Additionally, the reading frame of the carboxyl tail of the rhesus monkey β3AR is shifted relative to the human, resulting in divergent amino acids in this region of the protein (WALSTON et al. 1997). Such divergence is not seen for the β1AR and β2AR, and suggests little selective pressure to maintain specific sequences in the carboxyl tail region.

The human β3AR gene has been localized to a region on chromosome near 8p11.2 (NAHMIAS et al. 1991). Recent work has shown that codon 64 in the human β3AR gene is polymorphic. The predominant allele encodes tryptophan (frequency between 0.9 and 0.7, depending upon the population sampled), whereas a minority of individuals possess an allele that encodes arginine in this position. In 1995, three papers appeared demonstrating statistically significant associations of the arg64 polymorphism with indices of obesity and insulin resistance (WALSTON et al. 1995; WIDEN et al. 1995; CLEMENT et al. 1995). In Finns and Pima Indians the arg64 allele was associated with slightly earlier onset of NIDDM. However, the arg64 and trp64 alleles are equally rep-

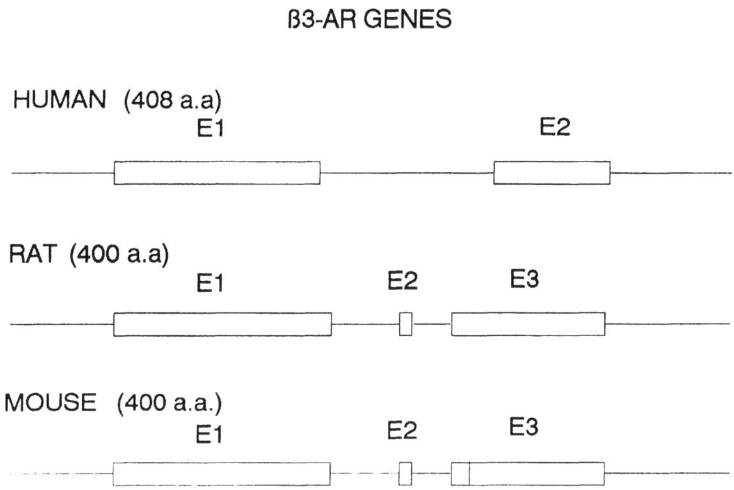

Fig. 1. Schematic representation of the intron/exon organization of the human and rodent $\beta 3$AR genes. E, exon. The mouse $\beta 3$AR gene contains an alternative splice site in the 3′ nontranslated region

resented in Pima Indians with NIDDM (Walston et al. 1995). The arg64 allele was not significantly associated with body mass index (BMI) or metabolic rate in Pima Indians. Clement et al. (1995) examined the arg64 allele in massively obese French patients and found that the allele was not over-represented in these subjects compared to normal controls. However, within this population of massively obese individuals, the rate of weight gain after age 20, and the maximal weight achieved, were slightly greater in subjects bearing the arg64 allele. Many more reports have appeared in the following 2 years, most of which show significant associations with indices of obesity, insulin resistance and dyslipidia (Elbein et al. 1996; Kurabayshi et al. 1996; Zhang et al. 1996; Urhammer et al. 1996; Fujisawa et al. 1996; Nagase et al. 1997; Higashi et al. 1997). In general, the associations are weak, and are by no means consistent across studies. Thus, some studies report significant association with insulin resistance, but no significant association obesity, while others have found the reverse. The meaning of these differences is unclear. Obesity and NIDDM are clearly polygenic disorders, and it seems likely that while the 64arg allele modifies these disorders, it is not a major determinant. The frequency of the allele varies significantly across various populations, and it is likely that the significance of the arg64 allele will be greater in some populations and lesser in others. In this regard, Pima Indians and Nauruans both have high rates of obesity and NIDDM, yet differ significantly in the frequency of the arg64 allele (Walston et al. 1995; Silver et al. 1996). Recent work indicates that the $\beta 3$AR is contained in a quantitative trait locus (QTL) that has been associated with elevated leptin levels in a population of Mexican Americans (Comuzzie et al.

1997), suggesting a relationship between β3AR and adipose tissue mass. The relationship between the putative QTL and the arg64 allele was examined very recently with paired sibling analysis, which demonstrated a significant association with body mass index, fat mass, and waist circumference (MITCHELL et al. 1998).

The genetic association studies cited above strongly suggest that the β3AR gene is active in humans and that it plays a role in human obesity. Nevertheless, the mechanism by which the 64arg β3AR contributes to obesity is unclear. Two papers have been published examining the functional activity of the arg64 and trp64 variants expressed in Chinese hamster ovary (CHO) and 293 (human embryonic kidney) cells. The first found no differences in the potency or efficacy of catecholamines and two classes of β3AR-selective agonists (CANDELORE et al. 1996). The second study (PIETRI-ROUXEL et al. 1997) examined catecholamine agonists as well as the aryloxypropanolamine CGP 12177 and found a reduction in efficacy (maximal effect) at the arg64 variant, without a change in potency (EC_{50}). The interpretation of the latter experiment is complicated somewhat by the fact that the responsiveness of CHO cells (but not 293 cells) to the direct adenylyl cyclase activator forskolin was strongly suppressed in cells expressing the arg64 variant. The maximal stimulation of cyclic AMP accumulation was also reduced in 293 cells expressing the arg64 variant, without changes in forskolin-stimulated activity. However, no allowance was made for the fact that 293 cells contain endogenous β2AR, making interpretation of the catecholamine data difficult. Curiously, CGP 12177 stimulated nearly twice as much cyclic AMP accumulation as did epinephrine and norepinephrine in 293 cells expressing the arg64 variant, whereas the reverse was true in CHO cells. It is noteworthy that arg is found in the homologous position in all species cloned to date (including rhesus monkey), and activation of human and rodent β3AR by catecholamine agonists is virtually indistinguishable (EMORINE et al. 1989; NAHMIAS et al. 1991; LIGGETT et al. 1992). It is highly likely that the ancestral β3AR gene encoded arg codon 64 and that the arg to trp mutation occurred relatively recently in evolution.

Relatively little work has been done on the beta β3AR promoter. In rats, transcription of the β3AR gene occurs at two widely-spaced promoters (GRANNEMAN and LAHNERS 1993). Within the proximal promoter transcription is initiated at multiple sites in a tissue-specific fashion. In brown fat, most transcripts are initiated just down-stream of a consensus TATA box and transcription initiator sequence, while transcription in white fat occurs here and just downstream of a GC box. Analysis of human transcription start sites in BAT and SK-N-MC cells indicates the presence of multiple start sites that are homologous to the rodent gene. Sequence analysis of the promoter regions of the rodent and human β3AR genes demonstrates a surprising divergence (GRANNEMAN et al. 1993; BROWN and MACHIDA 1994). For example, the proximal promoters of the rat and human β3AR are about 65% similar, whereas the β1AR promoters are more than 90% similar between these species. These data suggest that there are likely to be important species differences in the regulation of β3AR gene expression by hormones, second messengers and tissue-specific factors.

C. Tissue Distribution of β3AR mRNA

The comparative tissue distribution of β3AR mRNA in human and rodents is a critical issue given the demonstrated efficacy of β3AR-selective agonist in rodent obesity models. In rodents, β3AR mRNA is abundant in brown and white adipose tissue where levels are 8–20 times greater than β1AR mRNA levels (GRANNEMAN et al. 1992; COLLINS et al. 1994). β3AR mRNA levels in the gastrointestinal tract are considerable lower, being about 5%–10% of that found in adipose tissue (GRANNEMAN et al. 1991). Notably, β3AR transcripts are essentially absent from rodent muscle, liver, heart, and brain. The distribution of β3AR transcripts in humans has been somewhat controversial. Use of the highly sensitive, but non-quantitative, reverse-transcription/polymerase chain reaction has identified transcripts in numerous tissues including adipose tissue, gall bladder, ileum (KREIF et al. 1993; GRANNEMAN et al. 1993). In many cases, adipose tissue RNA samples containing β3AR transcripts also contain the mRNA encoding uncoupling protein, suggesting that β3AR transcripts might be derived from brown adipocytes contained within white fat depots. Quantitative analysis of β3AR transcripts by solution hybridization strongly supports the view that β3AR are expressed abundantly in human brown fat, but are virtually absent in white fat (GRANNEMAN 1995). Indeed, β3AR mRNA levels in human white fat are less than 2% of those found in mouse or rat white fat. Although not extensively studied, levels of β3AR transcripts in the gastrointestinal tract appear to be comparably low in rodent and man.

The relation between β3AR mRNA and protein levels is unclear. In rodent adipocytes, variations in β3AR mRNA levels are strongly correlated with changes in β3AR ligand binding and functional activity (FEVE et al. 1992, 1994; GRANNEMAN et al. 1992; EL HADRI et al. 1996, 1997). However, degradation of β3AR mRNA appears to depend upon translation of the protein (GRANNEMAN and LAHNERS 1994), so it is possible that steady-state levels of β3AR mRNA may not be predictive of β3AR protein levels across tissues. For example, the gastrointestinal tract contains very low levels of β3AR mRNA, yet functional responses mediated by β3AR are readily observed in this tissue (COHEN et al. 1995).

D. Signaling Properties of Recombinant β3AR

One major difficulty is assessing the signaling properties of the β3AR in native tissues is that this subtype is invariably co-expressed with other βAR subtypes. The molecular cloning of β3AR allowed detailed analysis of the molecular pharmacology of this subtype which has revealed unique signaling properties and important species differences.

I. Receptor-Ligand Interactions

Catecholamine agonists like norepinephrine have a relatively low binding affinity (>30 μmol/l) for the β3AR (EMORINE et al. 1989; NAHMIAS et al. 1991; FEVE et al. 1991; MUZZIN et al. 1991; GRANNEMAN et al. 1993). By contrast, the affinity of norepinephrine for β1AR is about 100 times greater than that for the β3AR, and this difference initially raised questions regarding the physiological relevance of the β3AR (EMORINE et al. 1991; CHAUDHRY et al. 1992). In intact cells, however, the potency of norepinephrine in stimulating cyclic AMP accumulation appears to be far greater than would be predicted on the basis of ligand binding studies (EMORINE et al. 1989; MUZZIN et al. 1991; NAHMAIS et al. 1991; Granneman et al., see below). Although this discrepancy is greatest in cells over-expressing the receptor, it is clearly present in adipocytes and SK-N-MC cells that natively express much lower levels (CHAUDHRY et al. 1994; ESBENSHADE et al. 1992). The reasons for the discrepancy between binding affinity and functional response are unclear; nevertheless the nuances of β3AR signaling properties have important implications for the development of drug screens. Low binding affinity of catecholamines seems to be an intrinsic property of the receptor in that it is found in both intact cells and in membrane preparations (EMORINE et al. 1989; NAHMAIS et al. 1991; FEVE et al. 1992; J.G. Granneman et al., unpublished). The high coupling efficiency of β3AR to adenylyl cyclase, however, is highly dependent upon the integrity of the cells. Thus, the K_{act} of norepinephrine in membrane adenylyl cyclase assays of transfected cells can be more than 100 times greater than the K_{act} in whole-cell cAMP accumulation assays (NAMHIAS et al. 1991; FEVE et al. 1992; GRANNEMAN et al. 1991; MUZZIN et al. 1991). In general, the potency (K_{act}) of catecholamines in membrane adenylyl cyclase assays appears to be very similar to binding affinity (K_i) determined in radioligand binding assays, while these values in turn are 50–1000 times greater than K_{act} determined in cAMP accumulation assays. Clearly, the assay method is an important consideration in the development of β3AR agonist screen, and techniques based upon ligand binding affinity or membrane adenylyl cyclase assay could greatly underestimate potency in the intact cell (WILSON et al. 1996).

II. Receptor/G-Protein Interactions

It is well established that β3AR couple to adenylyl cyclase and thereby stimulate adenylyl cyclase. As detailed above, the mode of coupling of β1AR and β3AR to cAMP generation differs, implying differential interactions with G_s. The GTP dependency of adenylyl cyclase stimulation differs for β1AR and β3AR in adipocytes (Fig. 2). Notably, activation of adenylyl cyclase by β3AR, but not β1AR, is markedly affected when activity of G_i is varied by GTP concentration and by pertussis toxin treatment (CHAUDHRY et al. 1994; BEGIN-HEICK 1995). The coupling of purified β2AR receptors to G_i has been unequivocally demonstrated with purified, reconstituted receptors and

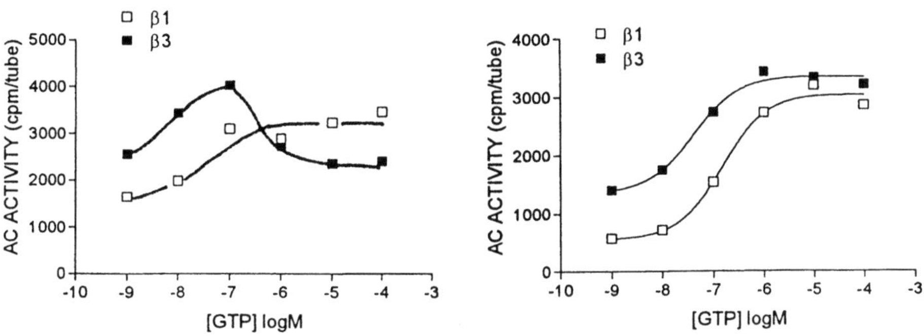

Fig. 2. Differential effects of GTP and pertussis toxin on the coupling of β1 and β3AR to cyclic AMP generation in white adipocyte membranes. *Left*: Stimulation of adenylyl cyclase activity (AC) via β3 AR, but not β1AR, is suppressed by GTP concentrations greater than 100 nmol/l. *Right*: GTP-dependent suppression of β3AR-stimulated AC is abolished in membranes from adipocytes treated with pertussis toxin indicating that GTP-induced suppression is mediated by Giα

G proteins (RUBENSTEIN et al. 1991), as well as in additional cellular systems (MARBACH et al. 1988; XIAO et al. 1995; ZHOU et al. 1997; DAAKA et al. 1997). The results of these experiments indicate that β3AR, but not β1AR, interact with G_i in adipocyte membranes (CHAUDHRY et al. 1994). The physiological significance of these results is presently unclear. It is likely that interactions with G_i involve cell-specific factors: the interaction appears to be present in 293 cells and adipocytes, but absent in CHO cells (PIETRI-ROUXEL et al. 1997; J.G. GRANNEMAN, unpublished). Heterotrimeric G_i proteins influence a variety of signaling pathways, including some that are independent of cAMP. Heterotrimeric G_i proteins are much more abundant than heterotrimeric G_s, and thus could be a significant source of βγ subunits following β3AR stimulation. βγ subunits are known to interact differentially with adenylyl cyclase isoforms (SUNAHARA et al. 1996) and it is possible that this plays a role in differences in the efficiency of the βAR subtypes to couple to cAMP generation. Additionally, G-protein βγ subunits activate several signaling pathways, including MAP kinase and PI-3 kinase (CRESPO et al. 1995; STOYANOV et al. 1995; LOPEZ-ILASACA et al. 1997). In this regard, β3AR have been found to activate protein kinase B in a cAMP-independent fashion (MOULE et al. 1997), and we have found that β3AR activate MAP kinase in transfected cells (J.G. GRANNEMAN, unpublished). These observations raise the possibility that, although fat cells contain both β1AR and β3AR, the intracellular signals generated by the subtypes may be qualitatively different.

E. Regulation of Receptor Activity and Expression

Given that β3AR agonists are currently being developed as therapeutics for obesity and diabetes, it is important to consider the regulation of this subtype by agonist exposure. It is well established that acute adrenergic stimulation of β1AR and β2AR leads to loss of responsiveness or desensitization (for review see LEFKOWITZ et al. 1998). Desensitization of the β2AR has been particularly well characterized and is thought to occur through the phosphorylation of multiple serine and threonine residues in the third cytoplasmic loop and carboxyl-terminal tail of the receptor by regulatory kinases which impairs the ability of the β2AR to couple to G_s and activate adenylyl cyclase. In contrast to β1AR and β2AR, the β3AR has fewer potential phosphorylation sites and thus would be expected to be resistant to agonist-induced desensitization (EMORINE et al. 1989; GRANNEMAN et al. 1991). Interestingly, although there is considerable species divergence in the primary amino of the carboxyl tail of the β3AR, none contain phosphorylation sites for known regulatory kinases. Several studies have shown that adipocyte β3AR are resistant to agonist-induced desensitization (GRANNEMAN 1992; CARPENE et al. 1993). The lack of β3AR desensitization after short-term agonist exposure has been confirmed in Chinese hamster fibroblasts (CHW) and murine Ltk^- (L cells) that have been transfected to express the recombinant human β3AR (NANTEL et al. 1993; LIGGETT et al. 1993).

Resistance to desensitization, however, may not be an intrinsic property of the β3AR. For example, it has been reported that β3AR responses are desensitized in rodent fundus and human SK-N-MC cells following agonist exposure (McLAUGLIN and MACDONALD 1991; ESBENSHADE et al. 1992). Recombinant rat and human β3AR undergo desensitization when expressed in 293 cells, but not when expressed in CHO cells. Furthermore, the magnitude of desensitization in 293 cells is similar for transfected β1AR and β3AR. These data indicate that cellular background strongly influences receptor desensitization, and suggests that alternative pathways, which may be independent of receptor phosphorylation, exist in these cells in which β3AR are susceptible to desensitization. The precise mechanism of uncoupling β3AR from adenylyl cyclase remains to be determined, but it does not appear to involve receptor down-regulation or sequestration (CHAUDHRY et al. 1995).

Most work examining agonist regulation of β3AR expression has been performed on rodent cells. This work demonstrates that the mRNAs encoding βAR subtypes are differentially regulated by agonist exposure. Thus, denervation of rat brown fat upregulates β3AR mRNA whereas treatment with exogenous βAR agonists down-regulates the message (GRANNEMAN and LAHNERS 1992; REVELLI et al. 1992). Under these conditions, the β1AR mRNA is not affected (GRANNEMAN and LAHNERS 1992). Isoproterenol treatment also down regulates β3AR mRNA in cultured adipocytes (GRANNEMAN and LAHNERS 1994; KLAUS et al. 1995). The down regulation produced by isoproterenol can be mimicked by 8-bromo-cyclic AMP, and appears to involve suppression of β3AR gene transcription (GRANNEMAN and LAHNERS 1994).

β3AR gene expression is highly sensitive to the nutritional status of the animal. β3AR gene expression is suppressed by insulin, and variations in β3AR expression and activity can be observed during the transition from fed to fasted state (HADRI et al. 1997) β3AR mRNA is dramatically down-regulated in leptin-deficient mice and rodents lacking the leptin receptor (BEGIN HEICK 1995; COLLINS et al. 1995; CHARON et al. 1995; ARBEENY et al. 1995). Exogenous leptin restores β3-AR expression in *ob/ob* mice, without affecting expression in lean animals (BRESLOW et al. 1997). These obese animals are rather unresponsive to acute stimulation by adrenergic agonists, and it is likely that decreased β3AR activity in adipose tissue contributes to the obese phenotype. Work in cultured adipocytes has shown that insulin and glucocorticoids dramatically suppress β3-AR gene transcription (FEVE et al. 1992, 1994), and it is likely that these hormones play a role in the massive suppression of β3AR expression seen in genetic obesity models (ONAI et al. 1995; GETTYS et al. 1997). Interestingly, chronic treatment of genetically obese rodents with β3AR agonists improves responsiveness to β3AR stimulation in part by *upregulation* of β3AR expression and activity (YOUNG et al. 1985; MUZZIN et al. 1989; COLLINS et al. 1997). These observations indicate that the genetic response to agonist stimulation is complex and regulation in normal and obese states can vary considerably.

Little is known about regulation of β3AR expression in man owing to the lack of appropriate cell models. In this regard dexamethasone, isoproterenol, and phorbol esters down-regulate the mouse β3AR mRNA in 3T3-F442A cells, but have no effect on human β3AR mRNA levels in SK-N-MC neuroepithelioma cells (GRANNEMAN and LAHNERS 1994). The reasons for this discrepancy are unclear. As noted above, there are numerous differences in the proximal promoters of the human and rodent genes (VANSPRONSEN et al. 1993; GRANNEMAN and LAHNERS 1994), suggesting that differences in cis-acting elements could be involved. On the other hand, β3AR are not expressed to any significant extent in human or rodent brain. Thus, expression in SK-N-MC appears to be ectopic, and the regulation in these cells could differ from sites of authentic expression. Recently, a transformed brown adipocyte cell line of human origin that express β3AR has been described (ZILBERFARB et al. 1997). These cells should be very useful in assessing the regulation of human β3AR in a physiologically-relevant cellular background. It will be particularly useful to learn whether cyclic AMP upregulates human β3AR, as it does in SK-N-MC cells, or down-regulates it, as occurs in rodent adipocytes (GRANNEMAN and LAHNERS 1994).

F. Pharmacology of β3AR-Selective Ligands

In general, two classes of β3AR-selective agonists have been developed. The first, developed by Beecham in the early 1980s, are unsubstituted and 3-chloro-phenethanolamines. The prototypic compound in this class is BRL 37344 and include CL 316243 (Fig. 3). BRL 37344 and CL 316243 were developed using rodent-based biological screens, and have moderate affinity and high efficacy at rodent β3AR (ARCH et al. 1984; BLOOM et al. 1992). BRL 37344

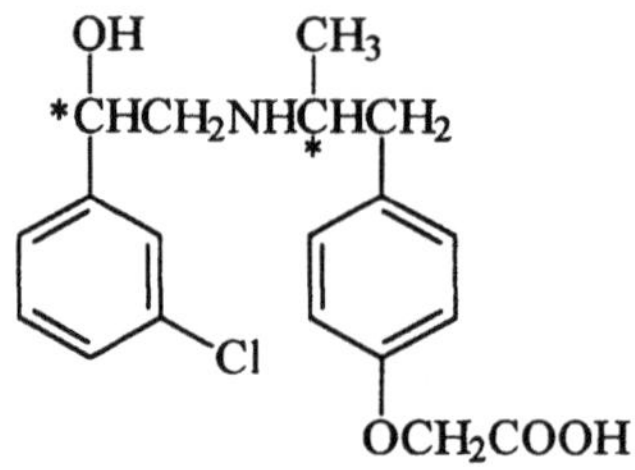

BRL 37344

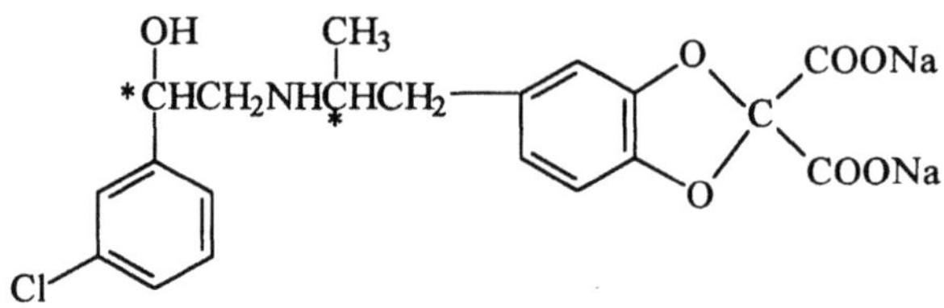

CL 316,243

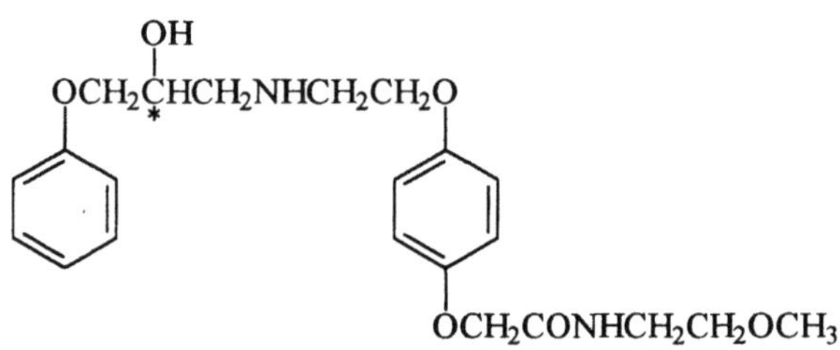

ICI D7114

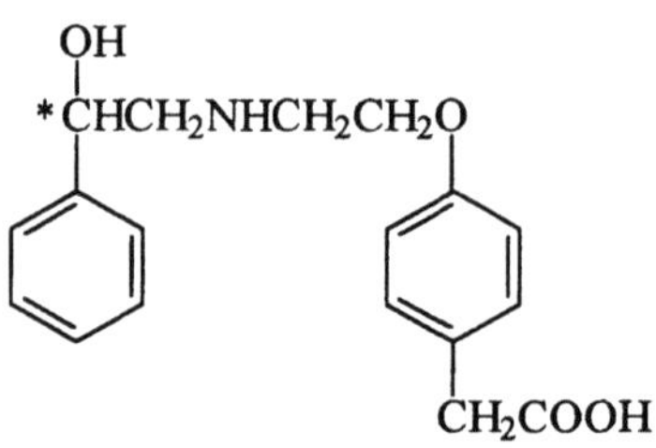

ZD 2079

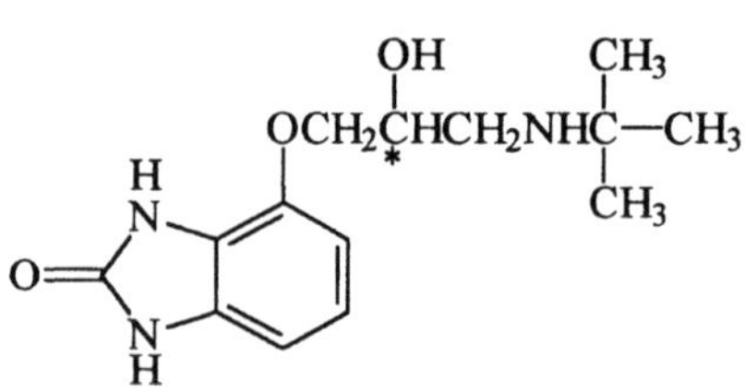

CGP 12177

* Asterisk indicates center of asymmetry

Fig. 3. Chemical structures of selective $\beta3$AR agonists discussed in the text

displays only moderate (sixfold) selectivity for β3AR over β2AR, whereas CL 316243 is highly selective for rodent β3AR and has been extremely useful in assessing β3AR function in various rodent models. Unfortunately, both BRL 37344 and CL 316243 have lower affinity and poor efficacy at the human β3AR (WILSON et al. 1996; ARCH and WILSON 1996).

A second class of β3AR agonists are arlyoxypropanolamines. The prototypic compound in this class is CGP 12177. Members of this class include certain antagonists of the β1AR and β2AR, like pindolol, as well as novel compounds that have been developed with the cloned human receptor. Much of the data regarding this class derives from experiments with CGP 12177, which has moderate affinity and efficacy at both human and rodent β3AR. The interaction of CGP 12177 with β3AR appears to differ from that of catecholamines and substituted phenethanolamines. First, the potency of CGP 12177 in stimulating cAMP accumulation is less than its ability to inhibit [125]ICYP binding or activate membrane adenylyl cyclase, whereas the opposite is true of catecholamine and phenethanolamine agonists. Second, certain mutations in the seventh transmembrane region of the β1AR and β3AR strongly affect the binding and activity of phenethanolamines like CL 316,243, yet have no effect on CGP 12177 action, indicating the site of interactions with the receptors differ (GRANNEMAN et al., in press). Recent work with β1AR strongly suggests that CGP12177 (and possibly other agents that are also antagonists of β1/2AR) interact with two sites on βAR and that the mode of that interaction differs from catecholamine and phenethanolamine agonists (see below).

G. Functional Effects of β3AR Agonists

I. Effects in Rodents

β3AR agonists are highly effective in reducing body weight in obese rodents without affecting food intake (ARCH et al. 1984; MEIER et al. 1984; YEN 1984; LANDE et al. 1993; YOSHIDA et al. 1994; SANTTI et al. 1994). The reduction in body weight results exclusively from the loss in body fat, with no reduction in carcass protein content (ARCH et al. 1989; CAWTHORNE et al. 1992). Since food intake is largely unaffected by β3AR agonists treatment, the loss in lipid results from dramatically increased thermogenesis induced by these agents. Virtually all of the increase in energy expenditure results from the oxidation of lipid substrates, as evidenced by the fact that methylpalmoxirate, an inhibitor of fatty acid oxidation, abolishes β3AR-induced thermogenesis (WILSON et al. 1986). The relative contribution of various tissues to the enhanced thermogenesis, however, is not entirely clear.

Many investigators have focused upon brown adipose tissue as a major site of β3AR action. BAT contains a high density of β3AR and β3AR agonists strongly activate thermogenesis in isolated brown adipocytes (ARCH et al. 1984; HOLLOWAY et al. 1992; ATGIE et al. 1997). Furthermore, chronic treatment with $\beta 3$AR agonists induces BAT hypertrophy and greatly increases its thermogenic capacity (HOLLOWAY et al. 1991; HIMMS-HAGEN et al. 1994; GHORBANI et al. 1997). In rodents and dogs chronic treatment with $\beta 3$AR agonists induces expression of BAT markers in depots that are usually considered WAT (CHAMPIGNY et al. 1991; HIMMS-HAGEN et al. 1994; GHORBANI et al. 1997). The ability to induce UCP1 expression in white fat depots appears to occur reliably in mice, although there appears to be large strain-specific variations (COLLINS et al. 1997). The induction of brown fat-specific markers probably represents the recruitment of existing brown adipocytes and not the proliferation from brown adipocyte precursors since CL 316243, unlike NE, does not increase DNA content of the tissue (HIMMS-HAGEN et al. 1994). The inability of CL 316243 to stimulate BAT proliferation probably results from the fact that $\beta 3$AR are expressed late in brown adipocyte differentiation and thus are unlikely to be present on brown adipocyte progenitor cells (KLAUS et al. 1995; EL HADRI et al. 1996).

Although brown fat is clearly involved in thermogenesis mediated by $\beta 3$AR agonists, it is far less certain whether brown fat is the sole, or even dominant, site of increased energy expenditure. Measurements of regional blood flow and oxygen content indicate that BAT provides only a small fraction of total thermogenesis induced by $\beta 3$AR agonists (THURLBY and ELLIS 1986; ARCH et al. 1989). Recent work with $\beta 3$AR gene knockout and transgenic expression has provided valuable insights into the sites of action of $\beta 3$AR agonists. Targeted disruption of the $\beta 3$AR gene completely abolishes the thermogenic effects of the highly-selective $\beta 3$AR agonist CL 316243 (SUSULIC et al. 1995). Transgenic re-expression of the $\beta 3$AR in brown adipose tissue (but not white adipose tissue) in the knockout mice restored only about 1/3 of the thermogenic response seen in wild type mice or in mice in which $\beta 3$AR expression was restored in both BAT and WAT (GRUJIC et al. 1997). These data strongly indicate that activation of white adipocyte $\beta 3$AR is required for the full thermogenic response. However, the role WAT plays in $\beta 3$AR-mediated thermogenesis is not clear. As stated above, the $\beta 3$AR-induced thermogenesis results from the oxidation of lipid, and it seems likely fatty acids that are mobilized from white fat are an important energy source. In addition, fatty acids are known to be effective uncouplers of oxidative phosphorylation (WOJTCZAK and SCHONFELD 1993), and the recent identification of uncoupling proteins 2 and 3 (FLEURY et al. 1997; GIMENO et al. 1997; BOSS et al. 1997; VIDAL-PUIG et al. 1997) suggests that tissues like skeletal muscle could be important sites of $\beta 3$AR-mediated thermogenesis. In this regard, $\beta 3$AR stimulation has been shown to increase expression of UCP3 in white adipose tissue (GONG et al. 1997), although the significance of this expression on agonist-induced thermogenesis is presently unclear. Finally, mounting evidence strongly indicates that white adipose tissue functions as an endocrine organ and it is possible that adipocyte-derived signals (fatty acids, leptin, etc.) play a role in coordinating the metabolic response to $\beta 3$AR stimulation.

The lipid-mobilizing and thermogenic effects of $\beta 3$AR agonists readily explain the antiobesity properties of these compounds. Interestingly, $\beta 3$AR agonists also enhance insulin sensitivity in several rodent models of type II diabetes. The insulin-sensitizing effects occur at doses that do not significantly affect body composition (ARCH et al. 1989; CAWTHORNE et al. 1992), suggesting that the antiobesity and antidiabetes effects might involve different mechanisms. One possibility is that chronic treatment with low doses of β3AR agonists reduce circulating levels of free fatty acids by augmenting their oxidation. The lowering of circulating free fatty acids and the possible increase in lipid storing capacity of white adipocytes could mediate the improvement in insulin sensitivity. In this regard, the effects of β3AR agonist treatment is reminiscent of the effects of cold exposure (SEYDOUX et al. 1982) and thiazolidinedione (PPARγ agonist) treatment (MERCER and TRAYHURN 1986; FUJIWARA et al. 1991). It is relevant to note that each of these treatments improves sensitivity much more in obese vs lean animals, and further suggests these treatments involve alterations in adipose tissue. In addition to effects on lipid metabolism, β3AR agonist treatment greatly increases glucose utilization in brown adipose tissue, without affecting use in skeletal muscle (YOUNG et al. 1985; DESOUSA et al. 1997). The relevance of adipocyte glucose utilization to the antiobesity/antidiabetes properties of β3AR agonists, however, is unclear (DANFORTH and HIMMS-HAGEN 1997).

II. Effects in Humans

In rodents, acute treatment with β3AR agonists increases metabolic rate by 40%–50%, an effect that increases somewhat during chronic dosing (THURLBY and ELLIS 1986; ARCH et al. 1989; GHORBANI et al. 1997). In contrast, the experience with first generation phenethanolamine β3AR agonists as agents that promote thermogenesis and insulin action in humans has been disappointing (ARCH and WILSON 1996). For example, while early work demonstrated that chronic treatment with BRL 26830 enhanced thermogenesis and increased weight loss in obese individuals subjected to a calorie-restricted diet (CONNACHER et al. 1988, 1992), it was unclear whether the thermogenic effects were due to actions at β2AR and/or β3AR (CONNACHER et al. 1990). Other compounds, such as ZD 2079 and ICI D7114, demonstrated no thermogenic activity when tested in man (GOLDBERG et al. 1995; TOUBRO and ASTRUP 1995). Several reasons, discussed below, have been cited for the relatively weak thermogenic effects of first generation β3AR agonists observed in humans, including species differences in the pharmacological sensitivity of the receptors, differences in drug metabolism, differences in the tissue distribution and abundance of the target receptors, and duration of treatment (ARCH and WILSON 1996). Thus, there remains considerable controversy regarding whether β3AR mediate thermogenic effects in man.

It is quite clear that many of the first generation β3AR-selective compounds that have failed in the clinic have low efficacy at the cloned human β3AR. To circumvent the problem of poor efficacy of selective β3AR agonists, WHEELDON et al. (1993) examined the metabolic effects of isoproterenol, a full β3-AR agonist, in the presence of β1- and β2AR blockade. These workers found that isoproterenol increased metabolic rate by 30%, half of which could be attributed to activation of β1AR. However, isoproterenol still increased metabolic rate by 9.5% in the presence of β1/β2AR blockade that was sufficient to abolish stimulation of heart rate and tremor. Similar results were obtained for BRL 35135, which increased metabolic rate by 8.9% in the presence of β1/β2AR blockade (WHEELDON et al. 1994). The results with BRL 35135 are somewhat surprising given that the efficacy of its active metabolite is very low compared to that of isoproterenol (ARCH and WILSON 1996). Lastly, LIU et al. (1995) reported that 40% of the thermogenic response to the indirect agonist ephedrine was resistant to β1/β2AR blockade. While these results suggest that β3AR mediate thermogenesis in man, it is unclear what tissues are thermogenically responsive to β3AR activation. The putative β3AR-mediated thermogenesis occurred in the absence of changes in plasma free fatty acids, plasma glucose or respiratory quotient. Thus, the metabolic profile differs considerably from that observed in rodents, and may indicate that the physiological mechanisms for the induced thermogenesis differ as well.

Perhaps more important than pharmacological responsiveness of the receptors are species differences in the density and distribution of β3AR. As mentioned above, activation of β3AR in white adipocytes appears to be required for 2/3 of the thermogenic response to β3AR agonist stimulation in mice (SUSLIC et al. 1997). However, there is very little evidence to support the existence of functional β3AR in human white adipocytes. Initial binding studies with [3H]CGP 12177 suggested that low affinity, high capacity binding sites in white fat cells could represent β3-AR (REVELLI et al. 1993). However, subsequent studies with the more selective [3H] SB206606 demonstrated that these sites were not β3AR, and that human white adipocytes did not contain detectable β3AR (DENG et al. 1996). These results are in full agreement with quantitative nuclease protection and Northern blot analysis which indicated the lack of β3AR mRNA in human white adipose tissue (GRANNEMAN et al. 1992; GRANNEMAN 1995; DENG et al. 1996).

Studies evaluating the actions of catecholamines and phenethanolamines on lipolysis also indicate that human white adipocytes contain few, if any, β3AR. It is well established that isoproterenol is a full agonist of the human β3AR (EMORINE et al. 1989; GRANNEMAN et al. 1993), yet the antagonism of isoproterenol-induced lipolysis by various blockers is consistent with an interaction with only β1/β2AR (BOUSQUET-MELOU et al. 1994; VAN LIEFDE et al. 1994; TAVERNIER et al. 1996). Additionally, the phenethanolamines BRL 37344 and CL 316,243 have little or no lipolytic activity in man, and when present the response is potently blocked by standard β-AR antagonists indicating weak interaction with β1/β2-AR (BOUSQUET-MELOU et al. 1994; ARCH and WILSON 1996; TAVERNIER et al. 1996; UMEKAWA et al. 1997). It is important to note that, although these compounds have relatively low efficacy at the human β3AR, their activities at the β1/β2AR are even lower (BLIN et al. 1994).

The strongest support for the existence of functional β3AR on human white fat cells comes from work with the aryloxypropanolamine CGP 12177, which was developed as a βAR antagonist and was subsequently shown to be a partial agonist of the human β3AR. The effects of CGP 12177 on adipocyte lipolysis have been mixed. In studies in where CGP 12177 was inactive (Bousquet-Melou et al. 1994; Van Liefde et al. 1994; Tavernier et al. 1996), the potency of isoproterenol was in the mid-nanomolar range (i.e., moderate to low receptor reserve). In the few studies that have shown partial agonism by CGP 12177, isoproterenol was 1000 times more potent (Lonnqvist et al. 1993). Subsequent work demonstrated that the efficacy of CGP 12177 is highly correlated with the potency of the non-selective agonist isoproterenol (Hoffstedt et al. 1995). Isoproterenol was potently blocked by propranolol (100 nmol/l shifted the isoproterenol curve by four orders of magnitude), whereas activation by CGP 12177 was more resistant. Together, the available data indicate that CGP 12177-mediated lipolysis is observable only in the presence of a large receptor reserve, consistent with a very low density of β3AR in adipose tissue. Indeed, if one assumes that CGP 12177 has an efficacy of 0.3–0.7 relative to isoproterenol (Granneman et al. 1993; Blin et al. 1994) and β1 and β3-AR couple equally to lipolysis, it can be estimated that β1/2AR are at least 100 times more abundant than β3-AR in white adipocytes.

The conclusion that CGP 12177-induced lipolysis is mediated via β3-AR, however, is not readily reconciled with the ability of propranolol to block completely isoproterenol-induced lipolysis with nanomolar potency. Additionally, BRL 37344 and CGP 12177 have similar equal efficacy at the cloned human β3AR (Granneman et al. 1993; Blin et al. 1994), yet BRL 37344 does not seem to induce lipolysis mediated via β3AR. It has been suggested that the superior numbers of β1/2AR interfere somehow with activation by catecholamines and phenethanolamines, but not with activation by CGP 12177 (Arch and Wilson 1996). Regardless, these observation suggest that CGP 12177 induces lipolysis via a mechanism that is distinct from isoproterenol and phenethanoloamine agonists.

As discussed above, the interaction of CGP 12177 with the β3AR differ from that of catecholamine and phenethanolamine agonists, and it is possible that the nature of the interaction affects the signaling pathways that are activated. For example, the ability of α2AR to activate or inhibit adenylyl cyclase depends upon the type of agonist occupying the receptor (Pepperl and Regan 1993; Eason et al. 1994). It is also possible that the actions of CGP 12177 in human adipocytes involve atypical interactions with the β1AR. Although commonly considered to be an antagonist, CGP 12177 was originally shown to have sympathomimetic effects (Staehelin et al. 1983). Recent work by Pak and Fishman (1996) has clearly shown that CGP 12177 activates β1AR. Indeed, we have found that, at similarly high levels of receptor expression, CGP 12177 activates β1 and β3AR to a similar degree. Interestingly, the interaction of CGP 12177 with the β1-AR is complex: at low concentrations it antagonizes isoproterenol-induced activation, while high concentrations reverse this antagonism (Fig. 4). This behavior is not predicted for simple partial agonism, and suggests that CGP 12177 acts at distinct antagonist and

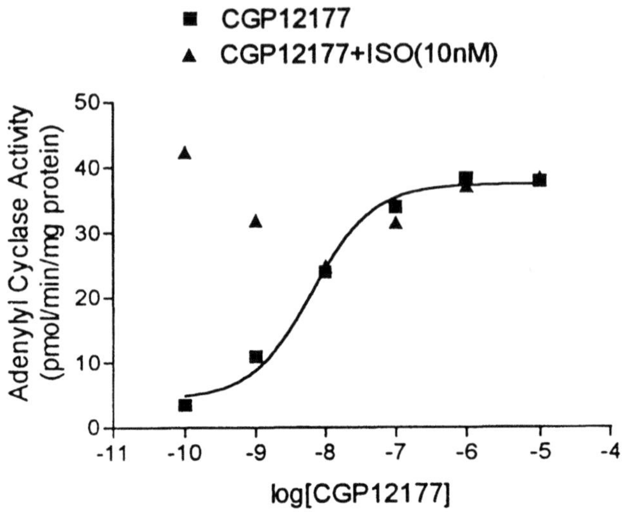

Fig. 4. Anomalous interaction of CGP 12177 with β1AR. CGP 12177 activates β1AR. In the presence of 10 nmol/l isoproterenol (ISO), CGP 12177 suppresses, then stimulates activity – behavior not predicted for simple partial agonism

agonist sites on the β1-AR. Significantly, propranolol weakly antagonizes activation of β1-AR by CGP 12177 (KONKAR and GRANNEMAN, in press). The basis of this phenomenon is under investigation, and could relate to the ability of this compound to assume folded (antagonist) and extended (agonist) conformations (BLIN et al. 1994). In any event, it is likely that β1AR contributes significantly to the effects of CGP 12177 in cells exhibiting large receptor reserve.

In contrast to white adipose tissue, human brown fat contains substantial β3AR mRNA and binding sites detected by 3H SB 206606 (BENSAID et al. 1993; GRANNEMAN et al. 1995; DENG et al. 1996). Indeed, transgenic expression of the human β3AR gene in mice results in expression in brown fat, but not in white adipose tissue (LOWELL and FLIER 1996). Nevertheless, reverse transcription/polymerase chain reaction (RT/PCR) analysis of human white adipose tissue depots has identified low levels of β3AR mRNA (KRIEF et al. 1993; GRANNEMAN et al. 1993). In several cases, the ability to amplify β3AR mRNA is correlated with the presence of UCP1 mRNA, a specific marker of brown adipocytes (KRIEF et al. 1993). These data indicate that, like rodents, white adipocyte depots contain a small population of brown adipocytes.

In rodents and dogs, chronic treatment with β3AR agonists elevates the expression of UCP1 in white adipose tissue depots, suggesting that these agents can recruit or reactivate dormant brown adipocytes contained within white fat depots (CHAMPIGNY et al. 1991; HIMMS-HAGEN et al. 1994; COLLINS et al. 1997). It is presently unknown whether this process results from the reactivation of dormant mature brown adipocytes, the differentiation of brown adipocyte progenitor cells, or the conversion of white adipocytes into brown. In any event, a key question for assessing the therapeutic potential of $\beta 3$AR agonists is whether a similar process can occur in humans. Recently, CHAMPIGNY and RICQUIER (1996) examined the ability of first generation $\beta 3$AR agonists to induce expression of UCP1 in human perirenal adipocytes that were differentiated in culture. These investigators found that phenethanolamine and aryloxypropanolamine agonists induced UCP1 expression. Although not rigorously tested, it is likely that these agents induced UCP1 expression via $\beta 3$AR based upon the concentrations that were effective. Whether such effects will be observed in vivo is unknown. It is clear that brown adipocytes can be recruited in adults by adrenergic stimulation (e.g., in patients with adrenal chromaffin tumors, GARUTTI and RICQUIER 1992). However, it is uncertain whether the pool of brown adipocytes containing $\beta 3$AR is sufficiently large to generate a meaningful biological response (ASTRUP et al. 1985). Given that $\beta 3$AR appear late in differentiation and that the highly-selective $\beta 3$AR agonist CL 316,243 does not promote cell hyperplasia in rodents (FEVE et al. 1992; HIMMS-HAGEN et al. 1994), it is doubtful that treatment with $\beta 3$AR agonists alone will expand the population of brown adipocytes. On the other hand, peroxisome proliferator activated receptor gamma (PPARγ) have been shown to promote the differentiation of brown adipocytes in vitro and increase the mass of brown fat in rodents (ROTHWELL et al. 1987; TAI et al. 1996). Recently, the PPARγ agonist BRL 49653 was recently found to promote UCP1 expression in human preadipocytes cultured from various "white" adipocyte depots, raising the exciting possibility that PPARγ agonists might promote $\beta 3$AR responsiveness in man (LENHARD et al. 1997).

H. Summary and Prospects of $\beta 3$AR as a Therapeutic Target in Man

Much progress has been made evaluating the regulation and functional properties of $\beta 3$AR. The relatively low affinity of this receptor for catecholamines and its resistance to desensitization indicated the $\beta 3$AR is likely to play an important role in signaling during periods of intense, sustained adrenergic stimulation as occurs in brown adipose tissue of many species, including man. Additionally, $\beta 3$AR are likely couple to pathways which are distinct from those activated by other βAR subtypes. It is thus likely that $\beta 3$AR serve unique signaling functions in adipocyte. Furthermore, expression of $\beta 3$AR is highly sensitive to the endocrine and nutritional status of animals, indicating that the receptor plays an important role in the control of energy balance. Together, these observations indicate that $\beta 3$AR occupy a unique signaling niche among the βAR subtypes.

While the efficacy of β3AR agonists in rodent models of obesity and NIDDM is truly impressive, it is still uncertain whether the β3AR will be a useful target for treatment of human obesity. Several factors have delayed unequivocal evaluation of the therapeutic potential of β3AR agonists. Most widely considered have been the significant issues of poor selectivity, efficacy, and/or bioavailability of first generation β3AR agonists (ARCH and WILSON 1996). It can be anticipated, however, that these obstacles can be overcome in part by employing the appropriate human-based screens. Perhaps more significant is whether humans express enough β3AR in the relevant tissues to generate a significant therapeutic effect. It is quite clear that the thermogenic effects of β3AR agonists in rodents require expression of β3AR in brown and white fat. The virtual absence of β3AR in human white fat and the relatively low levels of brown fat in adult humans indicate that the therapeutic effect of β3AR agonists in man will be either substantially less than that seen in rodents, or will be achieved by different mechanisms. Full evaluation of β3AR agonist may require chronic dosing, since repeated treatment in rodents increases thermogenic responsiveness and recruits brown adipocytes to an activated state. Very recent studies indicating that thiazolidindiones increase UCP 1 expression in cultured perirenal adipocytes raise the possibility that these agents might promote β3AR responsiveness in man.

References

Arbeeny CM, Meyers DS, Hillyer DE, Bergquist KE (1995) Metabolic alterations associated with the antidiabetic effect of beta 3-adrenergic receptor agonists in obese mice. Am J Physiol 268(4 Pt 1):E678–E684

Arch JR, Ainsworth AT, Cawthorne MA, Piercy V, Sennitt MV, Thody VE, Wilson C, Wilson S (1984) Atypical beta-adrenoceptor on brown adipocytes as target for anti-obesity drugs. Nature 309(5964):163–165

Arch JRS, Bywater RJ, Coney KA, Ellis RDM, Thurlby PL, Smith, SA, Zed C (1989) Influences on Body Composition and Mechanism of Action of the β-Adrenoceptor Agonist BRL26830A. In: Lardy H, Stratman F (eds) Hormones, Thermogenesis, and Obesity. Elsevier Science Publishing Co., London, pp 465–475

Arch JRS, Wilson S (1996) Prospects for β_3-adrenoceptor agonists in the treatment of obesity and diabetes. Int J Obes 20:191–199

Astrup A, Bulow J, Madsen J, Christensen NJ (1985) Contribution of BAT and skeletal muscle to thermogenesis induced by ephedrine in man. Am J Physiol 248(5 Pt 1):E507–E515

Atgie C, D'Allaire F, Bukowiecki LJ (1997) Role of beta 1- and beta 3-adrenoceptors in the regulation of lipolysis and thermogenesis in rat brown adipocytes. Am J Physiol 273(4 Pt 1):C1136–C1142

Bahouth SW, Malbon CC (1988) Subclassification of beta-adrenergic receptors of rat fat cells: a re-evaluation. Mol Pharmacol 34(3):318–326

Begin-Heick N (1995) Beta 3-adrenergic activation of adenylyl cyclase in mouse white adipocytes: modulation by GTP and effect of obesity. J Cell Biochem 58(4):464–473

Blaak EE, van Baak MA, Kempen KP, Saris WH (1993) Role of alpha- and beta-adrenoceptors in sympathetically mediated thermogenesis. Am J Physiol 264(1 Pt 1): E11–E17

Blin N, Camoin L, Maigret B, Donny Strosberg A (1993) Structural and conformational features determining selective signal transduction in the $\beta 3$-adrenergic receptor. Mol Pharmacol 44:1094–1104

Bloom JD, Dutia MD, Johnson BD, Wissner A, Burns MG, Largis EE, Dolan JA, Claus TH (1992) Disodium (R,R)-5-[2-[[2-(3-chlorophenyl)-2-hydroxyethyl]-amino]propyl]-1,3-benzodioxole-2,2-dicarboxylate (CL 316, 243). A potent beta-adrenergic agonist virtually specific for beta 3 receptors. A promising antidiabetic and antiobesity agent. J Med Chem 35(16):3081–3084

Boss O, Samec S, Paoloni-Giacobino A, Rossier C, Dulloo A, Seydoux J, Muzzin P, Giacobino J-P (1997) Uncoupling protein-3: a new member of the mitochondrial carrier family with tissue-specific expression. FEBS Lett 408:39–42

Breslow MJ, An Y, Berkowitz DE (1997) Beta-3 adrenoceptor (beta-3AR) expression in leptin treated OB/OB mice. Life Sci 61(1):59–64

Brown JA, Machida CA (1994) The 5' flanking region of the rat beta 3-adrenergic receptor gene: divergence with the human gene and implications for species-specific gene expression. DNA Seq 4(5):319–324

Candelore MR, Deng L, Tota LM, Kelly LJ, Cascieri MA, Strader CD (1996) Pharmacological characterization of a recently described human $\beta 3$-adrenergic receptor mutant. Endocrinology 137(6):2638–2641

Carpene C, Galitzky J, Collon P, Esclapez F, Dauzats M, Lafontan M (1993) Desensitization of beta-1 and beta-2, but not beta-3, adrenoceptor-mediated lipolytic responses of adipocytes after long-term norepinephrine infusion. J Pharmacol Exp Ther 265(1):237–247

Casteilla L, Muzzin P, Revelli JP, Ricquier D, Giacobino JP (1994) Expression of beta 1- and beta 3-adrenergic-receptor messages and adenylate cyclase beta-adrenergic response in bovine perirenal adipose tissue during its transformation from brown into white fat. Biochem J 297(Pt 1):93–97

Cawthorne MA, Sennitt MV, Arch JRS, Smith SA (1992) BRL 35135, a potent and selective atypical β-adrenoceptor agonist. Am J Clin Nutr 55:252S–257S

Champigny O, Ricquier D (1996) Evidence from in vitro differentiating cells that adrenoceptor agonists can increase uncoupling protein mRNA level in adipocytes of adult humans: an RT-PCR study. J Lipid Res 37:1907–1914

Champigny O, Ricquier D, Blondel O, Mayers RM, Briscoe MG, Holloway BR (1991) Beta 3-adrenergic receptor stimulation restores message and expression of brown-fat mitochondrial uncoupling protein in adult dogs. Proc Natl Acad Sci USA 88(23):10774–10777

Chapman BJ, Farquahar DL, Galloway SM, Simpson GK, Munro JF (1988) The effects of a new beta-adrenoceptor agonist BRL 26830A in refractory obesity. Int J Obes 12(2):119–123

Charon C, Krief S, Diot-Dupuy F, Strosberg AD, Emorine LJ, Bazin R (1995) Early alterations in the brown adipose tissue adenylate cyclase system of pre-obese Zucker rat fa/fa pups: decreased G-proteins and beta 3-adrenoceptor activities. Biochem J 312(Pt 3):781–788

Cimmino M, Bukowiecki LJ, Geloen A (1997) In situ lipolysis measured by in vivo microdialysis during acute cold exposure. Can J Physiol Pharmacol 75(9):1112–1115

Clémente K, Vaisse C, Manning BS, Basdevant A, Guy-Grand B, Ruiz J, Silver KD, Shuldiner AR, Froguel P, Donny Strosberg A (1995) Genetic variation in the β_3-adrenergic receptor and an increased capacity to gain weight in patients with morbid obesity. N Engl J Med 333(6):351–354

Cohen ML, Granneman JG, Chaudhry A, Schenck KW, Cushing DJ, Palkowitz AD (1995) Is the "atypical" beta-receptor in the rat stomach fundus the rat beta 3 receptor?. J Pharmacol Exp Ther 272(1):446–451

Collins S, Daniel KW, Petro AE, Surwit RS (1997) Strain-specific response to β_3-adrenergic receptor agonist treatment of diet-induced obesity in mice. Endocrinology 138(1):405–413

Collins S, Daniel KW, Rohlfs EM, Ramkumar V, Taylor IL, Gettys TW (1994) Impaired expression and functional activity of the beta 3- and beta 1-adrenergic receptors in adipose tissue of congenitally obese (C57BL/6J ob/ob) mice. Mol Endocrinol 8(4):518–527

Connacher AA, Bennet WM, Jung RT (1992) Clinical studies with the beta-adrenoceptor agonist BRL 26830A. Am J Clin Nutr 55(1 Suppl):258S–261S

Connacher AA, Bennet WM, Jung RT, Rennie MJ (1992) Metabolic effects of three weeks administration of the beta-adrenoceptor agonist BRL 26830A. Int J Obes Relat Metab Disord 16(9):685–694

Connacher AA, Jung RT, Mitchell PEG (1988) Weight loss in obese subjects on a restricted diet given BRL 26830A, a new atypical β adrenoceptor agonist. Brit Med J 296:1217–1220

Connacher AA, Lakie M, Powers N, Elton RA, Walsh EG, Jung RT (1990) Tremor and the anti-obesity drug BRL 26830A. Br J Clin Pharmacol 30(4):613–615

Crespo P, Cachero TG, Xu N, Gutkind JS (1995) Dual effect of β-adrenergic receptors on mitogen-activated protein kinase: Evidence for a $\beta\gamma$-Dependent activation and a $G\alpha_s$-cAMP-mediated inhibition. J Biol Chem 270(42):25259–25265

Daaka Y, Luttrell LM, Lefkowitz RJ (1997) Switching of the coupling of the beta 2-adrenergic receptor to different G proteins by protein kinase A. Nature 390(6655):88–91

Danforth Jr E, Himms-Hagen J (1997) Obesity and diabetes and the beta-3 adrenergic receptor. Eur J Endocrinol 136:362–365

Deng C, Moinat M, Curtis L, Nadakal A, Preitner F, Boss O, Assimacopoulos-Jeannet F, Seydoux J, Giacobino J-P (1997) Effects of β-Adrenoceptor subtype stimulation on *obese* gene messenger ribonucleic acid and on leptin secretion in mouse brown adipocytes differentiated in culture. Endocrinology 138(2):548–552

Deng C, Paoloni-Giacobino A, Kuehne F, Boss O, Revelli JP, Moinat M, Cawthorne MA, Muzzin P, Giacobino JP (1996) Respective degree of expression of beta 1-, beta 2-, beta 3-adrenoceptors in human brown and white adipose tissues. Brit J Pharmacol 118(4):929–934

de Souza CJ, Hirshman MF, Horton ES (1997) CL-316,243, a beta 3-specific adrenoceptor agonist, enhances insulin-stimulated glucose disposal in nonobese rats. Diabetes 48(8):1257–1263

Dolan JA, Muenkel HA, Burns MG, Pellegrino SM, Fraser CM, Pietri F, Donny Strosberg A, Largis EE, Dutia MD, Bloom JD, Bass AS, Tanikella TK, Cobuzzi A, Lai FM, Claus TH (1994) Beta 3- adrenoceptor selectivity of the dioxolane dicarboxylate phenethanolamines. J Pharmacol Exp Ther 269(3):1000–1006

Donny Strosberg A (1997) Structure and function of the β_3-adrenergic receptor. Annu Rev Pharmacol Toxicol 37:421–450

Eason MG, Jacinto MT, Liggett SB (1994) Contribution of ligand structure to activation of alpha 2-adrenergic receptor subtype coupling to Gs. Mol Pharmacol 45(4):696–702

Elbein SC, Hoffman M, Barrett K, Wegner K, Miles C, Bachman K, Berkowitz D, Shuldiner AR, Leppert MF, Hasstedt S (1996) Role of the beta 3-adrenergic receptor locus in obesity and noninsulin-dependent diabetes among members of Caucasian families with a diabetic sibling pair. J Clin End Met 81(12):4422–4427

Emorine LJ, Marullo S, Briend-Sutren M-M, Patey G, Tate K, Delavier-Klutchko C, Donny Strosberg A (1989) Molecular characterization of the human β_3-adrenergic receptor. Science 245:1118–1120.

Enocksson S, Shimizu M, Lonnqvist F, Nordenstrom J, Arner P (1995) Demonstration of an in vivo functional beta 3-adrenoceptor in man. J Clin Invest 95(5):2239–2245

Feve B, Baude B, Krief S, Strosberg AD, Pairault J, Emorine LJ (1992) Inhibition by dexamethasone of beta 3-adrenergic receptor responsiveness in 3T3-F442A adipocytes. Evidence for a transcriptional mechanism. J Biol Chem 267(22):15909–15915

Feve B, Elhadri K, Quignard-Boulange A, Pairault J (1994) Transcriptional down-regulation by insulin of the beta 3-adrenergic receptor expression in 3T3-F442A adipocytes: a mechanism for repressing the cAMP signaling pathway. Proc Natl Acad Sci USA 91(12):5677–5681

Fleury C, Neverova M, Collins S, Raimbault S, Champigny O, Levi-Meyrueis C, Bouillaud F, Seldin MF, Surwit RS, Ricquier D, Warden CH (1997) Uncoupling protein-2: a novel gene linked to obesity and hyperinsulinemia. Nat Genetics 15:269–272

Fujisawa T, Ikegami H, Yamato E, Takekawa K, Nakagawa Y, Hamada Y, Oga T, Ueda H, Shintani M, Fukuda M, Ogihara T (1996) Association of Trp64 Arg mutation of the beta3-adrenergic-receptor with NIDDM and body weight gain. Diabetologia 39(3):349–352

Fujiwara T, Wada M, Fukuda K, Fukami M, Yoshioka S, Yoshioka T, Horikoshi H (1991) Characterization of CS-045, a new oral antidiabetic agent, II. Effects on glycemic control and pancreatic islet structure at a late stage of the diabetic syndrome in C57BL/KsJ-db/db mice. Metabolism 40(11):1213–1218

Gettys TW, Watson PM, Seger L, Padgett M, Taylor IL (1997) Adrenalectomy after weaning restores beta 3-adrenergic receptor expression in white adipocytes from C57BL/6J-ob/ob mice. Endocrinology 138(7):2697–2704

Garruti G, Ricquier D (1992) Analysis of uncoupling protein and its mRNA in adipose tissue deposits of adult humans. Int J Obes Relat Metab Disord 16(5):383–390

Ghorbani M, Claus TH, Himms-Hagen J (1997) Hypertrophy of brown adipocytes in brown and white adipose tissues and reversal of diet-induced obesity in rats treated with a beta 3-adrenoceptor agonist. Biochem Pharmacol 54(1):121–131

Gimeno RF, Dembski M, Weng X, Deng N, Shyjan AW, Gimeno CJ, Iris F, Ellis SJ, Woolf EA, Tartaglia LA (1997) Cloning and characterization of an uncoupling protein homolog: A potential molecular mediator of human thermogenesis. Diabetes 46:900–906

Goldberg GR, Prentice AM, Murgatroyd PR, Haines W, Tuersley MD (1995) Effects on metabolic rate and fuel selection of a selective β-3 agonist (ICI D7114) in healthy lean men. Int J Obes 19:625–631

Gong DW, He Y, Karas M, Reitman M (1997) Uncoupling protein-3 is a mediator of thermogenesis regulated by thyroid hormone, beta 3-adrenergic agonists, and leptin. J Biol Chem 272(39):24129–24132

Granneman JG (1995) Why do adipocytes make the β_3 adrenergic receptor?. Cell Signal 7(1):9–15

Granneman JG, Lahners KN (1994) Analysis of human and rodent β_3-adrenergic receptor messenger ribonucleic acids. Endocrinol 135(3):1025–1031

Granneman JG, Lahners KN, Chaudhry A (1993) Characterization of the human β_3-adrenergic receptor gene. Mol Pharmacol 44:264–270

Granneman JG, Lahners KN, Chaudhry A (1991) Molecular cloning and expression of the rat β_3-adrenergic receptor. Mol Pharmacol 40:895–899

Granneman JG, Lahners KN, Rao DD (1992) Rodent and human β_3-adrenergic receptor genes contain an intron within the protein-coding block. Mol Pharmacol 42:964–970

Granneman JG, Lahners KN, Zhai Y (1998) Agonist interactions with chimeric and mutant β1- and β3-adrenergic receptors: Involvement of the seventh transmembrane region in conferring subtype specificity. Mol Pharmacol, in press

Grujic D, Susulic VS, Harper ME, Himms-Hagen J, Cunningham BA, Corkey BE, Lowell BB (1997) Beta 3-adrenergic receptors on white and brown adipocytes mediate beta 3-selective agonist-induced effects on energy expenditure, insulin secretion, and food intake. A study using transgenic and gene knockout mice. J Biol Chem 272(28):17686–17693

Hadri KE, Charon C, Pairault J, Hauguel-De Mouzon S, Quignard-Boulange A, Feve B (1997) Down-regulation of beta 3-adrenergic receptor expression in rat adipose

tissue during the fasted/fed transition: evidence for a role of insulin. Biochem J 323(Pt 2):359–364
Hadri KE, Courtalon A, Gauthereau X, Chambaut-Guerin AM, Pairault J, Feve B (1997) Differential regulation by tumor necrosis factor-alpha of beta 1-, beta 2-, and beta 3-adrenoreceptor gene expression in 3T3-F442A adipocytes. J Biol Chem 272(39):24514–24521
Hadri KE, Feve B, Pairault J (1996) Developmental expression and functional activity of beta 1- and beta 3-adrenoceptors in murine 3T3-F442A differentiating adipocytes. Eur J Pharmacol 297(1–2):107–119
Harms HH, Zaagsma J, Van der Wal B (1974) Beta-adrenoceptor studies. III. On the beta-adrenoceptors in rat adipose tissue. Eur J Pharmacol 25(1):87–91
Higashi K, Ishikawa T, Ito T, Yonemura A, Shige H, Nakamura H (1997) Association of a genetic variation in the beta 3-adrenergic receptor gene with coronary heart disease among Japanese. Biochem Biophys Res Comm 232(3):728–730
Himms-Hagen J, Cui J, Danforth Jr E, Taatjes DJ, Lang SS, Waters BL, Claus TH (1994) Effect of CL-316, 243, a thermogenic beta 3-agonist, on energy balance and brown and white adipose tissues in rats. Am J Physiol 266(4 Pt 2):R1371–R1382
Hoffstedt J, Shimizu M, Sjostedt S, Lonnqvist F (1995) Determination of beta 3-adrenoceptor mediated lipolysis in human fat cells. Obes Res 3:447–459
Holloway BR, Howe R, Rao BS, Stribling D (1992) ICI D7114: a novel selective adrenoceptor agonist of brown fat and thermogenesis. Am J Clin Nutr 55(Suppl 1):262S–264S
Holloway BR, Howe R, Rao BS, Stribling D, Mayers RM, Briscoe MG, Jackson JM (1991) ICI D7114 a novel selective beta-adrenoceptor agonist selectively stimulates brown fat and increases whole-body oxygen consumption. Br J Pharmacol 104(1):97–104
Kelly Moule S, Welsh GI, Edgell NJ, Foulstone EJ, Proud CG, Denton RM (1997) Regulation of protein kinase B and glycogen synthase kinase-3 by insulin and β-adrenergic agonists in rat epididymal fat cells: activation of protein kinase B by wortmannin-sensitive and -insensitive mechanisms. J Biol Chem 272(12):7713–7719
Klaus S, Muzzin P, Revelli JP, Cawthorne MA, Giacobino JP, Ricquier D (1995) Control of beta 3-adrenergic receptor gene expression in brown adipocytes in culture. Mol Cell Endocrinol 109(2):189–195
Krief S, Lonnqvist F, Raimbault S, Baude B, Van Spronsen A, Arner P, Strosberg AD, Ricquier D, Emorine LJ (1993) Tissue distribution of beta 3-adrenergic receptor mRNA in man. J Clin Invest 91(1):344–349
Kurabayashi T, Carey DGP, Morrison NA (1996) The β3-adrenergic receptor gene Trp64Arg Mutation is overrepresented in obese women: Effects on weight, BMI, abdominal fat, blood pressure, and reproductive history in an elderly Australian population. Diabetes 45:1358–1363
Landi M, Croci T, Manara L (1993) Similar atypical beta-adrenergic receptors mediate in vitro rat adipocyte lipolysis and colonic motility inhibition. Life Sci 53(18): PL297–PL302
Langin D, Portillo MP, Saulnier-Blache JS, Lafontan M (1991) Coexistence of three beta-adrenoceptor subtypes in white fat cells of various mammalian species. Eur J Pharmacol 199(3):291–301
Lefkowitz RJ, Pitcher J, Krueger K, Daaka Y (1998) Mechanisms of beta-adrenergic receptor desensitization and resensitization. Adv Pharmacol 42:416–420
Lelias JM, Kaghad M, Rodriguez M, Chalon P, Bonnin J, Dupre I, Delpech B, Bensaid M, LeFur G, Ferrara P, Caput D (1993) Molecular cloning of a human β3-adrenergic receptor cDNA. FEBS Lett 324(2):127–130
Lenhard JM, Klliewer SA, Paulik MA, Plunket KD, Lehmann JM, Weiel JE (1997) Effect of troglitazone and metformin on glucose and lipid metabolism: alterations of two distinct molecular pathways. Biochem Pharmacol 54:801–808

Liu YL, Toubro S, Astrup A, Stock MJ (1995) Contribution of beta 3-adrenoceptor activation to ephedrine-induced thermogenesis in humans. Int J Obes Relat Metab Disord 19(9):678–685
Lonnqvist F, Krief S, Strosberg AD, Nyberg S, Emorine LJ, Arner P (1993) Evidence for a functional beta 3-adrenoceptor in man. Br J Pharmacol 110(3):929–936
Lopez-Ilasaca M, Crespo P, Pellici PG, Gutkind JS, Wetzker R (1997) Linkage of G protein-coupled receptors to the MAPK signaling pathway through PI 3-kinase gamma. Science 275(5298):394–397
Lowell BB, Flier JS (1997) Brown Adipose Tissue, ß3-Adrenergic Receptors, and Obesity. Annu Rev Med 48:307–316
Marbach I, Shiloach J, Levitzki A (1988) Gi affects the agonist-binding properties of beta-adrenoceptors in the presence of Gs. Eur J Biochem 172(1):239–246
McLaughlin DP, MacDonald A (1991) Characterization of catecholamine-mediated relaxations in rat isolated gastric fundus: evidence for an atypical beta-adrenoceptor. Br J Pharmacol 103(2):1351–1356
Meier MK, Alig L, Burgi-Saville ME, Muller M (1984) Phenethanolamine derivatives with calorigenic and antidiabetic qualities. Int J Obes 8[Suppl 1]:215–225
Mercer SW, Trayhurn P (1986) Effects of ciglitazone on insulin resistance and thermogenic responsiveness to acute cold in brown adipose tissue of genetically obese (ob/ob) mice. FEBS Lett 195(1–2):12–16
Mitchell BD, Blangero J, Comuzzie AG, Almasy LA, Shuldiner AR, Silver K, Stern MP, MacCluer JW, Hixon JE (1998) A paired sibling analysis of the beta-3 adrenergic receptor and obesity in Mexican Americans. J Clin Invest 101:584–587
Mitchell TH, Ellis RD, Smith SA, Robb G, Cawthorne MA (1989) Effects of BRL 35135, a beta-adrenoceptor agonist with novel selectivity, on glucose tolerance and insulin sensitivity in obese subjects. Int J Obes 13(6):757–766
Mohell N, et al (1989) The beta-adrenergic radioligand [3H]CGP-12177, generally classified as an antagonist, is a thermogenic agonist in brown adipose tissue. Biochem J 261(2):401–405
Muzzin P, Revelli JP, Ricquier D, Meier MK, Assimacopoulos-Jeannet F, Giacobino JP (1989) The novel thermogenic beta-adrenergic agonist Ro 16–8714 increases the interscapular brown-fat beta-receptor-adenylate cyclase and the uncoupling-protein mRNA level in obese (fa/fa) Zucker rats. Biochem J 261(3):721–724
Nagase I, Aoki A, Yamamoto M, Yasuda H, Kado S, Nishikawa M, Kugai N, Akatsu T, Nagata N (1997) Lack of association between the Trp64 Arg mutation in the beta 3-adrenergic receptor gene and obesity in Japanese men: a longitudinal analysis. J Clin Endocrin Metabol 82(4):1284–1287.
Nagase I, Yoshida T, Kumamoto K, Umekawa T, Sakane N, Nikami H, Kawada T, Saito M (1996) Expression of uncoupling protein in skeletal muscle and white fat of obese mice treated with thermogenic ß3-adrenergic agonist. J Clin Invest 97(12):2898–2904
Nahmias C, Blin N, Elalouf J-M, Mattei MG, Strosberg AD, and Emorine LJ (1991) Molecular characterization of the mouse ß3-adrenergic receptor: relationship with the atypical receptor of adipocytes. EMBO Journal 10(12):3721–3727
Nougues J, Reyne Y, Champigny O, Holloway B, Casteilla L, Ricquier D (1993) The beta 3-adrenoceptor agonist ICI-D7114 is not as efficient on reinduction of uncoupling protein mRNA in sheep as it is in dogs and smaller species. J Anim Sci 71(9):2388–2394
Onai T, Kilroy G, York DA, Bray GA (1995) Regulation of beta 3-adrenergic receptor mRNA by sympathetic nerves and glucocorticoids in BAT of Zucker obese rats. Am J Physiol 269(3 Pt 2):R519–R526
Pak MD, Fishman PH (1996) Anomalous behavior of CGP 12177A on beta 1-adrenergic receptors. J Recept Signal Transduct Res 16(1–2):1–23
Pepperl DJ, Regan JW (1993) Selective coupling of alpha 2-adrenergic receptor subtypes to cyclic AMP-dependent reporter gene expression in transiently transfected JEG-3 cells. Mol Pharmacol 44(4):802–809

Piétri-Rouxel F, Lenzen G, Kapoor A, Drumare M-F, Archimbault P, Donny Strosberg A, Manning BS (1995) Molecular cloning and pharmacological characterization of the bovine ß3-adrenergic receptor. Eur J Biochem 230:350–358
Piétri-Rouxel F, Manning BS, Gros J, Donny Strosberg A (1997) The biochemical effect of the naturally occurring Trp64·Arg mutation on human β3-adrenoceptor activity. Eur J Biochem 247:1174–1179
Revelli JP, Muzzin P, Giacabino J-P (1992) Modulation in vivo of β-adrenergic receptor subtypes in rat brown adipose tissue by the thermogenic agonist RO 16-8714. Biochem J 286:743–746
Rothwell NJ, Stock MJ, Tedstone AE (1987) Effects of ciglitazone on energy balance, thermogenesis and brown fat activity in the rat. Mol Cell Endocrinol 51:253–257
Rubenstein RC, Linder ME, Ross EM (1991) Selectivity of the beta-adrenergic receptor among Gs, Gi's, and Go: assay using recombinant alpha subunits in reconstituted phospholipid vesicles. Biochemistry 30(44):10769–10777
Santti E, Rouvari T, Rouru J, Huupponen R, Koulu M (1994) Effect of chronic treatment with ICI D7114, a selective beta 3-adrenoceptor agonist, on macronutrient selection and brown adipose tissue thermogenesis in Sprague-Dawley rats. Pharmacol Toxicol 75(3–4):166–169
Seydoux J, Assimacopoulos-Jeannet F, Jeanrenaud B, Girardier L (1982) Alterations of brown adipose tissue in genetically obese (ob/ob) mice. I. Demonstration of loss of metabolic response to nerve stimulation and catecholamines and its partial recovery after fasting or cold adaptation. Endocrinology 110(2):432–438
Shimizu M, Blaak EE, Lonnqvist, Gafvels ME, Arner P (1996) Agonist and antagonist properties of beta 3-adrenoceptors in human omental and mouse 3T3-L1 adipocytes. Pharmacol Toxicol 78(4):254–263
Silver K, Walston J, Wang Y, Dowse G, Zimmet P, Shuldiner AR (1996) Molecular scanning for mutations in the beta 3-adrenergic receptor gene in Nauruans with obesity and noninsulin-dependent diabetes mellitus. J Clin End Met 81(11):4155–4158
Staehelin M, Simons P, Jaeggi K, Wigger N (1983) CGP-12177. A hydrophilic beta-adrenergic receptor radioligand reveals high affinity binding of agonists to intact cells. J Biol Chem 258(6):3496–3502
Stoyanov B, Volinia S, Hanck T, Rubio I, Loubtchenkov M, Malek D, Stoyanova S, Vanhaesebroeck B, Dhand R, Nurnberg B, et al (1995) Cloning and characterization of a G protein-activated human phosphoinositide-3 kinase. Science 269(5224):690–693
Strickland S, Loeb JN (1981) Obligatory separation of hormone binding and biological response curves in systems dependent upon secondary mediators of hormone action. Proc Natl Acad Sci USA 78(3):1366–1370
Sunahara RK, Dessauer CW, Gilman AG (1996) Complexity and diversity of mammalian adenylyl cyclases. Annu Rev Pharmacol Toxicol 36:461–480
Susulic VS, Frederich RC, Lawitts J, Tozzo E, Kahn BB, Harper ME, Himms-Hagen J, Flier JS, Lowell BB (1995) Targeted disruption of the beta 3-adrenergic receptor gene. J Biol Chem 270(49):29483–29492
Tai TAC, Jenrmann C, Brown KK, Oliver BB, MacGinnitie MA, Wilison WO, Brown HR, Lehmann JM, Kliewer SA, Morris DC, Graves RA (1996) Activation of the nuclear receptor peroxisome proliferator-activated receptor gamma promotes brown adipocyte differentiation. J Biol Chem 271:29909–29914
Tavernier G, Barbe P, Galitzky J, Berlan M, Caput D, Lafontan M, Langin D (1996) Expression of beta 3-adrenoceptors with low lipolytic action in human subcutaneous white adipocytes. J Lipid Res 37(1):87–97
Thurlby PL, Ellis RDM (1985) Differences between the effects of noradrenaline and the ß-adrenoceptor agonist BRL 28410 in brown adipose tissue and hind limb of the anaesthetized rat. Can J Physiol Pharmacol 64:1111–1114
Toubro S, Astrup A (19956) The selective β3 agonist ZD 2079 stimulates 24-hour energy expenditure through increased fidgeting. A 14 day, randomized placebo-controlled study in obese subjects. Int J Obes 19:070 (abstr)

Urhammer SA, Clausen JO, Hansen T, Pedersen O (1996) Insulin sensitivity and body weight changes in young white carriers of the codon 64 amino acid polymorphism of the beta 3-adrenergic receptor gene. Diabetes 45(8):1115–1120
Van Liefde I, Van Ermen A, Vauquelin G (1994) No functional atypical beta-adrenergic receptors in human omental adipocytes. Life Sci 54(12):PL209–PL214
Van Spronsen A, Nahmias C, Krief S, Briend-Sutren M-M, Donny Strosberg A, Emorine LJ (1993) The promoter and intron/exon structure of the human and mouse ß3-adrenergic-receptor genes. Eur J Biochem 213:1117–1124
Vidal-Puig A, Solanes G, Grujic D, Flier JS, Lowell BB (1997) UCP3: An uncoupling protein homologue expressed preferentially and abundantly in skeletal muscle and brown adipose tissue. Biochem Biophys Res Comm 235:79–82
Viguerie-Bascands N, Bousquet-Melou A, Galitzky J, Larrouy D, Ricquier D, Berlan M, Casteilla L (1996) Evidence for numerous brown adipocytes lacking functional beta 3-adrenoceptors in fat pads from nonhuman primates. J Clin Endocrinol Metab 81(1):368–375
Walston J, Lowe A, Silver K, Yang Y, Bodkin NL, Hansen BC, Shuldiner AR (1997) The beta 3-adrenergic receptor in the obesity and diabetes prone rhesus monkey is very similar to human and contains arginine at codon 64, Gene 188(2):207–213
Wheeldon NM, McDevitt DG, Lipworth BJ (1993) Do β_3-adrenoceptors mediate metabolic responses to isoprenaline. Quart J Med 86:595–600
Wheeldon NM, McDevitt DG, McFarlane LC, Lipworth BJ (1993) β-Adrenoceptor subtypes mediating the metabolic effects of BRL 35135 in man. Clin Sci 86: 331–337
Widén E, Lehto M, Kanninen T, Walston J, Shuldiner AR, Groop LC (1995) Association of a Polymorphism in the β_3-Adrenergic-Receptor Gene with Features of the Insulin Resistance Syndrome in Finns. N Engl J Med 333(6):348–351
Wilson S, Chambers JK, Park JE, Ladurner A, Cronk DW, Chapman CG, Kallender H, Browne MJ, Murphy GJ, Young PW (1996) Agonist potency at the cloned human beta-3 adrenoceptor depends on receptor expression level and nature of assay. J Pharmacol Exp Ther 279(1):214–221
Wilson S, Thurlby PL, Arch JRS (1986) Substrate supply for thermogenesis induced by the ß-adrenoceptor agonist BRL 26830A. Can J Physiol Pharmacol 65:113–119
Wojtczak L, Schonfeld P (1993) Effect of fatty acids on energy coupling processes in mitochondria. Biochem Biophys Acta 1183(1):41–57
Xiao RP, Ji X, Lakatta EG (1995) Functional coupling of the beta 2-adrenoceptor to a pertussis toxin-sensitive G protein in cardiac myocytes. Mol Pharmacol 47(2):322–329
Yen TT (1984) The antiobesity and metabolic activities of LY79771 in obese and normal mice. Int J Obes 8(1):69–78
Young P, Cawthorne MA, Smith SA (1985) Brown adipose tissue is a major site of glucose utilisation in C57B1/6 ob/ob mice treated with a thermogenic beta-adrenoceptor agonist. Biochem Biophys Res Commun 130(1):241–248
Zhang Y, Wat N, Stratton IM, Warren-Perry MG, Orho M, Groop L, Turner RC (1996) UKPDS 19: heterogeneity in NIDDM: separate contributions of IRS-1 and beta 3-adrenergic receptor mutations to insulin resistance and obesity respectively with no evidence for glycogen synthase gene mutations. Diabetologia 39(12):1505–1511
Zhou YY, Cheng H, Bogdanov KY, Hohl C, Altschuld R, Lakatta EG, Xiao RP (1997) Localized cAMP-dependent signaling mediates beta 2-adrenergic modulation of cardiac excitation-contraction coupling. Am J Physiol 272(3 Pt 2):H1611–H1618
Zilberfarb V, Piétri-Rouxel F, Jockers R, Krief S, Delouis C, Issad T, Donny Strosberg A (1997) Human immortalized brown adipocytes express functional β_3-adrenoceptor coupled to lipolysis. J Cell Sci 110:801–807

CHAPTER 15

Insulin Sensitization

C.F. BURANT

A. Introduction

Type 2 diabetes (formerly known as non-insulin-dependent diabetes mellitus (NIDDM) or adult onset diabetes) is characterized by peripheral insulin resistance, increased hepatic glucose production, and defects in insulin secretion from pancreatic β-cells (DEFRONZO 1988). In skeletal muscle, the tissue responsible for up to 90% of insulin-stimulated glucose disposal, both oxidative and non-oxidative glucose utilization is impaired (DEFRONZO 1988). The increase in hepatic glucose production is associated with increased levels of glucagon and an increased utilization of peripherally derived three carbon gluconeogenic precursors, the latter of which may be due to skeletal muscle insulin resistance (MOLLER 1993). The β-cell dysfunction is due to multiple defects including blunted responses to glucose, changes in insulin secretory patterns, and secretion of incompletely and partially processed insulin (POLONSKY 1995). Although the underlying cause of these defects in the majority of patients with NIDDM is unknown, there is a strong correlation with obesity, especially centripetal obesity (BODEN 1997).

It is not clear whether the positive energy balance associated with obesity interacts with a genetic predisposition in skeletal muscle to become insulin resistant or if factors associated with the expanded adipose tissue is the primary cause of resistance, independent of genetic risk. Support for the former comes from a number of observations. Studies have demonstrated that insulin resistance can be detected in offspring of NIDDM patients who are not obese (WARRAM et al. 1990). The rapid induction of insulin resistance with fat feeding (OAKES et al. 1997a) and the observation that even short term fasting markedly improves insulin sensitivity without weight change in susceptible individuals (PERSEGHIN et al. 1996) suggests factors other than just fat mass predispose to resistance. In addition, the demonstration that insulin resistance in skeletal muscle induced by genetic manipulation of mice can result in obesity (MOLLER et al. 1996) demonstrates that genetic factors can lead to obesity. The latter gains support from epidemiological studies which demonstrate that obesity is nearly always associated with insulin resistance (ERICKSON et al. 1989) and that weight gain and weight loss are associated with changes in insulin sensitivity (SIMS et al. 1973) to an even greater degree than that seen in short term weight changes. The expanded fat cell mass may elaborate one or more factors, either lipids and/or humoral factors, which result in insulin resistance (SPIEGELMAN and FLIER 1996). The reality of the situation is that both theories may be operative. Insulin resistance begets obesity and the expanded fat mass results in greater insulin resistance.

A number of interventions, both physiologic and pharmacologic, can result in improvement in insulin sensitivity and it is not surprising that the majority of interventions are targeted to the skeletal muscle or fat. Improvement in insulin sensitivity is associated with many salutary effects on the cardiovascular health of humans including decreases in insulin, triglyceride, free fatty acid and cholesterol levels, and improvements in the fibrinolytic system. Only recently have we come to understand some of the mechanisms underlying insulin resistance and insulin sensitization. However, there are still many unanswered questions and this is due in large part to our incomplete understanding of the ways by which insulin transduces binding to its receptor to the myriad of biological effects which insulin produces.

B. Molecular Mechanisms of Insulin Resistance

Insulin resistance can be defined as a concentration of insulin which gives less than the expected biological effect. Except for defined genetic causes, the exact cause of insulin resistance in humans or animal models is not clear. There are a number of correlative associations as described below, but none have proven to provide a complete molecular explanation of the metabolic abnormalities. The few candidate mediators seen in association with obesity have been extensively studied and provide some insight into the metabolic alterations in obesity.

I. Lipids

Expanded adipose tissue mass is associated with elevated levels of triglycerides and free fatty acids (FFA) and is seen in the majority of patients with obesity, insulin resistance, and NIDDM (MOLLER 1993; MCGARRY 1994; REAVEN 1995). The elevated circulating lipid levels and increased delivery of lipid to muscle have been implicated in the development of peripheral insulin resistance. The accumulation of intracellular triglyceride may lead to a preferential oxidation of fat vs glucose in skeletal muscle. The so-called Randale cycle (RANDALE et al. 1963), which was originally demonstrated in heart and diaphragm, hypothesizes that an increase in FFA metabolism is associated with increased accumulation of acetyl CoA in the mitochondria. This in turn would lead to a decrease in the NAD/NADH ratio and inhibit pyruvate dehydrogenase. The resulting increased mitochondrial citrate would "leak" into the cytoplasm with the allosteric inhibition of phosphofructokinase. The inhibition of this enzyme would lead to an increase in glucose-6-phosphate and allosteric inhibition of muscle hexokinase II, leading to decreased passive glucose transport because of the build up of intracellular free glucose.

The primacy of the Randale cycle in producing insulin resistance has been called into question. Using sophisticated NMR techniques, Shulman and coworkers (RODEN et al. 1996) have shown that the inhibition of glucose transport and/or phosphorylation, as reflected by decreasing glucose-6-phosphate levels, preceded the fall in glycogen synthesis in human subjects infused with a combination of intralipid and heparin to increase serum FFA levels. Subsequent studies by this group, using a similar experimental paradigm, demonstrated that IRS-1 associated PI3-kinase activity was not activated above basal levels after insulin/FFA infusion while subjects infused with insulin/glycerol showed a fourfold increase (BORKMAN et al. 1993). Free intracellular glucose was virtually undetectable, suggesting that the activation of glucose transport and not phosphorylation is inhibited by FFA infusion. However, the mechanism for the decrease in IRS-1 activation is not clear, as previous studies suggest that insulin receptor phosphorylation is not affected (SAHA et al. 1995) under a similar experimental paradigm.

Increased lipid delivery has been reported to change the phospholipid composition of membranes, and has also been associated with insulin resistance (BORKMAN et al. 1993) though this does not seem to explain the effects of acutely elevated FFA levels. Additional effects of increased delivery of FFA to muscle include alterations in the levels of key metabolic intermediates such as malonyl CoA (SAHA et al. 1995) and long chain acyl CoA (OAKES et al. 1997b). Accumulation of malonyl CoA in insulin resistance leads to an inhibition of CPT1 activity on the mitochondrial outer membrane (MCGARRY 1994). The result is an accumulation of cytosolic long chain acyl CoA levels that have been hypothesized to lead to inhibition of glycogen synthase or insulin signaling (OAKES et al. 1997b). The FFA hypothesis is confounded by the inconsistent improvement in insulin sensitivity by lowering free fatty acid levels pharmacologically (INOUE et al. 1995; MATSUI et al. 1997). The differing results could reflect the different models of insulin resistance or different human populations examined in the studies.

II. TNF-α

Studies in humans and animal models of obesity have identified changes in the expression and activity of key molecules involved in the insulin signaling pathway. Decreases in the number and the kinase activity of insulin receptors (GUMBINER et al. 1996), impairment in the activation of IRS-1 (SAAD et al. 1992), PI3 kinase (ZIERATH et al. 1997; ANAI et al. 1998) and protein kinase B (KROOK et al. 1997) have been observed. While the basis for the changes are, in general, unknown, a TNF-α mediated mechanism for the decreased activity in the initial steps of the insulin signaling cascade has been proposed. TNF-α, made and secreted by adipocytes, is elevated in a variety of experimental models of obesity (HOTAMISLIGIL and SPIEGELMAN 1994). The kinase activity of the insulin receptor in rats (MILES et al. 1997) or in 3T3-L1 adipocytes (HOTAMISLIGIL et al. 1996) treated with TNF-α are reduced, possibly by increased serine phosphorylation (HOTAMISLIGIL et al. 1996). Fat fed mice with genetic ablation of TNF-α production have increased kinase activity of the insulin receptor compared to controls and demonstrate increased insulin sensitivity (UYSAL et al. 1997). In addition, rats treated with neutralizing antisera or soluble TNF receptors demonstrate an amelioration of their insulin resistance. As will be described below, other interventions to decrease TNF-α action result in increased insulin sensitivity.

III. Glucotoxicity/Glucosamine

Hyperglycemia is a primary factor in the development of the complications of diabetes and decreases in average blood glucose have a profound effect to prevent complications in both type 1 (THE DIABETES CONTROL AND COMPLICATIONS TRIAL RESEARCH GROUP 1993) and type 2 diabetes (TURNER 1998). Hyperglycemia itself can cause insulin resistance. In Pima Indians, the levels of fasting glycemia are the primary determinant of insulin sensitivity (SAKUL et al. 1997). The defect is primarily in skeletal muscle (YKI-JARVINEN 1990) and is related to the degree of hyperglycemia.

Entry of glucose into the cell results in its phosphorylation to glucose-6-phosphate that has multiple metabolic fates. The enzyme glutamine:fructose-6-phosphate amidotransferase (GFAT) caries out the rate-limiting step of the hexosamine pathway (KORNFIELD 1967). Recent evidence suggests that the "hexosamine" pathway underlie the defect in glucose utilization associated with hyperglycemia. Hexosamines, such as glucosamine, when incubated with adipose tissue induces insulin resistance in fat cells (MARSHALL et al. 1991) and in skeletal muscle (ROBINSON et al. 1995). Infusion of glucosamine into rats resulted in a dose-dependent increase in insulin resistance of skeletal muscle (ROBINSON et al 1995). Finally, transgenic mice that overexpress GFAT specifically in skeletal muscle acquire severe insulin resistance (HERBERT et al. 1996). By a pathway that is unclear, glucosamine overproduction results in a disruption of the ability of insulin to cause translocation of GLUT4 to the cell surface (BARON 1995). It is through its anti-insulin action that the hexosamine pathway has been hypothesized to be a glucose sensor that allows the cell to sense and adapt to the prevailing glucose (YKI-JARVINEN 1998).

C. Exercise and Insulin Sensitivity

Exercise is clearly effective in increasing insulin sensitivity in animals and in humans. There appears to be two separable but related actions of exercise on insulin action. A single bout of exercise can result in an acute increase in insulin-independent glucose transport measurable during, and for a relatively short period after, exercise (NESHER et al. 1985; WALLBERG-HENRIKSSON and

Holloszy 1985; Wallberg-Henriksson and Constable 1988; Young et al. 1987, Douen et al. 1990). Like insulin, exercise/muscle contractions increase glucose transport by translocation of intracellular GLUT4 glucose transporters to the cell surface (Goodyear et al. 1990, 1991; Lund et al. 1995).

I. Acute Exercise

The signaling pathway leading to the exercise induced increase in glucose transporter translocation and increase glucose transport is unknown, though there is ample evidence that the pathway is independent of the insulin-stimulated, receptor mediated pathway. The effect of exercise/contractions on translocation/transport is additive to the maximal effect of insulin (Nesher et al. 1985; Zorzano et al. 1986; Henriksen et al. 1990; Gao et al. 1994; Lund et al. 1995). Insulin-stimulated glucose transport in muscle is inhibited by specific inhibitors of phosphatidylinositol 3-kinase, such as wortmanin, while transport/translocation stimulated by muscle contractions is insensitive to these inhibitors (Lund et al. 1995; Lee et al. 1995; Yeh et al. 1995). Unlike insulin, stimulation of muscle contractions in situ and exercise do not increase insulin receptor phosphorylation or tyrosine kinase activity, IRS phosphorylation, or PI 3-kinase activity (Treadway et al. 1989; Goodyear et al. 1995). Additionally, in many insulin resistant states the acute exercise-stimulated (but not insulin-stimulated) glucose transport and GLUT4 translocation is normal. This has been demonstrated in the obese, insulin-resistant Zucker rat (Brozinic et al. 1992) and in type 2 diabetic patients (Kennedy et al. 1999). Finally, hypoxia, a stimulus for glucose transport which is also independent of the insulin receptor mediated pathway, is also effective in increasing glucose transport in muscle strips from obese, insulin-resistant individuals and in type 2 diabetics (Azevedo et al. 1995).

The acute effect of exercise and hypoxia may be mediated by AMP-dependent protein kinase (AMPK). AMPK is thought to be a sensor of intracellular energy stores and is activated by increases in intracellular AMP. The riboside precursor of ZMP, 5-aminoimidazole-4-carboxamide 1-β-D-ribofuranoside (AICAR), can activate AMPK in cells leading to increased phosphorylation of known substrates for AMPK, including HMG-CoA reductase, Acyl-CoA carboxylase, and creatine kinase (Winder and Hardie 1999). Treatment of incubated skeletal muscle with AICAR results in increased glucose uptake and glucose transporter translocation (Hayashi et al. 1998). Similarly, the inclusion of 2 mmol/l AICAR into the perfusate of the rat hindlimb, resulted in inactivation of ACC, decreases in malonyl CoA levels, and a twofold increase in glucose uptake (Merril et al. 1997). Bergeron et al. (1999) used a euglycemic clamp in conscious rats to demonstrate that infusion of AICAR resulted in a more that twofold increase in glucose utilization. Uptake of the glucose analogue 2-deoxyglucose was also increased twofold in vivo in the soleus and gastrocnemius muscle. As with previous studies, this uptake was not associated with PI 3-kinase activation, again indicating a separate pathway from that of insulin.

A second effect of exercise, which becomes evident as the acute effect on glucose transport reverses, consists of a large increase in the sensitivity of glucose transport to stimulation by insulin (RICHTER et al. 1982; GARETTO et al. 1984; CARTEE et al. 1989; RICHTER et al. 1989). This effect is due to translocation of more GLUT4 glucose transporters to the cell surface for any given dose of insulin (HANSEN et al. 1998; THORELL et al. 1999). As with the acute stimulation of transport by exercise, the cellular mechanisms leading to enhanced translocation in response to submaximally effective stimuli are unknown. However, several studies have shown that steps in the insulin-signaling cascade leading to activation of PI 3-kinase are not enhanced following a bout of exercise. There is no change in insulin binding to its receptor (BONEN et al. 1984; ZORZANO et al. 1985; TREADWAY et al. 1989), insulin stimulation of receptor tyrosine kinase activity (TREADWAY et al. 1989; WOJTASZEWSKI et al. 1997), increase in insulin-stimulated tyrosine phosphorylation of IRS1 (HANSEN et al. 1998), or PI 3-kinase activity associated with IRS1 (GOODYEAR et al. 1995; WOJTASZEWSKI et al. 1997).

II. Exercise Training

In contrast to the acute effects of insulin, exercise training also results in increases in insulin sensitivity (OSHIDA et al. 1989; DEFRONZO et al. 1987; TOMINO 1989; DEVLIN et al. 1987) and can delay or prevent the onset of type 2 diabetes in those at high risk (HELMRICH et al. 1991). Using the hyperinsulinemic-euglycemic clamp, PERSEGHIN et al. (1996) used exercise training of 45 min on a stair climbing machine 4 days per week for 6 weeks in normal insulin sensitive subjects and a group of high risk, insulin-resistant relatives of type 2 diabetics. A 100% increase in insulin sensitivity was seen in both groups without significant change in body weight. Interestingly, the higher basal and glucose-stimulated insulin release in the insulin resistant subjects did not change after exercise training. The effect of training on insulin sensitivity has been proposed to be due to upregulation of glucose transporter number, changes in capillary density and increases in the number of red, glycolytic (type IIa) fibers (EBELING et al. 1993; HOUMARD et al. 1991).

In summary, exercise is an effective measure to increase glucose uptake in the absence of insulin and to increase insulin sensitivity both acutely and in the long term. The effect is independent of weight loss, changes in lipid parameters, or adiposity. Given its efficacy and the other cardiovascular benefits, exercise is recommended to be an integral part of any regimen in most patients with insulin resistance and type 2 diabetes.

D. Pharmacological Agents

While diet and exercise are efficacious in improving insulin sensitivity, their utility in the long term is limited primarily by patient compliance. Thus pharmacological therapy to improve insulin sensitivity is now a primary goal of the pharmaceutical industry. Prior to the introduction of the thiazolidinediones, pharmacological treatment of type 2 diabetes consisted of insulin, sulfonylureas, metformin, and α-glucosidase inhibitors. While improvements in glycemia are associated with improvements in insulin action, none of these agents are thought to be direct sensitizers to insulin action. The exception to this general rule is metformin, which may improve insulin sensitivity of the liver and, to a small extent, insulin sensitivity in skeletal muscle, though these effects may also be due to changes in glycemia (JOHNSON et al. 1993; STUMVOLL et al. 1996). The agents described are those that have clearly been shown to have a primary effect on changing insulin sensitivity in either animal models of insulin resistance or in humans.

I. PPARγ Activators

Thiazolidinedione compounds have been demonstrated to have potent anti-hyperglycemic actions in animals and humans. Ciglitazone was the first thiazolidinedione compound that was demonstrated to have anti-diabetic properties in animals (FUJITA et al. 1993) and, subsequently, many compounds with a core thiazolidine-2-4 dione structure have been described with similar actions in animal models (SALTIEL and OLEFSKY 1996) (Fig. 1). As will be described in detail below, thiazolidinediones are potent activators of the nuclear receptor PPARγ.

1. Structure and Distribution of PPAR Receptors

Peroxisomal Proliferator Activated Receptors (PPAR) are a family of nuclear hormone receptors designated PPARα, PPARβ (or δ), and PPARγ (EVANS 1988). The structure of PPAR is similar to other members of the superfamily of nuclear hormone receptors (Fig. 2). An n-terminal AF-1 domain can interact with specific coactivators and corepressors and is the most divergent among the PPAR subtypes (SCHOONJANS et al. 1996b; FORMAN et al. 1996). This is followed by the DNA binding domain with a nuclear localization signal and is highly conserved. The c-terminus of the receptors has the dimerization domain and a second domain which also interacts with specific nuclear factors (MANGELSDORF et al. 1995) in a ligand dependent fashion. The PPARs form obligate hetrodimers with the retinoid x receptor (RXR) (MANGELSDORF and EVANS 1995).

PPARα was originally discovered as the receptor for fibrates that cause peroxisomal proliferation and transformation in the liver of rats (ISSEMAR and GREEN 1990). PPARα is primarily expressed in liver, kidney, heart, and brown adipose tissue. The primary activity of PPARα is the regulation of fatty acid homeostasis (FRUCHART et al. 1999). Binding of the PPARα/RXR heterodimer to peroxisomal proliferator response elements (PPRE) results in activation or repression of multiple genes involved in fatty acid metabolism and include

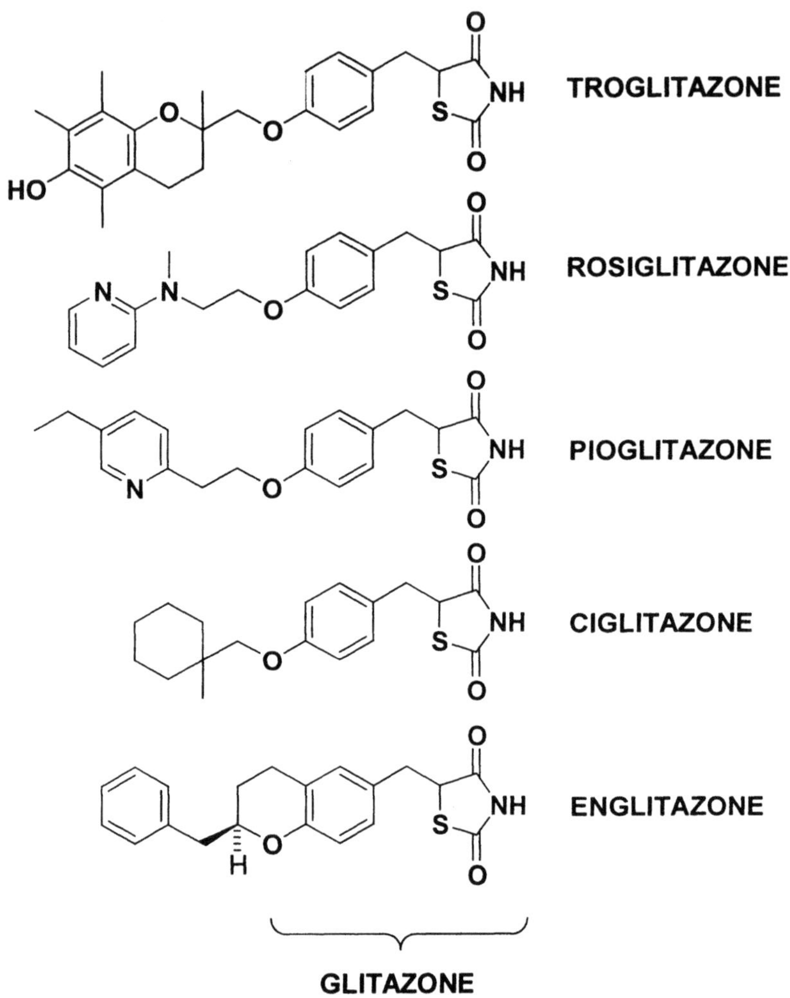

Fig. 1. Structure of the insulin sensitizing thiazolidinediones

peroxisomal enzymes which leads to peroxisomal proliferation (SCHOONJANS et al. 1996a). PPARβ/δ is expressed ubiquitously (AMRI et al. 1995). The exact role of PPARβ/δ in physiology is unclear though activation of the receptor results in elevated HDL levels (BERGER et al. 1999). PPARβ/δ has recently been implicated as a target of the APC gene and the unregulated expression of PPARβ/δ may be critical to the formation of colon tumors in APC-null mice (HE et al. 1999). PPARγ has its highest expression in fat cells, but is also found in a variety of other cell types including colonic epithelial cells, macrophages, skeletal muscle, and in several other tissues at lower levels (TONTONOZ et al. 1994b; AUBOEUF et al. 1997). As will be described, PPARγ likely functions as a regulator of lipid metabolism but also plays a role in insulin sensitivity.

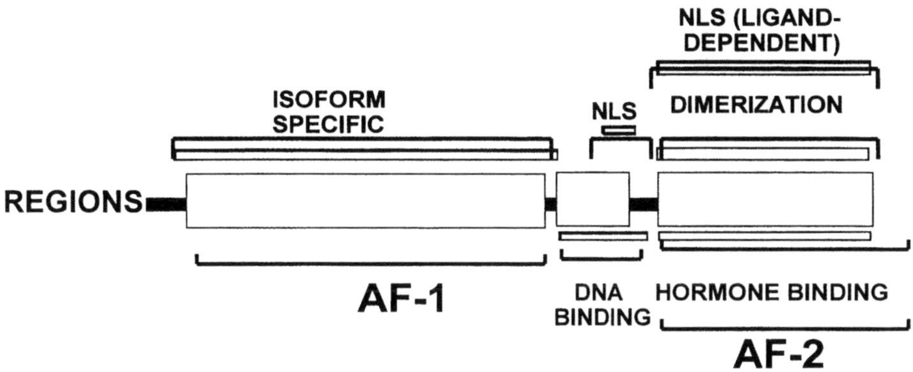

Fig. 2. Schematic structure of the PPAR receptor. *NLS*, nuclear localization signal; *AF-1* and *AF-2*, activation factor binding sites 1 and 2

PPARγ has three distinct 5' ends which arise from a single PPARγ gene and each is under the control of its own promoter (Zhu et al. 1995; Fajas et al. 1997). PPARγ1 and PPARγ3 differ only in their 5' untranslated region and give rise to the same protein whereas PPARγ2 has an additional 18 amino acids at the amino terminus of the protein (Elbrecht et al. 1996). PPARγ2 is expressed at high levels in adipose tissue and at low levels in the kidney. PPARγ1 and PPARγ3 are also high in adipose tissue and lower in other tissues such as intestine and kidney (Elbrecht et al. 1996). The level of PPAR receptor mRNA levels are influenced by weight with the level of PAPRγ2 increasing 75% in the adipose tissue of obese females while fasting decreased transcript levels (Bastard et al. 1999). In culture, the levels of PPARγ increased with insulin and glucocorticoid treatment (Wu et al. 1996). Leptin treatment can cause a decrease in triglyceride stores in fat without release of free fatty acids. This is accompanied by a marked reduction in PPARγ levels (Qian et al. 1998). PPARγ levels can also be regulated in human skeletal muscle cultures by acute insulin treatment (Park et al. 1998).

2. Role of PPARγ in Adipocyte Differentiation

Transfection of PPARγ into NIH 3T3 cells resulted in their differentiation into fat cells (Tontonoz et al. 1994a). The differentiation was enhanced by the addition of fatty acids and fatty acid derivatives including the prostaglandin PGJ2 to the medium (Forman et al. 1995). Similarly, when PPARγ is introduced into myoblasts, there is a trans-differentiation to an adipocyte-like phenotype with the cells expressing adipocyte markers and accumulating triglyceride (Hu et al. 1995). The thiazolidinedione class of insulin sensitizers were subsequently demonstrated to induce differentiation of fibroblasts into fat cells (Kletzien et al. 1992) and were shown to be high affinity ligands for PPARγ (Lehmann et al. 1995). During adipocyte differentiation, C/EBPβ and C/EBPδ are expressed early and induce the expression of PPARγ (Chawla et al. 1994). In turn, PPARγ can increase its own expression and the expression of C/EBPα (which can also increase PPARγ expression). C/EBPα and PPARγ increase the expression of multiple genes associated with adipocyte differentiation including aP2, lipoprotein lipase, and acyl CoA synthase (Tontonoz et al. 1994a; Wu et al. 1999). There does appear to be a C/EBP independent pathway for induction of PPARγ since in C/EBPβ / C/EBPδ double knockout mice PPARγ and C/EBPα expression continues (Tanaka et al. 1997).

3. Effect of Disruption of PPARγ gene

Homozygous disruption of the PPARγ receptor is lethal at embryonic day 10 due to defective trophoblasic fusion and decreased placental vascularization (Rosen et al. 1999; Barak et al. 1999). When the PPARγ null embryos are rescued by aggregation of tetraploid embryos, the pups survive to term but subsequently die due to lipodystrophy and multiple hemorrhages (Barak et al. 1999; Kubota et al. 1999). Mice with a heterozygous disruption of the PPARγ gene display interesting metabolic properties (Kubota et al. 1999). While on a standard chow diet, they have a normal weight gain. However, when placed on a high fat diet, the heterozygous animals have an attenuation of the weight gain seen in wild type animals. The heterozygous mice also show a resistance to increases in serum glucose levels after consuming a high fat diet. It should be noted that both wild type mice and heterozygous mice have a marked hyperglycemic response to an oral glucose tolerance test after feeding on the high fat diet. During the glucose tolerance test, the excursion in insulin levels is less in the heterozygous mice, again suggesting an increase in insulin sensitivity, though it also indicates a defective glucose sensing by the β-cell to hyperglycemia. Interestingly, treatment of these mice with pioglitazone resulted in increased insulin levels. This appears to be paradoxical since the activation of PPARγ by thiazolidinediones results in increased insulin sensitivity. The authors propose the results may be due to changes in adipose metabolism. Small fat cells may be metabolically more favorable to the animal by being metabolically more active and secreting more leptin (Okuno et al. 1998). Indeed, in the heterozygous PPARγ null mice, the average fat cell size is significantly smaller than in the wild type animal and, when placed in culture, synthesize more leptin. Treatment of the heterozygous mouse with pioglitazone increased the average size of the adipocyte from the heterozygous mouse. Treatment of mice with thiazolidinediones will also cause a decrease in leptin secretion (Kallen and Lazar 1996). While this may explain some of the results, further studies will be needed to clarify the metabolic consequences of disruption of the PPARγ gene.

4. Mutations of PPARγ in Humans

The human PPARγ gene is located on chromosome 3, band 3p25 (Beamer et al. 1997). An amino-acid variant (Pro12·Ala) was found in screening for mutations in the coding region of PPARγ2 gene. When expressed in cells, it was found that this variant had decreased the ability to stimulate gene transcription (RINGEL et al. 1999). The variant was not associated with diabetes, obesity, or hyperlipidemia in either type 1 or type 2 diabetics (MANCINI et al. 1999). Interestingly, when a group of severely obese patients (BMI > 30 kg/m^2) who had the Pro12·Ala variant were studied, they exhibited *increased* insulin sensitivity (KOCH et al. 1999). In contrast, a recent report demonstrates that a patients harboring inactivating mutations of PPARγ have decreased insulin sensitivity (BARROSO et al. 1999) and display many of the features of syndrome X including hypertension and obesity at an early age. The apparent decreased activity of PPARγ variants in humans associated with increased sensitivity is reminiscent of the heterozygous PPARγ-null mice described above.

PPARγ can be phosphorylated on Ser by MAP kinase and decreases its transcriptional activity (ADAMS et al. 1997). A Pro115Gln variant in PPARγ2 that decreases its ability to be phosphorylated on Ser has been proposed to lead to increases in adiposity (RISTOW et al. 1998). Transfection of this mutant receptor into fibroblasts accelerated the differentiation into adipocytes. The fasting insulin concentrations in these patients were lower than detected in patients with a matching BMI, suggesting an increase in insulin sensitivity. Reconciling the apparent increase in insulin sensitivity seen in patients with allelic variations of PPARγ which decrease activity with those variants which increase activity will need further investigation.

5. PPARγ Activators in Lipid and Glucose Homeostasis

As already mentioned, thiazolidinediones have been determined to be high affinity ligands for PPARγ (LEHMANN et al. 1995). These compounds have emerged as very effective therapeutic agents for the treatment of type 2 diabetes, primarily through the alleviation of insulin resistance (SALTIEL and OLEFSKY 1996). In general, there is a distinct relationship between the affinity of thiazolidinediones and nonthiazolidinediones for the PPARγ receptor, their potency in causing adipocyte differentiation, and in vivo potency in lowering blood glucose (LEHMANN et al. 1995). Because there is no effect on glucose in insulin deficient animals, the improvements in blood glucose in whole animals is not due to a peripheral insulin-mimetic affect (FUJII et al. 1997). In addition, there is no evidence that insulin secretion is stimulated by thiazolidinediones; however, there are reports of TZDs interacting with the β-cell KATP channel in a manner similar to that of the sulfonylureas (SUNAGA et al. 1999). The importance of this latter effect is unclear. There are also reports of thiazolidinedione effects which clearly are not PPARγ mediated (OKUNO et al. 1997; WANG et al. 1999). What the mechanism is for these latter effects remains to be discovered.

a) Effects in Animal Models of Diabetes

In the main, the blood glucose lowering effect of the thiazolidinediones appear to be due to marked increases in insulin sensitivity (SALTIEL and OLEFSKY

1996). In murine models, studies with hyperinsulinemic-euglycemic clamps have demonstrated both an increase in peripheral glucose disposal and a decrease in hepatic glucose production after thiazolidinedione treatments (SUGIYAMA et al. 1990; ZIERATH et al. 1998).

Zucker diabetic fatty (ZDF), a substrain of the *fa/fa* rat, have been used extensively as a model of type 2 diabetes. The *fa/fa* rat has a mutation in the leptin receptor and is hyperphagic (WHITE et al. 1997). It develops obesity and insulin resistance with age but does not become diabetic. In these animals there is a β-cell hypertrophy and increased basal levels of insulin secretion (CHAN et al. 1999). In the ZDF there is a second mutation that results in β-cell failure that occurs spontaneously in the male ZDF at 10–12 weeks of age (TOKUYAMA et al. 1995). The β-cell failure is associated with increased rates of apoptosis (PICK et al. 1998). Similar events are seen in the *db/db* mouse that also has a mutation affecting the function of the leptin receptor (CHEN et al. 1996). Treatment of the ZDF (STURIS et al. 1995) rat or *db/db* (FUJIWARA et al. 1991) mouse with thiazolidinediones prevents the onset of diabetes, prevents the β-cell hypertrophy and degranulation, and improves β-cell function. Troglitazone treatment in vitro (SHIMABUKURO et al. 1998) or in vivo (SREENAN et al. 1999) lowers islet associated TG in the ZDF rat which, in addition to its peripheral effect, may play a role in the improvement in β-cell function.

In addition to improvement of glucose homeostasis, treatment of the ZDF with thiazolidinediones is accompanied by marked lowering of serum triglyceride, free fatty acid, and cholesterol levels (STURIS et al. 1995; SREENAN et al. 1999). There is marked decrease in liver and skeletal muscle triglyceride levels (SREENAN et al. 1999), the latter of which has been correlated to decreased insulin sensitivity (PHILLIPS et al. 1996). The mechanism by which FFA and triglyceride levels are improved in thiazolidinedione treated animals is not clear. Pioglitazone has been demonstrated to improve peripheral clearance of triglycerides in fatty rats due to an increase in VLDL catabolism (KAUMI et al. 1996). This may be accomplished by induction of lipoprotein lipase (LEFEBVRE et al. 1997), FATP, and acyl CoA synthase (MARTIN et al. 1997) in adipocytes which would result in increased disposal of fat. Improvements in blood glucose may decrease hepatic TG synthesis as well (BODEN 1997).

How does a drug which has high levels of receptor in adipose tissue improve muscle insulin sensitivity? A number of possible explanations exist. First, enhancing FFA uptake into fat cells and lowering serum free fatty acid concentrations would result in decreased FFA uptake and utilization in skeletal muscle. Since the bulk of energy for skeletal muscle metabolism arises from FFA utilization, the final effect would be to improve skeletal muscle glucose utilization. This may be by decreasing the "Randale cycle" (RANDALE et al. 1963) or by decreasing the other proposed effects of FFA on skeletal muscle metabolism (see Sect. B.I.). A second possible effect is that thiazolidinediones affect the release of one or more cytokines (such as TNFs, Sect. B.II.) or other mediators that are released from fat and cause defective insulin signaling in skeletal muscle. Indeed, TNF release from adipocytes is inhibited by TZD treatment (SZALKOWSKI et al. 1995). In addition, there is a restoration of insulin signaling in culture cells (PERALDI et al. 1997) and in animals (MILES et al. 1997) treated with TNF-α by exposure to thiazolidinediones. Third, there might be direct effects of TZD on skeletal muscle and a number of studies suggest that this is a primary site of action of these drugs. Recent evidence suggests that, despite low levels of PPARγmessage, substantial levels of PPAR protein is expressed in skeletal muscle (PARK et al. 1998). In human studies, described below, the effect of troglitazone is to enhance disposal of glucose in skeletal muscle. In mice that were rendered "fatless" by diphtheria toxin expression mediated by the aP2 promoter, there is a marked attenuation of insulin sensitivity (BURANT et al. 1997). Treating these mice with troglitazone resulted in normalization of glucose tolerance and improvement in triglyceride levels without evidence of new fat growth. Indeed, preliminary results from humans with lipodystrophy and severe insulin resistance demonstrate that TZDs can also improve glucose tolerance (ARIOGLU et al. 1999). These results suggest that TZD action can be independent of fat tissue.

b) *Effects of Thiazolidinediones in Humans*

Troglitazone was the first thiazolidinedione to be approved for use in humans. The bulk of published studies concern the metabolic consequences of treatment with this agent though it is likely that many of the metabolic effects of troglitazone will be similar for the other "glitazones." Early studies in type 2 diabetics demonstrated that troglitazone had insulin sensitizing properties in humans. At a dose of 400 mg per day, lean (BMI of 22 kg/m^2) Japanese patients demonstrated decreases in insulin and glucose concentrations after three months of treatment (MIMURA et al. 1994). In a group of obese, type 2 diabetic patients again treated for three months with troglitazone, 75% showed glucose lowering. When these patients were analyzed with a hyperinsulinemic-euglycemic clamp, all patients responded with a mean improvement in insulin sensitivity of over 50% (IWAMOTO et al. 1991). Improvement was observed in both the patients responding to the drug with lowered fasting glucose and insulin and in those patients who did not respond. In the non-responding cohort, it may be that the improvement in insulin sensitivity afforded by troglitazone was not great enough to compensate for the relative deficiency in insulin secretion in these patients. In a separate study, hyperinsulinemic, euglycemic clamps were performed in a large cohort of patients treated with increasing doses of troglitazone in a multicenter trial (MAGGS et al. 1998). The study enrolled 93 obese patients (BMI 32 kg/m^2) who had failed diet therapy or were inadequately controlled on sulfonylureas. Patients were tested for glucose tolerance prior to treatment and again after six months. There was a decrease in fasting glucose and an increase in insulin sensitivity in these patients in a dose-response pattern from 100–600 mg/day. Patients treated with 400 mg/day or 600 mg/day had a 45% increase in insulin mediated glucose disposal. However, only the patients treated with 600 mg/day showed a decrease in endogenous glucose production. Treated patients also showed improvement in triglyceride, free fatty acid, and fasting C-peptide levels.

INZUCCHI et al. (1998) studied 29 obese patients, comparing 400 mg/day of troglitazone with 1000 mg twice per day of metformin. The patients were treated for three months in a blind fashion. Both troglitazone and metformin decreased fasting glucose by 20% and postprandial glucose by 25%. Troglitazone increased insulin sensitivity by 54% while metformin improved it by only 13%. Conversely, metformin had a greater effect on endogenous glucose production, decreasing it by 19% while troglitazone decreased it by 3%. After three months, troglitazone treated patients had metformin added and vice versa. After an additional three months of combination treatment, troglitazone addition increased glucose disposal by an additional 24% while metformin increased glucose disposal by 15%. These results confirm previous studies which demonstrate minimal effects of metformin on insulin sensitivity save that due to improved glycemia (STUMVOLL et al. 1995).

These exogenous insulin-clamp studies clearly show that troglitazone improves insulin action in humans similar to that seen in experimental animals. However, it is clear that the insulin action is not restored to normal in these patients and leaves open the possibility that more potent PPARγ activators, either of the thiazolidinedione class or nonthiazolidinediones, could lead to an even greater improvement in sensitivity and glycemic control in type 2 diabetics.

Two studies in patients treated with exogenous insulin also demonstrate the insulin sensitizing effects of troglitazone. In a study of 350 patients with poorly controlled type 2 diabetes, treated with at least 30 U of insulin daily, patients were randomly assigned to receive 200 mg or 600 mg of troglitazone or placebo daily for 26 weeks (SCHWARTZ et al. 1998). The adjusted mean glycosylated hemoglobin values decreased by 0.8% and 1.4%, respectively, in the group given 200 mg or 600 mg of troglitazone. The fasting serum glucose concentrations decreased by 35 mg/dl and 49 mg/dl despite decreases in the insulin dose of 11% and 29% in the troglitazone treated groups vs placebo. Total serum cholesterol, low-density lipoprotein cholesterol, and high-density lipoprotein cholesterol concentrations increased slightly and serum triglyceride concentrations decreased slightly in the troglitazone-treated patients.

The ability of troglitazone to reduce requirements for injected insulin while maintaining blood glucose levels was studied in a second 26-week double-blind trial performed in patients who had failed previous oral antidiabetic medication and took >30 U but <150 U of insulin daily (BUSE et al. 1998). Patients received 200 mg or 400 mg troglitazone once daily or matching placebo. Reflecting the insulin sensitization, insulin dose was reduced by 13 U, 30 U, and 41 U in the placebo, 200-mg, and 400-mg troglitazone groups, respectively. HgbA1c decreased 0.09% for placebo, 0.13% for 200 mg, and 0.41% for 400 mg troglitazone. The small decrease in HgbA1c reflects the protocol restrictions with insulin dose decreased by 25% for every 5 mg% decrease in blood glucose.

c) *Effects of Thiazolidinediones on Weight*

The effect of TZDs on weight in animals and humans are well documented. The question of the relationship between action of the drug in adipose tissue to promote increases in fat accumulation directly as opposed to an indirect mechanism secondary to improved metabolic parameters, including decreased glucosuria and decreased gluconeogenesis, is still controversial. Treatment of normal rats or mice with the thiazolidinedione, troglitazone, has no effect on body weight or food consumption (SREENAN et al. 1996). Similar results are seen when the nondiabetic, hyperphagic *fa/fa* mouse is treated with troglitazone (SREENAN et al. 1999). In contrast, the diabetic ZDF rat gains significantly more weight and eats less when troglitazone is admixed in the diet (SREENAN et al. 1996, 1999). The increase in weight was accompanied by marked improvement in metabolic parameters and decreased blood glucose. The decrease in food consumption may be due to the decrease in glycemia, a known stimulant to food intake in diabetes. OKUNO et al. (1998) demonstrated that troglitazone treatment results in a decrease in mRNA TNFα and leptin in the retroperitioneal and mesenteric fat pads of *fa/fa* rats. They also showed that these fat pads had a greater number of small adipocytes and decreased number of large adipocytes. They suggest that the smaller adipocytes are metabolically more active and secrete less cytokine, leading to an improvement in insulin sensitivity. In contrast to troglitazone, the thiazolidinedione, rosiglitazone, has been reported to induce hyperphagia and weight gain in *fa/fa* rats but not in lean controls (ELDERSHAW et al. 1995). In this study, the plasma leptin levels were twice as high in untreated fatty rats as in lean rats and were unaffected by rosiglitazone after treatment for 20 days. The reason for the differences between troglitazone and rosiglitazone is unclear, though differential action at the PPARγ receptor is possible.

In humans there is clear evidence for increases in weight associated with troglitazone therapy. Weight gain (which correlated with improved glycemic control) was observed during the double-blind phases of studies examining the glycemic effects of combinations of troglitazone with either sulfonylureas (HORTON et al. 1998) or with insulin (SCHWARTZ et al. 1998), but not of troglitazone monotherapy (M. Ghazzi and H. Foyt, personal communication) (Fig. 3). An integrated analysis of weight change in the double-blind phase of the studies showed that weight changes in patients treated with troglitazone monotherapy of all doses (mean –0.9 kg, median –1.1 kg) (Fig. 3C) were comparable to those in placebo-treated patients (mean –1.1 kg, median –0.5 kg) (Fig. 3A). However, a mean increase in weight was observed in patients treated with combination therapy of various doses of troglitazone with sulfonylurea or insulin (mean 2.8 kg, median 2.5 kg) (Fig. 3D). Weight gain among these patients correlated significantly with the improvement in HgbA1c levels (1.0–1.2 kg weight gain per 1% decrease in HgbA1c; $p = 0.0001$). Interestingly, in the small cohort of patients on sulfonylureas therapy alone, there was no correlation between improvement in HgbA1c and weight (Fig. 3B). In these

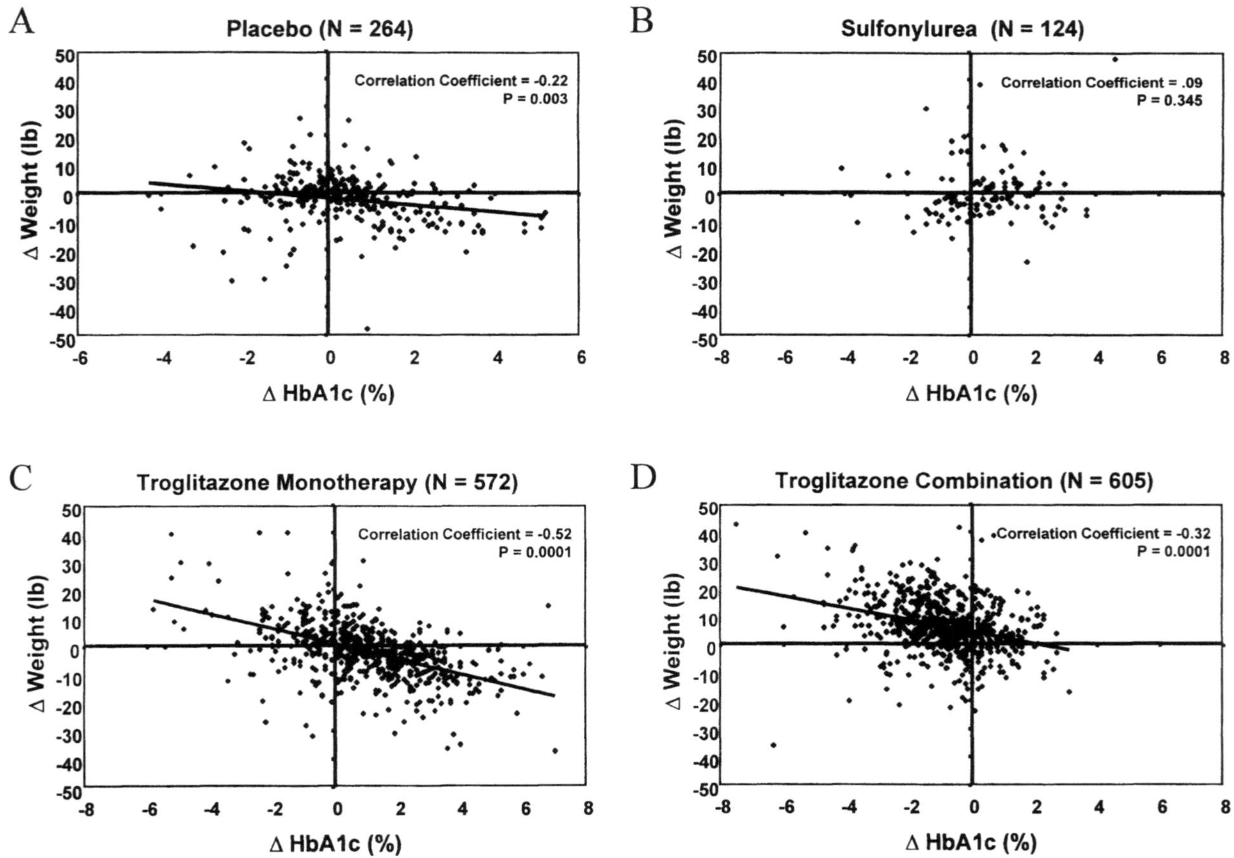

Fig. 3A–D. Relationship between weight gain and HgbA1c. Data from registration clinical trials of troglitazone was correlated for changes in weight against HgbA1c concentration: **A** patients on placebo; **B** patients taking glyburide (5–10 mg/day); **C** patients on troglitazone monotherapy (200–600 mg/day); **D** patients on a combination of troglitazone with either glyburide or insulin

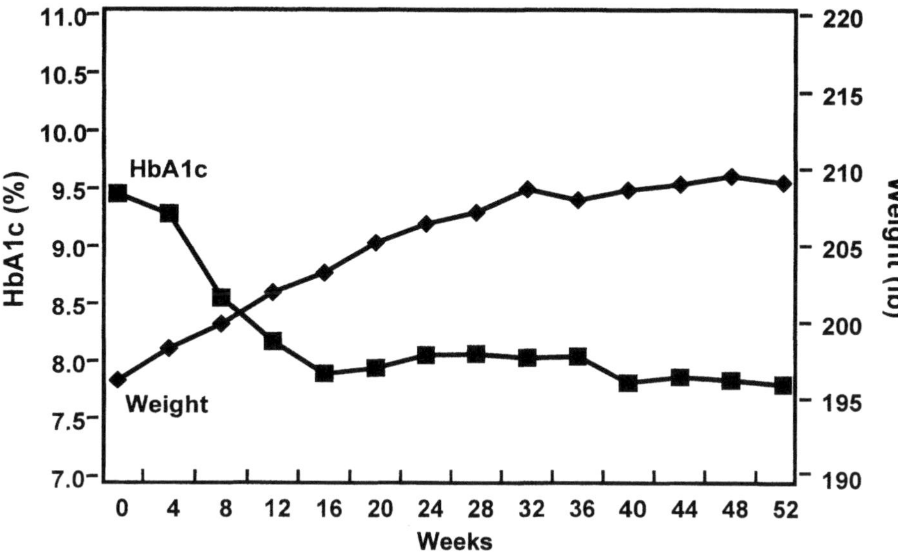

Fig. 4. Time course of HgbA1c and weight change in patients on a stable combination of 12 mg/day of glynase and 600 mg/day of troglitazone

studies, weight stabilized within 12–20 weeks and remained unchanged after that time; the time course for this increase in weight correlated with that seen for the decrease in HgbA1c (Fig. 4). The results of these human and experimental animal studies suggest, in aggregate, that the bulk of the weight gain is due to improvements in glycemic control.

While the weight gain may be due to improved glycemia, there is some suggestion that the pattern of fat distribution may change during troglitazone therapy. Kelly et al. (1999) treated type 2 diabetics with troglitazone (eight men, three women) or placebo (eight men, two women) matched for age, BMI, total body fat percentage by underwater weighing, and intra-abdominal fat (kilograms) by MRI. The patients were treated for 12 weeks. The body weight changes in the troglitazone group (+0.66 kg) and the placebo group (+0.25 kg) were not different nor were changes in total body fat. However, a decrease in intra-abdominal fat mass in the troglitazone-treated group (–0.47 kg) was significantly different from placebo treatment (–0.06 kg). Whether the decrease in intraabdominal fat mass contributes to the improvement in insulin sensitivity or is a reflection of changes in sensitivity is unclear.

Shimizu et al. (1998) treated 20 thinner, Japanese type 2 diabetic patients with 200 mg troglitazone twice daily for 12 weeks. There was improvement in glycemic control in all the patients with a significantly increased BMI. Serum levels of leptin was significantly reduced by troglitazone treatment. Interestingly, two-thirds of the patients complained of increased hunger after the start of troglitazone administration. However, the caloric intake and the relationship to body weight gain in this population was not documented.

II. RXR Agonists

The heterodimeric partner of the PPARγ receptor is the retinoid X receptor (MANGELSDORF and EVANS 1995). Binding of ligand to the RXR receptor results in the binding of coactivators to the RXR/ PPARγ heterodimer and may act to transactivate those genes activated with PPARγ ligand binding. When the high affinity, specific RXR agonists LGD100268 and LGD10069 were administered to diabetic, hyperglycemic *ob/ob* mice for 14 days, there was a decrease in glucose and insulin levels, indicating an insulin sensitizing response (MUKHERJEE et al. 1997). Oral glucose tolerance tests also improved as did serum triglyceride levels. Similar results were observed with LGD100268 in *db/db* mice (LENHARD et al. 1999). In this study, it was demonstrated that the effects of the RXR agonist were additive to two different PPARγ activators. However, humans treated with LGD10069 for indications other than diabetes showed significant side effects of therapy including hypertriglyceridemia (VU-DAC et al. 1998) and central hypothyroidism (SHERMAN et al. 1999). That there may be multiple side effects from RXR agonists should not be surprising since RXR is the heterodimeric partner to the thyroid hormone receptor, LXR, Vitamin D receptor, among others (MANGELSDORF et al. 1995). RXR can also homodimerize and function at the DR1 element of certain promoters after ligand binding (T. Leff, personal communication).

III. Vanadate

The trace metal vanadate can act as phosphotyrosine phosphatase inhibitors. Especially effective are peroxovanadate species. Peroxovanadate, administered to insulinopenic rats result in glucose lowering alone (MEYEROVITCH et al. 1987) and in combination with insulin. Indeed, peroxovanadates can increase the maximum stimulatory effects of insulin. It is also effective in lowering blood glucose levels in insulin resistant animal models of diabetes (MEYEROVITCH et al. 1991). Preliminary studies in humans with type 2 diabetes demonstrates the clinical efficacy of these compounds (HALBERSTAM et al. 1996). In a small group of patients, administration of vanadyl sulfate, 100 mg/day for 3 weeks, resulted in an 87% increase in insulin stimulated glucose disposal as assessed with a hyperinsulinemic-euglycemic clamp. The bulk of the effect was due to a decrease in endogenous glucose production (HALBERSTAM et al. 1996); however, there was also a significant increase in peripheral glucose disposal. Similar results were observed in a second study of diabetic patients (CUSI et al. 1997). A third study found that treatment with sodium metavanadate, 125 mg/day, did not change insulin sensitivity but did result in a decrease in insulin requirements (GOLDFINE et al. 1995). The utility of these compounds for treatment of type 2 diabetes will likely be limited by toxicological concerns. Vanadate accumulates in the kidneys, liver, and bone and may cause serious dysfunction (MONGOLD et al. 1990). In addition, the potential oncogenic potential of inhibition of phosphotyrosine phosphatases makes the long-term treatment with these agents potentially risky.

IV. Dichloroacetic Acid

Dichloroacetic acid (DCA) is an activator of pyruvate dehydrogenase (PDH) (BAUDRY et al. 1982). PDH activity is decreased in both type 1 and type 2 diabetes (STACPOOLE 1989). PDH activity is modulated by reversible phosphorylation carried out by a family of pyruvate dehydrogenase kinases (PDKs) (POPOV et al. 1997). PDK is allosterically inactivated by pyruvate, NADH, and ATP. DCA is a structural analogue of pyruvate that can bind to PDK and inactivate it similar to pyruvate. Inactivation of the kinase results in increased PDH enzymatic activity and increased glucose utilization. In animal models, DCA can increased glucose utilization in insulin resistant states (STACPOOLE and GREENE 1992). The long-term use of DCA is limited by the development of neuropathy in animals (KURLEMANN et al. 1995). Structural analogues that are metabolized in an alternative manner may have fewer side effects and prove to be useful therapeutic agents.

V. α-Lipoic Acid

Also know as thioctic acid, the racemic R/S α-lipoic acid has been used for a number of years in Germany as a treatment for neuropathy, perhaps through its antioxidant properties (ZIEGLER and GRIES 1997). Thioctic acid is a natural cofactor in a number of enzyme complexes of mitochondria, including pyruvate dehydrogenase and branch chain ketoacid dehydrogenase (OIZUMI and HAYAKAWA 1989). Long-term treatment of Zucker diabetic rats with thioctic acid was demonstrated to improve glucose uptake into skeletal muscle from these animals (BLACK et al. 1998). In recent studies, type 2 diabetics treated for 4 weeks had an improvement in oral glucose tolerance tests and in insulin stimulated glucose disposal. KONRAD et al. (1999) examined the effect of lipoic acid on insulin sensitivity and glucose effectiveness after oral glucose tolerance tests (OGTTs) and modified frequently sampled intravenous glucose tolerance tests (FSIGTTs) in ten lean and ten obese patients with type 2 diabetes. FSIGTT data were analyzed by minimal modeling technique to determine SI and SG before and after lipoic acid treatment 600 mg, twice a day, for 4 weeks. Lipoic acid treatment was associated with increased glucose effectiveness in both the lean and diabetic groups. Higher sensitivity to insulin and lower fasting glucose were measured in lean diabetic patients only. Serum lactate and pyruvate, before and after glucose loading, were approximately 45% lower in both the lean and obese diabetic patients after lipoic acid treatment.

The mechanism by which thioctic acid improved insulin sensitivity is unclear. The stimulation of PDH activity by thioctic acid has been show in both skeletal muscle and in liver (Sugden et al. 1995). The binding of PDK to the E2 subunit of PDH is partially through a lipoic acid moiety (Loffelhardt et al. 1995). It is possible that the disruption of binding of PDK to the PDH would result in decreased phosphorylation of PDH and increased activity levels. A report by Estrada et al. (1996) has demonstrated that the improvement in insulin-stimulated transport in 3T3LI adipocytes is wortmanin-sensitive and may increase the activity of PI3 kinase. The mechanism by which this occurs remains obscure.

VI. Dehydroepiandrosterone

Dehydroepiandrosterone (DHEA) is a C19 steroid that is formed naturally in the adrenal gland from pregnenolone, though up to 25% of DHEA and metabolites in men is formed in the testis. Its immediate metabolic products are androstenedione and testosterone as well as the inactive sulfur conjugate, DHEA-S, which is the most abundant circulating steroid in the serum of humans. Obesity and type 2 diabetes is associated with lower DHEA/DHEA-S levels, perhaps by increasing clearance (Nestler 1995), while weight loss study in men demonstrated it was associated with a 125% increase in DHEA-S levels during weight reduction (Jakubowicz et al. 1995). As number of studies have demonstrated the usefulness of DHEA in animal models of insulin resistance. In obese Zucker rats, DHEA administration decreased food intake and produced a decrease in the selection of fat as foodstuff (Porter and Svec 1995). The weight loss associated with DHEA administration was observed only in obese animals while lean rats demonstrated a small weight gain (Wright et al. 1993). Treatment of insulin-resistant mice with high doses of etocholanones and related analogues of DHEA reduces insulin resistance and hyperglycemia, and can preserve β-cell function without changing food consumption (Coleman 1985).

Nestler et al. (1988) treated five healthy men with 1600 mg/day of DHEA for 28 days. There was a significant decrease in body fat and an increased in muscle mass. Bates et al. (1995) has proposed that DHEA may augment insulin sensitivity in postmenopausal women and attenuate the age-associated decrease in insulin tolerance. The molecular mechanism of these compounds is unclear though it is thought they work through a nuclear receptor pathway (Kroboth et al. 1999). There is no evidence that they activate the PPARγ receptor, but they produce a number of metabolic effects similar to thiazolidinediones. Further work on DHEA and related compounds is clearly needed since these compounds are becoming popular as an over-the-counter diet supplements and are at times taken in large quantities.

VII. IGF-1

Insulin-like growth factor 1 is related to the insulin receptor structurally and functionally. It binds to insulin and proinsulin at about 10% affinity compared to that of insulin. IGF-1 uses many of the same insulin signaling pathways as insulin, but the IGF-1 signaling pathways do not appear to develop "resistance" in obesity or type 2 diabetes (DUNGER and ACERINI 1997).

Direct demonstration that IGF-1 can substitute for insulin action in humans comes from studies in patients with mutations in the insulin receptor that abrogates signaling (type A insulin resistance). When recombinant human insulin-like growth factor I was given intravenously to three women with type A insulin resistance, an immediate fall in blood glucose was observed. The markedly elevated insulin and C-peptide levels fell in a parallel manner to blood glucose (SCHOENLE et al. 1991). A similar study demonstrated that the effectiveness of IGF-1 can be maintained over longer periods with subcutaneous injections (VESTERGAARD et al. 1997).

In normal humans, recombinant human IGF-I (rhIGF-I) was has been shown to decrease serum levels of insulin and C-peptide in fasted subjects without affecting plasma glucose levels. In their study, ZENOBI et al. (1992a) examined the responses of normal volunteers to intravenous rhIGF-I infusions during a standard oral glucose tolerance test and meal tolerance test. Glucose tolerance remained unchanged during the hIGF-I infusions despite lowering of insulin and C-peptide levels due to an enhanced tissue sensitivity to insulin. hIGF-I did not alter the ratio between C-peptide and insulin, suggesting that the metabolic clearance of endogenous insulin, which could augment insulin responses, remained unchanged. The authors suggest that part of the effect may be due to suppressed growth hormone secretion. In insulin resistant type 2 diabetes (ZENOBI et al. 1992b), subcutaneous injections of rhIGF-I ($2 \times 120\,\mu g/kg$) Dadministration increased total IGF-I serum levels 5.3-fold. With fasting, glucose decreased by nearly 40% while insulin and C-peptide concentrations fell 60%–70% and triglyceride levels declined by 40%. Postprandial areas under the glucose, insulin, and C-peptide curve decreased during an oral glucose tolerance test to 77%, 52%, and 60% of control, respectively. In a similar study, MOSES et al. (1996) determined the effect of 6 weeks of recombinant human IGF-I (rhIGF-I) administration on insulin resistance and glycemic control in obese insulin-resistant patients with type II diabetes. In addition to improvements in glycemic parameters and glucose tolerance, they showed a significant decline in HgbA1c levels (10.4% to 8.1%).

The enthusiasm for IGF-1 in the long-term treatment of insulin resistant diabetics is tempered by the occurrence of side effects of the therapy. These include edema of the face and hands, weight gain, dyspnea, jaw tenderness, arthralgias and myalgias, fatigue, tachycardia, flushing, orthostatic hypotension, and local burning at the injection site (JABRI et al. 1994).

VIII. AMP-Activated Protein Kinase Agonists

As described above (Sect. C), the activation of AMP-activated protein kinase is thought to be the mediator of the acute effects of exercise or hypoxia in stimulating glucose transport (HAYASHI et al. 1998). The activation of AMPK can also decrease the activity of acyl CoA carboxylase which generates malonyl CoA from acetyl CoA. Malonyl CoA is an allosteric inhibitor of CPT1 (WINDER and HARDIE 1999) and decreases in the level of malonyl CoA in muscle cells have been associated with increased fat and glucose oxidation. Indeed, as described above, treatment of rats with the AMPK activator AICAR can increase glucose utilization and decrease hepatic gluconeogenesis (RUSSELL et al. 1999). The development of agents that are orally available and are specific activators of AMPK may prove to be useful agents in the treatment of insulin resistance.

References

Adams M, Reginato MJ, Shao DL, Lazar MA, Chatterjee VK (1997) Transcriptional activation by peroxisome proliferator-activated receptor gamma is inhibited by phosphorylation at a consensus mitogen-activated protein kinase site. J Biol Chem 272:5128–5132

Amri EZ, Bonino F, Ailhaud G, Abumrad NA, Grimaldi PA (1995) Cloning of a protein that mediates transcriptional effects of fatty acids in preadipocytes. Homology to peroxisome proliferator-activated receptors. J Biol Chem 270:2367–2371

Anai M, Funaki M, Ogihara T, Terasaki J, Inukai K, Katagiri H, Fukushima Y, Yazaki Y, Kikuchi M, Oka Y, Asano T (1998) Altered expression levels and impaired steps in the pathway to phosphatidylinositol 3-kinase activation via insulin receptor substrates 1 and 2 in Zucker fatty rats. Diabetes 47:13–23

Arioglu E, Duncan Morin, JL, Gottlieb N, Lieberman J, Sumnar A, Rother KI, Seberin N, Reitman M, Reynolds J, Taylor SI (1999) Efficacy of troglitazone in the treatment of lipoatrophic diabetes: preliminary results of an ongoing study. Diabetes 48 [Suppl 1]:403

Auboeuf D, Rieusset J, Fajas L, Vallier P, Frering V, Riou JP, Staels B, Auwerx J, Laville M, Vidal H (1997) Tissue distribution and quantification of the expression of mRNAs of peroxisome proliferator-activated receptors and liver X receptor-alpha in humans: no alteration in adipose tissue of obese and NIDDM patients. Diabetes 46:1319–1327

Azevedo JL, Carey JO, Pories WJ, Morris PG, Dohm GL (1995) Hypoxia stimulates glucose transport in insulin-resistant human skeletal muscle. Diabetes 44:695–698

Barak Y, Nelson MC, Ong ES, Jones YZ, Ruiz-Lozano P, Chien KR, Koder A, Evans RM (1999) PPAR gamma is required for placental, cardiac, and adipose tissue development. Mol Cell 4:585–595

Baron AD, Ahu J-S, Zhu J-H, Weldon H, Maianu L, Garvey WT (1995) Glucosamine induces insulin resistance in vivo by affecting GLUT4 translocation in skeletal muscle. J Clin Invest 96:2792–2801

Barroso I, Gurnell M, Crowley VEF, Agostini M, Schwabe W, Soos MA, Li Maslen, G, Williams TDM, Lweis H, Schafer AJ, Chatterjee VKK, O'Rahilly S (1999) Nature 402:880–883

Bastard JP, Hainque B, Dusserre E, Bruckert E, Robin D, Vallier P, Perche S, Robin P, Turpin G, Jardel C, Laville M, Forest C, Vidal H (1999) Peroxisome proliferator activated receptor-gamma, leptin and tumor necrosis factor-alpha mRNA expression during very low calorie diet in subcutaneous adipose tissue in obese women. Diabetes Metab Res Rev 15:92–98

Bates GW Jr, Egerman RS, Umstot ES, Buster JE, Casson PR (1995) Dehydroepiandrosterone attenuates study-induced declines in insulin sensitivity in postmenopausal women. Ann NY Acad Sci 774:291–293

Baudry M, Kessler M, Smith EK, Lynch G (1982) The regulation of pyruvate dehydrogenase activity in rat hippocampal slices: effect of dichloroacetate. Neurosci Lett 31:41–46

Beamer BA, Negri C, Yen CJ et al (1997) Chromosomal localization and partial genomic structure of the human peroxisome proliferator activated receptor-gamma (hPPARgamma) gene. Biochem Biophys Res Commun 233:756–759

Berger J, Leibowitz MD, Doebber TW, Elbrecht A, Zhang B, Zhou G, Biswas C, Cullinan CA, Hayes NS, Li Y, Tanen M, Ventre J, Wu MS, Berger GD, Mosley R, Marquis R, Santini C, Sahoo SP, Tolman RL, Smith RG, Moller DE (1999) Novel peroxisome proliferator-activated receptor (PPAR) gamma and PPARdelta ligands produce distinct biological effects. J Biol Chem 274:6718–6725

Bergeron R, Russell RB, Young LH, Ren JM, Marcucci M, Lee A, Shulman GI (1999) Effect of AMPK activation on muscle glucose metabolism in conscious rats. Am J Physiol 276:E938–E934

Black K, Qu X, Seale JP, Donnelly R (1998) Metabolic effects of thioctic acid in rodent models of insulin resistance and diabetes. Clin Exp Pharmacol Physiol 25:712–714

Boden G (1997) Role of fatty acids in the pathogenesis of insulin resistance and NIDDM. Diabetes 46:3–10

Bonen A, Tan MH, Watson-Wright WM (1984) Effects of exercise on insulin binding and glucose metabolism in muscle. Can J Physiol Pharmacol 62:1500–1504

Borkman M, Storlien LH, Pan DA, Jenkins AB, Chisholm DJ, Campbell LV (1993) The relations between insulin sensitivity and the fatty acid composition of skeletal-muscle phospholipids. N Engl J Med 238:238–244

Braissant O, Foufelle F, Scotto C, Dauca M, Wahli W (1996) Differential expression of peroxisome proliferator-activated receptors (PPARs): tissue distribution of PPAR-alpha, -beta, and -gamma in the adult rat. Endocrinology 137:354–366

Brozinick JT Jr, Etgen JG Jr, Yaspelkis BB III, Ivy JL (1992) Contraction-activated glucose uptake is normal in insulin-resistant muscle of the obese Zucker rat. J Appl Physiol 73:382–387

Burant C, Sreenan GS, Hirano K, Tai T-A C, Lohmiller J, Lukens J, Davidson NO, Ross S, Graves RA (1997) Troglitazone action is independent of adipose tissue. J Clin Invest 100:2900–2908

Buse JB, Gumbiner B, Mathias NP, Nelson DM, Faja BW, Whitcomb RW (1998) Troglitazone use in insulin-treated type 2 diabetic patients. The Troglitazone Insulin Study Group. Diabetes Care 21:1455–1461

Cartee GD, Young DA, Sleeper MD, Zierath J, Wallberg-Henriksson H, Holloszy JO (1989) Prolonged increase in insulin-stimulated glucose transport in muscle after exercise. Am J Physiol 256:E494–E499

Chan CB, MacPhail RM, Sheu L, Wheeler MB, Gaisano HY (1999) Beta-cell hypertrophy in fa/fa rats is associated with basal glucose hypersensitivity and reduced SNARE protein expression. Diabetes 48:997–1005

Chawla A, Schwarz EJ, Dimaculangan DD, Lazar MA (1994) Peroxisome proliferator-activated receptor (PPAR) gamma: adipose-predominant expression and induction early in adipocyte differentiation. Endocrinology 135:798–800

Chen H, Charlat O, Tartaglia LA, Woolf EA, Weng X, Ellis SJ, Lakey ND, Culpepper J, Moore KJ, Breitbart RE, Duyk GM, Tepper RI, Morgenstern JP (1996) Evidence that the diabetes gene encodes the leptin receptor: identification of a mutation in the leptin receptor gene in db/db mice. Cell 84:491–495

Coleman DL (1985) Antiobesity effects of etiocholanolones in diabetes (db), viable yellow (Avy), and normal mice. Endocrinology 117:2279–2283

Cusi D, Cukier S, Defronzo RA, Torres M (1997) Metabolic effects of treatment with banakyl sulfate in NDDM. Diabetes 46:34A

DeFronzo RA (1988) Lilly lecture. The triumvirate: β-cell, muscle, liver. A collusion responsible for NIDDM. Diabetes 37:667–687

De Fronzo RA, Sherwin RS, Kraemer N (1987) Effect of physical training on insulin action in obesity. Diabetes 26:1379–1386

Devlin JT, Hirschman M, Horton ED, Horton ES (1987) Enhanced peripheral and splanchnic insulin sensitivity in NIDDM after single bout of exercise. Diabetes 36:434–439

Douen AG, Ramlal T, Rastogi S, Bilan PJ, Cartee GD, Vranic M, Holloszy JO, Klip A (1990) Exercise induces recruitment of the "insulin-responsive glucose transporter". Evidence for distinct intracellular insulin- and exercise-recruitable transporter pools in skeletal muscle. J Biol Chem 265:13427–13430

Drener A, Laurent D, Marcucci M, Griffin ME, Dufour S, Cline GW, Slezak LA, Andersine DD, Hundal RS, Rothmand DL, Petersen DF, Shulman GI (1999) Effects of free fatty acids on glucose transport and IRS-1-associated phosphatidylinsoitol 3-kinase activity. J Clin Invest 103:253–259

Dunger DB, Acerini CL (1997) Does recombinant human insulin-like growth factor-1 have a role in the treatment of diabetes? Diabet Med 14:723–731

Ebeling P, Bourey R, Koranyi L (1993) Mechanism of enhanced insulin sensitivity in athletes: increased blood flow, muscle glucose transport protein (GLUT4) concentration and glycogen synthase activity. J Clin Invest 92:1623–1631

Elbrecht A, Chen Y, Cullinan CA et al (1996) Molecular cloning, expression and characterization of human peroxisome proliferator activated receptors gamma 1 and gamma 2. Biochem Biophys Res Commun 224:431–437

Eldershaw TP, Rattigan S, Cawthorne MA, Buckingham RE, Colquhoun EQ, Clark MG (1995) Treatment with the thiazolidinedione (BRL 49653) decreases insulin resistance in obese Zucker hindlimb. Horm Metab Res 27:169–172

Erickson J, Franssila-Kallunki A, Ekstrad A, Saloranta C, Widen E, Schalin C, Groop L. (1989) Early metabolic defects in persons at increased risk for non-insulin dependent diabetes mellitus. N Eng J Med 321:337–343

Estrada DE, Ewart HS, Tsakiridis T, Volchuk A, Ramlal T, Tritschler H, Klip A (1996) Stimulation of glucose uptake by the natural coenzyme alpha-lipoic acid/thioctic acid: participation of elements of the insulin signaling pathway. Diabetes 45: 1798–1804

Evans RM (1988) The steroid and thyroid hormone receptor superfamily. Science 240:889–895

Fajas L, Auboeuf D, Raspe E, Schoonjans K, Lefebvre AM, Saladin R, Najib J, Laville M, Fruchart JC, Deeb S, Vidal-Puig A, Flier J, Briggs MR, Staels B, Vidal H, Auwerx J (1997) The organization, promoter analysis, and expression of the human PPARgamma gene. J Biol Chem 272:18779–18789

Forman BM, Tontonoz P, Chen J, Brun RP, Spiegelman BM, Evans RM (1995) 15-Deoxy-delta 12, 14-prostaglandin J2 is a ligand for the adipocyte determination factor PPARgamma. Cell 83:803–812

Forman BM, Chen J, Evans RM (1996) The peroxisome proliferator-activated receptors: ligands and activators. Ann NY Acad Sci 804:266–275

Fruchart JC, Duriez P, Staels B (1999) Peroxisome proliferator-activated receptor-alpha activators regulate genes governing lipoprotein metabolism, vascular inflammation and atherosclerosis. Curr Opin Lipidol 3:245–257

Fujii M, Takemura R, Yamaguchi M, Hasegawa G, Shigeta H, Nakano K, Kondo M (1997) Troglitazone (CS-045) ameliorates albuminuria in streptozotocin-induced diabetic rats. Metabolism 46:981–983

Fujita T, Sugiyama Y, Taketomi S, Sohda T, Dawamatsu Y, Iwatsuka H, Suzuoki Z (1993) Reduction of insulin resistance in obese and/or diabetic animals by 3[-4-(1-methylcyclohexylmithoxy)benzyl]-thiazolidine-2,4-dione (ADD-3878, U63,287, ciglitazone), a new antidiabetic agent. Diabetes 32:804–810

Fujiwara T, Wada M, Fukuda K, Fukami M, Yoshioka S, Yoshioka T, Horikoshi H (1991) Characterization of CS-045, a new oral antidiabetic agent, II. Effects on glycemic control and pancreatic islet structure at a late stage of the diabetic syndrome in C57BL/KsJ-db/db mice. Metabolism 40:1213–1218

Gao J, Ren J-M, Gulve EA, Holloszy JO (1994) Additive effect of contractions and insulin on GLUT4 translocation into the sarcolemma. J Appl Physiol 77:1597–1601

Garetto LP, Richter EA, Goodman MN, Ruderman NB (1984) Enhanced muscle glucose metabolism in the rat: the two phases. Am J Physiol 246:E472–E475

Goldfine AB, Simonson DC, Folli F, Patti ME, Kahn CR (1995) Metabolic effects of sodium metavanadate in humans with insulin-dependent and noninsulin-dependent diabetes mellitus in vivo and in vitro studies. J Clin Endocrinol Metab 80:3311–3320

Goodyear LJ, Giorgino F, Balon TW, Condorelli G, Smith RJ (1995) Effects of contractile activity on tyrosine phosphoproteins and PI 3-kinase activity in rat skeletal muscle. Am J Physiol 268:E987–E995

Goodyear LJ, Hirshman MF, King PA, Horton ED, Thompson CM, Horton ES (1990) Skeletal muscle plasma membrane glucose transport and glucose transporters after exercise. J Appl Physiol 68:193–198

Gumbiner B, Mucha JF, Lindstrom JE, Rekhi I, Livingston JN (1996) Differential effects of acute hypertriglyceridemia on insulin action and insulin receptor autophosphorylation. Am J Physiol 270(3 Pt 1):E424–E429

Halberstam M, Cohen N, Shlimovich P, Rossetti L, Shamoon H (1996) Oral vanadyl sulfate improves insulin sensitivity in NIDDM but not in obese nondiabetic subjects. Diabetes 45:659–666

Hansen PA, Nolte LA, Chen MM, Holloszy JO (1998) Increased GLUT4 translocation mediates enhanced insulin sensitivity of muscle glucose transport after exercise. J Appl Physiol 85:1218–1222

Hayashi T, Hirshman MF, Kurth EJ, Winder WW, Goodyear LJ (1998) Evidence for 5'-AMP-activated protein kinase mediation of the effect of muscle contraction on glucose transport. Diabetes 47:1369–1373

He TC, Chan TA, Vogelstein B, Kinzler KW (1999) PPARdelta is an APC-regulated target of nonsteroidal anti-inflammatory drugs. Cell 99(3):335–345

Helmrich SP, Ragland DR, Leung RW, Paffenbarger RS (1991) Physical activity and reduced occurrence of non-insulin-dependent diabetes mellitus. N Engl J Med 325:147–152

Henriksen EJ, Bourey RE, Rodnick KJ, Koranyi L, Permutt MA, Holloszy JO (1990) Glucose transporter protein content and glucose transport capacity in rat skeletal muscles. Am J Physiol 259:E593–E598

Herbert LF, Daniels MC, Zhou J, Crook ED, Turner RL, Simmons ST, Neidigh JL, Shu J-S, Baron AD, McClain DA (1996) Overexpression of glutamine:fructose-6-phosphate amidotransferase in skeletal muscle of transgenic mice leads to insulin resistance. J Clin Invest 98:930–936

Horton ES, Whitehouse F, Ghazzi MN, Venable TC, Whitcomb RW (1998) Troglitazone in combination with sulfonylurea restores glycemic control in patients with type 2 diabetes. The Troglitazone Study Group. Diabetes Care 9:1462–1469

Hotamisligil GS, Peraldi P, Budavari A, Ellis R, White MF, Speigelman BM (1996) IRS-1-mediated inhibition of insulin receptor tyrosine kinase activity in TNF-a and obesity-induced insulin resistance. Science 271:665–668

Hotamisligil GS, Speigelman BM (1994) Tumor necrosis factor. A key component of the obesity-diabetes link. Diabetes 43:1271–1278

Houmard JA, Egan PC, Neufer OD (1991) Elevated skeletal muscle glucose transporter levels in exercise-trained middle-aged men. Am J Physiol 261:E437–E443

Hu E, Tontonoz P, Spiegelman BM (1995) Transdifferentiation of myoblasts by the adipogenic transcription factors PPAR gamma and C/EBPalpha. Proc Natl Acad Sci 21:9856–9860

Inoue I, Takahashi K, Katayama S, Harada Y, Negishi K, Itabashi A, Ishii J (1995) Effect of troglitazone (CS-045) and bezafibrate on glucose tolerance, liver glycogen synthase activity, and β-oxidation in fructose-fed rats. Metabolism 44:1626–1630

Inzucchi SE, Maggs DG, Spollett GR, Page SL, Rife FS, Walton V, Shulman GI (1998) Efficacy and metabolic effects of metformin and troglitazone in type II diabetes mellitus. N Engl J Med 338:867–872

Issemar I, Green S (1990) Activation of a member of the hormone receptor superfamily by peroxisome proliferators. Nature 347:645–680
Iwamoto Y, Kuzuya T, Matsuda A, Awata T, Kumakura S, Inooka G, Shiraishi I (1991) Effect of new oral antidiabetic agent CS-045 on glucose tolerance and insulin secretion in patients with NIDDM. Diabetes Care 14:1083–1086
Jabri N, Schalch DS, Schwartz SL, Fischer JS, Kipnes MS, Radnik BJ, Turman NJ, Marcsisin VS, Guler HP (1994) Adverse effects of recombinant human insulin-like growth factor I in obese insulin-resistant type II diabetic patients. Diabetes 43:369–374
Jakubowicz DJ, Beer NA, Beer RM, Nestler JE (1995) Disparate effects of weight reduction by diet on serum dehydroepiandrosterone-sulfate levels in obese men and women. J Clin Endocrinol Metab 80:3373–3376
Johnson AB, Webster JM, Sum C-F, Heseltine L, Argyraki M, Cooper BG, Taylor R (1993) The impact of metformin therapy on hepatic glucose production and skeletal muscle glycogen synthase activity in overweight type II diabetic patients. Metabolism 42:1217–1222
Kahn CR (1985) The molecular mechanism of insulin action. Ann Rev Med 36:429–451
Kallen CB, Lazar MA (1996) Antidiabetic thiazolidinediones inhibit leptin (ob) gene expression in 3T3-L1 adipocytes. Proc Natl Acad Sci 93:5793–5796
Kaumi T, Hirano T, Odaka H, Ebara T, Amano N, Hozumi T, Ishida Y, Yoshino G (1996) VLDL triglyceride kinetics in Wistar fatty rats, an animal model of NIDDM: effects of dietary fructose alone or in combination with pioglitazone. Diabetes 45:806–811
Kelley DE, Simoneau JA (1994) Impaired free fatty acid utilization in non-insulin dependent diabetes mellitus. J Clin Invest 94:2349–2356
Kelly IE, Han TS, Walsh K, Lean ME (1999) Effects of a thiazolidinedione compound on body fat and fat distribution of patients with type 2 diabetes. Diabetes Care 22:288–293
Kennedy JW, Hirshman MF, Gervino EV, Ocel JV, Forse RA, Hoenig SJ, Aronson D, Goodyear LJ, Horton ES (1999) Acute exercise induced GLUT4 translocation in skeletal muscle of normal human subjects and subjects with type 2 diabetes. Diabetes 48:1192–1197
King PA, Betts JJ, Horton ED, Horton ES (1993) Exercise, unlike insulin, promotes glucose transporter translocation in obese Zucker rat muscle. Am J Physiol 265:R447–R452
Kletzien RF, Foellmi LA, Harris PK, Wyse BM, Clarke SD (1992) Adipocyte fatty acid-binding protein: regulation of gene expression in vivo and in vitro by an insulin-sensitizing agent. Mol Pharmacol 42:558–562
Koch M, Rett K, Maerker E et al (1999) The PPAR gamma2 amino acid polymorphism Pro 12 Ala is prevalent in offspring of Type II diabetic patients and is associated to increased insulin sensitivity in a subgroup of obese subjects. Diabetologia 42:758–762
Konrad T, Vicini P, Kusterer K, Hoflich A, Assadkhani A, Bohles HJ, Sewell A, Tritschler HJ, Cobelli C, Usadel KH (1999) α-Lipoic acid treatment decreases serum lactate and pyruvate concentrations and improves glucose effectiveness in lean and obese patients with type 2 diabetes. Diabetes Care 22:280–287
Kornfeld R (1967) Studies on L-glutamine D-fructose 6-phosphate amidotransferase. I. Feedback inhibition by uridine diphosphate-*N*-acetylglucosamine. J Biol Chem 242:3135–3141
Kroboth PD, Salek FS, Pittenger AL, Fabian TJ, Frye RF (1999) DHEA and DHEA-S: A review. J Clin Pharmacol 39:327–348
Krook A, Kawano Y, Song XM, Efendic S, Roth RA, Wallbergg-Henriksson H, Zierath JR (1997) Improved glucose tolerance restores insulin-stimulated Akt kinase activity and glucose transport in skeletal muscle from diabetic Goto-Kakizaki rats. Diabetes 46:2110–2114

Kubota N, Terauchi Y, Miki H, Tamemoto H, Yamauchi T, Komeda K, Satoh S, Nakano R, Ishii C, Sugiyama T, Eto K, Tsubamoto Y, Okuno A, Murakami K, Sekihara H, Hasegawa G, Naito M, Toyoshima Y, Tanaka S, Shiota K, Kitamura T, Fujita T, Ezaki O, Aizawa S, Kadowaki T (1999) PPAR gamma mediates high-fat diet-induced adipocyte hypertrophy and insulin resistance. Mol Cell 4:597–609

Kurlemann G, Paetzke I, Moller H, Masur H, Schuierer G, Weglage J, Koch HG (1995) Therapy of complex I deficiency: peripheral neuropathy during dichloroacetate therapy. Eur J Pediatr 154:928–932

Lee AD, Hansen PA, Holloszy JO (1995) Wortmannin inhibits insulin-stimulated but not contraction-stimulated glucose transport activity in skeletal muscle. FEBS Letters 361:51–54

Lefebvre AM, Peinado-Onsurbe J, Leitersdorf I, Briggs MR, Paterniti JR, Fruchart JC, Fievet C, Auwerx J, Staels B (1997) Regulation of lipoprotein metabolism by thiazolidinediones occurs through a distinct but complementary mechanism relative to fibrates. Arterioscler Thromb Vasc Biol 17:1756–1764

Lehmann JM, Moore LB, Smith-Oliver TA, Wilkison WO, Willson TM, Kliewer SA (1995) An antidiabetic thiazolidinedione is a high affinity ligand for peroxisome proliferator-activated receptor gamma (PPAR gamma). J Biol Chem 270: 12953–12956

Lenhard JM, Lancaster ME, Paulik MA, Weiel JE, Binz JG, Sundseth SS, Gaskill BA, Lightfoot RM, Brown HR (1999) The RXR agonist LG100268 causes hepatomegaly, improves glycaemic control and decreases cardiovascular risk and cachexia in diabetic mice suffering from pancreatic beta-cell dysfunction. Diabetologia 42:545–554

Loffelhardt S, Bonaventura C, Locher M, Borbe HO, Bisswanger H (1995) Interaction of alpha-lipoic acid enantiomers and homologues with the enzyme components of the mammalian pyruvate dehydrogenase complex. Biochem Pharmacol 50:637–646

Lund S, Holman GD, Schmitz O, Pedersen O (1995) Contraction stimulates translocation of glucose transporter GLUT4 in skeletal muscle through a mechanism distinct from that of insulin. Proc Natl Acad Sci 92:5817–5821

Maggs DG, Buchanan TA, Burant CF, Cline G, Gumbiner B, Hsueh WA, Inzucchi S, Kelley D, Nolan J, Olefsky JM, Polonsky KS, Silver D, Valiquett TR, Shulman GI (1998) Metabolic effects of troglitazone monotherapy in type 2 diabetes mellitus. A randomized, double-blind, placebo-controlled trial. Ann Intern Med 128:176–185

Mancini FP, Vaccaro O, Sabatino L, et al. (1999) Pro12Ala substitution in the peroxisome proliferator-activated receptor- gamma2 is not associated with type 2 diabetes. Diabetes 48:1466–1468

Mangelsdorf DJ, Evans RM (1995) The RXR heterodimers and orphan receptors. Cell 83:841–850

Mangelsdorf DJ, Thummel C, Beato M, Herrlich P, Schutz G, Umesono K, Kastner P, Mark M, Chambon P, Evans RM (1995) The nuclear receptor superfamily: the second decade. Cell 83:835–839

Marshall S, Bacote V, Traxinger RR (1991) Discovery of a metabolic pathway mediating glucose-induced desensitization of the glucose transport system: role of hexosamine biosynthesis in the induction of insulin resistance. J Biol Chem 266:4706–4712

Martin G, Schoonjans K, Lefebvre AM, Staels B, Auwerx J (1997) Coordinate regulation of the expression of the fatty acid transport protein and acyl-CoA synthetase genes by PPARalpha and PPARgamma activators. J Biol Chem 272:28210–28217

Matsui H, Okumura K, Kawakami K, Hibino M, Toki Y, Ito T (1997) Improved insulin sensitivity by bezafibrate in rats: relationship to fatty acid composition of skeletal-muscle triglycerides. Diabetes 46:348–353

McGarry JD (1994) Disordered metabolism in diabetes: have we underemphasized the fat component? J Cell Biochem 55:29–38

Merril GG, Kurth EJ, Hardie DG, Winder WW (1997) AICA riboside increases AMP-activated protein kinase, fatty acid oxidation and glucose uptake in rat muscle. Am J Physiol 273: E1107–E1112
Meyerovitch J, Farfel Z, Sack J, Shechter Y (1987) Oral administration of vanadate normalizes blood glucose levels in streptozotocin-treated rats. Characterization and mode of action. J Biol Chem 262:6658–6662
Meyerovitch J, Rothenberg P, Shechter Y, Bonner-Weir S, Kahn CR (1991) Vanadate normalizes hyperglycemia in two mouse models of non-insulin-dependent diabetes mellitus. J Clin Invest 87:1286–1294
Miles PDG, Romeo OM, Higo K, Cohen A, Rafaat K, Olefsky JM (1997) TNF-a-induced insulin Resistance in vivo and its prevention by troglitazone. Diabetes 46:1678–1689
Mimura K, Umeda F, Hiramatsu S, Taniguchi S, Ono Y, Nakashima N, Kobayashi K, Masakado M, Sako Y, Nawata H (1994) Effects of a new oral hypoglycaemic agent (CS-045) on metabolic abnormalities and insulin resistance in type 2 diabetes. Diabet Med 11:685–691
Moller DE (1993) Insulin Resistance. Wiley, London
Moller DE, Chang PY, Yaspelkis BB, Flier JS, Wallberg-Henriksson H, Ivy JL (1996) Transgenic mice with muscle-specific insulin resistance develop increased adiposity, impaired glucose tolerance, and dyslipidemia. Endocrinol 137:2397–2405
Mongold JJ, Cros GH, Vian L, Tep A, Ramanadham S, Siou G, Diaz J, McNeill JH, Serrano JJ (1990) Toxicological aspects of vanadyl sulphate on diabetic rats: effects on vanadium levels and pancreatic β-cell morphology. Pharmacol Toxicol 67:192–198
Moses AC, Young SC, Morrow LA, O'Brien M, Clemmons DR (1996) Recombinant human insulin-like growth factor I increases insulin sensitivity and improves glycemic control in type II diabetes. Diabetes 45:91–100
Mukherjee R, Davies PJ, Crombie DL, Bischoff ED, Cesario RM, Jow L, Hamann LG, Boehm F, Mondon CE, Nadzan AM, Paterniti JR Jr, Heyman RA (1997) Sensitization of diabetic and obese mice to insulin by retinoid X receptor agonists. Nature 386:407–410
Nesher R, Karl IE, Kipnis DM (1985) Dissociation of effects of insulin and contraction on glucose transport in rat epitrochlearis muscle. Am J Physiol 249:C226–C232
Nestler JE (1995) Regulation of human dehydroepiandrosterone metabolism by insulin. Ann N Y Acad Sci 774:73–81
Nestler JE, Barlascini CO, Clore JN, Blackard WG (1988) Dehydroepiandrosterone reduces serum low density lipoprotein levels and body fat but does not alter insulin sensitivity in normal men. J Clin Endocrinol Metab 66:57–61
Oakes ND, Cooney GJ, Camilleri S, Chisholm DJ, Kraegen EW (1997) Mechanisms of liver and muscle insulin resistance induced by chronic high-fat feeding. Diabetes 46:1768–1774A
Oakes ND, Bell KS, Furler SM, Camilleri S, Saha AK, Ruderman NB, Chisholm DJ, Kraegen EW (1997) Diet-induced muscle insulin resistance in rats is ameliorated by acute dietary lipid withdrawal or a single bout of Exercise. Parallel relationship between insulin stimulation of glucose uptake and suppression of long-chain fatty Acyl CoA. Diabetes 46:2022–2028B
Oizumi J, Hayakawa K (1989) Liberation of lipoate by human serum lipoamidase from bovine heart pyruvate dehydrogenase. Biochem Biophys Res Commun 162:658–663
Okuno A, Ikeda K, Shiota M, Fujiwara T, Yoshioka S, Sugano T, Horikoshi H (1997) Acute effect of troglitazone on glucose metabolism in the absence or presence of insulin in perfused rat hindlimb. Metabolism 46:716–721
Okuno A, Tamemoto H, Tobe K, Ueki K, Mori Y, Iwamoto K, Umesono K, Akanuma Y, Fujiwara T, Horikoshi H, Yazaki Y, Kadowaki T (1998) Troglitazone increases the number of small adipocytes without the change of white adipose tissue mass in obese Zucker rats. Clin Invest 101:1354–1361

Oshida Y, Yamanouchi K, Hayamizu S, Sato Y (1989) Long-term mild jogging increase insulin action despite no influence on body mass index or VO2max. J Appl Physiol 66:2206–2210

Park KS, Ciaraldi TP, Abrams-Carter L, Mudaliar S, Nikoulina SE, Henry RR (1998) PPAR-gamma gene expression is elevated in skeletal muscle of obese and type II diabetic subjects. Diabetes 46:1230–1234

Peraldi P, Xu M, Spiegelman BM (1997) Thiazolidinediones block tumor necrosis factor-alpha-induced inhibition of insulin signaling. J Clin Invest 100:1863–1869

Perseghin G, Price TB, Petersen KF, Roden M, Cline GW, Gerow K, Rothman DL, Shulman GI (1996) Increased glucose transport-phosphorylation and muscle glycogen synthesis after exercise training in insulin-resistant subjects. New Eng J Med 335:1357–1362

Phillips DI, Caddy S, Ilic V, Fielding BA, Frayn KN, Borthwick AC, Taylor R (1996) Intramuscular triglyceride and muscle insulin sensitivity: evidence for a relationship in nondiabetic subjects. Metabolism 45:947–950

Pick A, Clark J, Kubstrup C, Levisetti M, Pugh W, Bonner-Weir S, Polonsky KS (1998) Role of apoptosis in failure of beta-cell mass compensation for insulin resistance and beta-cell defects in the male Zucker diabetic fatty rat. Diabetes 47:358–364

Polonsky KS (1995) Lilly Lecture. The beta-cell in diabetes: from molecular genetics to clinical research. Diabetes 44:705–717

Popov KM, Hawes JW, Harris RA (1997) Mitochondrial alpha-ketoacid dehydrogenase kinases: a new family of protein kinases. Adv Second Messenger Phosphoprotein Res 31:105–111

Porter JR, Svec F (1995) DHEA diminishes fat food intake in lean and obese Zucker rats. Ann N Y Acad Sci 774:329–331

Qian H, Hausman GJ, Compton MM, Azain MJ, Hartzell DL, Baile CA (1998) Leptin regulation of peroxisome proliferator-activated receptor-gamma, tumor necrosis factor, and uncoupling protein-2 expression in adipose tissues. Biochem Biophys Res Commun 246:660–667

Randale PJ, Garland PB, Hales CN, Newsholme EA (1963) The glucose fatty-acid cycle: its role in insulin sensitivity and the metabolic disturbances of diabetes mellitus. Lancet I:785–789

Reaven GM (1995) The fourth musketeer – from Alexandre Dumas to Claude Bernard. Diabetologia 38:3–13

Richter EA, Mikines KJ, Galbo H, Kiens B (1989) Effect of exercise on insulin action in human skeletal muscle. J Appl Physio 66:876–885

Richter EA, Garetto LP, Goodman MN, Ruderman NB (1982) Muscle glucose metabolism following exercise in the rat. Increased sensitivity to insulin. J Clin Invest 69:785–793

Ringel J, Engeli S, Distler A, Sharma AM (1999) Pro12Ala missense mutation of the peroxisome proliferator activated receptor gamma and diabetes mellitus. Biochem Biophys Res Commun 254:450–453

Ristow M, Muller-Wieland D, Pfeiffer A, Krone W, Kahn CR (1998) Obesity associated with a mutation in a genetic regulator of adipocyte differentiation. N Engl J Med 339:953–959

Robinson KA, Weinstein ML, Lindenmayer GE, Buse MG (1995) Effects of diabetes and hyperglycemia on the hexosamine synthesis pathway in rat muscle and liver. Diabetes 44:1438–1446

Roden M, Price TB, Perseghin G, Peterson KF, Rothman DL, Cline GW, Shulman GI (1996) Mechanism of free fatty acid-induced resistance in humans. J Clin Invest 97:2859–2865

Rosen ED, Sarraf P, Troy AE, Bradwin G, Moore K, Milstone DS, Spiegelman BM, Mortensen RM (1999) PPAR gamma is required for the differentiation of adipose tissue in vivo and in vitro. Mol Cell 4:611–617

Rothman DL, Shulman RG, Shulman GE (1992) ^{31}P nuclear magnetic resonance measurements of muscle glucose-6-phosphate: evidence for reduced insulin-dependent

glucose transport or phosphorylation activity in non-insulin-dependent diabetes mellitus. J Clin Invest 89:1069–1075
Russell RR III, Bergeron R, Shulman GI, Young LH (1999) Translocation of myocardial GLUT-4 and increased glucose uptake through activation of AMPK by AICAR. Am J Physiol 277(2 Pt 2):H643–H649
Saad MJ, Araki E, Miralpeix M, Rothenberg PL, White MF, Kahn CR (1992) Regulation of insulin receptor substrate-1 in liver and muscle of animal models of insulin resistance. J Clin Invest 90:1839–1849
Saha AK, Kurowski TG, Ruderman NB (1995) A malonyl CoA fuel-sensing mechanism in muscle: Effects of insulin, glucose and denervation. Am J Physiol 269:E95–E101
Sakul H, Pratley R, Cardon L, Ravussin E, Mott D, Bogardus C (1997) Familiality of physical and metabolic characteristics that predict the development of non-insulin-dependent diabetes mellitus in Pima Indians. Am J Hum Genet 60:651–656
Saltiel AR, Olefsky JM (1996) Thiazolidinediones in the treatment of insulin resistance and Type II diabetes. Diabetes 45:1661–1669
Schoenle EJ, Zenobi PD, Torresani T, Werder EA, Zachmann M, Froesch ER (1991) Recombinant human insulin-like growth factor I (rhIGF I) reduces hyperglycaemia in patients with extreme insulin resistance. Diabetologia 34:675–679
Schoonjans K, Staels B, Auwerx J (1996) Role of the peroxisome proliferator activated receptor (PPAR) in mediating effects of fibrates and fatty acids on gene expression. J Lipid Res 37:907–925A
Schoonjans K, Staels B, Auwerx J (1996) The peroxisomal proliferator activated receptors (PPARs) and their effects on lipid metabolism and adipocyte differentiation. Biochim Biophys Acta 1302:93–109B
Schwartz S, Raskin P, Fonseca V, Graveline JF (1998) Effect of troglitazone in insulin-treated patients with type II diabetes mellitus. Troglitazone and Exogenous Insulin Study Group. N Engl J Med 338:861–866
Sherman SI, Gopal J, Haugen BR, Chiu AC, Whaley K, Nowlakha P, Duvic M (1999) Central hypothyroidism associated with retinoid X receptor-selective ligands. N Engl J Med 340:1075–1079
Shimabukuro M, Zhou YT, Lee Y, Unger RH (1998) Troglitazone lowers islet fat and restores beta cell function of Zucker diabetic fatty rats. J Biol Chem 273:3547–3550
Shimizu H, Tsuchiya T, Sato N, Shimomura Y, Kobayashi I, Mori M (1998) Troglitazone reduces plasma leptin concentration but increases hunger in NIDDM patients. Diabetes Care 21:1470–1474
Sims EAH, Danforth, E Horton ES, Bray GA, Glennon JA, Salan LB (1973) Endocrine and metabolic effects of experimental obesity in man. Rec Prog Horm Res 29:457–496
Spiegelman BM, Flier JS (1996) Adipogenesis and obesity: Rounding out the big picture. Cell 87:377–389
Sreenan S, Keck S, Fuller T, Cockburn B, Burant CF (1999) Effects of troglitazone on substrate storage and utilization in insulin-resistant rats. Am J Physiol 276(6 Pt 1):E1119–E1129
Sreenan S, Sturis J, Pugh W, Burant CF, Polonsky KS (1996) Prevention of hyperglycemia in the Zucker diabetic fatty rat by treatment with metformin and troglitazone. Am J Physiol 271:E742–E774
Stacpoole PW (1989) The pharmacology of dichloroacetate. Metabolism 38:1124–1144
Stacpoole PW, Greene YJ (1992) Dichloroacetate. Diabetes Care 15:785–791
Stumvoll M, Nurjhan N, Perriello G, Dailey G, Gerich JE (1996) Metabolic effects of metformin in non-insulin-dependent diabetes mellitus. N Engl J Med 333:550–554
Sturis J, Pugh WL, Tang J, Polonsky KS (1995) Prevention of diabetes does not completely prevent insulin secretory defects in the ZDF rat. Am J Physiol 269(4 Pt 1):E786–E792
Sugden MC, Orfali KA, Holness MJ (1995) The pyruvate dehydrogenase complex: nutrient control and the pathogenesis of insulin resistance. J Nut 125:1746S–1752S

Sugiyama Y, Shimura Y, Ikeda H (1990) Effects of pioglitazone on hepatic and peripheral insulin resistance in Wistar fatty rats. Arzneimittelforschung 40:436–440

Sunaga Y, Inagaki N, Gonoi T, Yamada Y, Ishida H, Seino Y, Seino S (1999) Troglitazone but not pioglitazone affects ATP-sensitive K(+) channel activity. Eur J Pharmacol 381:71–76

Szalkowski D, White-Carrington S, Berger J, Zhang B (1995) Antidiabetic thiazolidinediones block the inhibitory effect of tumor necrosis factor-alpha on differentiation, insulin-stimulated glucose uptake, and gene expression in 3T3-L1 cells. Endocrinology 136:1474–1481

Tanaka T, Yoshida N, Kishimoto T, Akira S (1997) Defective adipocyte differentiation in mice lacking the C/EBPβ and/or C/EBPδ gene. EMBO J 16:7432–7443

The Diabetes Control and Complications Trial Research Group (1993) The effect of intensive treatment of diabetes on the development and progression of long-term complications in insulin-dependent diabetes mellitus. N Engl J Med 329:997–986

Thorell A, Hirshman MF, Nygren J, Jorfeldt L, Wojtaszewski JFP, Dufresne SD, Horton ES, Ljungqvist O, Goodyear LJ (1999) Exercise and insulin cause GLUT4 translocation in human skeletal muscle. Am J Physiol 277:E733–E741

Tokuyama Y, Sturis J, DePaoli AM, Takeda J, Stoffel M, Tang J, Sun X, Polonsky KS, Bell GI (1995) Evolution of beta-cell dysfunction in the male Zucker diabetic fatty rat. Diabetes 44:1447–1457

Tomino RP (1989) Effect of physical training on the insulin resistance of aging. Am J Physiol 256:E352–E356

Tontonoz P, Hu E, Spiegelman BM (1994) Stimulation of adipogenesis in fibroblasts by PPAR gamma 2, a lipid-activated transcription factor. Cell 79:1147–1156A

Tontonoz P, Hu E, Graves RA, Budavari AI, Spiegelman BM (1994) mPPARγ: Tissue-specific regulator of an adipocyte enhancer. Genes Dev 8:18779–18789B

Treadway JL, James DE, Burcel E, Ruderman NB (1989) Effect of exercise on insulin receptor binding and kinase activity in skeletal muscle. Am J Physiol 256: E138–E144

Turner RC (1998) The U.K. Prospective Diabetes Study. A review. Diabetes Care 3:C35–C38

Uysal KT, Wiesbrock SM, Marino MW, Hotamisligil GS (1997) Protection from obesity-induced insulin resistance in mice lacking TNF-a function. Nature 389:610–614

Vestergaard H, Rossen M, Urhammer SA, Muller J, Pedersen O. (1997) Short- and long-term metabolic effects of recombinant human IGF-I treatment in patients with severe insulin resistance and diabetes mellitus. Eur J Endocrinol 136:475–482

Vu-Dac N, Gervois P, Torra IP, Fruchart JC, Kosykh V, Kooistra T, Princen HM, Dallongeville, Staels B (1998) Retinoids increase human apo C-III expression at the transcriptional level via the retinoid X receptor. Contribution to the hypertriglyceridemic action of retinoids. J Clin Invest 102:625–632

Wallberg-Henriksson H, Constable SH, Young DA, Holloszy JO (1988) Glucose transport into rat skeletal muscle: interaction between exercise and insulin. J Appl Physiol 65:909–913

Wallberg-Henriksson H, Holloszy JO (1985) Activation of glucose transport in diabetic muscle: responses to contraction and insulin. Am J Physiol 249(3 Pt 1):C233–C237

Wang M, Wise SC, Leff T, Su TZ (1999) Troglitazone, an antidiabetic agent, inhibits cholesterol biosynthesis through a mechanism independent of peroxisome proliferator-activated receptor-gamma. Diabetes 48:254–260

Warram JH, Martin BC, Krolewski AS, Soeldner JS, Kahn CR (1990) Slow glucose removal rate and hyperinsulinemia precede the development of type II diabetes in the offspring of diabetic parents. Ann Int Med 113:909–915

White DW, Wang DW, Chua SC Jr, Morgenstern JP, Leibel RL, Baumann H, Tartaglia LA (1997) Constitutive and impaired signaling of leptin receptors containing the Gln·Pro extracellular domain fatty mutation. Proc Natl Acad Sci 94:10657–10662

Winder WW, Hardie DG (1999) AMP-activated protein kinase, a metabolic master switch: possible roles in Type 2 diabetes. AM J Physiol 277:E1–E10
Wojtaszewski JFP, Hansen BF, Kiens B, Richter EA (1997) Insulin signaling in human skeletal muscle. Time course and effect of exercise. Diabetes 46:1775–1781
Wright BE, Browne ES, Svec F, Porter JR (1993) Divergent effect of dehydroepiandrosterone on energy intakes of Zucker rats. Physiol Behav 53:39–43
Wu Z, Bucher NL, Farmer SR (1996) Induction of peroxisome proliferator-activated receptor gamma during the conversion of 3T3 fibroblasts into adipocytes is mediated by C/EBPbeta, C/EBPdelta, and glucocorticoids. Mol Cell Biol 16:4128–4136
Wu Z, Rosen ED, Brun R, Hauser S, Adelmant G, Troy AE, McKeon C, Darlington GJ, Spiegelman BM (1999) Cross-regulation of C/EBP alpha and PPAR gamma controls the transcriptional pathway of adipogenesis and insulin sensitivity. Mol Cell 3:151–158
Yeh J-I, Gulve EA, Rameh L, Birnbaum MJ (1995) The effects of wortmannin on rat skeletal muscle. Dissociation of signaling pathways for insulin-and contraction-activated hexose transport. J Biol Chem 270:2107–2111
Yki-Jarvinen H, Sahlin K, Ren JM, Koivisto VA (1990) Localization of the rate-limiting defect for glucose disposal in skeletal muscle of insulin-resistant Type 1 diabetic patients. Diabetes 39:157–167
Young DA, Wallberg-Henriksson H, Sleeper MD, Holloszy JO (1987) Reversal of the exercise-induced increase in muscle permeability to glucose. Am J Physiol 253:E331–E335
Zenobi PD, Graf S, Ursprung H, Froesch ER (1992) Effects of insulin-like growth factor-I on glucose tolerance, insulin levels, and insulin secretion. J Clin Invest 89:1908–1913A
Zenobi PD, Jaeggi-Groisman SE, Riesen WF, Roder ME, Froesch ER (1992) Insulin-like growth factor-I improves glucose and lipid metabolism in type 2 diabetes mellitus. J Clin Invest 90:2234–2241B
Zhu Y, Qi C, Korenberg JR, Chen XN, Noya D, Rao MS, Reddy JK (1995) Structural organization of mouse peroxisome proliferator-activated receptor gamma (mPPARgamma) gene: alternative promoter use and different splicing yield two mPPAR gamma isoforms. Proc Natl Acad Sci 92:7921–7925
Ziegler D, Gries FA (1997) Alpha-lipoic acid in the treatment of diabetic peripheral and cardiac autonomic neuropathy. Diabetes 46:S62–S66
Zierath JR, Ryder JW, Doebber T, Woods J, Wu M, Ventre J, Li Z, McCrary C, Berger J, Zhang B, Moller DE (1998) Role of skeletal muscle in thiazolidinedione insulin sensitizer PPARgamma agonist action. Endocrinology 139:5034–5041
Zierath JR, Houseknecht KL, Gnudi L, Kahn BB (1997) High-fat feeding impairs insulin-stimulated GLUT4 recruitment via an early insulin-signaling defect. Diabetes 46:215–223
Zorzano A, Balon TW, Garetto LP, Goodman MN, Ruderman NB (1985) Muscle α-aminoisobutyric acid transport after exercise: enhanced stimulation by insulin. Am J Physiol 248:E546–E552
Zorzano A, Balon TW, Goodman MN, Ruderman NB (1986) Additive effects of prior exercise and insulin on glucose and AIB uptake by rat muscle. Am J Physiol 251:E21–E26

Section IV
Strategies for Indentifying Future Targets

CHAPTER 16

New Obesity Targets: Molecular-Genetic and Transgenic Approaches

D. E. MOLLER and L.H.T. VAN DER PLOEG

A. Introduction

Several lines of evidence point to genetic factors as important determinants of obesity. First, the prevalence of obesity varies dramatically between different racial and ethnic groups; it is substantially higher in certain native American (e.g. Pima Indians) or South Pacific (e.g. Nauruans) populations (TURNER et al. 1993). Secondly, some studies have shown that twins reared apart are concordant for body mass index (BMI) (BOUCHARD et al. 1990), and thirdly, several rare syndromes of extreme obesity are inherited as either dominant or recessive single-gene disorders (BOUCHARD and PERUSSE 1993). These include the Prader–Willi (BUTLER 1990) and Bardet–Biedl (BEALES et al. 1997) syndromes, and recently described single cases of congenital leptin deficiency (MONTAGUE et al. 1997). In addition, a single patient with a mutation in prohormone convertase 1 (presumably leading to impaired central processing of regulatory peptides that control appetite or energy expenditure) has been described (JACKSON et al. 1997). Overall, about two thirds of the contribution to body mass index (BMI) seems to be genetically determined (BOUCHARD and PERUSSE 1993; STUNKARD et al. 1986). Although increased nutrient intake is a prominent cause of obesity, there is emerging evidence that suggests that energy expenditure per se is, at least in part, genetically determined (BOUCHARD et al. 1990). Moreover, clinically significant increases in adiposity may often occur when individuals with what was once a favored "thrifty genotype" (reduced satiety, more efficient fuel storage) are exposed to a sedentary lifestyle accompanied by increased food availability and advancing age (TURNER et al. 1993; WENDORF and GOLDFINE 1991). The prevalence of obesity in the United States is currently about 30%, representing a health care problem of enormous proportions (KUCZMARSKI et al. 1994). Even though accumulating data is providing insights into the control of BMI, our ability to therapeutically intervene in BMI disorders is limited since prolonged weight loss with current therapy cannot easily be accomplished (SAFER 1991; LEIBEL et al. 1995). This chapter will review genetic strategies (and recent attempts) to identify new targets for treatment of obesity. We will delineate developing models which attempt to explain the mode of action of several new targets.

The use of genetically engineered animals (transgenic – gene addition – and knockout approaches) as a means of testing the physiological relevance of particular gene products for energy homeostasis is also reviewed.

B. Molecular-Genetic Approaches for Identifying New Obesity Targets

I. Positional Cloning vs. Analysis of Candidate Genes

The search for specific genetic defects that contribute to obesity in humans is complicated by the fact that this heterogeneous disorder is usually polygenic and may be clinically apparent only when present in combination with appropriate environmental factors (BOUCHARD and PERUSSE 1993). Furthermore, physiological parameters including BMI, waist–hip ratio, percent body fat (or the size of specific adipose tissue depots), satiety–appetite, and metabolic rate are continuous variables within the population, so it may be difficult to distinguish affected from unaffected subjects.

"Positional cloning" is an important strategy for identifying disease-causing genetic variation. This involves using linkage analysis methods to assess coinheritance of the disease (or trait) with DNA markers spaced evenly across the genome (LANDER and SCHORK 1994). These methods have been successfully applied to classical monogenic disorders (e.g. cystic fibrosis), where disease genes can be eventually cloned based only on their chromosomal location (LANDER and SCHORK 1994). Since classical linkage analysis requires accurate estimates of parameters such as penetrance and mode of inheritance, its application is limited in the case of common forms of obesity (ALLISON and SCHORK 1997). A combination of the following strategies will facilitate the total genome approach to detection of new genes which confer an increased (or decreased) risk of obesity: (1) the implementation of refined techniques that are better suited to the study of complex traits (e.g. affected sib-pair analysis); (2) greater emphasis on the study of selected populations (e.g. Pima Indians) and discrete or extreme subphenotypes (e.g. central adiposity or lipoatrophy – see below). The use of multiple strategies and the application of data pooling techniques (such as meta analysis) will enhance the ability to identify obesity-related genes that may confer only a partial effect (ALLISON and SCHORK 1997).

Some progress is being made in identifying disease-causing genes for rare obesity-related syndromes such as Bardet–Biedl syndrome; in this case, several loci: 11q13, 16q21, and 15q22, have been mapped (BEALES et al. 1997) (see also Chapter (3). In contrast, the actual application of any genome-wide scanning methods to the identification of common obesity genes in humans is still in its infancy. NORMAN et al., (NORMAN et al. 1997) reported initial results from an ongoing study in Pima Indians – a population plagued by a very high prevalence of morbid obesity. Potential linkages between >600 markers and percent body fat were assessed by sib-pair analysis in 874 individuals. A region on chromosome 11(q21-q22) was implicated as a locus containing gene(s) that may influence adiposity (NORMAN et al. 1997). When serum leptin levels were used as a surrogate marker for body adiposity and a general pedigree-based variance component approach, a locus at 2p21 was implicated in Mexican–American subjects (COMUZZIE et al. 1997). Interestingly, the POMC gene is contained within this region of the genome.

An alternative strategy is focused on the study of "candidate genes" whose initial identification is often based on murine genetics and transgenic approaches. A rapidly expanding list of known – cloned – genes encode molecules that are implicated as components of in vivo systems that regulate energy intake, storage, and expenditure. Having a newly cloned candidate gene in hand allows the investigator to use one or more approaches to test for the presence of potential disease-causing mutations: (1) Polymorphic markers in or near the gene can be used to study affected pedigrees (linkage analysis or sib-pair analysis) or in population association studies to test for coinheritance or association of gene variants with the disease. This approach is exemplified by a recent study reported by Hasstedt et al (HASSTEDT et al. 1997) who failed to find linkage of obesity to any of nine candidate genes (some are described below) in families with apparent recessive inheritance of obesity and Type II diabetes. (2) With the polymerase chain reaction (PCR), relevant portions of the candidate gene can be isolated from affected subjects and be directly sequenced or screened for nucleotide variation with a "molecular scanning" approach (COTTON 1989). Scanning techniques such as single-strand conformation polymorphism (SSCP) allow for rapid determination of nucleotide variation with ≥90% sensitivity (VIDAL-PUIG and MOLLER 1994).

Although several potential mutations in a given candidate gene may be detected, the following criteria need to be fulfilled to prove the clinical significance of a putative mutation: (1) functional defects are evident after expression of the mutant gene product in an appropriate in vitro system; (2) co-segregation of the mutant allele with a relevant phenotype in affected pedigree(s); (3) the mutant allele is present in affected subjects in greater frequency than in well-matched normal controls. The distinction between functionally "silent" polymorphisms and pathogenic mutations is especially difficult, since defects involving more than one gene are likely to be necessary at least in certain cases.

It is critical to bear in mind that positional cloning and candidate gene approaches are complementary. Thus, tentative identification of a candidate chromosomal locus from a total genome search should be followed by a search for relevant candidate genes that have been previously mapped to the vicinity of the locus in question (LANDER and SCHORK 1994).

II. Genetic Mapping of Quantitative Trait Loci in Animals

Rather than using techniques applicable to finding genes that contribute to complex traits in the human population, many groups are now applying quantitative trait loci (QTL) mapping methods to relevant animal models that relate to obesity. These efforts go beyond the successful positional cloning of monogenic mouse obesity mutations such as *ob*, *db*, *fat*, *tubby* and *Agouti* (reviewed in chap. 17). The identification of new QTLs for increased adiposity relies on segregation of contributing alleles (in unknown genes) that occur in experimental crosses between divergent outbred lines or between a defined in-bred line and an outbred line (e.g. *Spretus* mice). Numerous investigators have performed experiments with mouse crosses such as between in-bred C57BL/6J mice and *Spretus* or between NZB and SM/J strains; a large number of potentially important obesity-related QTLs have been identified (reviewed by Fisler and Warden 1997; Chagnon and Bouchard 1996; Pomp 1997, and in Chapter 17). In most cases, the magnitude of the effect for any given QTL is modest, accounting for ≤10% of the phenotypic variation in body adiposity (Pomp 1997). Interestingly, the mouse UCP-2 locus (see below) is coincident with QTLs for obesity in several mouse models (Fleury et al. 1997). Despite the large recent effort to identify and further evaluate mouse QTLs for obesity, no defined genetic defects have emerged as clear contributors to polygenic variation in body adiposity. Recent refinement of a genetic map for the mouse that now includes 4006 markers will facilitate further efforts with this species (Lander and Dietrich 1996).

Although most attempts to identify loci that contribute to polygenic obesity have employed mouse models, some effort has been applied to other species. Andersson reported results of experiments with crosses between the European wild boar and the domesticated Large White pig (Andersson et al. 1994). F2 offspring were scored for measures of adiposity (body weight, abdominal fat mass, depth of subcutaneous fat). Markers were used at 1-cM intervals, to map several potential QTLs to chromosome 4.

III. Application of Molecular Genetics and Genetically Engineered Animals to the Analysis of Potential Candidate Genes for Obesity

As shown in Table 1, there is a broad and expanding array of candidate genes that may confer increased or decreased obesity susceptibility. These can be separated into genes involved in central regulation of appetite and thermogenesis, genes implicated in regulating peripheral energy expenditure, and those that relate to lipid storage and adipocyte differentiation. A fourth group consists of genes with currently unknown physiological mechanisms. These include cell adhesion markers (ICAM-1, MAC-1) and a gene that may be involved in apoptosis (*Tubby*). Transgenic (gene addition) and gene-knockout experiments (Tables 2 and 3, respectively) both represent an important means of gaining new insights into the role of specific proteins or tissues in the regulation of the in vivo energy balance. The spectrum of transgenic approaches employed to date have included overexpression of selected proteins in a ubiquitous or tissue-specific fashion, forced expression of antisense mRNA

Table 1. Selected candidate genes

Gene products	Proposed function(s)	References
Central regulation of satiety and thermogenesis		
Melanocortin receptor 4	Control of satiety and thermogenesis; partially explained by control of NPY	Fan et al. 1997
Leptin	Fat-derived hormone; Control of satiety and thermogenesis, numerous pathways affected, including MCR and NPY	Friedman 1997
Leptin receptor	Hypothalamic class I cytokine receptor	Lee et al. 1996
Melanocyte-concentrating hormone	Central regulation of food intake	Qu et al. 1996
Agouti	Antagonist of melanocortin receptors	Bultman et al. 1992
Agouti-related protein	Antagonist of melanocortin receptors MC3R, MC4R	Ollmann et al. 1997
fat f*at*	Carboxypeptidase E Δ (hormone-processing defects)	Naggert et al. 1995
Tubby t*ub/tub*	Probably phosphodiesterase defect (apoptosis proposed as mechanism)	Noben-Trauth et al. 1996
BRS-3	Regulation of metabolic rate	Ohki-Hamazaki et al. 1997
Afferent parasympathetic input		
CCK	Satiety inducing by action on vagal afferents and CNS	Liddle 1997
Enterostatin	Proteolytic product derived from pro-colipase; mechanism of action unknown, causes satiety	Erlanson-Albertsson and York 1997
GLP-1	Satiety, by unknown mechanism	Holst 1997
Peripheral thermogenesis		
$\beta 3$ adrenergic receptor	Stimulation of lipolysis, regulation of brown adipocyte function	Kim-Motoyama et al. 1997; Candelore et al. 1996
Uncoupling protein 1	Regulation of mitochondrial energy dissipation	Ricquier et al. 1986; Cassard-Doulcier et al. 1996

Table 1. (*Continued*)

Gene products	Proposed function(s)	References
Uncoupling protein 2	"	FLEURY et al. 1997; GIMENO et al. 1997
Uncoupling protein 3	"	BOSS et al. 1997b; VIDAL-PUIG et al. 1997; GONG et al. 1997
Adipocyte differentiation/function		
PPARγ	Transcriptional activation of adipogenesis	TONTONOZ et al. 1994
c/EBPα, c/EBPβ	Transcriptional activation of adipogenesis	WU et al. 1995; CORNELIUS et al. 1994
Perilipin	Regulation of lipolysis	GREENBERG et al. 1991
TNFα	Regulation of lipolysis, adipocyte differentiation	SPIEGELMAN and HOTAMISLIGIL 1993; VENTRE et al. 1997
Lipoprotein lipase	Regulation of fat accretion	WEINSTOCK et al. 1997
Pygmy, HMGC-1	Reduced fat-tissue development	BENSON and CHADA 1994
ICAM-1, MAC-1	Intercellular adhesion markers, regulation of fat tissue	DONG et al. 1997

directed towards a selected gene product, tissue ablation through targeted expression of diphtheria toxin, and standard targeted gene disruption (knock-out mice). The range of known candidate obesity genes and approaches used to characterize them are discussed below.

1. Central Regulation of Satiety and Thermogenesis

Several animal models with chemically induced hypothalamic lesions, result in an obese phenotype in rodents. Lesioning of the afferent vagus nerve can similarly affect food intake and BMI. Cross-talk between the CNS and the afferent vagus nerve thus contributes to the control of food intake in rodents (MACLEAN 1985). Several peripherally produced hormones, from pancreatic (insulin), intestinal (CCK, GLP-1, enterostatin, PP, PYY), or fat tissue (leptin) origins can also modulate food intake and thermogenesis, even when admin-

Table 2. Transgenic (gene addition) rodent models resulting in obesity-related phenotypes

Transgene	Site(s) of expression	Phenotype	References
Neurofilament promoter – antisense glucocorticoid receptor	Generalized plus hypothalamus	Obesity	PEPIN et al. 1992
GLUT4 glucose transporter	Adipose tissue	Mild obesity	SHEPHERD et al. 1993
Dominant-negative insulin receptor	Skeletal and cardiac muscle	Increased adiposity/reduced lean body mass	MOLLER et al. 1996
UCP-1 promoter – diphtheria toxin	Brown adipose tissue	Loss of brown fat, obesity	LOWELL et al. 1993
aP2 promoter – diphtheria toxin	Brown and white fat	Lipoatrophy	ROSS et al. 1993
PEPCK promoter-TGFβ1	Adipose tissue, liver, kidney	Lipoatrophy, lean on *ob/ob* background	CLOUTHIER et al. 1997
Agouti protein	Generalized overexpression	Obesity	KLEBIG et al. 1995
Agouti protein	Adipose tissue	Obese in response to insulin	MYNATT et al. 1997
Agouti-related protein	Generalized overexpression	Obesity	OLLMANN et al. 1997
Glycerol 3-phosphate dehydrogenase	Brown and white fat and other tissues	Brown fat hypertrophy; atrophic white fat	KOZAK et al. 1991

istered directly into the brain. Transport processes have been invoked to explain access of these hormones to the CNS. In addition, an imperfect blood–brain barrier at the hypothalamus may allow access of some of these hormones to CNS targets. Additional hormones and nutrients believed to be essential in the control of meal initiation, termination and thermogenesis include corticosterone, glucose, and tryptophan. Corticosterone significantly affects the sensitivity of central and peripheral pathways for food intake and metabolism.

Thus, as described below, a diverse set of neuropeptides, monoamines, and hormones from peripheral tissues, as well as nutrients, all serve essential roles in integrating signals that control satiety and thermogenesis (for a review, see BRAY and RYAN 1996). Determining the relative dominance of these components has significantly benefited from the use of mouse genetics.

Table 3. Targeted gene disruptions resulting in obesity-related phenotypes

Gene affected	Phenotype	References
Melanocortin receptor 4	Obesity, hyperinsulinemia	FAN et al. 1997
5H2C serotonin receptor	Hyperphagia, metabolic disorder, obesity	TECOTT et al. 1995
Neuropeptide Y	Increased sensitivity to obesity	ERICKSON et al. 1996a
NPY Y1	Obesity in males	PEDRAZZINI et al. 1997
MCH	Reduced body weight, hypophagia, and increased metabolic rate	SHIMADA et al. 1998; CHAMBERS et al. 1999; SAITO et al., 1999
$\beta 3$ adrenergic receptor	Normal with slightly increased sensitivity to diet-induced obesity	GRUJIC et al. 1997
Uncoupling protein 1	Normal with slightly increased sensitivity to diet-induced obesity	ENERBACK et al. 1997
Lipoprotein lipase	Reduced sensitivity to weight gain on *ob/ob* background	WEINSTOCK et al. 1997
Pygmy, HMGC-1	Reduced white fat stores, reduction in body size	ZHOU et al. 1995
C/EBPα Δ	Reduced glycogen and fat stores	WANG et al. 1995
RIIβ Δ (PKA regulatory)	Resistance to obesity	CUMMINGS et al. 1996
aP2 Δ	Remains insulin sensitive upon diet-induced obesity (low TNFα)	HOTAMISLIGIL et al. 1996
TNFα	Slightly reduced adiposity, improved insulin sensitivity	VENTRE et al. 1997
Bombesin receptor subtype-3	Obesity and thermogenic disorder	OHKI-HAMAZAKI et al. 1997
CCK B receptor	Sensitive to the action of CCK	LANGHANS et al. 1997; MIYASAKA et al. 1994
ICAM-1	Obese	DONG et al. 1997
MAC-1	Obese	DONG et al. 1997
PPARγ	Neonatal lethal in homozygotes; resistance to diet-induced obesity in heterozygotes	KUBOTA et al. 1999

a) Components of the Leptin Axis

Our ability to maintain a steady state BMI was believed to be controlled by a circulating satiety factor, initially identified in parabiosis experiments involving *ob/ob* and *db/db* mice (COLEMAN 1973). These experiments showed that the *ob/ob* mouse is defective in the production of a fat-derived hormone, leptin, which acts at a class I cytokine receptor. This receptor is defective in the *db/db* mouse (Zhang et al. 1994; Tartaglia et al. 1995; Lee et al. 1996). These findings provided significant support for the lipostat theory of BMI maintenance, which proposes that the size of the fat stores, and a fat-derived hormone, are major determinants of BMI control. However, sensitivity to leptin is inversely correlated to BMI, allowing for gradual weight gain (McGregor et al. 1996; Maffei et al. 1995; Caro et al. 1996). It is unclear whether leptin sensitivity is centrally controlled, by affecting leptin receptor signal transduction. Alternative explanations involve the interaction of leptin with binding proteins and saturable transport of leptin into the CNS, thus limiting the controlling effects that leptin can exert on food intake and metabolism (Van et al. 1997).

A mutation of the leptin receptor was also discovered in the Zucker fatty rat, where a glutamine 269 to proline mutation affects the efficiency of leptin receptor signaling (Phillips et al. 1996; Rosenblum et al. 1996; Chua et al. 1996). This also leads to resistance to the action of leptin and results in an obese phenotype. Leptin-receptor-expressing neurons are believed to exert part of their effects on the control of food intake by increasing the action of melanocortin 4 receptor (MC-4R) expressing neurons, by reducing the expression of ART (a CNS MC-4R antagonist), by reducing hypothalamic NPY expression, and by increasing sympathetic tone and thermogenesis (see next sections).

Recently, a mutation in the leptin gene was identified in two separate cases of extreme obesity in humans. This finding firmly established a role for leptin as a circulating satiety factor in humans (Montague et al. 1997). However, mutations involving leptin or its receptor are not currently thought to represent a common source of obesity susceptibility in humans (Caro 1997). Given the increased resistance to leptin in humans with increased BMI, it is unclear whether leptin administration will result in a fruitful therapy for severe obesity. It appears that overcoming resistance will be the key to the success of the leptin pathway as a therapeutic strategy for BMI control.

b) Melanocortins

The agouti mouse represents a naturally occurring obese rodent, with a late-life onset of obesity, which is not corticosterone-dependent (Cone et al. 1996; Bultman et al. 1992). The obesity results from ectopic expression of the agouti protein. As is the case with the naturally occurring Ay (yellow) mutant mouse, ubiquitous transgenic overexpression of agouti results in an obese phenotype (Klebig et al. 1995). Agouti is a paracrine antagonist of the melanocortin-1 receptor, a G-protein-coupled receptor of the hair follicle. MC-1R agonism, through its natural ligand (α-MSH) raises cAMP and the expression of the enzyme tyrosinase in the hairfollicle. Low levels of tyrosinase, which result from antagonism of MC-1R, result in reduced conversion of the hair color pigment pheaomelanin to eumelanin. As a result, a light (agouti) rather than black hair color results. The obese phenotype of the agouti mouse was ascribed to the expression of agouti in the brain, where it antagonizes MC-3R and -4R receptors. This conclusion was corroborated by the generation of an MC-4R knockout mouse which recapitulates the obese phenotype of the agouti mouse (HUSZAR et al. 1997).

Recently, an agouti-related protein was identified (AGRP encoded by the agouti-related transcript ART), which is exclusively expressed in neurons of the arcuate nucleus and in the adrenal gland (SHUTTER et al. 1997; OLLMANN et al. 1997). AGRP was shown to be a naturally occurring antagonist of MC-3R and -4R (OLLMANN et al. 1997; FONG et al. 1997). In addition, overexpression of ART mRNA in transgenic mice recapitulates the obese phenotype observed in the MC-4R knockout and the agouti mouse (OLLMANN et al. 1997).

Several other mutations act on the MCR pathways: A dominant allele of the extension locus encodes a mutation at MC-1R; this leads to its constitutive activity, resulting in a black coat color. In this mutant (*e*), extension renders the mouse resistant to the action of agouti for hair color, but not obesity. Two additional mutations, mahogany and mahoganoid (*mg, md*) also affect coat color and obesity. These act epistatic to agouti and thus suppress the effects of agouti on coat color and obesity, while acting upstream of the MCRs (MILLER et al. 1997). The identification of the products expressed by these two genes will shed further light on the pathways through which MCR and AGRP control BMI.

One of the natural ligands for the MCRs is a pro-opiomelanocortin (POMC) derived 13 amino acid peptide, α-MSH. α-MSH in its non-acetylated form is a weak agonist of MC-4R, while the acetylated form of α-MSH is a potent MC-4R agonist (SHIMIZU et al. 1989). ICV injection of desacetyl-MSH resulted in significantly increased weight gain, compared to the outcome of acetyl α-MSH administration. It is unclear if the level of acetylation serves a role in the control of BMI in vivo. The leptin receptor was shown to be expressed on POMC containing neurons (CHEUNG et al. 1997). It therefore appears possible that leptin may affect POMC production. In line with this finding, ICV administration SHU-9119 (an MC-3R, -4R antagonist) renders mice resistant to the action of leptin (SEELEY et al. 1997). However, BOSTON et al. recently showed that the effects of POMC and leptin are additive and independent, indicating that leptin may also act through independent pathways (BOSTON et al. 1997).

Agouti and MCRs expressed peripherally may also affect BMI. Although mice with transgenic fat-specific agouti overexpression appeared normal, they have increased sensitivity to obesity in response to chronic insulin administration (MYNATT et al. 1997). This finding has potential physiological implications since low levels of agouti expression have been detected in human adipose tissue. The MCRs expressed in fat include MC-1R, -2R and -5R and it is unclear which of these is responsible for the action of agouti on insulin sensitivity in this mouse model.

MCRs have also been implicated in human obesity, cortisol production, and hair and skin color. In the Quebec family study, evidence for linkage between MC-5R and body mass index, fat mass, and resting metabolic rate and between MC-4R and fat mass and % body fat was reported (CHAGNON et al. 1997). Mutations in MCRs in humans were also identified, providing a direct correlation between MCR deficiency and a phenotype in humans: (i) MC-2R was shown to be defective in hereditary glucocorticoid resistance (TSIGOS et al. 1993; WEBER et al. 1995); (ii) variants of the MC-1R, involved in skin pigmentation, were shown to be associated with red hair and fair skin in humans (VALVERDE et al. 1995; JACKSON 1997); (iii) Rare MC-4R mutations have been identified in humans: (1) a heterozygous deletion at the MC-4R (amino acid 211) results in a decreased EC50 for agonist ligands and obesity (YEO et al. 1998) and (2) a family with a homozygous MC-4R mutation that leads to truncation at TM5 resulted in obesity (VAISSE et al. 1998); and, finally, (iv) POMC deficient children exhibit early life obesity with light skin and red hair implicating the MC-4R receptor in obesity (KRUDE et al. 1998).

c) *Neuropeptide Y and Glucocorticoids*

NPY belongs to a family of neuropeptides. NPY (a 36 amino acid peptide) is one of the most abundant polypeptides in the mammalian brain, and controls a variety of behaviors, including food intake, anxiety, facilitation of learning and memory, and regulation of the cardiovascular and neuroendocrine systems (GERALD et al. 1996; ERICKSON et al. 1996a). NPY is widely expressed in the CNS with sites of action in the hypothalamus that relate to its control of food intake. The related neuropeptide, Peptide YY (PYY) is found in the intestine and the brain stem, while pancreatic polypeptide is absent from the brain. NPY mediates food intake, with a preference for carbohydrate meals in rodents. Continued administration of NPY in the CNS leads to hyperphagia, obesity, hyperinsulinemia, and decreased BAT energy expenditure (ZARJEVSKI et al. 1993). NPY activates several G-protein coupled receptors of which the NPY (Y) Y1 and Y5 receptors (GERALD et al. 1996) are a current focus of attention.

The role of NPY in food intake was explored in the generation of NPY knockout mice. NPY knock out mice are normal with respect to body mass (ERICKSON et al. 1996a). However, NPY null mice are more sensitive to the action of leptin (ERICKSON et al. 1996b). On the basis of the normal phenotype of the NPY knockout mouse, it is assumed that other neuropeptides may have partially compensated for the loss of NPY. Candidate genes that come to mind include peptide YY and pancreatic polypeptide, all of which can act at NPY receptors.

To define the receptors through which NPY may exert its effects, NPY related peptides with unique receptor selectivities, were injected ICV. Y1, Y5, and a thus far uncharacterized neuropeptide Yx receptor (Yx) have been implicated in the control of food intake and metabolic rate (GERALD et al. 1996; O'SHEA et al. 1997). The phenotype of Y1 and Y5 receptor knockout mice will prove useful in elucidating the role of neuropeptide Y receptors in animal physiology (PEDRAZZINI et al. 1997).

Corticosterone is essential for NPY's action on food intake in rodents and adrenalectomy prevents the effects of centrally administered NPY on food intake (CHEN and ROMSOS 1996; SAINSBURY et al. 1997). In attempting to assess the role of glucocorticoids in behavior, PEPIN et al. generated a novel obese mouse model through the expression of antisense glucocorticoid receptor mRNA, using the neurofilament promoter (PEPIN et al. 1992). Since circulating glucocorticoid levels were elevated, it is likely that feedback inhibition at the hypothalamic-pituitary level was attenuated by selectively reduced glucocorticoid receptor expression in the CNS.

d) Monoamines – 5HT2c

Serotonin and dopaminergic pathways are major regulators of food intake. In the mouse, a knockout of the 5HT2c receptor led to hyperphagia, indicating that 5HT2c agonists could be beneficial in the treatment of obesity (TECOTT et al. 1995). 5HT re-uptake inhibitors and agonists (e.g. dexfenfluramine) are known to cause decreased food intake and increased metabolism. The mode of action of 5HT re-uptake inhibitors and agonists was proposed to involve a suppression of NPY release and an increase in CRF release (DRYDEN et al. 1996; GRIGNASCHI et al. 1996; DRYDEN et al. 1995). The precise mechanisms by which 5HT re-uptake inhibitors and agonists control satiety and metabolism have not been resolved. Finally, allelic variants of the dopamine D2 receptor implicate the D2 receptor in the regulation of body weight in humans (COMINGS et al. 1993; BLUM et al. 1996).

e) Tubby *and* Fat/Fat *Mutations*

Several spontaneous mutations that lead to obesity in mice were characterized after their genetic mapping and cloning of the affected genes had taken place. The *tubby* mutation (initially identified as rd5; retinal degeneration-5) represents a spontaneously occurring mutation that affects BMI and sensory perception (NOBEN-TRAUTH et al. 1996). The gene disrupted by the *tubby* mutation was cloned and characterized, and was shown to be homologous to phosphodiesterases. The obesity in the *tubby* mouse, as is the case for the agouti mouse, occurs in late life and is not corticosterone dependent. The *fat* mutation in mice leads to the expression of a dysfunctional carboxy peptidase E, which is believed to affect neuropeptide and insulin expression (NAGGERT et al. 1995). As a result, the mice lack normally processed insulin; this hormonal defect might partially explain the obese phenotype of the *fat/fat* mouse.

2. Peripheral Thermogenesis

a) Uncoupling Proteins and BAT

Uncoupling protein (now known as UCP-1) is a major mediator of facultative thermogenesis which is expressed exclusively in brown adipose tissue. UCP-1 functions as a proton channel within the mitochondrial membrane that acts to uncouple oxidative phosphorylation, allowing for the dissipation of stored energy as heat (RICQUIER et al. 1986). UCP-1 expression is upregulated during states of increased thermogenesis, including cold exposure, thyroid hormone excess, and activation of the sympathetic nervous system (SILVA and RABELO 1997). Brown adipose tissue is scarce in humans and the role of UCP-1 is untested; however, UCP-1 remains a viable candidate gene. Indeed, a restriction fragment (*Bcl-I*) length polymorphism in the UCP-1 gene was shown to be associated with obesity (CLEMENT et al. 1996); however the *Bcl-1* polymorphism results from a single nucleotide change in the 5′ untranslated region and is unlikely to perturb UCP-1 function or expression (CASSARD-DOULCIER et al. 1996).

The in vivo role of brown adipose tissue (in mice) was clearly defined by UCP-1 promoter dependent expression of diphtheria toxin which resulted in selective ablation of this tissue (LOWELL et al. 1993). These mice developed substantial obesity associated with an increase in food intake. In contrast, overexpression of glycerol 3-phosphate dehydrogenase (GPDH), a lipogenic enzyme which is predominantly expressed in brown fat, resulted in hypertrophy of this tissue at the expense of reduced white fat (KOZAK et al. 1991). Surprisingly, targeted ablation of the UCP-1 gene in mice did not lead to an obese phenotype, although the mice exhibited increased sensitivity to cold exposure (ENERBACK et al. 1997).

The discrepancy between UCP-1 null mice and BAT-deficient mice suggested the existence of additional UCP isoforms. Recently, two other uncoupling proteins, UCP-2 (FLEURY et al. 1997; GIMENO et al. 1997) and UCP-3 (BOSS et al. 1997b; VIDAL-PUIG et al. 1997; GONG et al. 1997) were discovered, after partial cDNA sequences (EST's) had been deposited in the database. Recent experimental evidence suggests that both proteins can function to uncouple mitochondrial respiration (FLEURY et al. 1997; GONG et al. 1997). UCP-3 is predominantly expressed in skeletal muscle (BOSS et al. 1997b; VIDAL-PUIG et al. 1997), whereas UCP-2 is more widely expressed (including in brown and white adipose tissue) (BOSS et al. 1997a). Both UCP-2 and UCP-3 may be upregulated by fasting (MILLET et al. 1997). Importantly, UCP-3 expression in muscle is strongly induced by thyroid hormone; this suggests that it may have an important physiological role in mediating thermogenesis (GONG et al. 1997). Given the above findings, both UCP-2 and UCP-3 are obvious candidate genes for obesity. Their recent discovery and characterization illustrate the power of database mining as a means of identifying potential new therapeutic targets for obesity.

b) *β3 Adrenergic Receptors*

β3 adrenergic receptors are specifically expressed in adipose tissue and mediate thermogenesis; thus, agonists for this receptor promote brown adipose tissue accumulation, stimulate UCP1 expression, and induce weight loss in rodent models of obesity (CHARON et al. 1995; ARBEENY et al. 1995) (also reviewed in chap. 14). Analysis of this obvious candidate gene in humans revealed the presence of a common missense polymorphism ($Trp^{64} \rightarrow Arg$), which has been variably associated with obesity, insulin resistance, and dislipidemia (Clement et al. 1996; Kim-Motoyama et al. 1997). Importantly, different studies have come to divergent conclusions regarding the Arg^{64} allele – in some studies it was not associated with obesity (Hasstedt et al. 1997); in some studies it was associated with increased triglycerides, while in others the association was with decreased triglycerides (Kim-Motoyama et al. 1997). Thus, this "mutation" is of questionable physiological significance. Moreover, when the variant receptor was expressed in transfected cells and carefully characterized in comparison with wild-type, significant functional differences could not be detected (Candelore et al. 1996).

3. Adipocyte Differentiation/Function

Obesities are often the result of an increase in the amount of triglyceride stored intracellularly leading to adipocyte hypertrophy, while more severe obesities can also involve adipocyte hyperplasia. Interfering with adipogenesis could therefore provide an opportunity for the control of obesity. One interesting transgenic model which relates to adipocyte differentiation involved expression of diphtheria toxin under control of the adipocyte fatty acid binding protein (aP2) promoter. In contrast to obesity which occurs in brown fat deficient mice, these mice developed severe lipoatrophy (Ross et al. 1993). They were also resistant to obesity that can be induced by chemical means (monosodium glutamate). The lipoatrophy consequences that occurred included marked hypertriglyceridemia, fatty liver, and insulin-resistant diabetes (Ross et al. 1993). Similar to what was found for the human syndromes of lipoatrophy (Moller and O'Rahilly 1993), these observations serve as a warning that therapeutic measures for obesity that promote unrestrained lipolysis or adipocyte necrosis are unlikely to succeed as drugs. At present, a number of specific genes involved in adipogenesis are being mapped and characterized.

a) HMGI-C and Adipogenesis

The spontaneous *pygmy (pg)* mouse mutant exhibits growth retardation. Phenotypic analysis of the *pygmy* mouse revealed that the majority of tissues were reduced in weight by 40%. However, fat was disproportionately reduced to approximately one tenth of that of their wild-type littermates (Benson and Chada 1994). The *pygmy* gene was identified as a nuclear architectural factor belonging to the high-mobility group proteins, HMGI-C (Zhou et al. 1995). HMGs are DNA binding proteins that affect chromatin structure. HMGI-C belongs to the AT-Hook DNA binding domain proteins. They are believed to act by recruiting other DNA binding proteins and transcription factors. Fusion proteins of HMGI-C with other protein domains can result from DNA rearrangements, leading to the fusion of the DNA binding domain of HMGI-C with acidic serine- and threonine-rich transcriptional activation domains, causing benign tumors and lipomas (ASHAR et al. 1995). Knockout of the murine HMGI-C gene showed that HMGI-C is expressed in the mesenchyme before its overt differentiation during embryogenesis. Subsequent to the murine studies, the human homolog was mapped to chromosome 12q13–15. This is a genomic site which is consistently involved in chromosomal translocations in a number of mesenchymally derived tumors, including lipomas. Null mutants of HMGI-C produce mice that contain fewer fat cells, whereas aberrant HMGI-C proteins lead to a proliferation of fat cells. This data suggests that HMGI-C functions in adipogenesis and mesenchyme differentiation.

b) PPARγ, C/EBPα, and Adipogenesis

Several transcription factors have been identified that can promote adipogenesis. These include PPARγ, the C/EBPs, and ADD-1. PPARγ is a member of the nuclear hormone receptor superfamily belonging to the subgroup of peroxisome proliferator activated receptors. PPARγ activation is sufficient to upregulate fat-specific genes (BRUN et al. 1996). Although homozygous PPARγ null mice were found not to be viable, heterozygous null mice were resistant to obesity when challenged with a high fat diet (Kubota et al. 1999). C/EBPα, a basic leucine zipper transcription factor, also regulates key enzymes, as can be seen by the phenotype of C/EBPα knockout mice which have reduced liver glycogen and fat stores. These mice died from hypoglycemia within 8 hours after birth, since they failed to activate gluconeogenic enzymes including phosphoenolpyruvate carboxykinase (PEPCK) and glucose 6-phosphatase (WANG et al. 1995). Hepatocytes and adipocytes derived from C/EBPα null mice failed to accumulate lipid. These neonatal deficiencies may not accurately predict the potential phenotype of C/EBPα deficiency in adult life, which is expected to have less of an impact on glucose metabolism due to rescue by C/EBPβ.

c) Possible Role of TGFβ1 in Adipocyte Differentiation

In order to study the in vivo role of TGFβ1 in cell differentiation or tissue fibrosis, the phosphoenolpyruvate carboxykinase (PEPCK) promoter was used to drive overexpression of TGFβ1 in adipose tissue, liver, and kidney (CLOUTHIER et al. 1997). In addition to other abnormalities, these mice failed to develop normal amounts of white adipose tissue, presumably as a consequence of impaired adipocyte differentiation.

d) Regulation of Lipolysis and Nutrient Partitioning

Insulin controls the rate of lipolysis through modulating cAMP dependent protein kinase (PK) levels in adipocytes, by activation of the "adipocyte-specific" cGMP inhibited cAMP phosphodiesterase IIIB (PDEIIIB) (ERIKSSON et al. 1995). PDEIIIB is activated by PKA, in response to insulin, and suppresses the hormone-sensitive lipase which requires cAMP dependent PK for its phosphorylation and activation. The role of enzymes in this pathway is partially validated by the obesity-resistant phenotype of the RIIβ PKA knockout mouse (Cummings et al. 1996).

Important insights into how altered nutrient partitioning may affect body adiposity were obtained by selective overexpression of the insulin-responsive glucose transporter (GLUT4) in adipose tissue (aP2 promoter), which resulted in obesity (Shepherd et al. 1993) and the creation of muscle-specific insulin resistance through the expression of dominant negative insulin receptors under the control of the muscle creatine kinase promoter (these mice also developed mild obesity) (Moller et al. 1996).

4. Miscellaneous Potential Targets Identified by Gene Knockout Experiments in Mice

a) Bombesin

Bombesin is a tetradecapeptide which can activate several G-protein-coupled receptors (GPCRs) including the bombesin receptor subtype-3 (BRS-3). Inactivation of the BRS-3 in mice resulted in mild obesity and reduction of metabolic rate, indicating that BRS-3 is one of the targets of bombesin action (Ohki-Hamazaki et al. 1997).

b) Cell Adhesion Markers: ICAM I

Mice deficient in the expression of the intercellular adhesion molecule 1 (ICAM-1) became obese in late life (Dong et al. 1997), while diet-induced obesity has a more aggressive progression. Mice defective in the counter-receptor of ICAM-1, the leukocyte integrin aMb2 (Mac-1), also developed an obesity comparable to that observed in the ICAM-1 knockout mice. The authors propose that leukocytes may play a role in lipid metabolism and energy expenditure, by preventing excess body-fat deposition.

c) Melanin Concentrating Hormone (MCH) and SLC-1

Rat and human MCH are identical 19 amino acid cyclic peptides encoded in the C-terminal portion of the MCH prohormone gene, which also encodes the neuropeptides EI and GE. MCH has been implicated in food intake based on ICV peptide administration in rodents (Rossi et al. 1997). Knock out of the MCH prohormone gene, which encodes at least three neuropeptides, led to an interesting phenotype in mice that exhibited hypophagia, increased metabolic rate, and decreased body weight (Shimada et al. 1998). This phenotype may implicate MCH as a ligand involved in the control of body weight and food intake. The MCH peptide ligand was subsequently shown to specifically activate the orphan G-protein coupled receptor, SLC-1 (Chambers et al. 1999; Saito et al. 1999).

C. Future Directions

I. Novel Approaches Using Genetically Engineered Animals

As an alternative to transgenic expression of foreign proteins, adenovirus-mediated gene therapy is a rapidly developing method where systemic expression of secreted proteins can be rapidly achieved or where local administration into a given tissue can also be employed. The power of this approach is best illustrated by recent experiments where chronically elevated systemic leptin levels were achieved in rats (CHEN et al. 1996) or in *ob/ob* mice (MUZZIN et al. 1996), resulting in a phenotype of substantial weight loss. Traditional gene-targeting experiments are subject to the following constraints: (1) the gene of interest is usually completely deleted or inactivated; (2) gene knockout is achieved in all tissues; (3) gene knockout (and transgenic gene expression) is usually present throughout development; this precludes the analysis of effects in adult animals if the knockout (or transgene) impairs development. Novel approaches are being developed to circumvent some of these problems. Alternative "in–out" targeting strategies now allow for a selected gene to be replaced with a modified version, such as one encoding a subtle mutation (VALANCIUS and SMITHIES 1991). The introduction of *loxP* sites into a gene of interest followed by cross-breeding with transgenic mice expressing Cre recombinase in the tissue(s) of interest has resulted in the recent ability to generate mice with tissue-specific inactivation of a desired gene (BARINAGA 1994). Moreover, many investigators are attempting to develop inducible expression systems – such as those which employ tetracycline- or ecdysone-responsive promoters (BARINAGA 1994). In this way, healthy adult animals with genetically engineered, but normally silent, genetic changes can be derived. Following exposure to the inducing agent, selected gene(s) will be expressed and/or inactivated in pre-determined tissue(s) of interest.

II. Assessment of Gene Expression as an Approach for New Target Identification

As a method for discovering new therapeutic obesity targets, the differential display polymerase chain reaction technique can be used to identify potential differences in mRNA levels between two or more samples selected for comparison (LIANG et al. 1995). Differential display was used in the obesity research that determined that melanin-concentrating hormone is overexpressed in the hypothalamus of *ob/ob* mice (QU et al. 1996). More recently, large numbers of probes (cDNAs, oligonucleotides) corresponding to known genes, organized as microarrays on glass chips, are being applied as a way of monitoring altered gene expression patterns (SCHENA et al. 1996; HOHEISEL 1997). Current technology has allowed the use of 1,000–20,000 cDNA probes for the parallel assessment of gene expression levels (SCHENA et al. 1996). In theory, probes representing the entire genome could be handled simultaneously. This novel genomic technology will undoubtedly lead to the discovery of many new potential target proteins in the obesity field and other therapeutic areas.

References

Allison DB, Schork NJ (1997) Selected methodological issues in meiotic mapping of obesity genes in humans: issues of power and efficiency. Behavior Genetics 27:401–421

Andersson L, Haley CS, Ellegren H, Knott SA, Johansson M, Andersson K, Andersson-Eklund L, Edfors-Lilja I, Fedholm M, Hansson I, Hakansson J, Lundstrom K (1994) Genetic mapping of quantitative trait loci for growth and fatness in pigs. Science 263:1771–1774

Arbeeny CM, Meyers DS, Hillyer DE, Bergquist KE (1995) Metabolic alterations associated with the antidiabetic effect of $\beta 3$-adrenergic receptor agonists in obese mice. Am J Physiol 268:E678–E684

Ashar HR, Fejzo MS, Tkachenko A, Zhou X, Fletcher JA, Weremowicz S, Morton CC, Chada K (1995) Disruption of the architectural factor HMGI-C: DNA-binding AT hook. Cell 82:57–65

Barinaga M (1994) Knockout mice: Round two. Science 265:26–28

Beales PL, Warner AM, Hitman GA, Thakker R, Flinter FA (1997) Bardet Biedl syndrome: a molecular and phenotypic study of 18 families. J Med Genetics 34:92–98

Benson KF, Chada K (1994) Mini-mouse: phenotypic characterization of a transgenic insertional mutant allelic to pygmy. Genet Res 64:27–33

Blum K, Braverman ER, Wood RC, Gill J, Li C, Chen TJ, Taub M, Montgomery AR, Sheridan PJ, Cull JG (1996) Increased prevalence of the Taq I A1 allele of the dopamine receptor. Pharmacogenetics 6:297–305

Boss O, Samec S, Dulloo A, Seydoux J, Muzzin P, Giacobino J-P (1997) Tissue-dependent upregulation of rat uncoupling protein-2 expression in response to fasting or cold. FEBS Letters 412:111–114

Boss O, Samec S, Paoloni-Giacobino A, Rossier C, Dulloo A, Seydoux J, Muzzin P, Giacobino J-P (1997) Uncoupling protein-3: a new member of the mitochondrial carrier family with tissue-specific expression. FEBS Letters 408:39–42

Boston BA, Blaydon KM, Varnerin J, Cone RD (1997) Independent and additive effects of central POMC and leptin pathways on murine obesity. Science 278:1641–1644

Bouchard C, Perusse L (1993) Genetics of obesity. Annu Rev Nutrition 13:337–354

Bouchard C, Tremblay A, Despres JP (1990) The response to long-term overfeeding in identical twins. N Engl J Med 322:1477–1482

Bray GA, Ryan DH (1996) Molecular and genetic aspects of obesity. Baton Rouge: Louisiana State University Press, Pennington Center Nutrition Series, vol 5

Brun RP, Tontonoz P, Forman BM, Ellis R, Chen J, Evans RM, Spiegelman BM (1996) Differential activation of adipogenesis by multiple PPAR isoforms. Genes Dev 10:974–984

Bultman SJ, Michaud EJ, Woychik R (1992) Molecular characterization of the mouse agouti locus. Cell 71:1195–1204

Butler MG (1990) Prader-Willi syndrome: current understanding of cause and diagnosis. Am J Med Genet 35:319–332

Candelore MR, Deng LP, Tota LM, Kelly LJ, Cascieri MA, Strader CD (1996) Pharmacological characterization of a recently described human beta 3 adrenergic receptor mutant. Endocrinology 137:2638–2641

Caro JF, Kolaczynski JW, Nyce MR, Ohannesian JP, Opentanova I, Goldman WH, Lynn RB, Zhang PL, Sinha MK, Considine RV (1996) Decreased cerebrospinal-fluid/serum leptin ratio in obesity: a possible mechanism for leptin resistance. Lancet 348:159–161

Caro JF (1997) Evidence against either a premature stop codon or the absence of obese gene mRNA in human obesity. Scientist 11:11
Cassard-Doulcier AM, Bouillaud F, Chagnon M, Gelly C, Dionne FT, Oppert JM, Bouchard C, Chagnon Y, Ricquier D (1996) The Bcl 1 polymorphism of the human uncoupling protein (ucp) gene is due to a point mutation in the 5′-flanking region. Intl J Obesity 20:278–279
Chagnon YC, Bouchard C (1996) Genetics of obesity: advances from rodent studies. Trends in Genetics 12:441–444
Chagnon YC, Chen WJ, Perusse L, Chagnon M, Nadeau A, Wilkison WO, Bouchard C (1997) Linkage and association studies between the melanocortin receptors 4 and 5 genes and obesity-related phenotypes in the Quebec Family Study. Mol Med 3:663–673
Chambers J, Ames RS, Bergsma D, Muir A, Fitzgerald LR, Hervieu G, Dytko GM, Foley JJ, Martin J, Liu WS, Park J, Ellis C, Ganguly S, Konchar S, Cluderay J, Leslie R, Wilson S, Sarau HM (1999) Melanin-concentrating hormone is the cognate ligand for the orphan G-protein-coupled receptor SLC-1. Nature 400:261–265
Charon C, Dupuy F, Marie V, Bazin R (1995) Effect of the β-adrenoceptor agonist BRL-35135 on development of obesity in suckling Zucker (fa/fa) rats. Am J Physiol 268:E1039–E1045
Chen GX, Koyama K, Yuan X, Lee Y, Zhou YT, O'Doherty R, Newgard CB, Unger RH (1996) Disappearance of body fat in normal rats induced by adenovirus mediated leptin gene therapy. Proc Natl Acad Sci 93:14795–14799
Chen HL, Romsos DR (1996) Dexamethasone rapidly increases hypothalamic neuropeptide Y secretion. Am J Physiol 271:E151–E158
Cheung CC, Clifton DK, Steiner RA (1997) Proopiomelanocortin neurons are direct targets for leptin in the hypothalamus. Endocrinology 138:4489–4492
Chua SC, Jr., White DW, Wu-Peng XS, Liu SM, Okada N, Kershaw EE, Chung WK, Power-Kehoe L, Chua M, Tartaglia LA, Leibel RL (1996) Phenotype of fatty due to Gln269Pro mutation in the leptin receptor. Diabetes 45:1141–1143
Clement K, Ruiz J, Cassard-Doulcier A-M, Bouillaud F, Ricquier D, Basdevant A, Guy-Grand B, Froguel P (1996) Additive effect of A-G (3826) variant of the uncoupling protein gene and the Trp64Arg mutation of the β3-adrenergic receptor gene on weight gain in morbid obesity. Int J Obesity 20:1062–1066
Clouthier DE, Comerford SA, Hammer RE (1997) Hepatic fibrosis, glomerulosclerosis, and a lipodystrophy-like syndrome in PEPCK-TGFβ1 transgenic mice. J Clin Invest 100:2697–2713
Coleman DL (1973) Effects of parabiosis of obese with diabetes and normal mice. Diabetologia 9:294–298
Comings DE, Flanagan SD, Dietz G, Muhleman D, Knell E, Gysin R (1993) The dopamine D2 receptor (DRD2) as a major gene in obesity and height. Biochem Med Metab Biol 50:176–185
Comuzzie AG, Hixson JE, Almasy L, Mitchell BD, Mahaney MC, Dyer TD, Stern MP, Maccluer JW, Blangero J (1997) A major quantitative trait locus determining serum leptin levels and fat mass is located on human chromosome 2. Nature Genetics 15:273–276
Cone RD, Lu D, Koppula S, Vage DI, Klungland H, Boston B, Chen W, Orth DN, Pouton C, Kesterson RA (1996) The melanocortin receptors: agonists, antagonists, and the hormonal control of pigmentation. Recent Prog Horm Res 51:287–317
Cornelius P, MacDougald OA, Lane MD (1994) Regulation of adipocyte development. Annu Rev Nutr 14:99–129
Cotton RGH (1989) Detection of single base changes in nucleic acids. Biochem J 263:1–10
Cummings DE, Brandon EP, Planas JV, Motamed K, Idzerda RL, McKnight GS (1996) Genetically lean mice result from targeted disruption of the RII beta subunit of protein kinase A. Nature 382:622–626

Dong ZM, Gutierrez-Ramos JC, Coxon A, Mayadas TN, Wagner DD (1997) A new class of obesity genes encodes leukocyte adhesion receptors. Proc Natl Acad Sci USA 94:7526–7530
Dryden S, Wang Q, Frankish HM, Pickavance L, Williams G (1995) The serotonin (5-HT) antagonist methysergide increases neuropeptide Y. Brain Res 699: 12–18
Dryden S, Wang Q, Frankish HM, Williams G (1996) Differential effects of the 5-HT 1B/2C receptor agonist mCPP and the 5-HT1A agonist flesinoxan on hypothalamic neuropeptide Y in the rat: evidence that NPY may mediate serotonin's effects on food intake. Peptides 17:943–949
Enerback S, Jacobsson A, Simpson EM, Guerra C, Yamashita H, Harper M-E, Kozak LP (1997) Mice lacking mitochondrial uncoupling protein are cold-sensitive but not obese. Nature 387:90–97
Erickson JC, Clegg KE, Palmiter RD (1996) Sensitivity to leptin and susceptibility to seizures of mice lacking neuropeptide Y. Nature 381:415–421
Erickson JC, Hollopeter G, Palmiter RD (1996) Attenuation of the obesity syndrome of ob/ob mice by the loss of neuropeptide Y. Science 274:1704–1707
Eriksson H, Ridderstrale M, Degerman E, Ekholm D, Smith CJ, Manganiello VC, Belfrage P, Tornqvist H (1995) Evidence for the key role of the adipocyte cGMP-inhibited cAMP phosphodiesterase in the antilipolytic action of insulin. Biochim Biophys Acta 1266:101–107
Erlanson-Albertsson C, York D (1997) Enterostatin – a peptide regulating fat intake. Obes Res 5:360–372
Fan W, Boston BA, Kesterson RA, Hruby VJ, Cone RD (1997) Role of melanocortinergic neurons in feeding and the agouti obesity syndrome. Nature 385:165–168
Fisler JS, Warden CH (1997) Mapping of mouse obesity genes: a generic approach to a complex trait. J Nutrition 127:S1909–S1916
Fleury C, Neverova M, Collins S, Raimbault S, Champigny O, Levi-Meyrueis C, Bouillaud F, Seldin MF, Surwit RS, Ricquier D, Warden CH (1997) Uncoupling protein-2: a novel gene linked to obesity and hyperinsulinemia. Nature Genetics 15:269–272
Fong TM, Mao C, MacNeil T, Kalyani R, Smith T, Weinberg D, Tota MR, Van der Ploeg LH (1997) ART (protein product of agouti-related transcript) as an antagonist of MC-3 and MC-4 receptors. Biochem Biophys Res Commun 237:629–631
Friedman J (1997) Leptin, leptin receptors and the control of body weight. Eur J Med Res 2:7–13
Gerald C, Walker MW, Criscione L, Gustafson EL, Batzl-Hartmann C, Smith KE, Vaysse P, Durkin MM, Laz TM, Linemeyer DL, Schaffhauser AO, Whitebread S, Hofbauer KG, Taber RI, Branchek TA, Weinshank RL (1996) A receptor subtype involved in neuropeptide-Y induced food intake. Nature 382:168–171
Gimeno RE, Dembski M, Weng X, Deng N, Shyjan AW, Gimeno CJ, Iris F, Ellis SJ, Woolf EA, Tartaglia LA (1997) Cloning and characterization of an uncoupling protein homolog. Diabetes 46:900–906
Gong D-W, He Y, Karas M, Reitman M (1997) Uncoupling protein-3 is a mediator of thermogenesis regulated by thyroid hormone, beta3-adrenergic agonists, and leptin. J Biol Chem 272:24129–24132
Greenberg AS, Egan JJ, Wek SA, Garty NB, Blanchette-Mackie EJ, Londos C (1991) Perilipin, a major hormonally regulated adipocyte-specific phosphoprotein associated with the periphery of lipid storage droplets. J Biol Chem 266:11341–11346
Grignaschi G, Sironi F, Samanin R (1996) Stimulation of 5-HT2A receptors in the paraventricular hypothalamus. Brain Res 708:173–176
Grujic D, Susulic VS, Harper ME, Himms-Hagen J, Cunningham BA, Corkey BE, Lowell BB (1997) Beta3-adrenergic receptors on white and brown adipocytes mediate beta3-selective agonist-induced effects on energy expenditure, insulin secretion, and food intake. J Biol Chem 272:17686–17693

Hasstedt SJ, Hoffman M, Leppert MF, Elbein SC (1997) Recessive inheritance of obesity in familial non insulin dependent diabetes mellitus and lack of linkage to nine candidate genes. Am J Human Genetics 61:668–677
Hoheisel JD (1997) Oligomer chip technology. Trends Biotechnol 15:465–469
Holst JJ (1997) Enteroglucagon. Annu Rev Physiol 59:257–271
Hotamisligil GS, Johnson RS, Distel RJ, Ellis R, Papaioannou VE, Spiegelman BM (1996) Uncoupling of obesity from insulin resistance through a targeted mutation in aP2, the adipocyte fatty acid binding protein. Science 274:1377–1379
Huszar D, Lynch CA, Fairchild-Huntress V, Dunmore JH, Fang Q, Berkemeier LR, Gu W, Kesterson RA, Boston BA, Cone RD, Smith FJ, Campfield LA, Burn P, Lee F (1997) Targeted disruption of the melanocortin-4 receptor results in obesity. Cell 88:131–141
Jackson IJ (1997) Homologous pigmentation mutations in human, mouse and other model organisms. Hum Mol Genet 6:1613–1624
Jackson RS, Creemers JW, Ohagi S, Raffin-Sanson ML, Sanders L, Montague CT, Hutton JC, O'Rahilly S (1997) Obesity and impaired prohormone processing associated with mutations in the human prohormone convertase 1 gene. Nature Genet 16:303–306
Kim-Motoyama H, Yasuda K, Yamaguchi T, Yamada N, Katakura T, Shuldiner AR, Akanuma Y, Ohashi Y, Yazaki Y, Kadowaki T (1997) A mutation of the beta 3-adrenergic receptor is associated with visceral obesity but decreased serum triglyceride. Diabetologia 40:469–472
Klebig ML, Wilkinson JE, Geisler JG, Woychik RP (1995) Ectopic expression of the agouti gene in transgenic mice causes obesity, features of type II diabetes, and yellow fur. Proc Natl Acad Sci USA 92:4728–4732
Kozak LP, Kozak UC, Clarke GT (1991) Abnormal brown and white fat development in transgenic mice overexpressing glycerol 3-phosphate dehydrogenase. Genes and Dev. 5:2256–2264
Krude H, Biebermann H, Luck W, Horn R, Brabant G, Gruters A (1998) Severe early-onset obesity, adrenal insufficiency and red hair pigmentation caused by POMC mutations in humans. Nat Genet 19:155–157
Kubota N, Terauchi Y, Miki H, Tamemoto H, Yamauchi Y, Komeda K, Satoh S, Nakano R, Ishii C, Sugiyama T, Eto K, Tsubamoto Y, Okuno A, Murakami K, Sekihara H, Hasegawa G, Naito M, Toyoshima Y, Tanaka S, Shiota K, Kitamura T, Fujita T, Ezaki O, Aizawa S, Nagai R, Tobe K, Kimura S, Kadowaki T (1999) PPARγ mediates high-fat diet-induced adipocyte hypertrophy and insulin resistance. Mol Cell 4:597–609
Kuczmarski RJ, Flegal KM, Campbell SM, Johnson CL (1994) Increasing prevalence of overweight among US adults. JAMA 272:205–211
Lander E, Dietrich W (1996) A genetic map of the mouse with 4006 simple sequence length polymorphisms. Scientist 10:14
Lander ES, Schork NJ (1994) Genetic dissection of complex traits. Science 265:2037–2048
Langhans N, Rindi G, Chiu M, Rehfeld JF, Ardman B, Beinborn M, Kopin AS (1997) Abnormal gastric histology and decreased acid production in cholecystokinin-B/gastrin receptor-deficient mice. Gastroenterology 112:280–286
Lee GH, Proenca R, Montez JM, Carroll KM, Darvishzadeh JG, Lee JI, Friedman JM (1996) Abnormal splicing of the leptin receptor in diabetic mice. Nature 379:632–635
Leibel RL, Rosenbaum M, Hirsch J (1995) Changes in energy expenditure resulting from altered body weight. N Engl J Med 332:621–628
Liang P, Bauer D, Averboukh L, Warthoe P, Rohrwild M, Muller H, Strauss M, Pardee AB (1995) Analysis of altered gene expression by differential display. Methods Enzymology 254:304–321
Liddle RA (1997) Cholecystokinin cells. Annu Rev Physiol 59:221–242
Lowell BB, S.-Susulic V, Hamann A, Lawitts J, Himms-Hagen J, Boyer BB, Kozak LP, Flier JS (1993)Development of obesity in transgenic mice after genetic ablation of brown adipose tissue. Nature 366:740–742

MacLean DB (1985) Abrogation of peripheral cholecystokinin – satiety in the capsaicin. Regul Pept 11:321–333
Maffei M, Halaas J, Ravussin E, Pratley RE, Lee GH, Zhang Y, Fei H, Kim S, Lallone R, Ranganathan S et al (1995) Leptin levels in human and rodent: measurement of plasma leptin and ob. Nat Med 1:1155–1161
McGregor GP, Desaga JF, Ehlenz K, Fischer A, Heese F, Hegele A, Lammer C, Peiser C, Lang RE (1996) Radioimmunological measurement of leptin in plasma of obese and diabetic. Endocrinology 137:1501–1504
Miller KA, Gunn TM, Carrasquillo MM, Lamoreux ML, Galbraith DB, Barsh GS (1997) Genetic studies of the mouse mutations mahogany and mahoganoid. Genetics 146:1407–1415
Millet L, Vidal H, Andreelli F, Larrouy D, Riou J-P, Ricquier D, Laville M, Langin D (1997) Increased uncoupling protein-2 and -3 mRNA expression during fasting in obese and lean humans. J Clin Invest 100:2665–2670
Miyasaka K, Kanai S, Ohta M, Kawanami T, Kono A, Funakoshi A (1994) Lack of satiety effect of cholecystokinin (CCK) in a new rat model not expressing the CCK-A receptor gene. Neurosci Lett 180:143–146
Moller DE, Chang P-Y, Yaspelkis BB, Flier JS, Wallberg-Henriksson H, Ivy JL (1996) Transgenic mice with muscle-specific insulin resistance develop increased adiposity, impaired glucose tolerance, and dyslipidemia. Endocrinology 137:2397–2405
Moller DE, O'Rahilly S (1993) Congenital syndromes of severe insulin resistance. In: Moller DE (ed) Insulin Resistance. Chichester: John Wiley and Sons
Montague CT, Farooqi IS, Whitehead JP, Soos MA, Rau H, Wareham NJ, Sewter CP, Digby JE, Mohammed SN, Hurst-JA, Cheetham CH, Earley AR, Barnett AH, Prins JB, O'Rahilly S (1997) Congenital leptin deficiency is associated with severe early-onset obesity in humans. Nature 387:903–908
Muzzin P, Eisensmith RC, Copeland KC, Woo SLC (1996) Correction of obesity and diabetes in genetically obese mice by leptin gene therapy. Proc Natl Acad Sci 93:14804–14808
Mynatt RL, Miltenberger RJ, Klebig ML, Zemel MB, Wilkinson JE, Wilkison WO, Woychik RP (1997) Combined effects of insulin treatment and adipose tissue specific agouti expression on the development of obesity. Proc Natl Acad Sci USA 94:919–922
Naggert JK, Fricker LD, Varlamov O, Nishina PM, Rouille Y, Steiner DF, Carroll RJ, Paigen BJ, Leiter EH (1995) Hyperproinsulinaemia in obese fat/fat mice associated with a carboxypeptidase E mutation which reduces enzyme activity. Nat Genet 10:135–142
Noben-Trauth K, Naggert JK, North MA, Nishina PM (1996) A candidate gene for the mouse mutation tubby. Nature 380:534–538
Norman RA, Thompson DB, Foroud T, Garvey WT, Bennett PH, Bogardus C, Ravussin E (1997) Genomewide search for genes influencing percent body fat in Pima Indians: suggestive linkage at chromosome 11Q21 Q22. Am J Human Genetics 60:166–173
Ohki-Hamazaki H, Watase K, Yamamoto K, Ogura H, Yamano M, Yamada K, Maeno H, Imaki J, Kikuyama S, Wada E, Wada K (1997) Mice lacking bombesin receptor subtype-3 develop metabolic defects and obesity. Nature 390:165–169
Ollmann MM, Wilson BD, Yang YK, Kerns JA, Chen Y, Gantz I, Barsh GS (1997) Antagonism of central melanocortin receptors in vitro and in vivo by agouti-related protein. Science 278:135–138
O'Shea D, Morgan DG, Meeran K, Edwards CM, Turton MD, Choi SJ, Heath MM, Gunn I, Taylor GM, Howard JK, Bloom CI, Small CJ, Haddo O, Ma JJ, Callinan W, Smith DM, Ghatei MA, Bloom SR (1997) Neuropeptide Y induced feeding in the rat is mediated by a novel receptor. Endocrinology 138:196–202
Pedrazzini T, Kuenstner P, Seydoux J, Aubert J-F, Grouzmann E, Beerman F, Brunner H-R (1997) Lack of NPY-induced blood pressure response and impaired metabolism in NPY Y1 receptor-deficient mice. Regul Pept 71:214

Pepin M-C, Pothier F, Barden N (1992) Impaired type II glucocorticoid-receptor function in mice bearing antisense RNA transgene. Nature 355:725–728
Phillips MS, Liu Q, Hammond HA, Dugan V, Hey PJ, Caskey CJ, Hess JF (1996) Leptin receptor missense mutation in the fatty Zucker rat [letter]. Nat Genet 13:18–19
Pomp D (1997) Genetic dissection of obesity in polygenic animal models. Behavior Genetics 27:285–306
Qu D, Ludwig DS, Gammeltoft S, Piper M, Pelleymounter MA, Cullen MJ, Mathes WF, Przypek J, Kanarek R, Maratos-Flier E (1996) A role for melanin-concentrating hormone in the central regulation of feeding behaviour. Nature 380:243–247
Ricquier D, Bouillaud F, Toumelin P, Mory G, Bazin R, Arch J, Penicaud L (1986) Expression of uncoupling protein mRNA in thermogenic or weakly thermogenic brown adipose tissue. Evidence for a rapid β-adrenoreceptor-mediated and transcriptionally regulated step during activation of thermogenesis. J Biol Chem 261:13905–13910
Rosenblum CI, Tota M, Cully D, Smith T, Collum R, Qureshi S, Hess JF, Phillips MS, Hey PJ, Vongs A, Fong TM, Xu L, Chen HY, Smith RG, Schindler C, Van der Ploeg LH (1996) Functional STAT 1 and 3 signaling by the leptin receptor (OB-R). Endocrinology 137:5178–5181
Ross SR, Graves RA, Spiegelman BM (1993) Targeted expression of a toxin gene to adipose tissue: transgenic mice resistant to obesity. Genes Dev 7:1318–1324
Rossi M, Choi SJ, O'Shea D, Miyoshi T, Ghatei MA, Bloom SR (1997) Melanin-concentrating hormone acutely stimulates feeding, but chronic administration has no effect on body weight. Endocrinology 138:351–355
Safer DJ (1991) Diet, behavior modification, and exercise: a review of obesity. South Med J 84:1470–1474
Sainsbury A, Cusin I, Rohner-Jeanrenaud F, Jeanrenaud B (1997) Adrenalectomy prevents the obesity syndrome produced by chronic central neuropeptide Y infusion in normal rats. Diabetes 46:209–214
Saito Y, Nothacker HP, Wang Z, Lin SH, Leslie F, Civelli O (1999) Molecular characterization of the melanin-concentrating-hormone receptor. Nature 400:265–269
Schena M, Shalon D, Heller R, Chai A, Brown PO, Davis RW (1996) Parallel human genome analysis: microarray based expression monitoring of 1000 genes. Proc Natl Acad Sci 93:10614–10619
Seeley RJ, Yagaloff KA, Fisher SL, Burn P, Thiele TE, van Dijk G, Baskin DG, Schwartz MW (1997) Melanocortin receptors in leptin effects [letter]. Nature 390:349
Shepherd PR, Gnudi L, Tozzo E, Yang H, Kahn BB (1993) Adipose cell hyperplasia and enhanced glucose disposal in transgenic mice overexpressing GLUT4 selectively in adipose tissue. J Biol Chem 268:22243–22246
Shimizu H, Shargill NS, Bray GA, Yen TT, Gesellchen PD (1989) Effects of MSH on food intake, body weight and coat color of the yellow obese mouse. Life Sci 45:543–552
Shimada M, Tritos NA, Lowell BB, Flier JS, Maratos-Flier E (1998) Mice lacking melanin-concentrating hormone are hypophagic and lean. Nature 396:670–674
Shutter JR, Graham M, Kinsey AC, Scully S, Luthy R, Stark KL (1997) Hypothalamic expression of ART, a novel gene related to agouti, is up-regulated in obese and diabetic mutant mice. Genes Dev 11:593–602
Silva JE, Rabelo R (1997) Regulation of the uncoupling protein gene expression. Euro J Endocrinology 136:251–264
Spiegelman BM, Hotamisligil GS (1993) Through thick and thin: wasting, obesity, and TNFα. Cell 73:625–627
Stunkard AJ, Sorensen TI, Hanis C, Teasdale TW, Chakraborty R, Schull WJ, Schulsinger F (1986) An adoption study of human obesity. N Engl J Med 314:193–198
Tartaglia LA, Dembski M, Weng X, Deng N, Culpepper J, Devos R, Richards GJ, Campfield LA, Clark FT, Deeds J et al. (1995) Identification and expression cloning of a leptin receptor, OB-R. Cell 83:1263–1271

Tecott LH, Sun LM, Akana SF, Strack AM, Lowenstein DH, Dallman MF, Julius D (1995) Eating disorder and epilepsy in mice lacking 5-HT2c serotonin receptors. Nature 374:542–546
Tontonoz P, Hu E, Spiegelman BM (1994) Stimulation of adipogenesis in fibroblasts by PPARγ2, a lipid activated transcription factor. Cell 79:1147–1156
Tsigos C, Arai K, Hung W, Chrousos GP (1993) Hereditary isolated glucocorticoid deficiency is associated with abnormalities of the adrenocorticotropin receptor gene. J Clin Invest 92:2458–2461
Turner RC, Levy JC, Clark A (1993) Complex genetics of type 2 diabetes: thrifty genes and previously neutral polymorphisms. Q J Med 86:413–417
Valancius V, Smithies O (1991) Testing an "In–Out" targeting procedure for making subtle genomic modifications in mouse embryonic stem cells. Mol Cell Biol 11:1402–1408
Valverde P, Healy E, Jackson I, Rees JL, Thody AJ (1995) Variants of the melanocyte-stimulating hormone receptor gene are associated with red hair and fair skin in humans. Nat Genet 11:328–330
Van Heek M, Compton DS, France CF, Tedesco RP, Fawzi AB, Graziano MP, Sybertz EJ, Strader CD, Davis HR, Jr (1997) Diet-induced obese mice develop peripheral, but not central, resistance to leptin. J Clin Invest 99:385–390
Vaisse C, Clement K, Guy-Grand B, Froguel P (1998) A frameshift mutation in human MC4R is associated with a dominant form of obesity. Nat Genet 20:113–114
Ventre J, Doebber T, Wu M, MacNaul K, Stevens K, Pasparakis M, Kolias G, Moller DE (1997) Targeted disruption of the tumor necrosis factor-α gene: metabolic consequences in obese and non-obese mice. Diabetes 46:1526–1531
Vidal-Puig A, Moller DE (1994) Comparative sensitivity of alternative single-strand conformation polymorphism (SSCP) methods. Biotechniques 17:490–494
Vidal-Puig A, Solanes G, Grujic D, Flier JS, Lowell BB (1997) UCP3: an uncoupling protein homologue expressed preferentially and abundantly in skeletal muscle and brown adipose tissue. Biochem Biophys Res Comm 235:79–82
Wang ND, Finegold MJ, Bradley A, Ou CN, Abdelsayed SV, Wilde MD, Taylor LR, Wilson DR, Darlington GJ (1995) Impaired energy homeostasis in C/EBP alpha knockout mice. Science 269:1108–1112
Weber A, Toppari J, Harvey RD, Klann RC, Shaw NJ, Ricker AT, Nanto-Salonen K, Bevan JS, Clark AJ (1995) Adrenocorticotropin receptor gene mutations in familial glucocorticoid deficiency: relationships with clinical features in four families. J Clin Endocrinol Metab 80:65–71
Weinstock PH, Levak-Frank S, Hudgins LC, Radner H, Friedman JM, Zechner R, Breslow JL (1997) Lipoprotein lipase controls fatty acid entry into adipose tissue, but fat mass is preserved by endogenous synthesis in mice deficient in adipose tissue lipoprotein lipase. Proc Natl Acad Sci 94:10261–10266
Wendorf M, Goldfine ID (1991) Archaeology of NIDDM: Excavation of the "thrifty" genotype. Diabetes 40:161–165
Wu Z, Xie Y, Bucher NLR, Farmer SR (1995) Conditional ectopic expression of C/EBPβ in NIH-3T3 cells induces PPARγ and stimulates adipogenesis. Genes Devel 9:2350–2363
Yeo GS, Farooqi IS, Aminian S, Halsall DJ, Stanhope RG, O'Rahilly S (1998) A frameshift mutation in MC4R associated with dominantly inherited human obesity. Nat Genet 20:111–112
Zarjevski N, Cusin I, Vettor R, Rohner-Jeanrenaud F, Jeanrenaud B (1993) Chronic intracerebroventricular neuropeptide-Y administration to normal rats mimics hormonal and metabolic changes of obesity. Endocrinology 133:1753–1758
Zhang Y, Proenca R, Maffei M, Barone M, Leopold L, Friedman JM (1994) Positional cloning of the mouse obese gene and its human homologue. Nature 372:425–432
Zhou X, Benson KF, Ashar HR, Chada K (1995) Mutation responsible for the mouse pygmy phenotype in the developmentally regulated factor HMGI-C. Nature 376:771–774

CHAPTER 17

Identification of Genes Involved in Animal Models of Obesity

D.B. West, Y. Ma, A.A. Truett, and B. York

A. Introduction

It is clear that human obesity has a significant genetic component (for recent reviews see Perusse and Chagnon 1997; Comuzzie and Allison 1998; Chagnon, Perusse and Bouchard 1998; and other chapters in this volume). Finding and characterizing these genetic risk factors will provide us with a greater understanding of the etiology of this disease and will certainly lead to novel therapeutic approaches. The identification of polygenes underlying complex disorders in human populations is extremely difficult. Genetic heterogeneity, difficulties in precise phenotyping of the subjects, partial penetrance of the gene effects, age-dependent expression of the trait, and variable gene–gene and gene–environment interactions can all serve to make the human linkage and association studies difficult to interpret. These factors result in great difficulty in finding linkage and/or demonstrating an association between a trait and a genetic variant. Therefore, the sample size of the human cohorts necessary to complete these studies with reasonable statistical power is difficult and expensive to achieve and combining data across multiple studies using different ascertainment strategies and ethnic compositions may not be appropriate. There are many interesting linkage and association data being generated in human studies of obesity (see Chagnon, Perusse and Bouchard 1998). However, often the statistical significance is marginal, the positioning of the linkage is not precise, and it is unclear how to best follow up on these observations to actually identify the genes involved and to conclusively prove that they contribute to the trait.

An alternative, economical and effective approach to understanding the genetics of a complex disorder in humans is to utilize animal models. Obesity is a clear example in which the use of genetic models of obesity in rodents has led to remarkable insights into the pathophysiology of this disease and has led to the identification of new genes and proteins that can contribute to the regulation of body lipid content. This chapter provides a brief overview of the strategies being utilized in animal models, primarily in the mouse, in order to identify the genes and metabolic pathways involved in obesity.

The general approaches for identifying candidate genes involved in obesity in murine models are presented schematically in Fig. 1.

The approaches include gene identification in monogenic models of obesity. These models can be spontaneous mutants or novel mutations induced by mutagenesis programs. These monogenic mutants have been very powerful and successful at the identification of novel genes involved in regulating body lipid stores. Another avenue for identifying possible obesity genes in the rodent models is gene targeting. In this approach, candidate genes identified through other mechanisms, for example through fine mapping starting with linkage data, expression screening or structural homology are manipulated by transgenic methods to determine if the genes functionally affect lipid storage. Finally, linkage of genomic regions to body fat can be studied in complex models and used ultimately to narrowly define chromosomal regions containing obesity genes. The approach of using human–mouse synteny to utilize the power of mouse models to pursue the linkage identified in humans is one untested but promising component of this approach. Once putative obesity genes are identified, then they can be evaluated in animal models and ultimately, since the goal is to characterize the genetic basis for predisposition to human obesity, the candidate genes can be evaluated in human populations for allelic variants that might be associated with obesity.

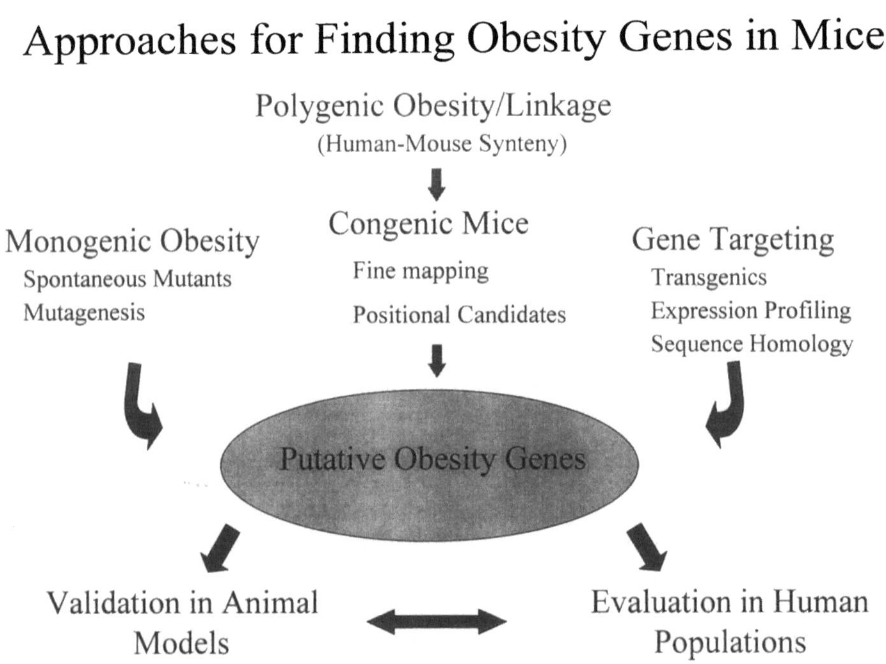

Fig. 1. Approaches for finding obesity genes in mice

B. Mendelian Models of Obesity in Mice

The monogenic models of murine obesity have been extraordinarily useful in demonstrating both the utility of the single gene obesity models for identifying novel obesity genes as well as illustrating the various approaches for pulling out these genes from the mouse models. For a recent review see CHUA (1997).

I. The Five Classical Mutants in the Mouse

There are five Mendelian obesity mutations in the mouse that have now been identified and cloned over the past 6 years. These include the *ob* (now *lep*) mutation, *db* (now *lepr*), *fat* (now cpe^{fat}), and *tubby* (*tub*) and *Agouti* (*A*) mutations. The identification of these genes was done through a variety of approaches that illustrate the versatility of mouse models for gene identification.

The cloning of the *ob* gene has resulted in the greatest degree of recent interest in terms of potential therapeutic development, although it was not the first mouse obesity gene cloned. This gene was positionally cloned by FRIEDMAN'S group in late 1994 (ZHANG et al. 1994) and given the name leptin from the Greek root leptos meaning "thin". Positional cloning relies upon fine chromosomal localization, usually based upon the linkage of the monogenic trait with molecular markers in crosses segregating for the phenotype. Once the position has been determined within less than a centimorgan (cM) of genetic distance (approximately 2 megabases of DNA in the mouse), then a variety of molecular methods are available for identifying the putative disease gene.

On the basis of studies by Coleman, in which *ob* mice were surgically interconnected (i.e., connected so that extracellular fluid could be exchanged between paired mice) to *db* mice and normal mice, it was inferred that the *ob* gene coded for a secreted protein, while the *db* gene coded for the receptor of this secreted protein (COLEMAN 1973). Indeed, this proved to be the case. The *ob* or *leptin* gene codes for a secreted protein expressed primarily in adipocytes but also expressed in some fetal tissues and in the stomach and placenta. It is correlated with body-fat content and thus may be an important signal mediating the regulation of body lipid stores (CONSIDINE et al. 1996). Subsequent to this discovery, TARTAGLIA et al. identified the *db* gene as the leptin receptor (*lepr*), through an expression-cloning strategy using a recombinant leptin to probe an expression library derived from a tissue (brain choroid plexus) shown histochemically to have a significant number of leptin receptors with relatively high and specific binding (TARTAGLIA et al. 1995). The *lepr* gene is expressed in a significant variety of tissues, including the hypothalamus and numerous alternative splice variants have been identified. The *lep–lepr* pathway is an important signaling pathway, identified only through the use of mouse mutants showing massive obesity. Subsequent studies with recombinant leptin have shown that it can reduce food intake and body weight in mice (PELLEYMOUNTER et al. 1995) and more recently in humans (HEYMSFIELD et al. 1998). However, in the human studies there is a significant heterogeneity in the response to this agent, suggesting that only specific subpopulations of obese patients may be responsive to treatment with leptin or leptin analogs. The role of leptin in normal physiology is complex and likely involves reproduction, hematopoiesis and other functions in addition to the regulation of body weight (for reviews see FLIER 1998; KIESS et al. 1998; CAMPFIELD et al. 1998).

Another example of a mouse obesity model leading to the identification of a signaling pathway with potentially important therapeutic applications is the *Agouti* mouse. *Agouti* is a coat color gene in the mouse for which there are multiple alleles including several (e.g., A^y and A^{vy}) that are dominant and result in massive obesity. For a review see SIRACUSA (1994). This gene was originally cloned by BULTMAN et al., using an inversion mutant in the mouse disrupting a previously cloned gene affecting limb development (the limb deformity locus) and also affecting the coat color *Agouti* gene (BULTMAN, MICHAUD and WOYCHIK 1992). Although the inversion mutation did not result in massive obesity it did disrupt the *Agouti* gene because coat color was affected. Since one gene was already cloned, it was possible to walk from the known sequence into the *Agouti* coding region and identify the gene. The A^y mutation results in the *Agouti* gene being put under the promoter control of another gene (*Raly*) and causes ubiquitous overexpression (MICHAUD et al. 1993). For a significant period of time it was unclear how this overexpression resulted in obesity since normally the *Agouti* gene is only (or primarily) expressed in the skin and its apparent function is the modulation of coat color. Two observations led to the pathway by which *Agouti* overexpression produced obesity. First, it was shown that the *Agouti* protein is an antagonist of the melanocyte-stimulating-hormone receptor (*Mc1r*) expressed in the skin (LU et al. 1994). By unknown mechanisms, *Agouti* not only antagonizes the endogenous ligand for this receptor, α-MSH, but also had specific independent actions although it binds to the same receptor (OLLMANN et al. 1998). This suggested that the *Agouti* gene might produce obesity by acting at melanocortin receptors outside the hair follicle. Second, it was observed that an *Agouti*-like gene in humans (*Agouti*-related transcript or *ART*; the mouse homolog is *Agrp* for *Agouti*-related protein) is expressed in the brain (SHUTTER et al. 1997). Subsequent studies suggested that ART might be an endogenous antagonist of other melanocortin receptors found in the brain (*Mc3r* or *Mc4r*) (OLLMANN et al. 1997). A knockout of the *Mc4* receptor in the mouse producing massive obesity has provided strong evidence that the *Agrp–Mc4r* pathway is important in body-weight regulation (see below; HUSZAR et al. 1997).

The other two classic Mendelian mouse mutations producing obesity have not led to the discovery of such clear pathways involved in the regulation of body-fat stores. However, the identification of these genes also illustrates the various approaches used with the mouse to identify obesity genes. The *fat* gene was identified by NAGGERT et al., (1995) by a positional candidate approach. The gene was fine-mapped to a location on proximal mouse chromosome 8 that placed it close to a prohormone-processing gene (carboxy peptidase E; *Cpe*). The investigators speculated that this gene could be important in the obesity phenotype since it is ubiquitously expressed in neuroendocrine tissues and is involved in the processing of a variety of hormones already postulated as candidate genes for obesity. Sequencing of the *Cpe* gene in the *fat* mouse confirmed a point mutation in the open reading frame of this gene and subsequent work has clearly determined that this mutation leads to the physiological derangement producing obesity. However, it is not understood whether the resulting obesity is due to the defective processing of a single hormone or multiple hormones affected by the mutation. It is most likely that the obesity is a pleotrophic effect of the Cpe^{fat} mutation; it is therefore not clear from the currently understood physiological effects of the Cpe^{fat} mutation which appropriate step or steps in multiple pathways should be intervened with therapeutics.

NAGGERT and colleagues identified the *tub* mutation by positional cloning in 1996 (NOBEN-TRAUTH et al. 1996) and the identification of the *tub* gene was published almost simultaneously by another group (KLEYN et al. 1996). The function of the *tub* gene is still not known and the mechanism responsible for the mutation-induced obesity remains uncertain. The *tub* gene mutation might act by producing specific neuronal apoptosis since the *tub* gene phenotype includes retinal degeneration and the gene belongs to a family of genes associated with abnormal ocular development (NORTH et al. 1997).

II. Mutations in Human Homologs of the Murine Obesity Genes

Of course, once these five mouse monogenic obesity genes were identified, it initiated a significant effort to determine if polymorphisms or mutations in the human homologs of these genes (or genes involved in these pathways) are involved in the etiology of human obesity. Rare recessive mutants of the human leptin gene (*LEP*) (STROBEL et al. 1998) and human leptin receptor gene (*LEPR*) (CLEMENT et al. 1998) have been found to produce massive obesity in humans. A mutation in humans in the proopiomelanocortin (POMC) pathway leading to α-MSH production has been reported to produce obesity in a single family, implicating that the human brain *ART–MC4R* pathway is important for body-weight regulation (KRUDE et al. 1998). Similarly, a rare mutation in an enzyme in the prohormone-processing pathway (prohormone-converting enzyme 1; *PC1*, JACKSON et al. 1997) has been identified in humans and this mutation produces obesity. However, it is now clear that allelic variants in the human homologs of the mouse obesity genes are not likely to contribute significantly to common forms of human obesity. This does not minimize the importance of exploiting the metabolic pathways identified, through the cloning of these genes in the mouse, for the development of therapeutic approaches for the treatment of obesity.

Thus, the identification of three of the five classic Mendelian mutations in the mouse has led to two heretofore unknown signaling pathways important for energy metabolism, for which pharmacotherapies are being developed. The methods employed for identifying these obesity genes are representative of the various approaches used in the mouse for the identification of any disease gene. These approaches include the positional candidate (*fat*) and positional cloning approaches (*lep* and *tub*), the use of mutagenesis to create mutations leading to gene identification (*Agouti*), as well as functional screens, based upon inferred mechanisms (e.g., *lepr*: the assay was based upon the identification of tissues containing putative receptors with immunohistological techniques).

III. Other Spontaneous Mouse Mutants

There are a number of other single-gene obesity models in the mouse that have not yet been cloned. These mutants may provide further new insights into the possible mechanisms producing or modulating obesity. For example, the mahogany (*mg*) and mahoganoid (*md*) mutations are coat-color mutations that modulate the *Agouti* allele effects on coat color. These coat-color mutations have also recently been shown to affect body weight/fat in the mouse and are likely modifiers of *Agrp* function/activity in the brain (MILLER et al. 1997). Several other spontaneously occurring Mendelian mutations have also been found to affect body-fat content. For example, the twirler (*Tw*) mutation mapped to mouse chromosome 18 has a pronounced phenotype of head shaking and circling behavior and is also associated with late-onset obesity (LYON 1958). The "adult obesity and diabetes" mutation (*Ad*) on mouse chromosome 7 was described as a spontaneous semidominant mutation resulting in spontaneous obesity (WALLACE 1975). However, this mutant appears to be extinct. A spontaneous maternal duplication in the region of chromosome 7 of the mouse, syntenic to the human chromosomal region containing the Prader Willi/Angelman syndrome, a syndrome with an obesity phenotype, has also been reported. This duplication is associated with an altered growth rate (CATTANACH et al. 1997). Thus, other "Mendelian" mutants exist or have occurred in the mouse that have marked effects on body-fat content and are also available for cloning. This suggests that mutagenesis programs (described below) may be a useful strategy for identifying new genes controlling adiposity, since such a strategy might reveal additional mutations in genes and pathways having large effects on body-fat content.

The cloning of genes causing monogenic disorders was difficult in the past because it primarily relied upon positional information and ultimately fine positioning of the gene by examining recombinant animals for the phenotype (e.g., *lep*, *tub*). This is a time-consuming and difficult process. It is clear that we are now entering an era where sequence and mapping information for both genes and expressed sequences will be abundant. This will greatly facilitate the process of pulling out positional candidates and will significantly accelerate the identification of both the genes underlying Mendelian traits as well as the genes involved in complex phenotypes (COLLINS 1995).

C. Gene Targeting

Another approach for identifying genes involved in the regulation of body fat is to develop new models by transgenic approaches to either overexpress these candidates ubiquitously or in specific tissues, or to knock them out. These mutations can either be targeted based upon known sequence and function of the gene, or can be random (see below for random mutagenesis approaches). Although the topic of transgenic models of obesity is covered in another chapter of this text, some of the results will be briefly reviewed here and discussed in terms of how they have elucidated our understanding of the pathogenesis of obesity.

At recent count there are at least 20 knockout or transgenic animals, with genes that have an effect on body fat and result in either an increase or a decrease in body lipid stores. Most of these genes were targeted on the basis of known or inferred roles of these genes and their products in energy or lipid metabolism.

I. Transgenics Increasing Body Fat

A number of transgenic or knockout animals that increase body-fat content have been developed. For example, the importance of brown-fat thermogenesis in maintaining normal body weight was evaluated by the use of a brown-adipose-tissue-specific suicide gene (driven by the brown-adipose-specific uncoupling-gene (*Ucp1*) promoter. This construct produced an endotoxin (diphtheria toxin) and resulted in ablation of the tissue (HAMANN, FLIER and LOWELL 1998). The animals became obese. The obesity produced in the brown-fat-ablated mice is eliminated when the animals are maintained at thermoneutrality, suggesting that hyperphagia induced in order to maintain core body temperature may be one factor in producing obesity in this model (MELNYK et al. 1997). Interestingly, a targeted knockout of the brown-adipose-tissue-specific uncoupling protein (*Ucp1*), had no discernable body-fat phenotype. This was unexpected since the essential function of brown adipose tissue is thermogenesis, a function primarily dependent upon *Ucp1* function. These *Ucp1* knockout animals did not thermoregulate well in the cold (ENERBACK et al. 1997), suggesting that there are compensatory mechanisms that may be activated in brown adipose tissue when *Ucp1* is lacking, allowing the animal to function normally except in extremes of cold.

A number of other transgenic or knockout models have also been used to evaluate the role of specific candidate genes in the regulation of body weight. For example, the recent melanocortin 4 receptor knockout (Huszar et al. 1997) tested the hypothesis that the *Mc4r* gene is an important gene in body-weight regulation, and this was confirmed by the resulting obesity in these animals. Similarly, ubiquitous overexpression of the *Agouti* gene (Klebig et al. 1995) and *Agouti*-related transcript (*Art*) (Graham et al. 1997) confirmed that the obesity associated with the A^y and other dominant *A* mutations was due to the ubiquitous overexpression of the *Agouti* gene. Furthermore, the *Art* transgenic phenotype confirmed that the brain-specific homolog of *Agouti* might be the endogenous ligand in the central nervous system melanocortin pathway important for the regulation of body weight and body fat. Another transgenic mouse producing an increased body-fat content is the adipose-specific overexpression of the glucose transporter 4 (*Glut4*) gene (Shepherd et al. 1993). This transgene results in obesity due to hyperplastic adipose tissue development although the precise mechanism by which this transgene may be stimulating preadipocyte differentiation remains unknown.

Transgenic and knockout approaches have also been useful for exploring the role of adrenergic receptors in the etiology of obesity. For example, the β-3 adrenoreceptor knockout results in an animal that is susceptible to high-fat-diet-induced obesity (Susulic et al. 1995). This susceptibility to obesity is, presumably, due to the decreased ability to activate brown-fat thermogenesis; a function largely regulated through β-3 adrenoreceptor sympathetic activation.

One caveat for the use of knockout mice to identify genes involved in energy homeostasis is that nonconditional knockouts, i.e., knockouts that are constitutive and result in loss of a functional gene product as early as in the single-cell, fertilized ovum, can result in developmental adaptations obscuring the true role of a gene product in the phenotype. For example, the recent knockout of the neuropeptide Y gene (*Npy*) results in no observable effect on body fat and does not prevent the development of obesity in either genetic or dietary models, suggesting that this gene is not important for normal energy homeostasis (Hollopeter, Erickson and Palmiter 1998). This contradicts significant evidence from other studies showing the importance of this neurotransmitter in the regulation of feeding behavior and energy homeostasis (e.g., Woods et al. 1998). Therefore, the lack of a phenotype effect of a transgene or knockout may not be definitive. Conditional gene targeting will be necessary for defining the true roles of specific genes in obesity and will circumvent early developmental adaptations that mitigate the effects of constitutive gene manipulations. Furthermore, the characterization of allelic series of genes with numerous mutations will be an important approach for understanding the genetics of complex disorders. Altering patterns of tissue expression, producing increased or decreased functionality, and altering protein processing and interactions leading to multiple effects will be important for fully characterizing the role of a specific gene in a complex metabolic pathway.

II. Trangenics Decreasing Body Fat

One must, of course, be cautious when interpreting the effect of any genetic manipulations that reduce growth and body lipid content, since such a phenotype could be attributable to adverse effects on development, or could be due to physical impairments unrelated to normal regulation of body lipid stores.

The adipocyte has been targeted in a number of white-adipose-tissue-specific gene manipulations to further understand the role of specific metabolic pathways in the control of lipid stores. For example, overexpression of the β-1 adrenergic receptor in white adipocytes results in less adipose-lipid content, suggesting that the tonic activation of adrenergic-mediated lipolysis may be an important factor in ultimately determining lipid stores (SOLOVEVA et al. 1997). Overexpression of *Ucp1* in white adipocytes also results in decreased adipose lipid stores to suggest that methods for increasing mitochondrial uncoupling in white adipose tissue could be a reasonable therapeutic target (KOPECKY et al. 1996). Although white adipose tissue does not express *Ucp1*, it does express another putative uncoupling protein, namely *Ucp2*, which could serve as a target for manipulation of lipid stores (FLEURY et al. 1997). Similarly, the ubiquitous overexpression of glycerol 3-phosphate dehydrogenase (KOZAK et al. 1991) increases brown adipose tissue size but significantly decreases white adipose lipid stores perhaps due to effects on adipose tissue development.

Of particular interest are studies of the skeletal-muscle overexpression of lipoprotein lipase, resulting in a significant reduction in adipose-lipid stores when the animals are placed on a high-fat diet (JENSEN et al. 1997). By overexpressing lipoprotein lipase in skeletal muscle, this might cause the muscle to preferentially take up and burn fatty acids and shunt lipid calories from storage to utilization. Of similar interest is the report that a disruption of the RII beta subunit of protein kinase A (PKA) results in a significant reduction in adipose lipid stores despite normal energy intake (CUMMINGS et al. 1996). Since adrenergic stimulation of lipolysis relies upon a cAMP-mediated PKA-dependent mechanism, these results suggest that disruption of this signaling pathway might lead to chronic activation of lipolysis.

The total ablation of white and brown adipose tissue has been achieved in mice by coupling the aP2 promotor to the diphtheria toxin minigene (BURANT et al. 1997). These mice are markedly hyperlipidemic, as might be expected, but are also hyperinsulinemic and hyperglycemic; suggesting that alterations in the normal disposal of lipid (by incorporation into stored lipids) may be a contributing factor in the etiology of obesity-associated type 2 diabetes.

III. Unexpected Obesity in Transgenic Mice

There have been other transgenic mice that have had significant unexpected effects on body fat. For example, a knockout of the *Nhlh2* gene, a transcription factor found in the brain, results in hypogonadism, decreased circulating gonadotropins and late-onset obesity (GOOD et al. 1997). This gene is clearly involved in the regulation of the hypothalamo–pituitary axis during development, but the specific mechanism by which it affects body weight is unknown. Targeted knockouts of the metallothionine 1 and 2 (*Mt1* and *Mt2*) genes, known to modulate zinc and copper clearance, also result in a late-onset obesity through an unknown mechanism (BEATTIE et al. 1998). Finally, knockouts of the intercellular adhesion molecule-1 molecule or the counterreceptor for this molecule, leukocyte integrin alpha *Mbeta2* (*Mac-1*), both become obese (DONG et al. 1997). These genes code for proteins mediating leukocyte adhesion; it suggests that leukocytes may in some way be involved in modulating lipid transport and storage. Such unexpected findings point to new and interesting mechanisms that may affect the regulation of body weight and lipid stores.

IV. Gene Dosage Effects in Transgenics

Combining knockouts and or transgenics can also lead to interesting inferences regarding the role of specific pathways in the regulation of body-fat stores. Although this has not yet been done for obesity, an excellent example of this approach is provided in the report combining heterozygous knockouts for both the insulin-receptor-substrate 1 (*Irs1*) and the insulin-receptor (*Ir*) gene in the mouse (BRUNING et al. 1997). Mice heterozygous for either single knockout have minimal or no phenotype, while mice heterozygous for both knockouts show profound insulin resistance and hyperinsulinemia. This model illustrates the importance of gene interactions in producing a phenotype, and also that the genetic modulations of complex phenotypes are probably due to the interaction of allelic variants in several genes. That complex diseases are the result of significant gene–gene interactions is certainly not a novel concept (see RISCH et al. 1993), and there are ample examples of these interactions in the mouse models of polygenic obesity (see below).

D. Mutagenesis Approaches

Mutagenesis approaches are also available, to screen either genome-wide, or in a relatively narrow chromosomal region, for genes affecting body-fat content. These methods can be used without any *a priori* knowledge of the specific genes involved in the obesity phenotype. For example, the classical method of ethyl-nitrosourea (ENU) induced mutagenesis is being used to screen genome-wide for a variety of phenotypes (BROWN and PETERS 1996; ROGERS et al. 1997), but to our knowledge has not been used yet to identify novel obesity mutations in a systematic way. One disadvantage of ENU mutagenesis is that one must either screen for gene-dosage effects (i.e., the effects of a heterozygous point mutation and therefore focus on dominant effects) or breed the mutations to homozygosity to find recessive phenotype effects. The location of the point mutation must then be determined by classical genetic techniques. Therefore, ENU mutagenesis may not be ideal for high-throughput and rapid gene identification.

Other techniques for nontargeted generation of mutations for the obesity phenotype are likely to be more efficient and rapid at identifying novel genes. For example, mutagenesis techniques producing large inversions or translocations have been used to screen for various physical and behavioral phenotypes and could be used to identify novel obesity genes (CULIAT et al. 1997). The advantage of this technique is that the general chromosomal location of the disrupted gene can be determined histochemically by chromosomal staining and provides at least some information about the gene location and allows potentially more rapid gene identification. Still other approaches can be used to screen large chromosomal regions for the presence of genes affecting a phenotype. For example, one can produce hemizygous deleted animals with the loxP/Cre method, deleting large sections of individual chromosomes, and then use a chemical mutagenesis approach to probe against these hemizygous deletions to identify genes affecting body lipid stores (JUSTICE et al. 1997). Hemizygous deletions produced in embryonic stem cells by irradiation could also serve as the source of mice carrying regional deletions against which one could then target with mutagenesis to cover a specific genomic region (YOU, BROWNING and SCHIMENTI 1997). Such approaches will prove useful in following up linkage data from mouse models or congenic mice used to identify chromosomal regions containing obesity genes (see below). Similarly, gene-dosage effects in large chromosomal regions can be tested with large-scale constructs (e.g., yeast artificial chromosomes) to create transgenic animals (SMITH and RUBIN 1997). The advantage of this approach is that if the YAC transgene affects the phenotype, then one has an immediate physical reagent for cloning the gene responsible. The disadvantages of this technique are that only some genes are likely to show a dosage effect on a phenotype or the effect may be subtle and require careful physiological characterization for detection. Finally, new methods of gene targeting requiring minimal information about the gene can be used to either knock out or generate specific point mutations (KHREBTUKOVA et al. 1998). This method has the advantage of being used on a genome-wide scale to target multiple genes of, for example, a specific functional class carrying common sequence motifs.

E. Genes in Complex Genetic Models of Obesity

Over the past several years, polygenic models have also been exploited to map the genes controlling body-fat content in the mouse and in the rat. The basic approach is known as quantitative trait locus, or QTL, analysis (LANDER and

BOTSTEIN 1989). In this approach one identifies mouse inbred strains differing significantly in a quantitative trait; intercrosses these strains to produce a hybrid F1 generation, and then backcrosses onto the parental lines or intercrosses the F1, to produce an N2 or F2 population, respectively. Due to independent assortment of the chromosomes and recombination during meiosis, the genes controlling the complex trait, such as body-fat content, are randomized. The resulting population contains a distribution of genotypes, and a phenotype ranging between the extremes of the parental phenotype and sometimes exceeding the phenotype of the parental strains. Using molecular markers, one can map these genes and find their chromosomal location. This approach has been applied over the past several years to map a number of genetic loci for obesity, body weight, or obesity-related traits as shown in Fig. 2.

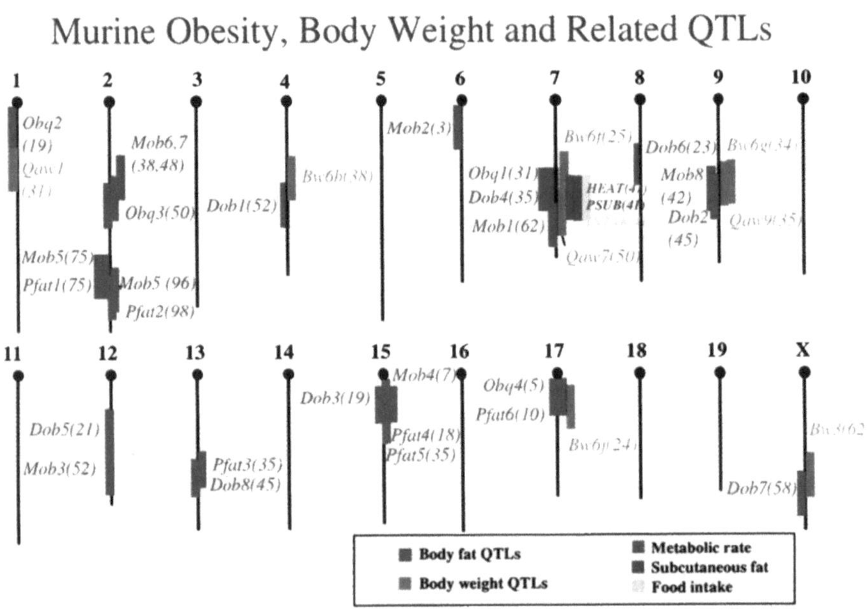

Fig. 2. A chromosomal cartoon of the 19 mouse autosomes and the x chromosome presenting murine obesity, body weight and related QTLs. The major body fat QTLs are shown by name as described in the text of the chapter and in Table 1. In addition, body weight and other obesity-related QTLs are portrayed which colocalize with QTLs for body fat alone. All of the murine body weight QTLs are not shown on this figure. Each QTL is represented by a vertically oriented bar which is scaled to cover approximately 20cM of the mouse genome. This generally covers the 95% confidence interval for most but not all of these QTLs. The numbers in parenthesis indicated the peak of the QTL in cM from the centromere. QTLs for body fat, body weight, metabolic rate, subcutaneous fat, and food intake are color coded and given the colors red, blue, purple, green and yellow respectively.

I. QTLs for Body Fat

There are now a number of crosses in the mouse in which body fat segregates, and the genes controlling this phenotype have been mapped. Warden et al. (1995) used a backcross between Mus Spretus and C57BL/6J mice to identify four chromosomal regions controlling body fat including loci on chromosomes 6 near the *ob* gene, chromosome 7 near the *tub* gene, chromosome 12 and chromosome 15. This was a model of spontaneous obesity, since the animals were not fed a particularly high-fat diet. These genetic loci were named *Mob1-4* for multigenic obesity. They displayed clear nonadditive interactions with the major locus on chromosome 15. Another interesting feature of this cross is that the spretus alleles were usually associated with greater body-fat content although the spretus mouse was the strain with the lower body fat.

West and colleagues (1994a,b; 1995; York 1995 and 1997) have identified a total of eight chromosomal locations linked to sensitivity to dietary obesity in an F2 cross between AKR/J and SWR/J mice. The body-fat content of the AKR/J strain markedly increases on a high-fat diet while the SWR/J mice remain lean (West et al. 1992). These genetic loci accounted for the majority of the genetic differences between these strains although no single locus accounts for more than 15% of the total variance in body fat in the F2 population. In addition to identifying autosomal and X-linked transmission, this group identified transmission of the phenotype across generations, most likely due to the inheritance of mitochondrial DNA from the maternal grandparent (York et al. 1997). These genetic loci have been identified as *Dob1-8* for dietary obesity loci and are found on chromosomes 4, 7, 8, 9, 12, 13, 15, and X.

In a more limited study on an F2 population from C57BL/6J and CAST/Ei mice, York et al. (1996) identified three chromosomal regions linked with body fat on distal chromosome 2, on chromosome 7, and on chromosome 15. Two of these regions (chromosome 7 and 15) were close to or precisely located at QTLs previously identified in the AKR/J◊SWR/J F2 mapping population described above. The linkage on chromosome 2 was in the same region as found by Lembertas in a separate cross (see below).

Taylor et al. (1996, 1997) used two separate crosses and the strategy of DNA pooling to identify a total of four genetic loci affecting body-fat content. In their initial report, a cross between 129/Sv and the EL/Suz strains was used to identify genetic loci on chromosome 7 and on chromosome 1 near previously mapped body-fat genetic loci. The chromosome 7 locus was of particular interest since it accounted for more than 12 percent of the phenotypic variance in the F2 population. In another cross of AKR/J and C57L/J mice, an additional two QTLs linked to body fat were identified on chromosomes 2 and 17. These QTLs were called *Obq1–4* for obesity QTLs.

More recently, Lembertas et al. (1997) reported a QTL, in a cross between NZB and SM mice, which maps to distal chromosome 2 (called *Bfq1* for body fat QTL 1) for which a congenic mouse strain also has an obesity phenotype. This region is of significant interest for two reasons: first, numerous mouse crosses have identified QTLs for body fat in this region (see Fig. 2), and second, there is evidence from several human linkage studies for a major gene affecting body fat in the human region syntenic to this mouse distal chromosome 2 region (i.e., human 20q).

Mehrabian et al. (1998) recently used a cross between C57BL/6J and CAST/Ei to map 4 chromosomal regions linked to body fat in an F2 population. Two of these QTLs map to proximal chromosome 2 while a third maps to distal chromosome 2 near the QTL mapped in the same cross by York et al. (1995). A fourth body-fat QTL was mapped by Mehrabian to chromosome 9 near the *Do2* locus. The differences in results in the two C57BL/6J × CAST/Ei crosses could be attributed to environmental differences in the phenotyping but also to the fact that the York et al. study utilized a relatively small mapping population without significant statistical power.

Brockmann et al. (1998) mapped body-fat QTLs in an F2 cross between Du6 and DUKs mice, which were selected lines divergent in body weight. The four identified QTLs were mapped to chromosomes 3, 4, 11, and 13,with only one of these loci co-localizing to an area previously linked in the mouse to body-fat content (the *Dob1* locus from the AKR/J◊SWR/J cross mapped to chromosome 4).

Finally, Pomp et al. (1997) used a cross between lines of mice selected for rapid growth (M16I) and the lean CAST/Ei strain to identify six QTLs for body fat called Pfat1–6. Two of these are on chromosome 2 (Pfat 1,2); two were on chromosome 15 (Pfat4,5) and QTLs for body fat were also found on chromosomes 13 (Pfat3) and 17 (Pfat6).

Of significant interest is the observation that the QTLs for body fat tend to cluster (see Fig. 2) with particular hot spots on distal chromosome 2, mid to distal chromosome 7 and proximal 15. This could be attributed to the same gene with allelic variants segregating in the multiple crosses or could be attributed to clusters of functionally related genes. As we learn more about the polygenic basis of obesity in these animal models, these alternative possibilities will be resolved.

II. QTLs for Body Weight and Obesity-Related Traits

There are additional QTLs that have been mapped for body weight and for obesity-related traits, such as insulin levels, in the mouse. For example, Seldin et al. (1994) mapped a QTL for circulating insulin levels to mid-chromosome 7, for which there is both data from several other murine studies identifying linkage for body fat or body weight, and for which the syntenic region in humans has also been linked to body fat (see Table 1 and Fig. 2). Moody et al. (1997) also recently reported the mapping of QTLs for energy intake, fatness and heat generation, done by direct calorimetry on a cross between a mouse line selected for high heat loss (HEAT) and an inbred line known for low heat loss. Heat loss QTLs were found on distal chromosome 1 and mid chromosome 2 and 3. On mid-chromosome 7 a number of QTLs were mapped, albeit with marginal significance scores, including QTLS for heat loss (HEAT), body fatness (PSUB) and energy intake (INTAKE). These middle chromosome 7 QTLs for obesity-related traits correspond to multiple QTLs for body fat and body weight reported by other investigators.

The QTLs for body weight are problematic since body weight (and growth rate) is a very complex trait and reflects the growth of lean body mass, accumulation of body mineral content, as well as the accumulation of body lipid stores. Nonetheless, it is of interest that some of the QTLs for body weight co-localize with some of the murine QTLs for body fat. Similarly, the QTLs for body weight/growth in the mouse are also reported to be syntenic with a number of human loci for body mass index (BMI) (see Table 1). This might not be surprising, since BMI is a complex estimate of body fatness which relies upon height and weight, and is more likely to reflect lean body mass and growth parameters than absolute body-fat content. The linkage data for body weight/growth rate from studies by Cheverud et al. (1966), Keightly et al. (1995); Lembertas et al (1966); Collins et al. (1993) and Rance et al. (1977a,b) are summarized in Table 1. Of particular note are the results from Cheverud (1996) and Keightly (1996), who both used relatively large mapping populations and found multiple loci with small effects. Of particular interest is that these loci correspond with fairly high frequency to the linkage data from human studies of obesity in which BMI is the measure of body fat. In addition, these body weight QTLs correspond in some cases with the QTLs for direct measures of body fat (either percent carcass lipid or adipose depot weights), which one might expect since adult body weight and body fatness are correlated in the rodent.

III. Following Up on Linkage Data

Once linkage is obtained in a chromosomal region for a specific trait in the mouse, one must then work on strategies to identify the genes controlling the phenotype. This is presently the bottleneck in identifying the genes in these complex animal models. Performing the linkage studies to identify general chromosomal regions affecting the trait is relatively easy compared with the next step of narrowing the linkage region and actually pulling out and identifying the gene involved in the trait. There are several approaches for doing this. One strategy is the positional candidate approach. Unfortunately the gene maps show the chromosomal location of less than 10% of the genes in the human genome; and even fewer are represented in the mouse. Even with such limited mapping data, there are often many potential candidates in the 95% confidence interval at a specific linked location. The reason for this is that the linkage data from F2 or N2 progeny from mouse intercrosses provide only a relatively crude estimation of location. This is because of the limited number of recombination events occurring in a specific population of several hundred to a thousand animals used for mapping and the fact that genes influencing the phenotype either on the same chromosome, or other chromosomes, make

Table 1. The table rows are organized by mouse chromosome, in sequential order (1–19 autosomes plus the X chromosome). Human linkage data for obesity or obesity-related traits are organized in each row to correspond with the mouse chromosomes. The human chromosomal region showing evidence of linkage or association in each study is indicated. The vertical columns represent the mouse linkage data from QTL studies for body fat, body weight, or insulin. Within each cell is indicated the chromosomal position of the mouse peak for linkage, and the name of the QTL where that is available. Mouse linkage data that are syntenic with reported human linkages are shown in bold typeface

	Human chromosome[a]	Anchor gene or marker[b]	Syntenic region(s) in mouse[c]	Mouse chromosome	Reference[d]					
					Monogenic obesity	WEST et al. (1994)	YORK et al. (1996)	WARDEN et al. (1995)	LEMBERTAS et al. (1997)	TAYLOR and PHILLIPS (1996)
Cross[e]						SWR × AKR	B6 × Cast	BSB	NZB × SM	129 × EL
Phenotype[f]					Fat	Fat	Fat	Fat	Fat	Fat
Human obesity locus[g]										
Morbid obesity in a Canadian kindred (MURRAY et al. 1994)	1q31	Synteny	*MMU1; 72–81 cM*	1						12 cM (Obq1)
Body fat QTL in French Canadians (LEMBERTAS et al. 1997)	20q11.2; 45–55 cM	Synteny	*MMU2; 65–80 cM*	2	88 cM (A)		84 cM		66.5 cM (Bfq1) (Mob5)	
BMI and sum of six skinfolds (BORECKI et al. 1994)	20q12-q13	ADA	MMU2; 94 cM	2						
Body fat and insulin in French Canadians (CHAGNON et al. 1997a)	1p32-p22	Synteny	*MMU3 (50–56) and (76–82)*; MMU4 (43–59); MMU5 (56 cM)	3						
Body fat and insulin in French Canadians (CHAGNON et al. 1997a)	1p32-p22	Synteny	MMU3 (50–56) and (76–82); *MMU4 (43–59)*; MMU5 (56 cM)	4	47 cM (db)	48 cM (Do1)				
Body fat and insulin in French Canadians (CHAGNON et al. 1997a)	1p32-p22	Synteny	MMU3 (50–56) and (76–82); MMU4 (43–59); *MMU5 (56 cM)*	5						
BMI and sum of six skinfolds (BORECKI et al. 1994)	7q33	Synteny	*MMU6; 11–14 cM*	6	10 cM (ob)			3 cM (Mob2)		
Obesity traits in Mexican Americans (DUGGIRALA et al. 1996)	7q32	LEP	*MMU6; 10 cM*	6	10 cM (ob)					
Extreme obesity (REED et al. 1996)	7q32	LEP	*MMU6; 10 cM*	6	10 cM (ob)					
Extreme obesity (CLEMENT et al. 1996a)	7q32	LEP	*MMU6; 10 cM*	6	10 cM (ob)					
Resting metabolic rate in French Canadians (BOUCHARD et al. 1997)	11; 78–83 cM	UCP2	*MMU7; 47 cM*	7	52 cM (tub) 74 cM (Ad)	37 cM (Do4)	57 cM	62 cM (Mob1)		33 cM (Obq2)
Morbid obesity in French Caucasians (HANI et al. 1997)	11p15	SUR	*MMU7, 41 cM*	7						
Serum leptin level in Mexican Americans (COMUZZIE et al. 1997)	8; 62–64 cM	ADRB3	MMU8; 10 cM	8	32 cM (fat)	23 cM (Do6)				
Percent body fat in Pima Indians (NORMAN et al. 1997)	11q21-q22	Synteny	*MMU9; 3–30 cM*	9		45 cM (Do2)				
				10						

Taylor and Phillips (1997)	Pomp (1997)	Pomp (1997)	Mehrabian et al. (1998)	Brockmann et al. (1998)	Seldin et al. (1994)	Cheverud et al. (1996)	Keightly et al. (1996)	Dragani et al. (1995)	Lembertas et al. (1996)	Collins et al. (1993)	Rance et al. (1997a,b)
AKR × C57L	M16i × CAST	High-heat × low-heat	B6 × CAST	DU6 × DUKs	B6 × AJ	LG × SM	B6 × DBA	F1 hybrid × B6	B6 × Spret	Qb × B6	High and low selected
Fat	Fat	Fat and other measures	Fat	Fat	Insulin	Weight	Weight	Weight	Weight	Weight	lines Weight
		129 cM (Heat) 59 cM (Fat)				53 cM	76 cM				
39 cM (Obq3)	72 cM (Pfat1) 98 cM (Pfat2)	80 cM (Heat)	48 cM (Mob6) 38 cM (Mob7) 96 cM (Mob5)			86 cM					
		41 cM (Heat)		46 cM		44 cM 93 cM					
				51–55 cM		29 cM	26 cM				
							35 cM 60 cM				
						62 cM	22 cM				
					50 cM	41 cM	25 cM				
		42 cM (Heat) (Fat) (Intake)				41 cM					
			42 cM (Mob8)			35 cM	32 cM				
										56 cM	

Table 1. (*Continued*)

	Human chromosome[a]	Anchor gene or marker[b]	Syntenic region(s) in mouse[c]	Mouse chromosome	Reference[d]					
					Monogenic obesity	West et al. (1994)	York et al. (1996)	Warden et al. (1995)	Lembertas et al. (1997)	Taylor and Phillips (1996)
				11						
BMI (Bailey-Wilson et al. 1993)	2p25	ACP1	*MMU12; 6–15 cM*	12		24 cM (Do5)		53 cM (Mob3)		
Serum leptin level in Mexican Americans (Comuzzie et al. 1997)	2p23	POMC	MMU12; 2–5 cM	12						
				13						
Percent body fat, sum of six skinfolds, in French Canadians (Borecki et al. 1994)	13q14.1–14.2	ESD	*MMU14; 41 cM*	14						
				15		21 cM (Do3)	21 cM	7 cM (Mob4)		
				16						
Percent body fat in Pima Indians (Norman et al. 1995)	6p21.2	TNFA	*MMU17; 19 cM*	17						
BMI, fat mass, RMR in French Canadians (Chagnon et al. 1997b)	18p11	MC5R	*MMU18;* 43 cM	18						
Morbid obesity (Clement et al. 1996b)	5q31-q32	GRL	MMU18; 20 cM	18						
				19						
				X		58 cM (Do7)				

[a] Human chromosomal region where linkage with body fat or body-fat-related trait was observed.
[b] Human gene used to establish synteny between mouse and human chromosomal regions for linkage.
[c] Syntenic region on specific mouse chromosomes. *Underlining* indicates evidence for shared linkage for obesity and related traits between syntenic regions. For some human linkage data the syntenic regions are represented on several mouse chromosomes.
[d] References for monogenic obesity models can be found in the appropriate section of the chapter. Citations for each of the mouse QTL studies are listed in the reference list for the chapter.
[e] Specific mouse intercross used for mapping studies.
[f] Major phenotype mapped in this cross.
[g] Citations are listed in the reference list for this chapter.

precise mapping difficult. However, some statistical approaches have been proposed for the use of the original linkage data to further refine the map position (Zeng 1994). These are as yet untested, since there are only a few reports in mouse linkage studies for a phenotype resulting in the identification of positional candidates likely to be controlling the trait. None of these reports involves body-fat content.

There are a number of strategies for narrowing down the chromosomal interval that use classical genetics; this was recently reviewed by Darvasi et al. (1998). These approaches encompass two general strategies; one is to set up crosses generating more recombinations in the region of interest and to then focus on these populations (e.g., advanced intercross lines) to more finely map the genes. The other approach is to develop congenic mice in which a segment from one strain (the donor strain) has been introduced by breeding

Taylor and Phillips (1997)	Pomp (1997)	Pomp (1997)	Mehrabian et al. (1998)	Brockmann et al. (1998)	Seldin et al. (1994)	Cheverud et al. (1996)	Keightly et al. (1996)	Dragani et al. (1995)	Lembertas et al. (1996)	Collins et al. (1993)	Rance et al. (1997a,b)
				6 cM		1 cM	45 cM			62 cM	
						11 cM					
	35 cM (Pfat3)			0 cM		25 cM 41 cM	59 cM 0 cM				
	18 cM (Pfat4) 35 cM (Pfat5)										
5 cm (Obq4)	10 cM (Pfat6)						14 cM				
						46 cM					
								7 cM (Bw1) 34 cM (Bw2); 61–70 cM (Bw3)	17 cM		8, 24 cM

into a uniform inbred background strain. If one starts with the same two strains used for the initial mapping studies, then the probability is increased that the resulting congenic line will demonstrate a phenotype effect. Theoretically, one can then set up backcrosses, identify recombinants, and by phenotyping these subcongenics narrow the interval to less than 1 cM (approximately 2megabases in the mouse). The primary limitation of this approach is that it takes time to do the breeding in order to develop these congenic lines. However, recent genetic approaches utilizing genetic markers scattered about the genome (so-called speed congenics) has accelerated this process (Wakeland et al. 1997; Markel et al. 1997). In addition, one can take pre-existing congenics, i.e., congenics developed for other purposes, and simply screen them for the phenotype of interest if the introgressed segment is located in a chromosomal region which by linkage (in human or mouse) has been found to be linked to the trait.

This was the approach taken by Lembertas et al. (1997), to identify a congenic line affecting body fat on mouse chromosome 2 where linkage for body-fat content had been previously found. Similarly, Fisler and Warden (1997) screened pre-existing congenics on mouse chromosome 7 to identify specific congenic lines affecting body fat. Although these congenics were developed to examine the effect of some of the histocompatibility loci on tissue rejection, it also appears that the same regions and strains were harboring genes with different alleles affecting body fat. Yet another approach is to develop congenics only in the region of interest and to allow the rest of the genome to independently assort (Bennett et al. 1997). This has the advantage that these "congenic" lines can be developed quickly. However, one limitation of this approach is that it might allow the confirmation of a QTL in a limited number of breeding cycles but it is likely to miss some congenic effects due to the large variation of the phenotype attributable to other loci still segregating in the rest of the genome.

Once a relatively narrow chromosomal region has been identified, then one must still identify the gene affecting the phenotype. However, by reducing the region to a1 cM or less interval, either by fine-mapping or by developing subcongenics, this greatly facilitates the process, since one can now sift through several megabases of DNA containing from 50 to several hundred genes, instead of 50 or more megabases of DNA containing literally thousands of genes. Positional candidates have indeed been identified for QTLs, and mutations in these positional candidates were confirmed to be responsible for the trait. For example, for multiple intestinal neoplasia in the mouse, a modifier of tumor incidence was mapped and called *Mom1* (for modifier of multiple intestinal neoplasia). A positional candidate, phospholipase A2 (*Pla2s*) was subsequently shown to contain a mutation in the *Mom1* mice resulting in a functional knockout of this gene (Dietrich et al. 1993; and MacPhee et al. 1995). Other less clear examples of positional candidates for QTL effects in mice have also been reported. For example, the interleukin 9 gene (*Il9*) has been proposed as the positional candidate underlying a differential airway responsiveness QTL (Nicolaides et al. 1997), on the basis of the significantly reduced expression of this gene in the strain of mouse having a hyporesponsive bronchial response. Similarly, the δ2-opoid receptor, *Oprd1*, has been proposed as the gene underlying a QTL for differential nociception in the mouse, on the basis of the differential response to specific opiate antagonists (Mogil et al. 1997). However, a sequence variation or differential expression has not been shown in this gene to provide proof that this is in fact the gene affecting the quantitative trait.

With the mouse, once a sequence polymorphism in a positional candidate is observed, one option is then to knock in the allele into the appropriate background to determine if the phenotype is recovered. Similarly, knockouts of the candidate gene, tissue-specific overexpression and conditional expression are all valuable tools available in the mouse to assess the functional role of the candidate in a specific phenotype.

F. Human–Mouse Synteny

One approach to accelerate the discovery of the genes involved in human obesity is the approach of using syntenic mouse models. Following the identification of linkage in human sib-pair or family studies in a specific chromosomal region, one can determine if there is linkage found in the same region syntenic (i.e., a chromosomal region which contains the mouse homologs in a conserved linkage order) in the mouse for the same or a related trait. Then one can use the mouse to accelerate the discovery of the genes controlling the trait in human populations. There are several examples now of human–mouse synteny for complex disease traits. For example, a region on human chromosome 20 linked to BMI in a family study is homologous to a region on mouse chromosome 2 with genetic linkage for body fat (LEMBERTAS et al. 1997). In addition, there is a congenic mouse (a mouse with a chromosomal segment from one strain introduced by breeding into a different inbred strain) that shows a body-fat phenotype in this region. CASTELLANI et al., (1998) found another example of homology, or synteny for linkage in a complex trait, by evaluating a recombinant congenic mouse panel, to show that a region on mouse chromosome 3 has a dramatic controlling influence over triglyceride levels, probably attributable to a spontaneous mutation. This mouse locus is syntenic with the genetic locus on human chromosome 1 for familial combined hyperlipidaemia and it is possible that the same gene is involved in controlling this trait in these two different species (PAJUKANTA et al. 1998).

Table 1 provides a partial list of the human regions shown to be linked to BMI or other measures of body fat and the mouse regions that are syntenic and have also been linked to obesity or obesity-related phenotypes such as insulin levels or body weight. This list covers the majority of the published mouse data describing linkage; however, there are numerous additional human studies, including family linkage studies, sib-pair analyses, and association analyses providing evidence for genes in chromosomal regions not represented by syntenic findings in the mouse. Of particular interest are the multiple reports of linkage on human chromosome 7 near the leptin structural gene (see Table 1). These multiple reports from independent laboratories strongly argue for a gene in this vicinity affecting body fat in humans. However, an analysis of sequence polymorphisms in the *LEP* gene and their association with body fat or related intermediate phenotypes (OKSANEN et al. 1997), as well as markers closely linked to the LEP gene (e.g., GOTODA et al. 1997; BRAY, BOERWINKLE and HANIS 1996), have not provided any convincing evidence for a common allelic variant in that gene being associated with human obesity. This suggests that a gene in relative close proximity, or regulatory regions at some distance from the *LEP* gene, may be important in this region for body-fat regulation. Similarly, there are now multiple reports of linkage on human 20q11; this suggests that this region is of importance in the regulation of body fat in humans. The availability of an existing mouse congenic in this region, with a body-fat phenotype (LEMBERTAS et al. 1997), may provide a reagent for rapid identification of this gene. One problem with the human–mouse synteny approach is the limited information on conserved linkage groups between human and mouse, although the quality of this human–mouse synteny data is rapidly improving (CARVER and STUBBS 1997).

This approach of using a mouse model to exploit and follow up linkage data in humans is as yet unproven and untested. For any complex disorder, it is likely that there is a set of genes in which allelic variations influence the trait. There are two primary issues determining whether the mouse–human synteny approach will work to accelerate gene discovery. The first is whether the same set of genes is controlling the phenotype in both species. This, of course, will depend upon the similarity in the metabolic pathways controlling the trait in mouse and human, and it is likely to be dependent on the specific trait or disease. For example, the set of genes controlling body lipid stores in human and mouse might have considerable overlap; while the set of genes contributing to cancer susceptibility might be very different in the two species. In addition, even if there is significant overlap in the set of genes controlling the trait in two distinct species, it is likely that allelic variants in specific genes will have developed in one species but perhaps not in the other. Simply showing linkage in homologous regions in two species for the same trait is no assurance that the same genes in these regions will be responsible for the phenotype. As our understanding and knowledge of the genetic and metabolic basis of disease develops, we will have more empirical information showing whether this approach will work.

G. Candidate Gene Assessment

I. Candidate Genes Based Upon Function

As we described above, the evaluation of the role of candidate genes, based solely on physiology, has generally not led to the identification of genes involved in the etiology of obesity, with rare exceptions (e.g., the *Cpe* gene). However, in the case of *Cpe*, the interval containing the obesity gene had been narrowed significantly, so that only a limited number of positional candidates, previously mapped to this proximal chromosome 8 location, could be the gene. Obesity is such a complex phenotype, with many contributing components such as energy expenditure, physical activity, diet selection, energy intake, lipid metabolism, sympathetic nervous system activity, etc., that there are many candidate genes, based upon a functional role in energy homeostasis or the regulation of feeding behavior. Certainly, screening all of these possible physiological candidates for sequence mutations is not feasible at present, nor have the limited efforts to date been productive. The human obesity genetics literature is replete with studies of allelic variants of functional candidates leading to case control association studies to determine if the sequence polymorphism is involved in the phenotype. A recent example of this approach is the evaluation of a mutation in the β-3 adrenoreceptor originally described by Shuldiner and colleagues (e.g., CLEMENT et al. 1995; WIDEN et al. 1995). Although the original data were quite promising and showed an association between a specific allelic variant and indirect measures of body fat and age of onset of diabetes, many subsequent studies have not consistently found the same association and presently the data are inconclusive. This inconsistency in the results may be due to the presence or absence of informative alleles in the different ethnic groups compared, subtle differences in phenotype, or lack of statistical power in some reports. Combining linkage data with positional candidates is certainly a more promising approach, since one can then restrict the potential candidates by positional means and have a greater likelihood of identifying sequence polymorphisms in positional candidates that are *functionally* involved in disease etiology.

II. Sequence Homology

Another approach for evaluating candidate genes is one based upon sequence homology with genes of known function. However, only a fraction of the genes for which we now have sequence have associated functions. The expressed sequence tag (EST) data bases now contain approximately 10^6 sequences, often including both 5′ and 3′ sequences, and likely include the majority of genes expressed in relatively high copy numbers. Although there are considerably more EST sequences in the human data sets, recent efforts by a consortium of investigators has been rapidly providing mouse EST sequences and much of this EST sequence data is now publicly available (see the IMAGE consortium home page: http://www-bio.lbnl.gov/bbrp/image/image.html). However, there are some limitations in these data bases. The EST data bases do not cover all tissues (e.g., libraries from adipose tissue are minimally represented), do not cover many developmental stages, and are likely to not include low copy number transcripts, despite attempts at library normalization. Furthermore, sequences starting from the 3' noncoding region may not provide enough open reading frame sequences for effective homology searching.

Despite these caveats, identifying possible obesity genes on the basis of homology searches with genes of known function is an important approach that has led to two notable success stories. FLEURY et al. (1997) found a human sequence partially homologous to the human uncoupling protein 1 (*UCP1*), a gene coding for a brown-adipose-tissue-specific uncoupling protein which is very important for energy homeostasis in rodents. Subsequent analysis of this novel gene, now called *UCP2*, has shown that it is expressed in multiple tissues, including brown adipose tissue, skeletal muscle and smooth muscle, and when transfected into yeast systems, it appears to alter mitochondrial transmembrane conductance. Furthermore, manipulations of energy intake and diet have effects on *Ucp2* expression in diet-sensitive and diet-responsive mouse strains that would be predicted if the gene product had a role in energy homeostasis (SURWIT et al. 1998). Recent findings in human populations also suggest that a polymorphism at or near this gene may be associated with differences in energy expenditure (WALDER et al. 1998), although these results should certainly be considered preliminary. Finally, a third putative uncoupling protein has been identified in mammalian systems (now called *Ucp3*) and this also may play a role in energy homeostasis (Boss et al. 1997).

Another example of using sequence homology to identify putative obesity genes, or genes involved in regulating feeding behavior or energy metabolism, was the identification of the *Agouti*-related transcript (ART) described above (SHUTTER et al. 1997). This gene was originally identified on the basis of sequence homology with the *Agouti* gene, and the sequence homology, along with expression localized to the brain, focused attention on it as a possible gene product whose function might be mimicked by the overexpression of *Agouti* in the dominant obesity mutations. Although the total amino acid sequence homology between ART and the agouti signaling protein (ASP) is relatively low (25%), the conserved spatial organization of the cysteine residues within the two genes provided the key clue that they were related peptides. This illustrates the need to utilize both secondary and tertiary structure analyses when using sequence homology to infer structure/function relationships.

Yet another example of sequence homology to pull out putative obesity genes (or genes involved in functions involving energy homeostasis) is the identification of the orexin gene (SAKURAI et al. 1998). The strategy utilized by these investigators was more indirect, but relied upon identifying G-protein-coupled orphan receptors, based upon the common 7 transmembrane motif and found in high numbers in EST databases. In this example, the orphan receptors were cloned, expressed in cells and then the transfected cells were treated with HPLC-derived extracts of hypothalamic tissue. A specific orphan receptor was found that responded (as assessed by intracellular calcium accumulation) to a specific fraction of hypothalamic extract and subsequent microsequencing led to the identification of the orexin peptide. Reverse genetics then led to the gene. This gene is expressed in the ventral medial hypothalamus; the full functional role in the regulation of feeding behavior remains to be elucidated. However, it is an example of a functional evaluation of candidates based upon homology.

With a relatively small proportion of gene sequences having associated functional information, it is perhaps surprising that approaches based solely on sequence homology have been successful. However, with additional functional information being added to the databases, this approach will certainly grow more useful with time. In addition, as the analytical methods are developed, allowing the comparison of predicted secondary and tertiary protein structure (BENNER 1995; DUNBRACK et al. 1997), the approach of homology analyses will certainly lead to even more successful identification of candidates involved in energy homeostasis.

H. Gene Expression Profiling and Proteomics

A plethora of expression profiling approaches are now available with most designed to evaluate relatively large number of transcripts based upon sequences derived from publicly available sequence data bases, especially the expressed-sequence tag (EST) databases. The approaches included chip-based technologies (e.g., SAPOLSKI and LIPSHUTZ 1996; LOCKHART et al. 1996), high-density arrays on glass slides or other solid supports (e.g., SHENA et al. 1995, 1996; CHEN et al 1998), sequence-based approaches such as the serial analysis of gene expression (SAGE; e.g., VELCULESCUE et al. 1995; WANG and ROWLEY 1998), quantitative RT-PCR (ZAMORANO et al. 1996) and other approaches. All of these methods have their relative strengths and weaknesses with the primary concerns being sensitivity to detect the presence of transcripts with low copy number and the ability to detect relatively small changes in gene expression. These expression-profiling methods have the potential ability to screen large numbers of genes to examine the effects of disease state, developmental stage, environmental perturbations, response to pharmacotherapy, etc., on gene expression in any tissue of interest. Often multiple differentially expressed transcripts are detectable through these approaches and the necessary follow-up to confirm the differential expression using an independent technique can be the bottleneck. Similarly, since expression levels do not necessarily correlate with the activity or abundance of the translated product, follow-up "proteomics" to assess the presence and/or activity of the protein product are necessary. These proteome approaches include 2-dimensional gel electrophoresis under a variety of conditions to separate the proteins in tissues, specific cell types, or even isolated cells (HUMPHEREY-SMITH et al. 1997) and follow-up mass spectroscopy (WHEELER et al. 1996). The accumulation of proteome data bases characterizing the normal and diseased protein content in specific tissues (CELIS et al. 1998) will be a useful complement to the analogous gene expression profiling of these same or similar tissues.

Although we are not aware of any novel obesity genes identified by these methods yet, the approach holds great promise to characterize the key molecular events involved in the etiology and pathophysiology of this disease state. The key tissues to evaluate include white and brown adipose tissue, skeletal muscle, and the central nervous system. At the present time, the EST databases do not carry a complete representation of the genes expressed uniquely in adipose tissues. However, this deficiency will be rapidly corrected. One group has been very successful at evaluating cDNA libraries from different human adipose depot sites, by direct sequencing of these libraries. These studies have resulted in the identification of unique differences in expression patterns among different depots (MAEDA et al. 1997; SHIMOMURA et al. 1996) and have further identified cDNAs which are uniquely expressed in adipose tissue (MAEDA et al. 1997; KURIYAMA et al. 1997; and MAEDA 1996). The identification and further characterization of these genes will certainly facilitate our understanding of the normal physiology of adipose tissue, and subsequent sequence analysis of these genes may lead to allelic variants important in genetic predisposition to disease.

I. Summary

In this chapter we have attempted to broadly represent the strategies currently being used to identify novel obesity genes or genes involved in pathways (controlling feeding behavior, energy expenditure, lipid metabolism, etc.) that might serve as targets for drug development for the treatment of obesity. The approaches are many, but share a common reliance upon the developing genomics resources in both human and mammalian systems such as the mouse. Additional mapping information of genomic regions linked to body fat or intermediate phenotypes will accelerate the identification of those chromosomal regions commonly harboring obesity disease genes with large effects. As more genes/transcripts are finely mapped, this will facilitate the evaluation of positional candidate genes. As the physical maps are filled in, this will provide the reagents for more rapid cloning of obesity genes, on the basis of fine mapping. As full-length sequences and DNA sequences are added to the existing data bases, and more functional information is tied to sequence, the electronic evaluation of putative obesity candidate genes will be aided. Finally, as the technologies for utilizing expression arrays to evaluate the changes in gene expression, as well as proteomics methodologies, are developed, the functional evaluation of candidate genes in the etiology of obesity will also certainly be facilitated.

References

Bailey-Wilson JE, Wilson AF, Bamba V (1993) Linkage analysis in a large pedigree ascertained due to essential familial hypercholesterolemia. Genet Epidemiol 10(6):665–669

Beattie JH, Wood AM, Newman AM, Bremner I, Choo KH, Michalska AE, Duncan JS, Trayhurn P (1998) Obesity and hyperleptinemia in metallothionein (-I and -II) null mice. Proc Natl Acad Sci 95(1):358–363

Benner SA (1995) Predicting the conformation of proteins from sequences. Progress and future progress. J Mol Recognit 8(1–2):9–28

Bennett B, Beeson M, Gordon L, Johnson TE (1997) Quick method for confirmation of quantitative trait loci. Alcohol Clin Exp Res 21(5):767–772

Borecki IB, Rice T, Perusse L, Bouchard C, Rao DC (1994) An exploratory investigation of genetic linkage with body composition and fatness phenotypes: the Quebec Family Study. Obes Res 2:213–219

Boss O, Samec S, Paoloni-Giacobino A, Rossier C, Dulloo A, Seydoux J, Muzzin P, Giacobino JP (1997) Uncoupling protein-3: a new member of the mitochondrial carrier family with tissue-specific expression. FEBS Lett 408(1):39–42

Bouchard C, Perusse L, Chagnon YC, Warden C, Ricquier D (1997) Linkage between markers in the vicinity of the uncoupling protein 2 gene and resting metabolic rate in humans. Hum Mol Genet 6(11):1887–1889

Bray MS, Boerwinkle E, Hanis CL (1996) OB gene not linked to human obesity in Mexican American affected sib pairs from Starr County, Texas. Hum Genet 98(5):590–595

Brockman GA, Haley CS, Renne U, Knott SA, Schwerin M (1998) Quantitative trait loci affecting body weight and fatness from a mouse line selected for extreme high growth. Genetics 150:369–381

Brown SDM, Peters J (1996) Combining mutagenesis and genomics in the mouse – closing the phenotype gap. Trends Genet 12:433–435

Bruning JC, Winnay J, Bonner-Weir S, Taylor SI, Accili D, Kahn CR (1997) Development of a novel polygenic model of NIDDM in mice heterozygous for IR and IRS-1 null alleles. Cell 88(4):561–572

Bultman SJ, Michaud EJ, Woychik RP (1992) Molecular characterization of the mouse *Agouti* locus. Cell 71(7):1195–1204

Burant CF, Sreenan S, Hirano K, Tai TA, Lohmiller J, Lukens J, Davidson NO, Ross S, Graves RA (1997) Troglitazone action is independent of adipose tissue. J Clin Invest 100(11):2900–2908

Campfield LA, Smith FJ, Burn P (1998) Strategies and potential molecular targets for obesity treatment. Science 280(5368):1383–1387

Carver EA, Stubbs L (1997) Zooming in on the human–mouse comparative map:genome conservation re-examined on a high-resolution scale. Genome Res 7(12):1123–1137

Castellani LW, Weinreb A, Bodnar J, Goto AM, Doolittle M, Mehrabian M, Demant P, Lusis AJ (1998) Mapping a gene for combined hyperlipidaemia in a mutant mouse strain. Nat Genet 18:374–377

Cattanach BM, Barr JA, Beechey CV, Martin J, Noebels J, Jones J (1997) A candidate model for Angelman syndrome in the mouse. Mamm Genome 8(7):472–478

Celis JE, Ostergaard M, Jensen NA, Gromova I, Rasmussen HH, Gromov P. Human and mouse proteomic databases: novel resources in the mouse protein universe. FEBS Lett 430(1):64–72

Cheverud JM, Routman EJ, Duarte FAM, van Swinderen B, Cothran K, Perel C (1996) Quantitative trait loci for murine growth. Genetics 142:1305–1319

Chagnon YC, Perusse L, Lamothe M, Chagnon M, Nadeau A, Dionne FT, Gagnon J, Chung WK, Leibel RL, Bouchard C (1997a) Suggestive linkages between markers on human 1p32-p22 and body fat and insulin levels in the Quebec Family Study. Obes Res 5(2):115–121

Chagnon YC, Chen WJ, Perusse L, Chagnon M, Nadeau A, Wilison WO, Bouchard (1997b) Linkage and association studies between the melanocortin receptors 4 and 5 genes and obesity-related phenotypes in the Quebec Family Study. Mol Med 3(10):663–673

Chagnon YC, Perusse L, Bouchard C (1998) The human obesity gene map: the 1997 update. Obes Res 6(1):76–92

Chen JJ, Wu R, Yang PC, Huang JY, Sher YP, Han MH, Kao WC, Lee PJ, Chiu TF, Chang F, Chu YW, Wu CW, Peck K (1998) Profiling expression patterns and isolating differentially expressed genes by cDNA microarray system with colorimetry detection. Genomics 51(3):313–324

Chua SC (1997) Monogenic models of obesity. Behav Genet 27(4):277–284

Clement K, Vaisse C, Manning BS, Basdevant A, Guy-Grand B, Ruiz J, Silve KD, Shuldiner AR, Froguel P, Strosberg AD (1995) Genetic variation in the beta 3 adrenergic receptor and an increased capacity to gain weight in patients with morbid obesity. N Engl J Medicine 333(6):352–354

Clement K, Philippi A, Jury C, Pividal R, Hager J, Demenais F, Basdevant A, Guy-Grand B, Froguel P (1996a) Candidate gene approach of familial morbid obesity: linkage analysis of the glucocorticoid receptor gene. Int J Obes Relat Metab Disord 20(6):507–512

Clement K, Garner C, Hager J, Philippi A, LeDuc C, Carey A, Harris TJ, Jury C, Cardon LR, Basdevant A, Demenais F, Guy-Grand B, North M, Froguel P (1996b) Indication for linkage of the human OB gene region with extreme obesity. Diabetes 45(5):687–690

Clement K, Vaisse C, Lahlou N, Cabrol S, Pelloux V, Cassuto D, Gourmelen M, Dina C, Chambaz J, Lacorte JM, Basdevant A, Bougneres P, Lebouc Y, Froguel P, Guy-Grand B (1998) A mutation in the human leptin receptor gene causes obesity and pituitary dysfunction. Nature 392(6674):398–401

Collins AC, Martin ICA, Kirkpatrick BW (1993) Growth quantitative trait loci (QTL) on mouse chromosome 10 in a Quackenbush-Swiss◊C57BL/6J backcross. Mammal Genome 4:454–458

Coleman DL (1973) Effects of parabiosis of obese with diabetes and normal mice. Diabetologia 9(4):294–298

Collins FS (1995) Positional cloning moves from perditional to traditional. Nature Genet 9:347–350

Comuzzie AG, Hixson JE, Almasy L, Mitchell BD, Mahaney MC, Dyer TD, Stern MP, MacCluer JW, Blangero J (1997) A major quantitative trait locus determining serum leptin levels and fat mass is located on human chromosome 2. Nat Genet 15(3):273–276

Comuzzie AG, Allison DB (1998) The search for human obesity genes. Science 280(5368):1374–1377

Considine RV, Sinha MK, Heiman ML, Kriauciunas A, Stephens TW, Nyce MR, Ohannesian JP, Marco CC, JcKee LJ, Bauer TL et al (1996) Serum immunoreactive-leptin concentrations in normal-weight and obese humans. N Engl J Med 334(5):292–295

Culiat CT, Carver EA, Walkowicz M, Rinchik EM, Cacheiro NL, Russell LB, Generoso WM, Stubbs L (1997) Induced mouse chromosomal rearrangements as tools for identifying critical developmental genes and pathways. Reprod Toxicol 11(2–3): 345–351

Cummings DE, Brandon EP, Planas JV, Motamed K, Idzerda RL, McKnight GS (1996) Genetically lean mice result from targeted disruption of the RII beta subunit of protein kinase A. Nature 15(382):622–626

Darvasi A (1998) Experimental strategies for the genetic dissection of complex traits in animal models. Nat Genet 18(1):19–24

Dietrich WF, Lander ES, Smith JS, Moser AR, Gould KA, Luongo C, Borenstein N, Dove W (1993) Genetic identification of Mom-1, a major modifier locus affecting Min-induced intestinal neoplasia in the mouse. Cell 75:631–639

Dong ZM, Gutierrez-Ramos JC, Coxon A, Mayadas TN, Wagner DD (1997) A new class of obesity genes encodes leukocyte adhesion receptors. Proc Natl Acad Sci 94(14):7526–7530

Dragani TA, Zenb Z-B, Canzian F, Gariboldi M, Ghilarducci MT, Manenti G, Pierotti MA (1995) Mapping of body weight loci on mouse chromosome X. Mammal Genome 6:778–781

Duggirala R, Stern MP, Mitchell BD, Reinhart LJ, Shipman PA, Uresanda OC, Chung WK, Leibel RL, Hales CN, O'Connell P, Blangero J (1996) Quantitative variation in obesity-related traits and insulin precursors linked to the OB gene region on human chromosome 7. Am J Hum Genet 59(3):694–703

Dunbrack Rl, Gerloff DL, Bower M, Chen X, Lichtarge O, Cohen FE (1997) Meeting review: the Second meeting on the critical assessment of techniques for protein structure prediction (CASP2). Fold Des 2(2):R27–R42

Enerback S, Jacobsson A, Simpson EM, Guerra C, Yamashita H, Harper ME, Kozak LP (1997) Mice lacking mitochondrial uncoupling protein are cold-sensitive but not obese. Nature 387(6628):90–94

Fisler JS, Warden CH (1997) Mapping of mouse obesity genes: a generic approach to a complex trait. J Nutr 127(9):1909S–1916 S

Fleury C, Neverova M, Collins S, Raimbault S, Champigny O, Levi-Meyrueis C, Bouillaud F, Seldin MF, Surwit RS, Ricquier D, Warden CH (1997) Uncoupling protein-2: a novel gene linked to obesity and hyperinsulinemia. Nat Genet 15(3):269–272

Flier JS (1998) Clinical review 94: What's in a name? In search of leptin's physiological role. J Clin Endocrinol Metab 83(5):1407–1413
Good DJ, Porter FD, Mahon KA, Parlow AF, Westphal H, Kirsch IR (1997) Hypogonadism and obesity in mice with a targeted deletion of the Nhlh2 gene. Nat Genet 15(4):397–401
Gotoda T, Manning BS, Goldstone AP, Imrie H, Evans AL, Strosberg AD, McKeigue PM, Scott J, Aitman TJ (1997) Leptin receptor gene variation and obesity: lack of association in a white British male population. Hum Mol Genet 6(6):869–876
Graham M, Shutter JR, Sarmiento U, Sarosi I, Stark KL (1997) Overexpression of Agrt leads to obesity in transgenic mice. Nat Genet 17(3):273–274
Hamann A, Flier JS, Lowell BB (1998) Obesity after genetic ablation of brown adipose tissue. Z Ernahrungswiss 37 [Suppl 1]:1–7
Hallopeter G, Erickson JC, Palmiter RD (1998) Role of neuropeptide Y in diet-, chemical- and genetic-induced obesity of mice. Int J Obes Relat Metab Disord 22(6):506–512
Hani EH, Clement K, Velho G, Vionnet N, Hager J, Philippi A, Dina C, Inoue H, Permutt MA, Basdevant A, North M, Demenais F, Guy-Grand B, Froguel P (1997) Genetic studies of the sulfonylurea receptor gene locus in NIDDM and in morbid obesity among French Caucasians. Diabetes 46(4):688–694
Humphery-Smith I, Cordwell SJ, Blackstock WP (1997) Proteome research: complementarity and limitations with respect to the RNA and DNA worlds. Elecrophoresis 18(8):1217–1242
Huszar D, Lynch CA, Fairchild-Huntress V, Dunmore JH, Fang Q, Berkemeier LR, Gu W, Kesterson RA, Boston BA, Cone RD, Smith FJ, Campfield LA, Burn P, Lee F (1997) Targeted disruption of the melanocortin-4 receptor results in obesity in mice. Cell 88(1):131–141
Heymsfield S, Greenberg A, Fujioka K, Dixon R, Kushner R, Hunt T, Lubina J, Patane J, Self B, McCamish M (1998) Weight and body composition changes in lean and obese subjects treated with recombinant methionyl human leptin (rL). Poster presented at the 8th International Congress on Obesity, August, 1998
Jackson RS, Creemers JW, Ohagi S, Raffin-Sanson ML, Sanders L, Montague CT, Hutton JC, O'Rahilly S (1997) Obesity and impaired prohormone processing associated with mutations in the human prohormone convertase 1 gene. Nat Genet 16(3):303–306
Jensen DR, Schlaepfer IR, Morin CL, Pennington DS, Marcell T, Ammon SM, Gutierrez-Hartmann A, Eckel RH (1997) Prevention of diet-induced obesity in transgenic mice overexpressing skeletal muscle lipoprotein lipase. Am J Physiol 272(2 Pt 2):R683–R689
Justice MJ, Zheng B, Woychik RP, Bradley A (1997) Using targeted large deletions and high-efficiency N-ethyl-N-nitrosourea mutagenesis for functional analyses of the mammalian genome. Methods 13(4):423–436
Keightley PD, Hardge T, May L, Bulfield G (1995) A genetic map of quantitative trait loci for body weight in the mouse. Genetics 142:227–235
Kuriyama H, Kawamoto S, Ishida N, Ohno I, Mita S, Matsuzawa Y, Matsubara K, Okubo K (1997) Molecular cloning and expression of a novel human aquaporin from adipose tissue with glycerol permeability. Biochem Biophys Res Commun 241(1):53–58
Khrebtukova I, Michaud EJ, Foster CM, Stark KL, Garfinkel DJ, Woychik RP (1998) Utilization of microhomologous recombination in yeast to generate targeting constructs for mammalian genes. Mutat Res 401(1–2):11–25
Kiess W, Blum WF, Aubert ML (1998) Leptin, puberty and reproductive function: lessons from animal studies and observations in humans. Eur J Endocrinol 138(1):26–29
Klebig ML, Wilkinson JE, Geisler JG, Woychik RP (1995) Ectopic expression of the agouti gene in transgenic mice causes obesity, features of type II diabetes, and yellow fur. Proc Natl Acad Sci 92(11):4728–4732

Kleyn PW, Fan W, Kovats SG, Lee JJ, Pulida JC, Wu Y, Berkemeier LR, Misumi DJ, Holmgren L, Charlat O, Woolf EA, Tayber O, Brody T, Shu P, Hawkins F, Kennedy B, Baldini L, Ebeling C, Alperin GD, Deeds J, Lakey ND, Culpepper J, Chen H, Glucksmann-Kuis MA, Moore KJ (1996) Identification and characterization of the mouse obesity gene tubby: a member of a novel gene family. Cell 85(2):281–290

Kopecky J, Hodny Z, Rossmeisl M, Syrovy I, Kozak LP (1996) Reduction of dietary obesity in aP2-Ucp transgenic mice: physiology and adipose distribution. Am J Physiol 270(5 Pt1):E768–E775

Kozak LP, Kozak UC, Clarke GT (1991) Abnormal brown and white fat development in transgenic mice overexpressing glycerol 3-phosphate dehydrogenase. Genes Dev 5(12 A):2256–2264

Krude H, Biebermann H, Luck W, Horn R, Brabant G, Gruters A (1998) Severe early-onset obesity, adrenal insufficiency and red hair pigmentation caused by POMC mutations in humans. Nat Genet 19(2):155–157

Lander ES, Botstein D (1989) Mapping mendelian factors underlying quantitative traits using RFLP linkage maps. Genetics 121:185–199

Lembertas AV, Fisler JS, Warden CH, Wen P-Z, Xia Y-R, Lusis AJ (1996) A locus on the x chromosome is linked to body length in mice. Mammal Genome 7:171–173

Lembertas AV, Perusse L, Chagnon YC, Fisler JS, Warden CH, Purcell-Huynh DA, Dionne FT, Gagnon J, Nadeau A, Lusis AJ, Bouchard C (1997) Identification of an obesity quantitative trait locus on mouse chromosome 2 and evidence of linkage to body fat and insulin on the human homologous region 20q. J Clin Invest 100(5):1240–1247

Lockhard DJ, Dong H, Byrne MC, Follettie MT, Gallo MV, Chee MS, Mittmann M, Wang C, Kobayashi M, Horton H, Brown EL (1996) Expression monitoring by hybridization to high-density oligonucleotide arrays. Nat Biotechnol 14(13):1675–1680

Lu D, Willard D, Patel IR, Kadwell S, Overton L, Kost T, Luther M, Chen W, Woychik RP, Wilkinson WO et al (1994) Agouti protein is an antagonist of the melanocyte-stimulating-hormone receptor. Nature 371(6500):799–802

Lyon MF (1958) Twirler: a mutant affecting the inner ear of the house mouse. J Embryol Exp Morphol 6:105–116

Maeda K, Okubo K, Shimomura I, Funahashi T, Matsuzawa Y, Matsubara K (1996) cDNA cloning and expression of a novel adipose specific collagen-like factor, apM1 (adipose most abundant gene transcript 1). Biochem Biophys Res Commun 221(2):286–289

Maeda K, Okubo K, Shimomura I, Mizuno K, Matsuzawa Y, Matsubara K (1997) Analysis of an expression profile of genes in human adipose tissue. Gene 190(2):227–235

Markel P, Shu P, Ebelikng C, Carlson GA, Nagle DL, Smutko JS, Moore KJ (1997) Theoretical and empirical issues for marker-assisted breeding of congenic mouse strains. Nat Genet 17(3):280–284

MacPhee M, Chepenick KP, Liddell RA, Nelson KK, Siracusa LD, Buchberg AM (1995) The secretory phospholipase A2 gene is a candidate for the Mom1 locus, a major modifier of ApcMin-induced intestinal neoplasia. Cell 81:957–966

Melnyk A, Harper ME, Himms-Hagen J (1997) Raising at thermoneutrality prevents obesity and hyperphagia in BAT-ablated transgenic mice. Am J Physiol 272(4):R1088–R1093

Mehrabian M, Wen PZ, Fisler J, Davis RC, Lusis AJ (1998) Genetic loci controlling body fat, lipoprotein metabolism, and insulin levels in a multifactorial mouse model. J Clin Invest 101(11):2485–2496

Murray JD, Bulman DE, Ebers GC, Lathrop GM, Rice GPA (1994) Linkage of morbid obesity with polymorphic microsatellite markers on chromosome 1q31 in a three-generation Canadian kindred. Am J Hum Genet 55:A197

Michaud EJ, Bultman SJ, Stubbs LJ, Woychik RP (1993) The embryonic lethality of homozygous lethal yellow mice (Ay/Ay) is associated with the disruption of a novel RNA-binding protein. Genes Dev 7(7 A):1203–1213

Miller KA, Gunn TM, Carrasquillo MM, Lamoreux ML, Galbraith DB, Barsh GS (1997) Genetic studies of the mouse mutations mahogany and mahoganoid. Genetics 146(4):1407–1415

Moody DE, Pomp D, Nielsen MK. (1997) Isolation of quantitative trait loci (QTL) controlling energy balance in mice. Proceedings of the 6th World Congress on Genetics of Applied Livestock Production 26:300–303

Mogil JS, Richards SP, O'Toole LA, Helms ML, Mitchell SR, Belknap JK (1997) Genetic sensitivity to hot-plate nociception in DBA/2J and C57BL/6J inbred mouse strains: possible sex-specific mediation by delta2-opoid receptors. Pain 70(2–3):267–277

Naggert JK, Fricker LD, Varlamov O, Nishina PM, Rouille Y, Steiner DF, Carroll RJ, Paigen BJ, Leiter EH. (1995) Hyperproinsulinaemia in obese fat/fat mice associated with a carboxypeptidase E mutation which reduces enzyme activity. Nat Genet 10(2):135–142

Nicolaides NC, Holroyd KJ, Ewart SL, Eleff SM, Kiser MB, Dragwa CR, Sullivan CD, Grasso L, Zhang LY, Messler CJ, Shou T, Kleeberger SR, Buetow KH, Levitt RC (1997) Interleukin 9: a candidate gene for asthma._Proc Natl Acad Sci 94(24): 13175–13180

Noben-Trauth K, Naggert JK, North MA, Nishina PM (1996) A candidate gene for the mouse mutation tubby. Nature 380(6574):534–538

Norman RA, Bogardus C, Ravussin E (1995) Linkage between obesity and a marker near the tumor necrosis factor-alpha locus in Pima Indians. J Clin Invest 96(1):158–162

Norman RA, Thompson DB, Foroud T, Garvey WT, Bennett PH, Bogardus C, Ravussin E (1997) Genomewide search for genes influencing percent body fat in Pima Indians: suggestive linkage at chromosome 11q21–22. Am J Hum Genet 60(1):166–173

North MA, Naggert JK, Yan Y, Noben Trauth K, Nishina PM (1997) Molecular characterization of TUB, TULP1 and TULP2, members of the novel tubby gene family and their possible relation to ocular diseases. Proc Natl Acad Sci 94(7):3128–3133

Oksanen L, Kainulainen K, Heiman M, Mustajoki P, Kauppinen-Makelin R, Kontula K (1997) Novel polymorphism of the human ob gene promoter in lean and morbidly obese subjects. Int J Obes Relat Metab Disord 21(6):489–494

Ollmann MM, Wilson BD, Yang YK, Kerns JA, Chen Y, Gantz I, Barsh GS (1997) Antagonism of central melanocortin receptors in vitro and in vivo by agouti-related protein. Science 278(5335):135–138

Ollmann MM, Lamoreux ML, Wilson BD, Barsh GS (1998) Interaction of agouti protein with melanocortin receptor in vitro and in vivo. Genes Dev 12(3):316–330

Pajukanta P, Nuotio I, Terwilliger JD, Porkka KVK, Ylitalo K, Pihlajamaki J, Suomalainen AJ, Syvanen A-C, Lehtimaki T, Viikari JSA, Laakso M, Taskinen M-R, Ehnholm C, Peltonen L (1998) Linkage of familial combined hyperlipidaemia to chromosome 1q21–q23. Nat Genet 18:369–373

Pelleymounter MA, Cullen MJ, Baker MB, Hecht R, Winters D, Boone T, Collins F (1995) Effects of the obese gene product on body weight regulation in ob/ob mice. Science 269(5223):540–543

Perruse L, Chagnon YC (1997) Summary of human linkage and association studies. Behav Genet 27(4):359–372

Pomp D (1997) Genetic dissection of obesity in polygenic animal models. Behav Genet 27(4):285–306

Rance KA, Hill WG, Keightley PD (1997) Mapping quantitative trait loci for body weight on the X chromosome in mice. 1. Analysis of a reciprocal F2 population. Genet Res 70(2):117–124

Rance KA, Heath SC, Keightley PD (1997) Mapping quantitative trait loci for body weight on the X chromosome in mice. 2. Analysis of congenic backcrosses. Genet Res 70(2):125–133
Reed DR, Ding Y, Xu W, Cather C, Green ED, Price RA (1996) Extreme obesity may be linked to markers flanking the human OB gene. Diabetes 45(5):691–694
Risch N, Ghosh S, Todd JA (1993) Statistical evaluation of multiple-locus linkage data in experimental species and its relevance to human studies: application to nonobese diabetic (NOD) mouse and human insulin-dependent diabetes mellitus (IDDM). Am J Hum Genet 53(3):702–714
Rogers DC, Fisher EMC, Brown SDM, Peters J, Hunter AJ, Martin JE (1997) Behavioral and functional analysis of mouse phenotypes: SHIRPA, a proposed protocol for comprehensive phenotype assessment. Mamm Genome 8:711–713
Sakurai T, Amemiya A, Ishii M, Matsuzaki I, Chemelli RM, Tanaka H, Williams SC, Richardson JA, Kozlowski GP, Wilson S, Arch JR, Buckingham RE, Haynes AC, Carr SA, Annan RS, McNulty DE, Liu WS, Terrett JA, Elshourbaby NA, Bergsma DJ, Yanagisawa M (1998) Orexins and orexin receptors: a family of hypothalamic neuropeptides and G protein-coupled receptors that regulate feeding behavior. Cell 92(4):573–585
Sapolsky RJ, Lipshutz RJ (1996) Mapping genomic library clones using oligonucleotide arrays. Genomics 33(3):445–456
Schena M, Shalon D, Davis RW, Brown PO (1995) Quantitative monitoring of gene expression patterns with a complementary DNA microarray. Science 270(5235): 467–470
Schena M, Shalon D, Heller R, Chai A, Brown PO, David RW (1996) Parallel human genome analysis: microarray-based expression monitoring of 1000 genes. Proc Natl Acad Sci 93(20) 10614–10619
Seldin MF, Mott D, Bhat D, Petro A, Kuhn CM, Kingsmore SF, Bogardus C, Opara E, Feinglos MN, Surwit RS (1994) Glycogen synthase: a putative locus for diet-induced hyperglycemia. J Clin Invest 94:269–276
Shepherd PR, Gnudi L, Tozzo E, Yang H, Leach F, Kahn BB (1993) Adipose cell hyperplasia and enhanced glucose disposal in transgenic mice overexpressing GLUT4 selectively in adipose tissue. J Biol Chem 268(30):22243–22246
Shimomura I, Funahashi T, Takahashi M, Maeda K, Kotani K, Nakamura T, Yamashita S, Miura M, Fukuda Y, Takemura K, Tokunaga K, Matsuzawa Y (1996) Enhanced expression of PAI-1 in visceral fat: possible contributor to vascular disease in obesity. Nat Med 2(7):800–803
Shutter JR, Graham M, Kinsey AC, Scully S, Luthy R, Stark KL (1997) Hypothalamic expression of ART, a novel gene related to agouti, is up-regulated in obese and diabetic mutant mice. Genes Dev 11(5):593–602
Siracusa LD (1994) The agouti gene: turned on to yellow. Trends Genet 10(12):423–428
Smith DJ, Rubin EM (1997) Functional screening and complex traits: human 21q22.2 sequences affecting learning in mice. Hum Mol Genet 6(10):1729–1733
Soloveva V, Granves RA, Rasenick MM, Spiegelman BM, Ross SR (1997) Transgenic mice overexpressing the beta 1-adrenergic receptor in adipose tissue are resistant to obesity. Mol Endocrinol 11(1):27–38
Strobel A, Issad T, Camoin L, Ozata M, Strosberg AD (1998) A leptin missense mutation associated with hypogonadism and morbid obesity. Nat Genet 18(3):213–215
Surwit RS, Wang S, Petro AE, Sanchis D, Raimbault S, Ricquier D, Collins S (1998) Diet-induced changes in uncoupling proteins in obesity-prone and obesity-resistant strains of mice. Proc Natl Acad Sci 95(7):4061–4065
Susulic VS, Frederich RC, Lawitts J, Tozzo E, Kahn BB, Harper ME, Himms-Hagen J, Flier JS, Lowell BB (1995) Targeted disruption of the beta 3-adrenergic receptor gene. J Biol Chem 270(49):29483–29492
Tartaglia LA, Dembski M, Weng X, Deng N, Culpepper J, Devos R, Richards GJ, Campfield LA, Clark FT, Deeds J et al (1996) Identification and expression cloning of a leptin receptor, OB-R. Cell 83(7):1263–1271

Taylor BA, Phillips SJ (1996) Detection of obesity QTLs on mouse chromosomes 1 and 7 by selective DNA pooling. Genomics 34(3):389–398

Taylor BA, Phillips SJ (1997) Obesity QTLs on mouse chromosomes 2 and 17. Genomics 43(3):249–257

Velculescu VE, Zhang L, Vogelstein B, Kinzler KW (1995) Serial analysis of gene expression. Science 270(5235):484–487

Wakeland E, Morel L, Achey K, Yui M, Longmate J (1997) Speed congenics: a classic technique in the fast lane (relatively speaking). Immunol Today 18(10):472–477

Walder K, Norman RA, Hanson RL, Schrauwen P, Neverova M, Jenkinson CP, Easlick J, Warden CH, Pecqueur C, Raimbault S, Ricquier D, Silver MHK, Shuldiner AR, Solanes G, Lowell BB, Chung WK, Leibel RL, Pratley R, Ravussin E (1998) Association between uncoupling protein polymorphisms (UCP2-UCP3) and energy metabolism/obesity in Pima Indians. Hum Mol Genet 7(9):1431–1435

Wallace ME, MacSwiney FJ (1975) Warfarin resistance and a new gene for obesity. Mouse News Lett 53:20

Wang SM, Rowley JD (1998) A strategy for genome-wide gene analysis: integrated procedure for gene identification. Proc Natl Acad Sci 95(20):11909–11914

Warden CH, Fisler JS, Shoemaker SM, Wen P-Z, Svenson KL, Pace MJ, Lusis AJ (1995) Identification of four chromosomal loci determining obesity in a multifactorial mouse model. J Clin Invest 95:1545–1552

West DB, Boozer CN, Moody DL, Atkinson RL (1992) Dietary obesity in nine inbred mouse strains. Am J Physiol 262:R1025–R1032

West DB, Truett GE, Goudey-Lefevre J, York B (1995) Loci on chromosomes 4, 7, 9, 12 and 15 control a significant proportion of the dietary obesity phenotype in the mouse. FASEB J 9:A722

West DB, Waguespack J, York B, Goudey-Lefevre J, Price RA (1994) Genetics of dietary obesity in SWR/J and AKR/J mice: segregation of the trait and identification of a linked locus on chromosome 4. Mamm Genome 5:546–552

West DB, Goudey-Lefevre, York B, Truett GE (1994) Dietary obesity linked to genetic loci on chromosomes 9 and 15 in a polygenic mouse model. J Clin Invest 94:1410–1416

Wheeler CH, Berry SL, Wilkins MR, Corbett JM, Ou K, Gooley AA, Humphery-Smith I, Williams KL, Dunn MJ (1996) Characterisation of proteins from two-dimensional electrophoresis gels by matrix-assisted laser desorption mass spectrometry and amino acid compositional analysis Electrophoresis 17(3):580–587

Widen E, Lehto M, Kanninen T, Walston J, Shuldiner AR, Groop LC (1995) Association of a polymorphism in the beta 3-adrenergic-receptor gene with features of the insulin resistance syndrome in Finns. N Engl J Med 333(6):348–351

Woods SC, Seeley RJ, Porte D Jr, Schwartz MW (1998) Signals that regulate food intake and energy homeostasis. Science 280:1378–1383

York B, Lei K, West DB (1996) Sensitivity to dietary obesity linked to a locus on chromosome 15 in CAST/Ei◊C57BL/6J F2 intercross. Mammal Genome 7:677–681

York B, Lei K, West DB (1997) Inherited non-autosomal effects on body fat in F2 mice derived from an AKR/J◊SWr/J cross. Mammal Genome 8:726–730

York B, Lei K, West DB (1995) A quantitative trait locus for regional fat on chromosome 8 in the mouse. Obesity Res [Suppl 3]:388S

You Y, Browning VL, Schimenti JC (1997) Generation of radiation-induced deletion complexes in the mouse genome using embryonic stem cells. Methods 13(4):409–421

Zamorano PL, Mahesh VB, Brann DW (1996) Quantitative RT-PCR for neuroendocrine studies. A minireview. Neuroendocrinology 63(5):397–407

Zhang Y, Proenca R, Maffei M, Barone M, Leopold L, Friedman JM (1994) Positional cloning of the mouse obese gene and its human homologue. Nature 372(6505):425–432

Zeng Z-B (1994) Precision mapping of quantitative trait loci. Genetics 136:1457–1468

CHAPTER 18

Primates in the Experimental Pharmacology of Obesity

B. C. HANSEN

A. Comparative Analysis of the Non-human Primate as a Model for Human Obesity

I. Obesity in Monkeys: Definition and Assessment

Obesity in monkeys is a disease of maturity, with no reports to date of spontaneous juvenile onset obesity in any primate species. Monkeys are sexually mature at about the age of 4 years, and continue increasing in lean body mass and growth to about 7 years. From 7 years on some monkeys spontaneously and gradually develop excess body fatness, while others, maintained under identical conditions, remain lean all their lives. Interestingly, this adult-onset obesity develops while monkeys are being maintained on ad libitum feeding of a diet which is "ideal" in composition – low in fat, negligible in cholesterol, reasonable in protein, and high in fiber. The usual chow diet contains about 13% of calories as fat, 18% as protein, and 69% as digestible carbohydrates, with cholesterol content of about 83 mg/kg diet. Composition of the diet is, therefore, not a significant factor in the development of spontaneous adult onset obesity in non-human primates. Obesity develops in free ranging monkeys (SCHWARTZ et al. 1993), of both sexes (SCHWARTZ et al. 1992), as well as in those maintained in a laboratory environment (JEN et al. 1985; KEMNITZ et al. 1986; HAMILTON et al. 1978).

Where monkeys have been observed in environments protected from predators with ad libitum food provided, obesity has been found in most primate species, including *Macaca mulatta* (rhesus monkeys), *Macaca fascicularis* (cynomolgus monkeys), *Macaca nemistrina* (pig-tailed macaques), African green monkeys, and many others. Since excess weight is age-dependent, the monkeys must have been observed in their middle age range.

In humans, overweight and obesity are defined by somewhat arbitrary thresholds referred to as the body mass index (BMI) chosen to identify persons with increased health risk due to excess body fat (for example, overweight in humans as been identified as a BMI of ≥25 and obesity as a BMI ≥ 30 kg/m^2) (NIH-NHLBI 1998). A similar index of fatness was defined in 1985 to apply to monkeys, and designated the Obesity Index Rh (for rhesus), or rhesus BMI (JEN et al. 1985). It is defined as the weight (kg) divided by the crown rump length (cm) squared (kg/cm^2), and was selected on the basis of its high correlation with per cent body fat and absence of correlation with height or length. The crown rump length is used for height because of increased measurement error and interobserver variation in measuring crown to heel length (height) in a reclining monkey.

II. Body Composition and Fat Content

Normal lean healthy adult male rhesus monkeys typically have body fat contents of 13.2 ± 5.4%, while a typical group of male monkeys with middle-aged onset obesity and with similar crown rump lengths have an average body fat content of 37.8 ± 6.6% (JEN et al. 1985). These two groups differed only slightly and non-significantly in lean body mass (lean vs obese groups: 8.8 ± 1.8 kg vs 10.9 ± 2.8 kg). Body fat percentages as high as 50% have been observed in laboratory-maintained ad libitum fed male rhesus monkeys. Females also show increased fatness with middle age and multiparous status (WALKER et al. 1984).

Body composition of monkeys has been measured principally by the tritiated water dilution method (PACE et al. 1947; WALIKE et al. 1977) which provides valid assessment of total body fat. Recently, the Dual Energy X-ray Absorptiometry (DXA or DEXA) method has been adapted for use in non-human primates (HUDSON et al. 1996).

III. Body Fat Distribution

As obesity develops, both sexes show increases in subcutaneous fat on the upper arms and thighs, but the greatest increase in fat for both sexes is in the abdominal or truncal region (KEMNITZ et al. 1989). Both visceral and subcutaneous adipose tissue depots increase. In humans, increased abdominal circumference carries additional health risk, and has been identified as clinically significant using waist circumference thresholds of 35 inches for women and 40 inches for men. As in humans, the abdominal circumference at about the level of the umbilicus provides an excellent indicator of fatness. For example in the above two groups of male monkeys, the lean animals had an abdominal circumference of 34.8 ± 6.2 cm, while the obese animals had a circumference of 57.5 ± 8.4 cm. Waist circumferences of greater than 45 cm in male rhesus monkeys carry increased risk of diabetes and its associated complications. Skin fold thickness has also been used to assess fat distribution; however, as in humans these measurements are subject to significant error. The abdominal skin fold thickness has proved to be a reliable indicator of fatness in monkeys, with abdominal skin folds of 9.82 ± 1.82 mm observed in lean males, and 22.74 ± 5.8 mm found in the above noted obese group (JEN et al. 1985).

IV. Obesity in Humans and Monkeys Compared

The characterization of obesity in humans and non-human primates shows extensive similarities which go well beyond its middle age onset and typically abdominal fat distribution. Many of the characteristics of primate obesity are summarized in Table 1.

V. Experimental Production of Obesity in Primates

Obesity has, in the past, been induced experimentally by several methods, including ventromedial hypothalamic lesions, high fat diet feeding (Christopher-Hennings et al. 1995), and forced over-feeding (Jen et al. 1984). Today such methods for inducing obesity are generally not used, except in use of high fat diets for the investigation of atherosclerosis, since monkeys at middle age develop obesity with sufficient frequency to be available as a natural model, rather than through experimental manipulations.

VI. The Natural History of Obesity in Non-human Primates

1. Longitudinal Metabolic Progression from Lean to Obese, With or Without Subsequent Development of Type 2 Diabetes

Long term prospective studies of many monkeys as they mature from young adulthood to advanced age have led to the identification of a series of phases through which monkeys move in the course of developing obesity with or without type 2 diabetes (Hansen et al. 1985, 1986, 1990; Bodkin et al. 1989). It should be noted that some monkeys stay lean all their lives, even under ad libitum fed laboratory maintained conditions. Other monkeys begin to develop obesity some years after reaching sexual maturity (after age 4). It has not been possible to identify among lean young adult monkeys which ones will go on to become obese in middle age, and which of those will develop spontaneous type 2 diabetes, a frequent but not universal occurrence. On average

Table 1. Comparison of human to non-human primate obesity

	Humans	Monkeys
Frequency in adults	>50%	>50%
Age of peak weight	40–60 years	15–20 years
Excess fat location	Central	Central
Insulin resistance/Hyperinsulinemia	Almost always	Almost always
Diabetes-% obese	>90%	>90%
Dyslipidemia	Increased frequency	Increased frequency
Hypertension	Increased frequency	Increased frequency
Hyperleptinemia	Present in most	Present in most

peak body weight for *Macaca mulatta* (rhesus monkeys) is reached between age 15 and 20 years, equivalent in humans to approximately age 45–60 years. Some develop a "simple" obesity in which they show little or no tendency to move toward overt diabetes, while most who become obese show movement across the sequence of phases in the direction of diabetes. Obese monkeys who show progression across the defined phases become overtly diabetic at an average age of 18. There is, however, a wide range in the age of onset of diabetes in obese monkeys. The type 2 diabetes diagnostic threshold of 140 mg/dl fasting plasma glucose (with impaired glucose tolerance) has been observed to be attained by individual monkeys as early as 10 years of age and as late as age 29 years. Since the average age of death in these laboratory maintained rhesus is around 23 years, it is clear that some monkeys who are on the "diabetes trajectory," based on longitudinal within animal studies, die before they reach overt diabetes. The mean changes in body weight, glucose, and insulin during the progression from lean to severe diabetes are shown in Fig. 1.

As shown in Fig. 2, obese rhesus monkeys develop increased hepatic glucose production which only becomes significant as elevated plasma glucose levels appear (HANSEN et al. 1985, 1986, 1990; BODKIN et al. 1989). In addition they progressively show insulin resistance or failure of insulin to induce normal glucose disposal rates at the whole body level (peripheral glucose uptake rate during a euglycemic hyperinsulinemic clamp or M).

Lipoprotein concentrations across the phases from normal young lean adult to obese to type 2 diabetes are shown in Fig. 2. Although not all diabetic animals develop dyslipidemia, many develop very highly elevated triglyceride levels (>500 mg/dl or 5.65 mmol/l) (HANNAH et al. 1991). These disturbances in lipoprotein concentrations have been observed in some obese monkeys well before the onset of clinically overt diabetes. This heterogeneity is consistent with observations in obese humans, and the abnormalities in diabetic monkeys, while similar to those seen in patients with diabetes, may be even more pronounced in non-human primates than those in humans. Some of the obese insulin resistant monkeys also develop dyslipidemia, consisting of increased plasma triglyceride concentrations, and gradually declining HDL cholesterol levels as also shown in Fig. 2.

VII. Insulin/Insulin Receptor/Hyperinsulinemia in Non-human Primates

Circulating plasma insulin levels in lean young adult normal monkeys generally range from 30 μU/ml to 70 μU/ml, about 8–10-fold higher than levels observed in humans. When monkeys become obese and hyperinsulinemic, fasting circulating insulin levels can range as high as 1000 μU/ml or greater. As shown above, in those progressing toward overt diabetes, insulin levels usually rise to the 300–500 μU range (HANSEN et al. 1985, 1986) during the early phases

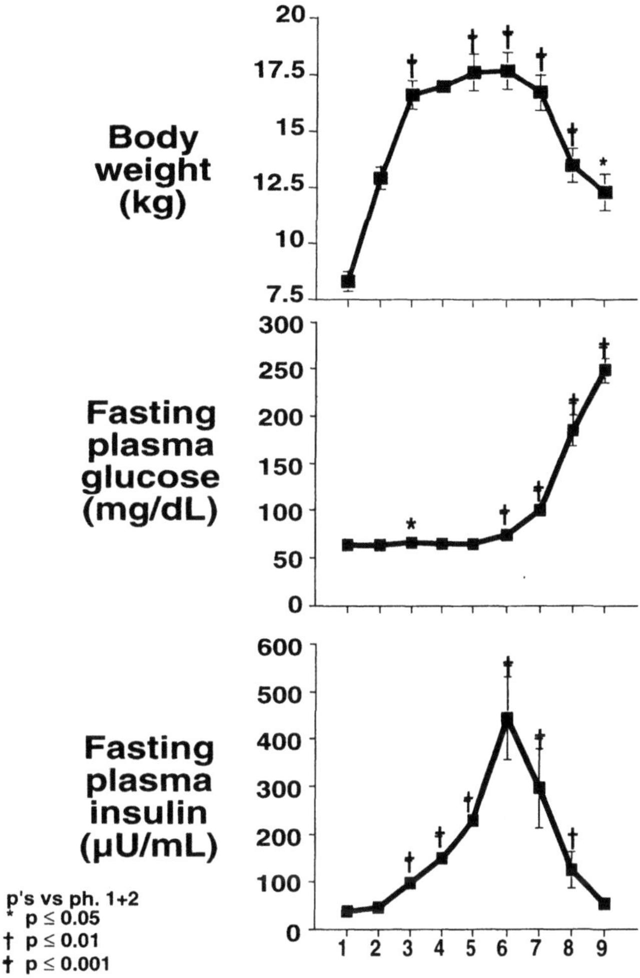

Fig. 1. Phases in the progression from lean to severely diabetic. Average body weight, fasting plasma glucose, and fasting plasma insulin are shown for phases 1–9. Phase 1 – Lean adult monkeys under the age of 10 and under 10 kg body weight. Phase 2 – lean and/or obese monkeys over the age of 10 showing no metabolic changes indicative of or associated with the development of insulin resistance or diabetes. Phases 3, 4, and 5 – a progressive period of increasing fasting plasma insulin and increasing β cell response to glucose usually accompanied by continuing weight gain and increasing adiposity through the early part of this progression. Phase 6 – a period when fasting plasma insulin is still on the rise, but the β cell response to glucose has begun to decline from the greatly elevated responsiveness of the earlier hyperinsulnemic period. Phase 7 – impaired glucose tolerance and the transition to diagnosis of overt type 2 diabetes. Phase 8 – meets or exceeds the diagnostic threshold for diabetes, including elevated fasting plasma glucose and impaired glucose tolerance. Phase 9 – severe advanced type 2 diabetes requiring insulin treatment to prevent weight loss and to maintain health

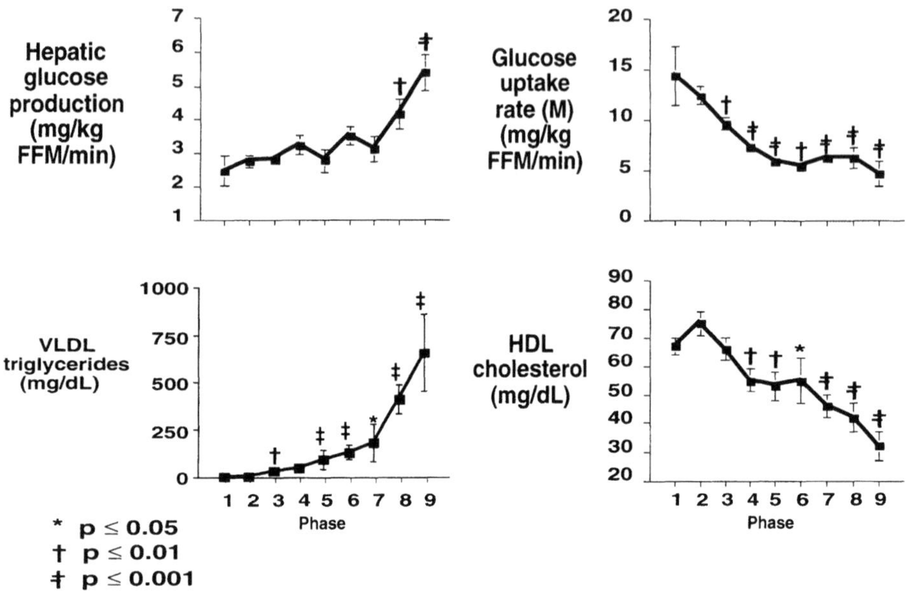

Fig. 2. Changes in hepatic glucose production, insulin sensitivity (insulin-stimulated glucose uptake rate), VLDL triglycerides, and HDL cholesterol in monkeys progressing from lean to obese to type 2 diabetes. Phases are as identified above

while glucose remains normal, and then insulin levels gradually decline as overt diabetes is approached in phases 8 and 9 as indicated in Fig. 3.

Both proinsulin and the insulin receptor of rhesus monkeys have been sequenced. Insulin in the monkey is identical in amino acid sequence to human insulin (NAITHANI et al. 1984). There is a single amino acid substitution in the C peptide molecule. The insulin receptor shows 99% amino acid identity to the human receptor (HUANG et al. 1993). Site directed mutagenesis to examine the functional consequences of the two non-conservative amino acid substitutions showed no evidence for impairment in receptor function induced by these two differences between human and monkey insulin receptor structure (FAN et al. 1995).

Basal secretion of insulin (and of C peptide) in monkeys oscillates with a periodicity of about 10–12 min per cycle, as shown in Fig. 4. This unique secretory process was first discovered in studies of monkeys because of their high plasma insulin levels and the consequent large amplitude of the oscillations. They were subsequently identified in humans as well. During the development of obesity with hyperinsulinemia, the amplitude of the oscillations grows. As fasting insulin levels fall the periodicity of insulin secretion also becomes disrupted and irregular. The effects of this disturbance in secretory rhythm are not understood.

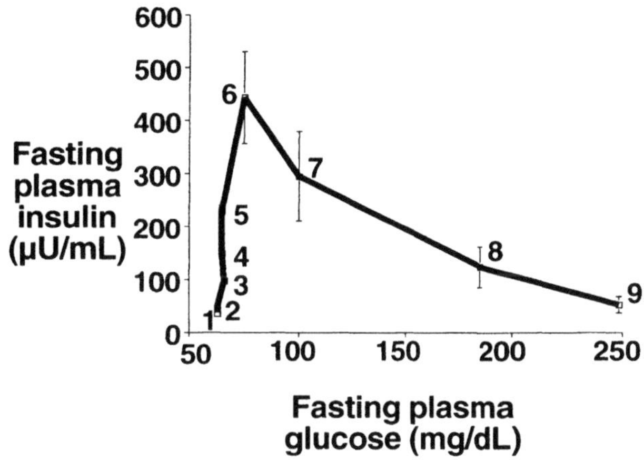

Fig. 3. The interactions of changing fasting plasma insulin levels and fasting plasma glucose levels during the progression to overt diabetes (between phases 7 and 8)

VIII. Characterization of Adipose Tissue in Monkeys

1. Fat Cell Size and Number

The adipose tissue of lean and obese monkeys has been the subject of a number of studies. Obese monkeys have been shown to have increased fat cell size compared to lean monkeys of similar crown rump height: lean: 0.61 ± 0.29 μg lipid per cell vs obese: 1.29 ± 0.54 μg lipid per cell. Fat cell numbers differed between the groups to an even greater extent than fat cell size: lean: 2.2 × 10^9 cells vs 6.1 × 10^9 cells (JEN et al. 1985)

2. Cellular Events in the Progression of Obesity to Diabetes

Adipocytes exhibit extraordinary changes in glucose metabolism during the development of obesity, and during the transition to overt type 2 diabetes that are seldom appreciated. Changes in the insulin responses of adipocytes during the metabolic changes associated with the development of diabetes were observed in both cross sectional and in longitudinal studies of monkeys during the progression from normal to obese, and from obesity to diabetes (HANSEN et al. 1988). In adipocytes isolated from subcutaneous abdominal adipose tissue, glucose utilization was evaluated by conversion of ^{14}C-glucose to CO_2 and lipids in the absence and in the presence of various concentrations of insulin. Two major changes were identified. The first was the increase – not decrease – in insulin sensitivity demonstrated by these cells obtained across time during the transition from lean to significantly obese. Obesity was clearly associated with enhanced adipocyte glucose utilization and lipid synthesis and enhanced sensitivity of adipocytes to insulin, relative to lean normal controls.

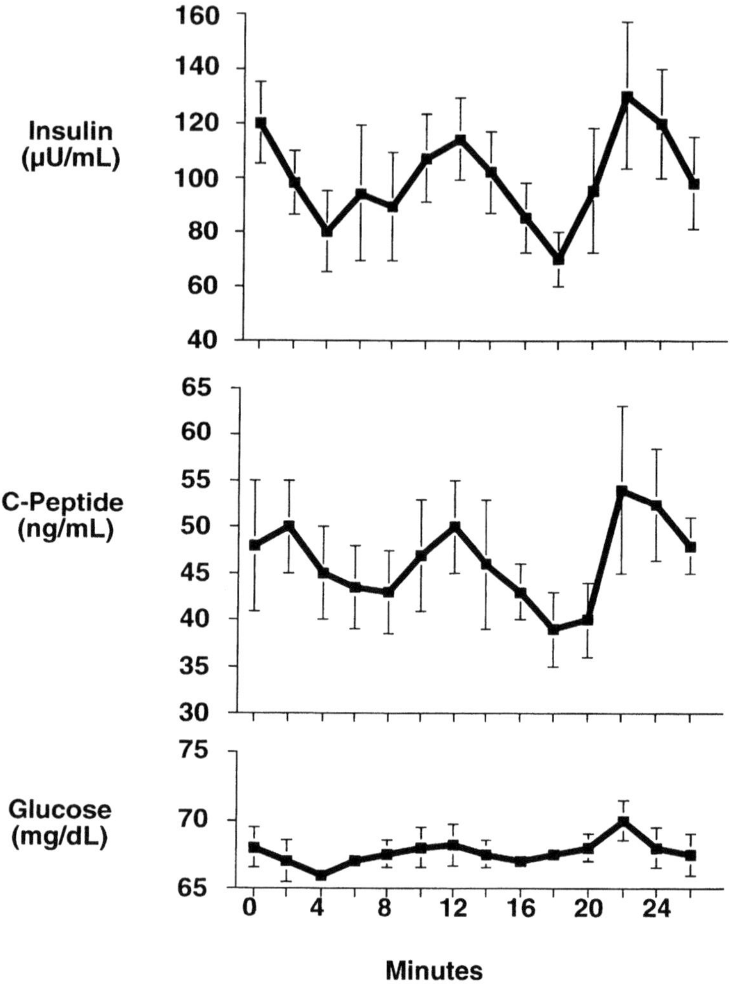

Fig. 4. Basal insulin and C peptide secretion with a periodicity of 10–12 min per cycle, and amplitude of ca. ±25% in monkeys

The second and later event was the major reduction in basal and in insulin-stimulated glucose utilization by adipocytes as hyperinsulinemic monkeys progressed into type 2 diabetes. Then, despite further worsening of glucose tolerance and increasing severity of diabetes, there was only a modest or negligible further impairment in basal and insulin-stimulated glucose oxidation, as shown in Fig. 5.

Snitker and colleagues reported related observations in an examination of in vitro studies of human adipose tissue (Snitker et al. 1998). They observed that basal in vitro adipose tissue lipolysis was negatively correlated with

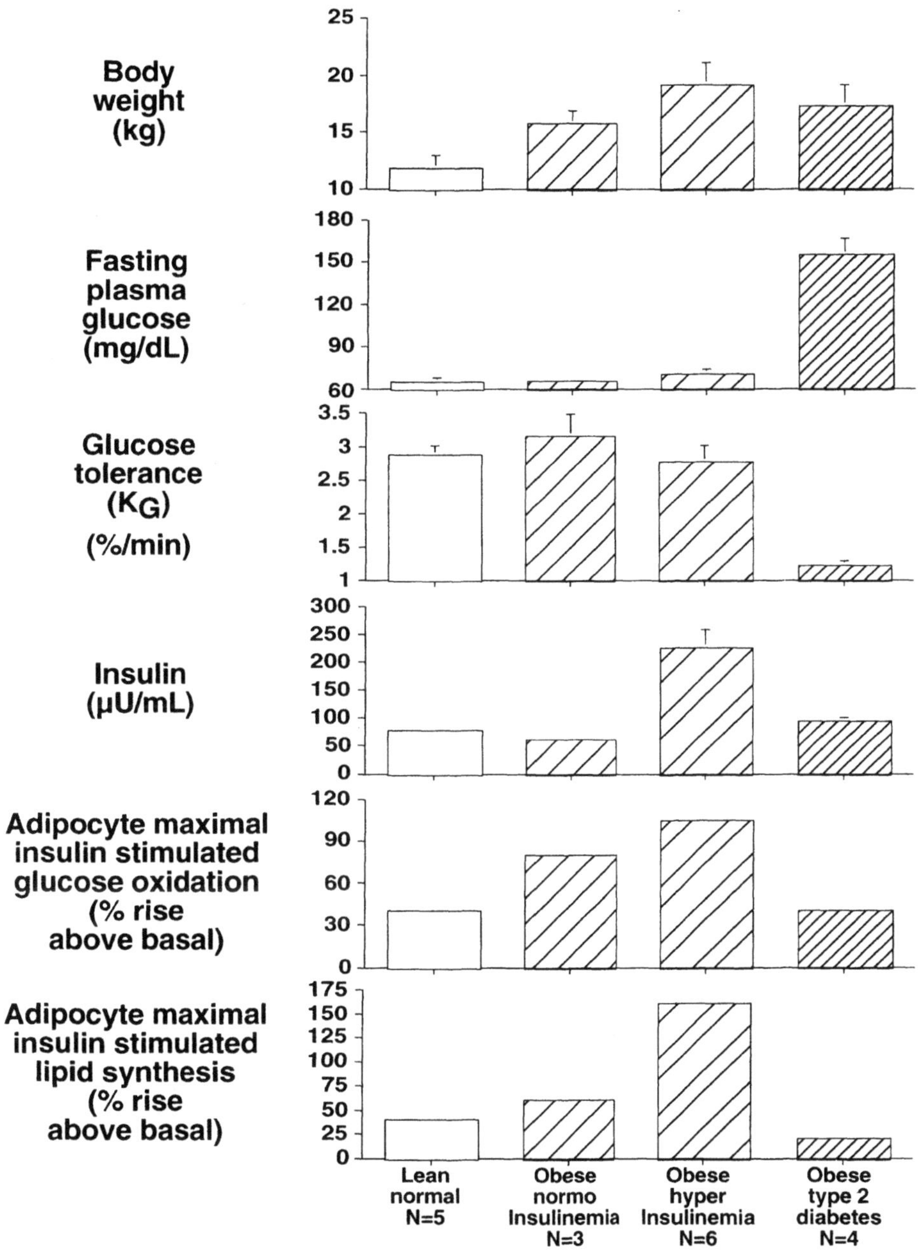

Fig. 5. Insulin action on adipocyte glucose oxidation and lipid synthesis in four groups of monkeys. The increase in insulin action on isolated adipocytes from obese hyperinsulinemic ("insulin resistant") monkeys is notable. Redrawn from HANSEN et al. (1988)

subsequent weight gain (low basal lipolysis was predictive of increased weight gain), as was insulin's ability to suppress lipolysis (increased insulin action was predictive of increased weight gain). Thus, in humans it has been confirmed that features related to enhanced (not reduced) insulin action on lipid storage in adipocytes are present in individuals who are developing obesity, just as observed in monkeys. We suspect that, if followed up further, this enhanced sensitivity to the antilipolytic effect of insulin will decline as those human subjects approach overt diabetes.

B. Pharmacologic Treatment of Obesity and the Insulin Resistance Syndrome: Applications in Primates

I. Brown Adipose Tissue, The β Adrenergic Receptors and Uncoupling Proteins (UCPs)

Brown adipose tissue in non-human primates is similar in amount and distribution to that found in humans. It has been identified by histological and immunological techniques in axillary, interscapular, subscapular, and cervical fat deposits in young adult male and female monkeys (KATES et al. 1990). UCP1, a member of the mitochondrial anion carrier family, is principally expressed in brown adipose tissue, and is induced by exposure to cold and by β adrenergic agonists (GONG et al. 1997). Since neither monkeys nor humans have significant brown adipose tissue in the adult, and since the UCP1 knock out in mice did not result in obesity, the role of brown adipose tissue in primate and human obesity has been questioned. In young baboons, heterogeneity of these tissues was noted, with UCP1, believed to be specific for brown adipose tissue, detected in the several fat deposits examined (VIGUERIE-BASCANDS et al. 1996). In cynomolgus monkeys, adipose tissue from axillary and visceral sites was shown to contain some brown adipocytes by morphological examination and by the presence of UCP1 mRNA. Omental white adipose tissue was found by RT-PCR to express higher amounts of $\beta 3$ adrenoceptors than subcutaneous tissue in humans and monkeys (FISHER et al. 1998). The $\beta 1$ and $\beta 2$ receptors were, however, expressed at levels detectable by Northern blots. Despite the difference in the number of these types of receptors expressed in omental adipose tissue, the high selectivity of some agonists for the $\beta 3$ receptor, and their ability to stimulate lipolysis selectively provides support for their potential as antiobesity compounds as discussed below. The normal physiologic or functional role of brown adipose tissue and of the $\beta 3$ receptor in adult humans and adult monkeys remains to be clarified.

Recent discoveries of new mitochondrial uncoupling proteins which are important in the regulation of thermogenesis, UCP2 which is abundant in white adipose tissue, and UCP3 which is highly expressed in muscle and rodent brown adipose tissue, and is expressed in a broader array of tissues than the first described UCP1 (found mainly in brown adipose tissue), have spurred interest in these as additional potential therapeutic targets for body weight regulation (GONG et al. 1997).

II. The β3 Adrenergic Receptor

The β3 adrenergic receptor makes an attractive candidate for therapeutic targeting in obesity because of its unique distribution, primarily in adipose tissue, and its functional activation by norepinephrine from the sympathetic nervous system terminals.

The primary evidence for the role of the adrenergic receptors was derived from rodents, in which the activation of β3 receptors was effective in activating brown adipose tissue and promoting weight loss. The human β3 adrenoreceptor was cloned in 1989 (EMORINE et al. 1989). The responses of the human and rodent β3 receptors to pharmacological stimuli were subsequently found to differ markedly. These species differences in potency as well as in specificity of responses to β3 agonists led us and others to explore the possibility that the non-human primate might make a better model for the examination of β3 agonists for their potential therapeutic usefulness for humans.

The entire sequence of the β3 receptor was determined in one rhesus monkey, and variations in parts of the sequence were examined in fifteen additional monkeys (WALSTON et al. 1997). In the first exon there were 19 amino acid differences from the human sequence (95% identity). Only 15 amino acid differences are present throughout the coding region prior to the C terminus region. At base pair 1135 a deletion of a single cytosine was identified which produced a frame shift resulting in 26 of the 30 C terminus amino acids differing from the human sequence, and this deletion was present in all monkeys examined. The C terminal has no PKA or βARK sites, and thus the frame shift is not likely to alter kinase mediated desensitization. There were also an additional ten amino acids coded in exon 2 of the rhesus, as shown in Fig. 6. In summary, these results show a total amino acid identity between humans and monkeys of 89%, with most of the difference from humans in the C terminal. Other species show even greater variability in the C terminus region, as shown in Table 2.

1. The β3 Receptor Variant ($Trp^{64}Arg$): Role in Obesity

A number of studies in humans have identified a possible association between a specific common variant of the β3 receptor, arginine substituted for $tryptophan^{64}$ ($Trp^{64}Arg$), and a wide range of characteristics related to obesity and/or to diabetes. We therefore examined this relationship in a group of non-human primates, including normal lean, normal obese, obese with hyperinsulinemia and insulin resistance, impaired glucose tolerant, and overtly diabetic monkeys, anticipating some distribution of the variant across these, and were surprised to discover that all 15 monkeys in which PCR analyses were

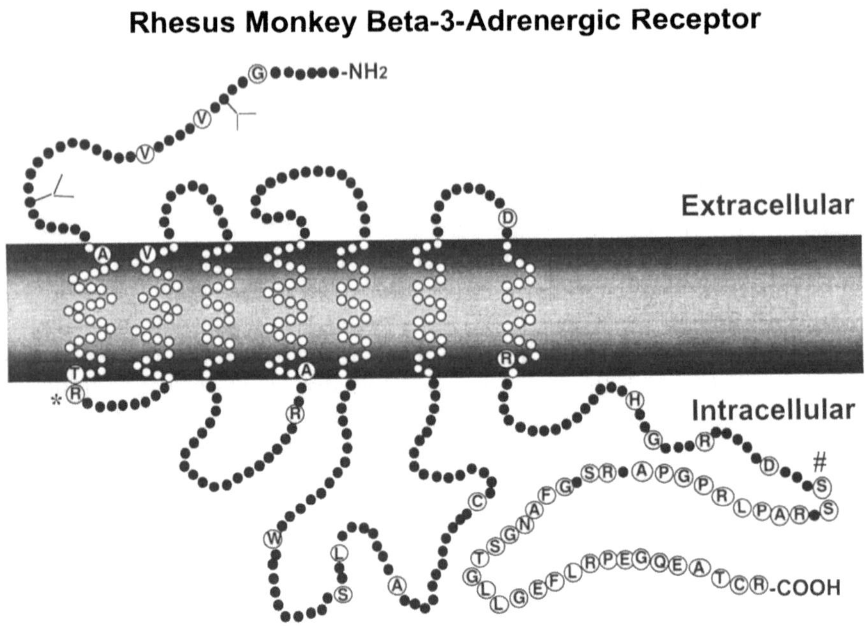

Fig. 6. Structure of the $\beta 3$ receptor in rhesus monkeys, showing the 45 amino acid differences and 10 additional amino acids compared to the human sequence. The *asterisk* indicates the tryptophan to arginine substitution. The # indicates the site of cytosine deletion

Table 2. Percentage of amino acid identity between human and rhesus monkey and between human and mouse β_3 adrenergic receptor by domain

Domain	Rhesus	Mouse
N – Term amino-terminal end of protein		
N – Term	91	89
Transmembrane domains		
1	93	93
2	96	89
3	100	100
4	96	93
5	100	95
6	100	86
7	100	95
Extra cellular loops		
1	100	91
2	100	95
3	91	91
Intra cellular loops		
1	89	89
2	95	95
3	92	83
C – Term Carboxy-terminated end of protein		
C – Term	52	35

performed were homozygous for this $Trp^{64}Arg$ variant, regardless of lean or obese or diabetic status. At least in the non-human primate, this variant does not differentiate those with or without obesity or a propensity to diabetes (WALSTON et al. 1997). If the relationship in humans continues to be supported, one would interpret this to suggest that the monkeys are in the resistant/obese portion of the human distribution. It may, however, be that the earlier associations of Arg at codon 64 found in humans with a higher incidence of obesity or diabetes will not be supported by larger studies or that the contribution of this variant is minor in the overall picture of obesity. This may be one of many gene contributions to obesity and diabetes, most of which have yet to be identified.

III. $\beta 3$ Adrenergic Receptor Agonists

The β adrenergic receptor activity in response to agonists has been examined in several primate species, marmoset, baboon, and macaque, and compared to rat, dog, and human (BOUSQUET-MELOU et al. 1994). Regardless of the relative amounts of the various β adrenoceptors, non-selective agonists such as isoproterenol or norepinephrine had similar lipolytic properties across the species studied. They differed significantly, however, in their responses to various selective agonists.

ENOCKSSON et al., in examining human subcutaneous white adipose tissue, demonstrated the coexistence of β_1, β_2, and β_3 functional adrenoceptor subtypes, and provided direct in vivo evidence for a functional $\beta 3$ adrenoceptor in human subcutaneous adipose tissue. Based on stimulation of lipolysis by three specific adrenoceptor agonists including CGP12177, they surmised that the dual role observed for the $\beta 2$ adrenoceptor in both mobilizing lipids and in stimulating blood flow might make it the predominant form in vivo in man (ENOCKSSON et al. 1995). They cautioned, however, against generalizing from in vitro to in vivo when investigating lipolysis in man, due to differences observed in relative potency of various agonists with these two methods. The relative efficacy of agonists to stimulate lipolysis via the several receptor types also differed between in vivo and in vitro studies (TAVERNIER et al. 1996).

Strikingly different results were obtained with the administration of $\beta 3$ selective agonists in vivo in humans compared to rodents, and thus, the primate model has become of high importance in the evaluation of the potential of new agents for efficacy in humans. An earlier agent, BRL-37344, was only weakly lipolytic in baboon, macaque, and human, but highly active in rat and dog white adipose cells. Similarly, CL-316243, ICI-D7114, and SR-58611A, all $\beta 3$ agonists in rat, dog, and marmoset, were inactive in baboons, macaques, and humans. Recent developments have focused upon human specific highly selective $\beta 3$ agonists, and have emphasized the importance of evaluation of such agents in species as close in functional regulation to humans as possible.

Meyers (1997) has used CGP 12177 to show that such an agent could have positive properties in humans, since, in cynomolgus monkeys increased lipolysis and oxygen consumption in brown adipocytes were observed. Another $\beta 3$ agonist, L-755507, was found to have >400-fold selectivity for activation of the $\beta 3$ receptor vs the $\beta 1$ or $\beta 2$ receptors of both humans and monkeys (Fisher et al. 1998). L-755507 was shown to increase metabolic rate in rhesus monkeys under acute conditions of intravenous drug administration (3 mg/kg) (Fisher et al. 1998). Increased lipolysis was also observed. Chronic exposure to this agent resulted in increased expression of UCP1 in brown adipose tissue of rhesus monkeys, but no increase in UCP2 expression in axillary brown adipose tissue (Fisher et al. 1998). Like the preceding agent, another $\beta 3$ agonist, L757,793, administered intravenously, stimulates lipolysis and produces a maximum increase in metabolic rate of approximately 25%, which equates to an approximate 3% increase in 24 h energy expenditure (MacIntyre 1998). L757,793 also produced increased UCP1 expression, but no increase in UCP2 expression, in axillary brown adipose tissue. In our unpublished studies, chronic dosing with the latter compound in monkeys did not produce weight loss or behavioral changes, probably due to the relatively short duration of action. If these metabolic changes were to be similarly observed in humans, with successful achievement of a more sustained duration of action, and without any compensatory increase in food intake, the weight loss produced might average 0.5 kg per month to 1 kg per month, potentially useful as a component of a weight loss program.

Cowan et al. 1998) have reported the effects of the human selective $\beta 3$ agonist GR5261X (dosed at 2 mg/kg, t.i.d. for 4 weeks) in six obese prediabetic monkeys (Cowan 1998). This agonist significantly reduced non-esterified fatty acids (NEFA) and LDL cholesterol, as shown in Fig. 7, as well as plasma insulin levels and triglycerides. No significant reduction in weight was observed with this dosing regimen and level, and there was no effect on food intake. There were no apparent side effects at any time during the 4 weeks of dosing. In anesthetized monkeys, acute dosing of GR5261 (0.1–3 mg/kg, i.v., $n = 4$) stimulated lipolysis as measured by increases in NEFA levels without affecting heart rate or blood pressure.

The $\beta 3$ agonist, CL214613 was found to increase UCP3 levels in adipose tissue, while starvation increased UCP3 mRNA expression in muscle and decreased it in brown adipose tissue. Expression levels and activity in non-human primates have not yet been reported, but can be expected to reflect these early reports of expression in human tissues. The low levels of $\beta 3$ adrenoceptors may mitigate against a significant effect in weight regulation in humans and monkeys, however, and even small stimulation of $\beta 3$ adrenoceptors on white adipose tissue appears to induce an increase in the output of insulin from the β cells by some unknown signaling pathway (Grujic et al. 1997). That this signal is not likely to be leptin is suggested by the decrease in circulating leptin levels with $\beta 3$ agonist treatment (Mantzoros et al. 1996). Could this signal be involved in the obesity-related enhancement of insulin output both basally and

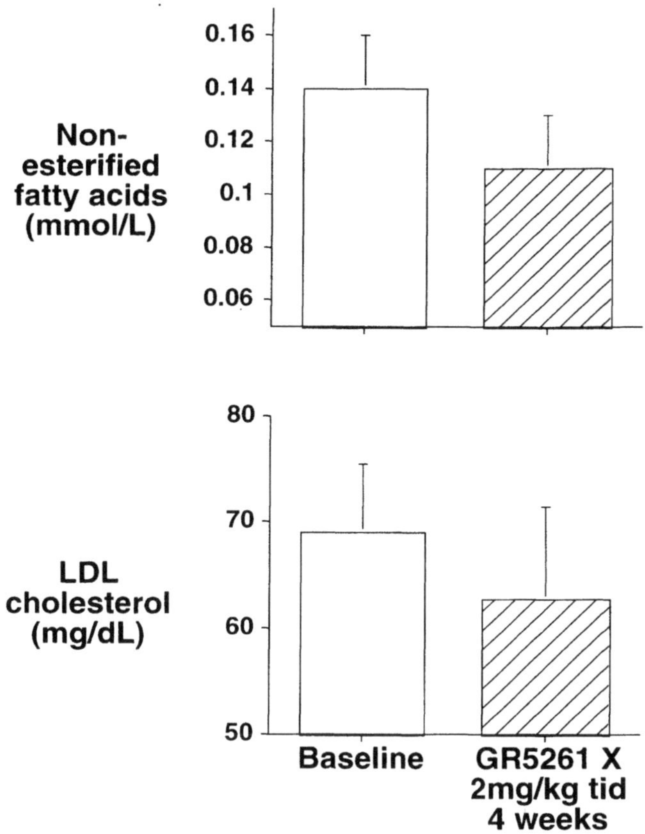

Fig. 7. Lowering of nonesterified free fatty acids (NEFA) and LDL cholesterol in monkeys by the β3 agonist, GR5261X, administered at a dose of 2 mg/kg, t.i.d. for 4 weeks

under glucose stimulation? This is an open question awaiting determination of the nature of the signal being generated.

The low level of expression of the β3 adrenoceptor and its variability, as well as the modest effectiveness of the selective agents so far reported, leave open to further research the question of the long term potential of this route to antiobesity action for humans and for non-human primates.

IV. Circulating Factors in the Regulation of Appetite and Body Weight – A New (Old) Question

1. Circulating Factors as Examined by Cross Circulation Experiments

The cross circulation studies of ob/ob and db/db mice suggested the possibility of a circulating factor which might profoundly affect body weight regulation, and a receptor-sensor of that factor, defects in both of which could produce dysregulation of body weight (COLEMAN 1978). Many years later these hypotheses concerning the causes of obesity and diabetes in two strains of mice were confirmed (ZHANG et al. 1994; TARTAGLIA et al. 1995). Today the important questions relate to whether these rodent mutations are related directly or indirectly to the causes of human obesity. Interestingly, studies in primates did not suggest this to be the case.

An early study of monkeys examined the possibility of a potent circulating factor controlling appetite regulation. In that study pairs of healthy young adult monkeys were cross circulated at the very rapid rate of about 200 ml/min, with one monkey allowed to feed freely and the other fully starved. Interestingly, the feeding monkey did not increase its food intake (although not prevented from doing so by gastrointestinal capacity), despite the fact that, conceptually, this monkey was "eating for two." This study suggested an extraordinarily powerful regulation of intake by neural input from the liver and/or gastrointestinal track to the CNS (WALIKE et al. 1972), an area of study claiming increasing attention today. That study also found that the monkey that did not receive food, whose appetite was examined in brief latency-to-feed tests, did not appear to be satiated by any circulating factor being received from the freely feeding monkey. This raises the important question of the power of endogenously derived circulating factors to regulate central nervous system control of food intake, and suggests the importance of neural signals from the periphery in this control system.

V. Ob Gene/Ob Receptor, and Circulating Leptin

In monkeys, as in humans, defects in the *ob* or *leptin* gene do not appear to be playing a significant role in obesity, with a few rare exceptions identified in several humans. Cloning of the *leptin* gene of the rhesus monkey has been performed, and revealed a predicted protein of 167 amino acids, as in humans, and with 91% identity to the human form (84% identity to mouse) (HOTTA et al. 1996). The expression of the *leptin* gene was examined in both lean and spontaneously obese monkeys and was found to be significantly correlated with body weight and body fatness in monkeys, and with fasting plasma insulin levels in the non diabetic group (HOTTA et al. 1996).

The ob or leptin receptor has also been cloned and its expression examined in rhesus monkeys (HOTTA et al. 1998). As in humans, there are long and short forms of the receptor, with the long form expressed more highly in liver, adipose tissue, hypothalamus, and choroid plexus, and other forms expressed in possibly all tissues. The expression of the long form, thought to be the signaling form of the receptor, was not related to body weight, plasma insulin, or glucose levels, or to plasma leptin levels (HOTTA et al. 1998).

Circulating levels of leptin have been examined extensively in monkeys (as well as in other species including humans).

Circulating leptin levels are, on average increased in obese humans and monkeys (BODKIN et al. 1996); however, it should be noted that within the obese groups there are substantial numbers of obese individuals without elevated leptin levels, and the levels are best represented by a logarithmic function since there is very high and to date unexplained variability in both the ob gene expression and the circulating leptin levels in the obese groups. That leptin levels are reduced in starvation suggests that although leptin is not acutely regulated by insulin (HOTTA et al. 1996), it is regulated by declining energy or nutrient flow.

The observation that both obese humans and obese monkeys have generally elevated circulating leptin levels has suggested to some that there may be resistance to circulating leptin. For this to be the case, however, those without such resistance must be found to be affected by administration of leptin. To date, this does not appear to be the case, at least in regard to a weight or appetite regulating role for leptin. More recently, many other potential functions of leptin have been identified, and only further research will sort out which features are involved in any physiological effects.

1. Is Leptin Truly a Satiety Hormone, and Does It Modulate Weight or Fatness in Non-human Primates?

Although leptin administration to rodents with defects in the *ob* gene has been shown to reduce the obesity of this model, such effects have been inconsistent and difficult to confirm in the rhesus monkey (RHODES et al. 1998). A prior report suggested possible but highly variable effects in monkeys which have not been substantiated in subsequent studies (WAGNER et al. 1996). Central administration of leptin to monkeys has, however, produced an inhibition of food intake and activation of the sympathetic nervous system (TANG-CHRISTENSEN et al. 1998). Although a brief report of the effects of leptin administration to humans has suggested a small positive effect on weight loss (GREENBERG 1998), the parameters of the study, together with the minimal efficacy data from administration to monkeys, suggest that leptin itself is unlikely to be a therapeutic agent for the treatment of obesity. This does not, however, negate the possibility that an analogue or a competitor, working at the Ob receptor could have potential effects, although such actions have not yet been reported.

VI. PPARγ and the Thiazolidinediones

1. Adipose Tissue Expression of Peroxisome Proliferator Activated Receptors (PPARs) and their Amino Acid Sequences in Monkeys

The peroxisome proliferator-activated receptor γ (PPAR γ) is a member of the family of nuclear hormone receptors identified as important in the differentiation of adipocytes from preadipocytes. As such they make attractive candidates for possible therapeutic targeting. In fact, the receptor was first identified based on its binding of an exogenous ligand – a thiazolidinedione, and was subsequently found to bind endogenous prostaglandin J2 with high affinity.

PPARγ forms a heterodimer with retinoid X receptor α. Its isoforms, PPARγ1 and γ2, have been cloned and their expression in adipose tissue and in other tissues of obese and lean monkeys has been determined (HOTTA et al. 1998). The PPARγ2 protein was 99% identical between monkeys and humans, and was found mainly in adipose tissue. The PPARγ1 receptor was found in various tissues, but most abundantly in adipose tissue. The ratio of PPARγ2 to total PPARγ was closely related to adiposity, as shown in Fig. 8, however, total PPARγ was not related to obesity. The expression of this adipocyte differentiation related receptor was closely coordinated with the expression of C/EBPα, LPL, and GLUT4.

2. Thiazolidinediones

Thiazolidinediones are a new class of antidiabetic agent with the principal action identified as increasing insulin sensitivity, but these agents have effects on many tissues, including the β cell of the pancreas, endothelial cells of the vasculature, and other targets. The thiazolidinedione class of pharmaceutical agents does not produce weight loss in humans or in monkeys; however, such agents may have a role in the treatment of obesity due to their well established insulin sensitizing actions. Short term studies of several different thiazolidinediones and related agents have shown positive results to improve insulin sensitivity.

Administration of a thiazolidinedione, Pioglitazone, to obese insulin resistant rhesus monkeys resulted in a dose response related reduction in plasma insulin levels by 64% at a dose of 3.0 mg/kg/day, and a reduction in triglyc-

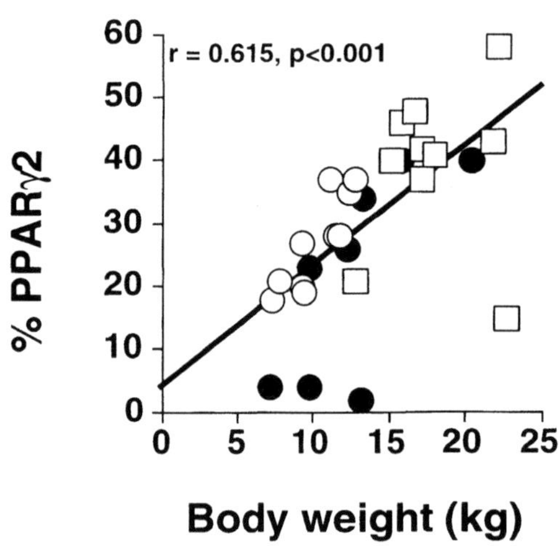

Fig. 8. Relative proportion of PPARγ2 to total PPARγ in relation to body weight and fatness. *Open circles* are normal monkeys, *open squares* are obese, and *filled circles* are diabetic (type 2 DM)

erides of 44% (KEMNITZ et al. 1994). Interestingly, although an improvement in insulin sensitivity, as measured by the minimal model, was reported, no change in glucose tolerance was observed.

Another thiazolidinedione, R-102380, was studied in rhesus monkeys (unpublished data). When 1 mg/kg was administered daily to type 2 diabetic monkeys, glucose was reduced by about 10% (unpublished observations). Importantly, this agent was most effective in lowering glucose in the mildly diabetic subjects, and did not affect glucose in those with a fasting glucose greater than 220 mg/dl. Insulin requiring type 2 diabetic individuals with diabetes of long duration may not prove to gain the positive effects expected in those in the pre-diabetic and early stages of diabetes. These monkeys were advanced in their diabetes, and no longer had elevated fasting insulin levels. Insulin was reduced by R-102380, with lesser effects in those whose insulin had already declined the most due to the natural progression of the disease. Triglycerides were significantly reduced.

R-102380 was also administered chronically for 6 weeks or more to monkeys who were not yet overtly diabetic (FPG < 140 mg/dl), but who were significantly insulin resistant and obese (unpublished data). In this group euglycemic hyperinsulinemic clamps with liver, muscle, and adipose tissue biopsies were performed at the start and end of each clamp, with this procedure carried out in the predosing baseline period and again at the end of 6 weeks of treatment. Additional effects of this thiazolidinedione on these insulin resistant non diabetic monkeys were noted. Those monkeys with initially elevated plasma glucose levels showed a decrease in glucose, while those with near normal glucose levels were unaffected by R-102380. Insulin levels in these hyperinsulinemic monkeys were significantly reduced. Triglyceride levels decreased, with the greatest effects on those with the most elevated basal triglycerides. Although the euglycemic clamp procedure did not demonstrate a significant increase in whole body glucose uptake rate, significant changes were observed in enzymes of glycogen metabolism. Specifically, skeletal muscle glycogen synthase activation during insulin stimulation was improved by R-102380. This enhanced insulin action was specific to skeletal muscle and was not seen in adipose tissue or liver glycogen synthase activities. Subcutaneous abdominal and omental adipose tissue expression of a wide range of genes considered potential candidates for alteration under the effects of an insulin sensitizer were also examined. Several trends were noted, but no significant alterations in either basal or insulin stimulated gene expression could be consistently identified in either subcutaneous or omental adipose tissue.

The Diabetes Prevention Program, an ongoing large scale multicenter trial aimed at slowing or halting the progression of impaired glucose tolerance to overt type 2 diabetes, had as one of its treatment arms the use of another thiazolidinedione, Troglitazone, with the hypothesis that this agent might delay the transition from obesity to diabetes by enhancing insulin action. Dual concerns about the power of this approach to prevent diabetes and possible idiopathic risks halted its use in this major diabetes prevention trial. Troglitazone remains in clinical use and its long term effect to prevent or delay diabetes is still to be determined. The potential benefits of long term improvement in insulin sensitivity are now under study. Non-human primates may offer a unique opportunity to establish this indication.

VII. Lipid Abnormalities and Pharmacologic Agents: PPARα Ligands and Fibrates

Lipid abnormalities are common in obese humans and in obese non-human primates. Pharmacologic approaches to altering the course of dyslipidemia have been carried out in rhesus monkeys, including published reports of the antidyslipidemic actions of a nicotinic acid analogue, Acipimox and a fibrate, Fenofibrate.

The mechanism by which a nicotinic acid analogue, Acipimox, acts to lower triglycerides was examined in a group of moderately hypertriglyceridemic insulin resistant nondiabetic rhesus monkeys. Acute lowering of free fatty acids (FFA) following an 8 mg/kg dose of Acipimox given with a meal was observed at 4 h, but this 30% decline in FFA was no longer present at 6 h after dosing (HANNAH et al. 1995). After 8 weeks of chronic administration, Acipimox (16 mg/kg q.i.d.) produced a 30% decline in triglyceride levels, as well as a decrease in LDL cholesterol concentration. Overnight fasted levels of FFA were not altered by chronic Acipimox treatment. FFA transport was increased, while FFA fractional clearance was unchanged. Both production and clearance of VLDL triglycerides were increased by Acipimox; however, clearance was relatively more affected. No decrease in insulin sensitivity was observed during this 8-week study (HANNAH et al. 1995).

Fibrates are activators of the peroxisome proliferator-activated receptor-alpha (PPAR-α), a member of the nuclear hormone receptor super family involved in gene transcription for enzymes of lipid metabolism. Fibrates have been shown to produce peroxisome proliferation in the liver of rodents, including increased peroxisomal fatty acid beta-oxidation and microsomal omega-hydroxylation leading to hepatomegaly with prolonged exposure. Non-human primates appear to be resistant to this peroxisome proliferative effect. Although it has been thought historically that non-human primates might be resistant to the lipid lowering effects of fibrates, this may be due to the nature of the previously used protocols in which dyslipidemia was experimentally induced in normal young adult monkeys through high fat diet feeding of susceptible species. In spontaneously obese, insulin resistant rhesus monkeys with naturally occurring dyslipidemia (increased triglycerides and reduced HDL cholesterol), fibrates have shown the same lipid lowering effects as have been observed in humans. In a recent report, fenofibrate (30 mg/kg b.i.d) was administered for 8 weeks to spontaneously dyslipidemic rhesus monkeys who were ingesting a low fat (20%), very low cholesterol, high fiber chow diet (WINEGAR

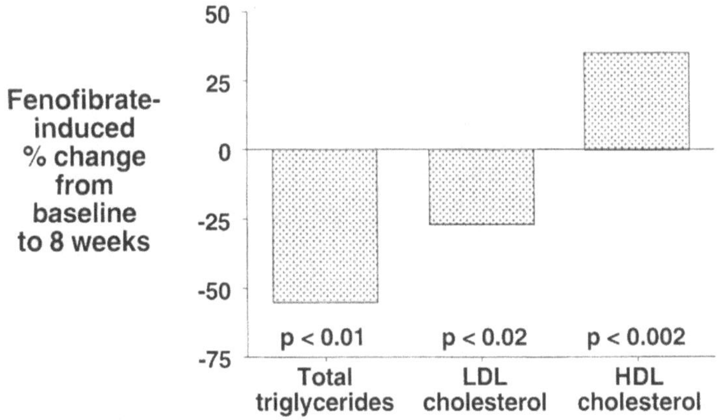

Fig. 9. Effects of 8 weeks of treatment with Fenofibrate (30 mg/kg b.i.d.) on serum triglycerides, LDL cholesterol, and HDL cholesterol levels in rhesus monkeys. Redrawn from WINEGAR et al. 1998

et al. 1998). Triglycerides were reduced by 55%, and LDL cholesterol by 27%, while HDL cholesterol increased 35%, as shown in Fig. 9. Apolipoprotein B-100 was reduced by 70% and apolipoprotein CIII was lowered by 29%. An improvement in insulin sensitivity was suggested by the 40% reduction in fasting plasma insulin levels, without change in body weight or food intake. The glucose levels in these normoglycemic monkeys were unchanged by fenofibrate. Furthermore, there were no changes in serum liver enzyme levels. The naturally occurring dyslipidemia of the obese rhesus monkey appears to be an ideal model for the examination of novel hypolipidemic agents with potential for human use.

VIII. Hypertension and ACE Inhibitors

Hypertension is common in obese monkeys, affecting 20%–30% of adults with insulin resistance and increased BMI. Captopril and other antihypertensive agents have been examined in rhesus monkeys (*Macaca mulatta*).

Captopril, a specific competitive ACE inhibitor, reduces the conversion of angiotensin I to angiotensin II, with its blood pressure lowering effects thought to be due to this suppression of the renin-angiotensin-aldosterone system. Because of concern that angiotensin I-converting enzyme (ACE) inhibitors might promote the progression to overt type 2 diabetes, a study was undertaken in obese hypertensive pre-diabetic monkeys. Captopril produced significant and consistent decreases in both systolic and diastolic blood pressure, as shown in Fig. 10 (BODKIN et al. 1995). During this study there was a decrease in fasting plasma glucose with no reduction in glucose tolerance, and no

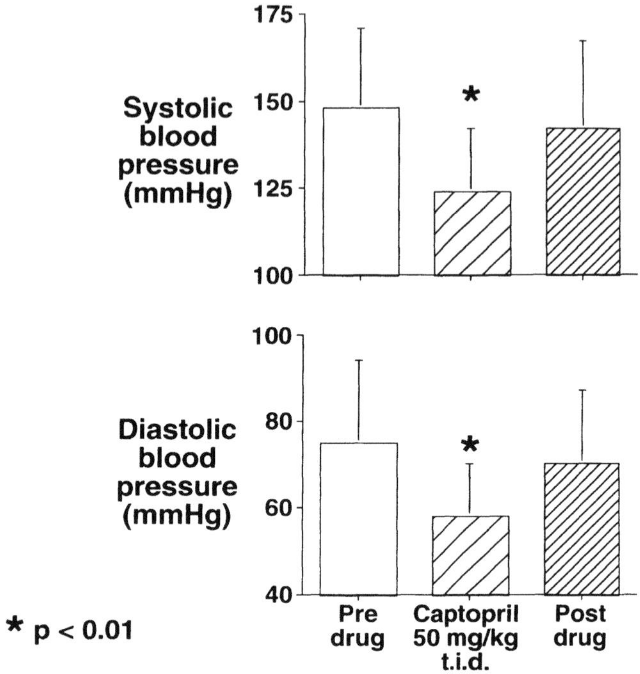

Fig. 10. Hypertension in prediabetic insulin resistant rhesus monkeys mitigated by administration of the ACE inhibitor, Captopril (50 mg/kg, t.i.d.). Redrawn from BODKIN et al. 1995

adverse effects on β cell insulin secretion in response to glucose. Thus no effects were noted to affect adversely the prediabetic state.

IX. Inositol Phosphoglycan Mediators in Insulin Action

A number of studies have suggested that D-chiroinositol metabolism may be altered in obese subjects with insulin resistance, and in diabetic subjects (KENNINGTON et al. 1990; OSTLUND et al. 1993). These observations draw on prior evidence that D-chiroinositol might act as an intracellular mediator of insulin action (LARNER et al. 1988). It is possible that the action of D-chiroinositol is through its incorporation into a mediator precursor which is subsequently processed to an active mediator containing chiroinositol. Myoinositol has also been implicated in disturbances in metabolism. Both D-chiroinositol [0.5 mg/kg (ORTMEYER et al. 1995) or 1.5 g/kg (ORTMEYER 1996)] and myoinositol [1.5 g/kg (ORTMEYER 1996)], when added to a meal, produced an improvement in glucose disappearance without any elevation in insulin output. In fact with the addition of myoinositol, post prandial insulin was significantly reduced as shown in Fig. 11. Further studies of these substrates are of interest for their potential to improve insulin sensitivity.

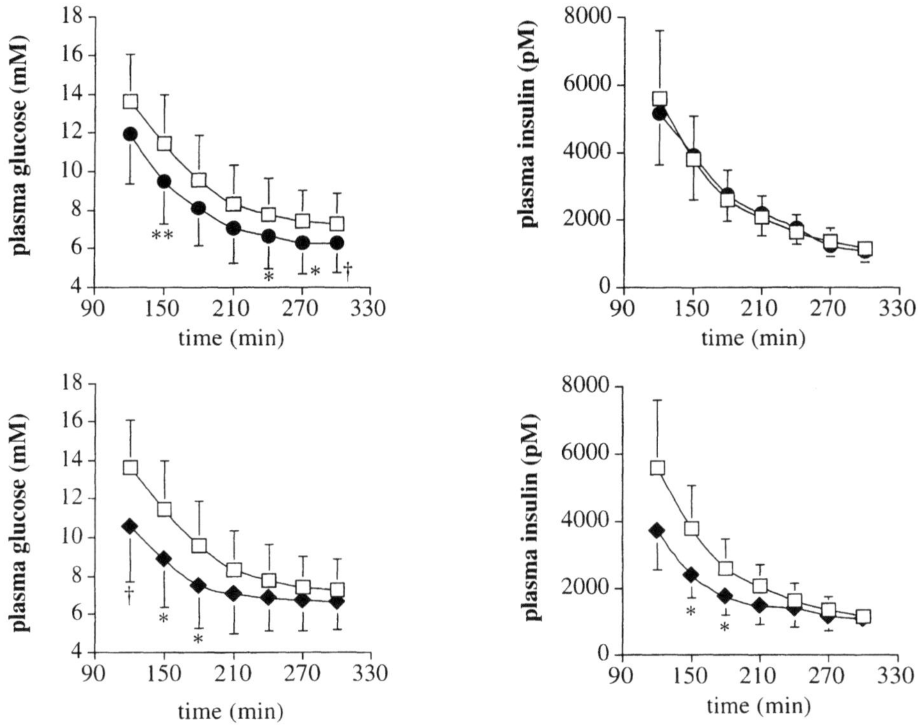

Fig. 11. Effects on post prandial plasma glucose and insulin with either 1.5 g/kg D-chiroinositol added to the meal (*top two panels*) or 1.5 g/kg myoinositol added to the meal (*bottom two panels*). *Open symbols* are without the inositol supplement, and *solid symbols* are with the agents. Redrawn from ORTMEYER 1996

X. Gastrointestinal Insulinotropic Agents

1. Glucagon-Like Peptides (GLP-1)

Obese rhesus monkeys, as they approach overt diabetes, lose the first phase β cell response to glucose, as noted above, and this is also true for humans. Several insulinotropic hormones have been identified and examined as possible pharmacological agents to reverse this decline and, therefore, to restore acute insulin release (BYRNE et al. 1998). GLP-1 is a potent insulin secretogogue which could be important in the therapeutic promotion of insulin secretion in those obese persons with declining β cell function (progressing toward diabetes). The hormone itself is a natural enteric peptide, however, it is rapidly inactivated by dipeptidylpeptidase IV and is rapidly cleared from plasma. GLP-1 has been shown to be the most potent of the endogenous incretins (FEHMANN et al. 1992), and lowers blood glucose in type 2 diabetic subjects (GUTNIAK et al. 1992). In rodents, GLP-1 also slows gastric emptying (DEORE et al. 1997). Both GLP-1 and Exendin-4 are potent agonists of the pancreatic

β cell GLP-1 receptor, with Exendin-4 having a longer duration of action (MONTROSE-RAFIZADEH et al. 1998).

In rhesus monkeys, incremental doses of continuous intravenous GLP-1 infusion (10, 20, 40, and 60 ng/kg per min), stepped up progressively during a continuous intravenous infusion of 20% dextrose and 8.5% amino acids (5 or 10 ml/kg per h), have been shown to produce increasing levels of circulating insulin in three of four monkeys, and increased plasma glucagon levels in two of the four monkeys, and to dampen the rise in glucose induced by the continuous intravenous nutrient infusion as shown in Fig. 12.

2. Exendin-4

The GLP-1 receptor agonist, exendin-4, is a 39 amino acid peptide, isolated from the venom of the Gila monster (*Heloderma suspectum*), and this agent is under investigation for possible use in type 2 diabetes. Because it amplifies

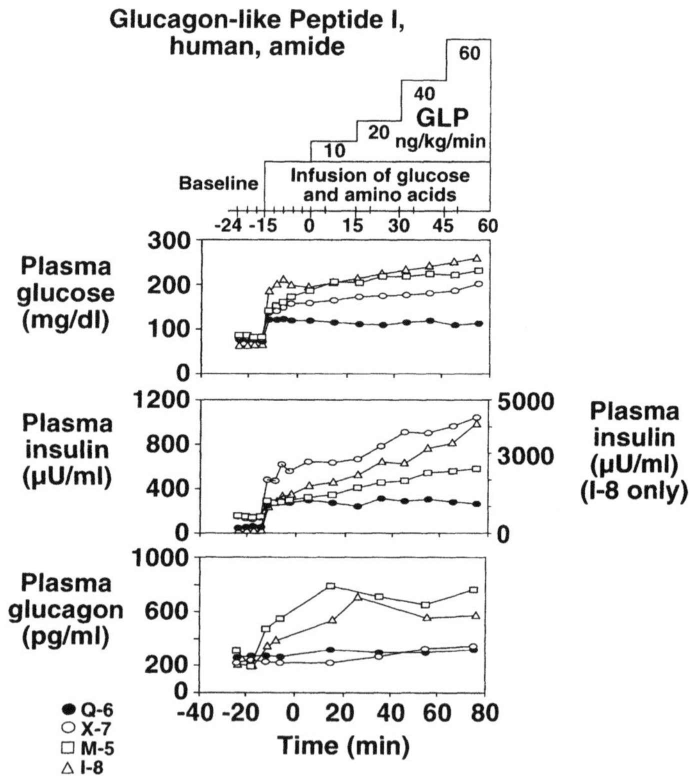

Fig. 12. Effects of GLP-1 administered intravenously at doses of 10, 20, 40, and 60 ng/kg per min during the continuous infusion of intravenous glucose and amino acids (5 ml/kg per h 20% dextrose and 8.5% amino acids–monkeys Q-6, X-7, and M-5, and 2.5 ml/kg per h monkey I-8)

β cell insulin secretion, its optimal application could prove to be as a candidate drug for use in obese patients whose β cell response to glucose is falling, but who have not yet developed overt type 2 diabetes. Exendin-4 has been found to bind to GLP-1 receptors, and to have a longer duration of action than GLP-1. Plasma glucose lowering in *db/db* mice led to dose response testing in obese type 2 diabetic non-human primates. Subcutaneous administration of Exendin-4 in doses up to 100 μg/kg to rhesus monkeys resulted in a significant mean maximum fall in plasma glucose from predosing levels to 120 min post dosing of more than 40% (Bodkin et al. 1998). There was a modest but variable increase in insulin levels. Although blood pressure increased in rats by an inotropic effect, in rhesus monkeys there was no change in blood pressure or heart rate during these dose ranging studies.

Other GLP-1 receptor agonists are currently under development and may prove to be useful to the obese insulin resistant patient for their insulinotropic effects.

C. Summary

Non-human primates exhibit a wide range of obesity related characteristics which are shared with those of humans. This is particularly true in the areas of fat distribution, insulin resistance, diabetes, and metabolism. The features of obesity in primates which appear to be involved in the pathogenesis of the disease, and which provide potential therapeutic targets, include the central distribution of fat, the adipose tissue secretory products and their receptors, the adipose specific $\beta 3$ adrenoreceptors, the β cell of the pancreas, incretins from the gastrointestinal tract, the PPAR family of nuclear receptors, and potential mediators of insulin action. In addition, frequent consequences of obesity or aspects of the insulin resistance syndrome which accompany obesity include hypertension and dyslipidemia. Obesity is a disease for which many new and unique pharmaceutical approaches will be found in the years to come, and primate models will continue to be key to determining long term efficacy and to examining mechanisms of action of new therapeutic agents.

References

Bodkin NL, Hansen BC (1995) Antihypertensive effects of captopril without adverse effects on glucose tolerance in hyperinsulinemic rhesus monkeys. J Med Primatol 24:1–6

Bodkin NL, Hansen BC, Young AA (1998) Dose-response for glucose lowering by Exendin-4 in diabetic rhesus monkeys (Macaca mulatta). Diabetes 47:A93

Bodki NL, Metzger BL, Hansen BC (1989) Hepatic glucose production and insulin sensitivity preceding diabetes in monkeys. Am J Physiol 256:E676–E681

Bodkin NL, Nicolson M, Ortmeyer HK, Hansen BC (1996) Hyperleptinemia: Relationship to adiposity and insulin resistance in the spontaneously obese rhesus monkey. Horm Metab Res 28:674–678

Bousquet-Melou A, Galitzky J, Carpene C, Lafontan M, Berlan M (1994) β-Adrenergic control of lipolysis in primate white fat cells: A comparative study with nonprimate mammals. Am J Physiol: Reg Int Comp Physiol 267:R115–R123

Byrne M, Ulrich W, Katschinske M, Goke B (1998) GLP-1 improves first phase insulin secretion without altering insulin sensitivity in subjects with impaired glucose tolerance. Diabetes 47:A192

Christopher-Hennings J, Kurzman ID, Haffa, AL, Kemnitz JW, Macewen EG (1995) The effect of high fat diet and dehydroepiandosterone (DHEA) administration in the rhesus monkey. In Vivo 9:415–420

Coleman DL (1978) Obese and diabetes: Two mutant genes causing diabetes-obesity syndromes in mice. Diabetologia 14:141–148

Cowan C et al. (1998) submitted

DeOre K, Greig NH, Holloway HW, Wang Y, Perfetti R, Egan JM (1997) The effects of GLP-1 on insulin release in young and old rats in the fasting state and during an intravenous glucose tolerance test. J Gerontol 52 A:B245–B249

Emorine LJ, Marullo S, Briend-Sutren M-M, Patey G, Tate K, Delavier-Klutchko C, Strosberg AD (1989) Molecular characterization of the human β3-adrenergic receptor. Science 245:1118–1121

Enocksson S, Shimizu M, Lonnqvist F, Nordenstrom J, Arner P (1995) Demonstration of an in vivo functional B-adrenoceptor in man. J Clin Invest 95:2239–2245

Fan Z, Kole H, Bernier M, Huang Z, Accilli D, Hansen BC, Shuldiner AR (1995) Molecular mechanism of insulin resistance in the spontaneously obese and diabetic rhesus monkey: site directed mutagenesis of the insulin receptor. The Endocrine Society 77:180

Fehmann H-C, Goke R, Goke B (1992) Glucagon-like peptide-1 (7–37)(7–36) amide is a new incretin. Mol Cell Endocrinol 85:C39–C44

Fisher MH, Amend AM, Bach TJ, Barker JM, Brady EJ, Candelore MR, Carroll D, Casieri MA, Chiu S-HL, Deng L, Forrest MJ, Hegarty-Friscino B, Guan X-M, Hom GJ, Hutchins JE, Kelly LJ, Mathvink RJ, Metzger JM, Miller RR, Ok HO, Parmee ER, Saperstein R, Strader CD, Stearns RA, Thompson GM, Tota L, Vicaario PP, Weber AE, Woods JW, Wyvratt MJ, Zafian PT, MacIntyre DE (1998) A selective human beta 3 adrenergic receptor agonist increases metabolic rate in rhesus monkeys. J Clin Invest (in press)

Gong W, He Y, Karas M, Reitman M (1997) Uncoupling protein-3 is a mediator of thermogenesis regulated by thyroid hormone, β3-adrenergic agonists, and leptin. J Biol Chem 272:24129–24132

Greenberg AS (1998) American Diabetes Association (oral communication)

Grujic D, Susuli, VS, Harper M-E, Himms-Hagen J, Cunningham BA, Corkey BE, Lowell BB (1997) β3-adrenergic receptors on white and brown adipocytes mediate β3-selective agonist-induced effects on energy expenditure, insulin secretion, and food intake. J Biol Chem 272:17686–17693

Gutniak M, Orskov C, Holst J. J, Ahren B, Efendic S (1992) Antidiabetogenic effect of glucagon-like peptide-1 (7–36) amide in normal subjects and patients with diabetes mellitus. New Engl J Med 326:1316–1322

Hamilton CL, Ciaccia P (1978) The course of development of glucose intolerance in the monkey (*Macaca mulatta*). J Med Primatol 7:165–173

Hannah JS, Bodkin NL, Paidi MS, Anh-Le N, Howard BV, Hansen BC (1995) Effects of acipimox on the metabolism of free fatty acids and VLDL triglyceride. Acta Diabetologia 32:279–293

Hannah JS, Verdery RB, Bodkin NL, Hansen BC, Le N-A, Howard BV (1991) Changes in lipoprotein concentrations during the development of noninsulin dependent diabetes mellitus in obese rhesus monkeys (*Macaca mulatta*). J Clin Endocrinol Metab 72:1067–1072

Hansen BC, Bodkin NL (1985) Beta cell hyperresponsiveness to glucose precedes both fasting hyperinsulinemia and reduced glucose tolerance. (Abstract) Diabetes 34 [Suppl 1]:8A

Hansen BC, Bodkin NL (1986) Heterogeneity of insulin responses: phases in the continuum leading to non-insulin-dependent diabetes mellitus. Diabetologia 29:713–719

Hansen BC, Bodkin NL (1990) β-cell hyperresponsiveness: earliest event in development of diabetes in monkeys. Am J Physiol 259 (Regulatory Integrative Comp Physiol 28):R612–R617

Hansen, B. C, Jen, K.-L, Schwartz, J (1988) Changes in insulin responses and binding in adipocytes from monkeys with obesity progressing to diabetes. Int J Obes 12:391–401

Hotta K, Gustafson TA, Ortmeyer HK, Bodkin NL, Hansen BC (1998) Monkey leptin receptor mRNA: Sequence, tissue distribution, and mRNA expression in the adipose tissue of normal, hyperinsulinemic and type 2 diabetic rhesus monkeys. Obes Res 6:353–360

Hotta K, Gustafson TA, Ortmeyer HK, Bodkin NL, Nicolson,MA, Hansen BC (1996) Regulation of *obese (ob)* mRNA and plasma leptin levels in rhesus monkeys: Effects of insulin, body weight and diabetes. J Biol Chem 271:25327–25331

Hotta K, Gustafson TA, Yoshioka S, Ortmeyer HK, Bodkin NL, Hansen BC (1998) Relationships of PPAR$\gamma$and PPARγ2 mRNA levels to obesity, diabetes and hyperinsulinemia in rhesus monkeys. Int J Obesity 22:1000–1010

Huang Z, Hansen BC, Shuldiner AR (1993) Characterization of the insulin receptor gene in the rhesus monkey, a diabetes-prone species. Exp Clin Endocrinol 101:358–359

Hudson JC, Baum ST, Frye DM, Roecker EB, Kemnitz JW (1996) Age and sex differences in body size and composition during rhesus monkey adulthood. Aging 8:197–204

Jen K-LC, Hansen BC (1984) Feeding behavior during experimentally induced obesity in monkeys. Physiol Behav 33:863–869

Jen K-LC, Hansen BC, Metzger BL (1985) Adiposity, anthropometric measures, and plasma insulin levels of rhesus monkeys. Int J Obes 9:213–224

Kates A-L, Park IRA, Himms-Hagen J, Mueller RW (1990) Thyroxine 5′-deiodinase in brown adipose tissue of the cynomolgus monkey *Macaca fascicularis*. Biochem Cell Biol 68:231–237

Kemnitz J, Francke, G (1986) Characteristics of spontaneous obesity in male rhesus monkeys. Physiol Behav 38:477–483

Kemnitz JW, Elson DF, Roecker EB, Baum ST, Bergman RN, Meglasson MD (1994) Pioglitazone increases insulin sensitivity, reduces blood glucose, insulin, and lipid levels, and lowers blood pressure, in obese, insulin-resistant rhesus monkeys. Diabetes 43:204–211

Kemnitz JW, Goy RW, Flitsch TJ, Lohmiller JJ, Robinson JA (1989) Obesity in male and female rhesus monkeys: fat distribution, glucoregulation, and serum androgen levels. J Clin Endocrinol Metab 69:287–293

Kennington AS, Hill CR, Craig J, Bogardus C, Raz I, Ortmeyer HK, Hansen BC, Romero G, Larner J (1990) Low urinary *chiro*-inositol excretion in non-insulin-dependent diabetes mellitus. N Engl J Med 323:373–378

Larner J, Huang LC, Schwartz CFW, Oswald AS, Shen T-Y, Kinter M, Tang G, Zeller K (1988) Rat liver insulin mediator which stimulates pyruvate dehydrogenase phosphatase contains galactosamine and d-chiroinositol. Biochem Biophys Res Comm 151:1416–1426

Larner J, Huan, LC, Tang G, Suzuki S, Schwartz CFW, Romero G, Roulidis Z, Zeller K, Shen TY, Oswald AS, Luttrell L (1988) Insulin mediators: structure and function. Cold Spring Harb Symp Quant Biol 53:965–971

MacIntyre DE (1998) Human β3 adrenergic receptor (β3AR) agonists: Acute and chronic effects in dogs and non human primates. Int J Obesity in press

Mantzoros CS, Qu D, Frederich RC, Susulic VS, Lowell BB, Maratos-Flier E, Flier JS (1996) Activation of beta3 adrenergic receptors suppresses leptin expression and

mediates a leptin-independent inhibition of food intake in mice. Diabetes 45:909–914
Montrose-Rafizadeh C, Yang H, Pritchette L, Eng J (1998) Distinct activation of GLP-1 receptor by Exendin-4 and GLP-1. Diabetes 47:A192
Naithani VK, Steffens GJ, Tager HS (1984) Isolation and amino-acid sequence determination of monkey insulin and proinsulin. Hoppe-Seyler's Z Physiol Chem 365:571–575
Ortmeyer HK (1996) Dietary myoinositol results in lower urine glucose and in lower postprandial plasma glucose in obese insulin resistant rhesus monkeys. Obesity Research 4:569–575
Ortmeyer HK, Larner J, Hansen BC (1995) Effects of D-chiroinositol added to a meal on plasma glucose and insulin in hyperinsulinemic rhesus monkeys. Obes Res 3:605S–608S
Ostlund REJ, McGill JB, Herskowitz I, Kipnis DM, Santiago JV, Sherman WS (1993) D-*chiro*-Inositol metabolism in diabetes mellitus. Proc Natl Acad Sci 90:9988–9992
Pace N, Kline L, Schachman HK, Harfenist M (1947) Studies on body composition. IV. Use of radioactive hydrogen for measurement in vivo of total body water. J Biol Chem 168:459–469
Rhodes L, Beall M, McGowan E, Tota M (1998) Rhesus macaques treated with human recombinant leptin (hOB) achieve high hOB blood levels, but show no effects on short term food intake. Int J Obesity 22:in press (Abstract # 653)
Schwartz SM, Kemnitz JW (1992) Age-and gender-related changes in body size, adiposity, and endocrine and metabolic parameters in free-ranging rhesus macaques. Am J Physical Anthro 89:109–121
Schwartz SM, Kemnitz JW, Howard CF (1993) Obesity in free-ranging rhesus macaques. Int J Obes 17:1–10
Snitker S, Foley JE, Ravussin E (1998) Low *in vitro* adipocyte lipolysis predicts weight gain in non-diabetic Pima Indians. Diabetes 47:A13
Tang-Christensen M, Havel PJ, Jacobs R, Larsen PJ, Cameron JL (1998) Central administration of human leptin inhibits food intake and activates the sympathetic nervous system in rhesus macaques. Int J Obesity 22:(In press) (Abstract # 587)
Tartaglia LA, Dembski M, Weng X, Deng N, Culpepper J, Devose R, Richards GJ, Campfield LA, Clark FT, Deeds J, Muir C, Sanker S, Moriarty A, Moore KJ, Smutko JS, GG, M, Woolf EA, Monroe CA, Tepper RI (1995) Identification and expression cloning of a leptin receptor, OB-R. Cell 83:1263–1271
Tavernier G, Barbe P, Galitzky J, Berlan M, Caput D, Lafontan M, Langin D (1996) Expression of $\beta3$-adrenoceptors with low lipolytic action in human subcutaneous white adipocytes. J Lipid Res 37:87–97
Viguerie-Bascands N, Bousquet-Melou A, Galitzky J, Larrouy D, Ricquier D, Berlan M, Casteilla L (1996) Evidence for numerous brown adipocytes lacking functional $\beta3$-adrenoceptors in fat pads from non-human primates. J Clin Endocrinol Metab 81:368–375
Wagner JD, Jayo MJ, Cefalu WT, Hardy VA, Rankin SE, Toombs CF (1996) Recombinant human leptin (rHuLeptin) reduces body weight and body fat and improves insulin sensitivity in non-human primates. Obes Res 4:27S
Walike BC, Goodner CJ, Koerke, DJ (1977) Assessment of obesity in pigtailed monkeys *(Macaca nemestrina)*. J Med Primatol 6:151–162
Walike BC, Smith OA (1972) Regulation of food intake during intermittent and continuous cross circulation in monkeys (*Macaca mulatta*). J Comp Physiol Psychol 80:372–381
Walker ML, Schwartz SM, Wilson ME, Musem, PI (1984) Estimation of body fat in female rhesus monkeys. Am J Physical Anthro 63:323–329
Walston J, Lowe A, Silver K, Yang Y, Bodkin NL, Hansen BC, Shuldiner AR (1997) The $\beta3$-adrenergic receptor in the obesity and diabetes prone rhesus monkey is very similar to human and contains arginine at codon 64. Gene 188:207–213

Winegar DA, Brown PJ, Wilkison WO, Lewis MC, Orr RJ, Tong WQ, Brown HR, Bodkin NL, Hansen BC (1998) Effects of fenofibrate on lipid parameters in obese rhesus monkeys. The Endocrine Society 216
Zhang Y, Proenca R, Maffei M, Barone M, Leopold L, Friedman JM (1994) Positional cloning of the mouse *obese* gene and its human homologue. Nature 372:425–432

Index

MIX
Papier aus verantwortungsvollen Quellen
Paper from responsible sources
FSC® C105338

FSC
www.fsc.org

If you have any concerns about our products,
you can contact us on
ProductSafety@springernature.com

In case Publisher is established outside the EU,
the EU authorized representative is:
Springer Nature Customer Service Center GmbH
Europaplatz 3, 69115 Heidelberg, Germany

Printed by Libri Plureos GmbH
in Hamburg, Germany